Actinobacteria and Myxobacteria

Actinobacteria and Myxobacteria —Important Resources for Novel Antibiotics

Editor

Joachim Wink

MDPI • Basel • Beijing • Wuhan • Barcelona • Belgrade • Manchester • Tokyo • Cluj • Tianjin

Editor
Joachim Wink
Helmholtz Centre for Infection Research (HZI)
Germany

Editorial Office
MDPI
St. Alban-Anlage 66
4052 Basel, Switzerland

This is a reprint of articles from the Special Issue published online in the open access journal *Microorganisms* (ISSN 2076-2607) (available at: https://www.mdpi.com/journal/microorganisms/special_issues/Actinobacteria_Myxobacteria_Antibiotics and https://www.mdpi.com/journal/microorganisms/special_issues/actinobacteria_myxobacteria_antibiotics_2).

For citation purposes, cite each article independently as indicated on the article page online and as indicated below:

LastName, A.A.; LastName, B.B.; LastName, C.C. Article Title. *Journal Name* **Year**, *Article Number*, Page Range.

ISBN 978-3-03943-529-6 (Pbk)
ISBN 978-3-03943-530-2 (PDF)

Cover image courtesy of Manfred Rohde, Helmholtz Centre for Infection Research.

Contents

About the Editor

Joachim Wink (Dr.) completed his Ph.D. in 1985 at Frankfurt University. He then went to the pharmaceutical industry and started his career at the Hoechst AG where he was responsible for strain collection and specialized in the cultivation and taxonomic characterization of Actinobacteria and Myxobacteria. During this period, he was responsible for the strain library within the pharmaceutical research and a number of screening projects with Hoechst Marion Russel, Aventis and Sanofi. In the year 2005, he did his habilitation at the Carolo Wilhelma University of Braunschweig and in 2012, he went to the Helmholtz Centre for Infection Research in Braunschweig where he founded the working group of the strain collection with its focus on Myxobacteria. Here, he is now working on the isolation and taxonomic characterization of Myxobacteria and Actinobacteria as well as the analysis of their secondary metabolites with a focus on the antibiotic active ones. The Compendium of Actinobacteria on the homepage of the German Culture Collection is a permanent actualized working tool for people working with Actinobacteria, which was prepared by him. He has published more than 60 papers on secondary metabolites and the taxonomy of the producing microorganisms in reputed journals, a number of reviews as well as book chapters and more than 35 patents. He is a member of the editorial board of a number of international journals.

 *microorganisms*

Editorial

Special Issue: "Actinobacteria and Myxobacteria—Important Resources for Novel Antibiotics"

Joachim Wink

Microbial Strain Collection, Helmholtz Centre for Infection Research GmbH (HZI), Inhoffenstrasse 7, 38124 Braunschweig, Germany; Joachim.Wink@helmholtz-hzi.de

Received: 22 September 2020; Accepted: 22 September 2020; Published: 24 September 2020

The history of our antibiotics is inseparably connected to microorganisms as producers. In particular, microorganisms with large genomes (often more than 8 MB) like many Actinobacteria and the *Myxococcales* show the highest potential for secondary metabolite formation. In addition, an important factor seems to be the differentiation process which is also found in both of these bacterial groups. Many of the clinically used antibiotics like the cephalosporins, anthrycylines, macrolides, glycopeptides, lipopeptides and aminoglycosides are originally products of Actinobacteria [1]. The producers belong to a number of different genera like *Streptomyces, Amycolatopsis, Micromonospora* and *Actinoplanes.* The phylum Actinobacteria belongs to the Gram-positive bacteria with a high GC content. They can be mainly found in soil but there are also pathogenic and saprophytic species. In particular, the mycelium-forming genera show characteristic differentiation by forming endospores that can be arranged in spore chains or sporangia.

Besides, the Actinobacteria members of the *Myxococcales* were first reported to show bacteriolytic effects in 1946 [2] but it took until the 1980s for the first antibiotic with high potential for market development, Sorangicin, to be isolated [3]. Like Actinobacteria, most of the members of the *Myxococcales* live in soil, they belong to the Gram-negative bacteria and form fruiting bodies during their differentiation process. With the knowledge of the genome information, it is now clear that they also harbor a large potential for the production of secondary metabolites [4,5].

The isolation of novel Actinobacteria and Myxobacteria still leads to new genetic potential for the identification and isolation of bioactive compounds, especially antibiotics. With more and more understanding of this genetic information, we also see a huge number of genes for which we do not know the resulting product and the induction of these silent genes is one of the challenges [6].

This issue gathers 16 papers including 11 articles, 4 reviews and 1 communication. Six of them describe novel species or isolates and their characterization by the use of a polyphasic approach [7–12] as well as their secondary metabolites. Two of the articles describe Actinobacteria from uncommon habitats like the Western Ghats region in India [13] and endophytic ones [14]. The study of the induction of secondary metabolites by use of the OSMAC approach is the subject of one other article [15]. The other two articles include the identification and heterologous expression of an antibiotic gene cluster [16] and the biological activity of some volatile secondary metabolites [17]. In the communication, the authors report the unexplored biosynthetic potential in Myxobacteria [18]. In the review section, we find reports on polyketide biosynthesis in *Streptomyces* [19], the role of elicitors in antibiotic biosynthesis [20], antiviral compounds from Myxobacteria [21] and an overview on the role of Myxobacteria as secondary metabolite producers [22]. Altogether, in my view, this is a balanced snapshot of this impressive research field.

Funding: This research received no external funding.

Acknowledgments: I would like to thank all authors who contributed their excellent papers to this Special Issue. I thank the reviewers for their help in improving the papers to the highest standard of quality. I am also grateful to all members of the *Microorganisms* Editorial Office for giving me this opportunity, and for their continuous support in managing and organizing this Special Issue.

Conflicts of Interest: The author declares no conflict of interest.

References

1. Omura, S. *The Search for Bioactive Compounds from Microorganisms*; Springer: New York, NY, USA, 1992.
2. Oxford, A.E.; Singh, B.N. Factors contributing to the bacteriolytic effect of species of Myxococci upon viable eubacteria. *Nature* **1946**, *158*, 745. [CrossRef] [PubMed]
3. Irschik, H.; Jansen, R.; Gerth, K.; Höfle, G.; Reichenbach, H. The sorangicins, novel and powerful inhibitors of eubacterial RNA polymerase isolated frommyxobacteria. *J. Antibiot.* **1987**, *40*, 7–13. [PubMed]
4. Weissman, K.J.; Müller, R. A brief tour of myxobacterial secondary metabolism. *Bioorg. Med. Chem.* **2009**, *17*, 2121–2136. [CrossRef] [PubMed]
5. Plaza, A.; Müller, R. Myxobacteria: Chemical diversity and screening strategies. In *Natural Products: Discourse, Diversity and Design*; Goss, R., Carter, G., Osbourne, A., Eds.; Wiley-Blackwell: Hoboken, NJ, USA, 2014.
6. Udwary, D.W.; Zeigler, L.; Asolkar, R.N.; Singan, V.; Lapidus, A.; Fenical, W.; Jensen, P.R.; Moore, B. Genome sequencing reveals complex secondary metabolomein the marine actinomycete *Salinispora tropica*. *Proc. Natl. Acad. Sci. USA* **2007**, *104*, 10376–10381. [CrossRef]
7. Singh, R.; Dubey, A.K. Isolation and Characterization of a New Endophytic Actinobacterium *Streptomyces californicus* Strain ADR1 as a Promising Source of Anti-Bacterial, Anti-Biofilm and Antioxidant Metabolites. *Microorganisms* **2020**, *8*, 929. [CrossRef] [PubMed]
8. Wang, H.; Sun, T.; Song, W.; Guo, X.; Cao, P.; Xu, X.; Shen, Y.; Zhao, J. Taxonomic Characterization and Secondary Metabolite Analysis of NEAU-wh3-1: An Embleya Strain with Antitumor and Antibacterial Activity. *Microorganisms* **2020**, *8*, 441. [CrossRef]
9. Ling, L.; Han, X.; Li, X.; Zhang, X.; Wang, H.; Zhang, L.; Cao, P.; Wu, Y.; Wang, X.; Zhao, J.; et al. A Streptomyces sp. NEAU-HV9: Isolation, Identification, and Potential as a Biocontrol Agent against *Ralstonia solanacearum* of Tomato Plants. *Microorganisms* **2020**, *8*, 351. [CrossRef] [PubMed]
10. Yu, Z.; Han, C.; Yu, B.; Zhao, J.; Yan, Y.; Huang, S.; Liu, C.; Xiang, W. Taxonomic Characterization, and Secondary Metabolite Analysis of *Streptomyces triticiradicis* sp. nov.: A Novel Actinomycete with Antifungal Activity. *Microorganisms* **2020**, *8*, 77.
11. Liu, C.; Zhuang, X.; Yu, Z.; Wang, Z.; Wang Yi Guo, X.; Xiang, W.; Huang, S. Characterization of *Streptomyces sporangiiformans* sp. nov., a Novel Soil Actinomycete with Antibacterial Activity against *Ralstonia solanacearum*. *Microorganisms* **2019**, *7*, 360.
12. Sujarit, K.; Mori, M.; Dobashi, K.; Shiomi, K.; Pathom-aree WLumyong, S. New Antimicrobial Phenyl Alkenoic Acids Isolated from an Oil Palm Rhizosphere-Associated Actinomycete, *Streptomyces palmae* CMU-AB204T. *Microorganisms* **2020**, *8*, 350. [CrossRef] [PubMed]
13. Siddharth, S.; Vittal, R.R.; Wink, J.; Steinert, M. Diversity and Bioactive Potential of Actinobacteria from Unexplored Regions of Western Ghats, India. *Microorganisms* **2020**, *8*, 225. [CrossRef] [PubMed]
14. Liu, C.; Zhuang, X.; Yu, Y.; Wang, Z.; Wang, Y.; Guo, X.; Xiang, W. Community Structures and Antifungal Activity of Root-Associated Endophytic Actinobacteria of Healthy and Diseased Soybean. *Microorganisms* **2019**, *7*, 243. [CrossRef] [PubMed]
15. Almeida, E.L.; Kaur, N.; Jennings, L.K.; Carrillo Rincón, A.F.; Jackson, S.A.; Thomas, O.P.; Dobson, A.D.W. Genome Mining Coupled with OSMAC-Based Cultivation Reveal Differential Production of Surugamide A by the Marine Sponge Isolate *Streptomyces* sp. SM17 When Compared to Its Terrestrial Relative *S. albidoflavus* J1074. *Microorganisms* **2019**, *7*, 394. [CrossRef]
16. Myronovskyi, M.; Rosenkränzer, B.; Stierhof, M.; Petzke, L.; Seiser, T.; Luzhetskyy, A. Identification and Heterologous Expression of the Albucidin Gene Cluster from the Marine Strain *Streptomyces Albus* Subsp. *Chlorinus NRRL B-24108*. *Microorganisms* **2019**, *7*, 394.
17. Siddharth, S.; Vittal, R.R. Evaluation of Antimicrobial, Enzyme Inhibitory, Antioxidant and Cytotoxic Activities of Partially Purified Volatile Metabolites of Marine *Streptomyces* sp.S2A. *Microorganisms* **2018**, *6*, 72. [CrossRef] [PubMed]

18. Gregory, K.; Salvador, L.A.; Akbar, S.; Adaikpoh, B.I.; Stevens, D.C. Survey of Biosynthetic Gene Clusters from Sequenced Myxobacteria Reveals Unexplored Biosynthetic Potential. *Microorganisms* **2019**, *7*, 181. [CrossRef] [PubMed]

19. Risdian, C.; Mozef, T.; Wink, J. Biosynthesis of Polyketides in *Streptomyces*. *Microorganisms* **2019**, *7*, 124. [CrossRef] [PubMed]

20. Tyurin, A.P.; Alferova, V.A.; Korshun, V.A. Chemical Elicitors of Antibiotic Biosynthesis in Actinomycetes. *Microorganisms* **2018**, *6*, 52. [CrossRef] [PubMed]

21. Mulwa, L.S.; Stadler, M. Antiviral Compounds from Myxobacteria. *Microorganisms* **2018**, *6*, 73. [CrossRef] [PubMed]

22. Mohr, K.I. Diversity of Myxobacteria—We Only See the Tip of the Iceberg. *Microorganisms* **2018**, *6*, 84. [CrossRef] [PubMed]

 *microorganisms*

Article

Isolation and Characterization of a New Endophytic Actinobacterium *Streptomyces californicus* Strain ADR1 as a Promising Source of Anti-Bacterial, Anti-Biofilm and Antioxidant Metabolites

Radha Singh and Ashok K. Dubey *

Department of Biological Sciences & Engineering, Netaji Subhas Institute of Technology, New Delhi 110078, India; dharana.radha@gmail.com
* Correspondence: adubey.nsit@gmail.com or akdubey@nsut.ac.in

Received: 24 May 2020; Accepted: 5 June 2020; Published: 19 June 2020

Abstract: In view of the fast depleting armamentarium of drugs against significant pathogens, like methicillin-resistant *Staphylococcus aureus* (MRSA) and others due to rapidly emerging drug-resistance, the discovery and development of new drugs need urgent action. In this endeavor, a new strain of endophytic actinobacterium was isolated from the plant *Datura metel*, which produced secondary metabolites with potent anti-infective activities. The isolate was identified as *Streptomyces californicus* strain ADR1 based on 16S rRNA gene sequence analysis. Metabolites produced by the isolate had been investigated for their antibacterial attributes against important pathogens: *S. aureus*, MRSA, *S. epidermis*, *Enterococcus faecium* and *E. faecalis*. Minimum inhibitory concentration (MIC_{90}) values against these pathogens varied from 0.23 ± 0.01 to 5.68 ± 0.20 µg/mL. The metabolites inhibited biofilm formation by the strains of *S. aureus* and MRSA (Biofilm inhibitory concentration [BIC_{90}] values: 0.74 ± 0.08–4.92 ± 0.49 µg/mL). The BIC_{90} values increased in the case of pre-formed biofilms. Additionally, the metabolites possessed good antioxidant properties, with an inhibitory concentration (IC_{90}) value of 217.24 ± 6.77 µg/mL for 1, 1-diphenyl-2-picrylhydrazyl (DPPH) free radical scavenging. An insight into different classes of compounds produced by the strain ADR1 was obtained by chemical profiling and GC-MS analysis, wherein several therapeutic classes, for example, alkaloids, phenolics, terpenes, terpenoids and glycosides, were discovered.

Keywords: endophytic actinobacteria; *Streptomyces* sp., anti-*S. aureus*; anti-MRSA; anti-biofilm

1. Introduction

The emergence of drug resistance among pathogens has assumed alarming proportions in recent times, causing rapid depletion in the current armamentarium of drugs to fight such infections [1]. This has posed a serious threat to human health globally and required the rapid development of new and effective drugs at a pace faster than resistance to achieve desirable outcomes in the treatment of infectious diseases. Some of these pathogens have gained exceptional notoriety due to their tremendous ability to adapt, to evade the host immune response and to develop drug resistance. This has led World Health Organization to enlist significant human pathogens under critical, high and medium priority categories [2]. *Enterococcus faecium* (vancomycin resistant) and *Staphylococcus aureus* (methicillin resistant) are considered as high priority pathogens for whom new antibiotics are required the most urgently. Infections involving such pathogens are often associated with biofilms, which are responsible for multi-fold increase in drug-resistance of the pathogens [3,4]. Therefore, it is highly desirable that the new antibiotics possess anti-biofilm activities: disruption of pre-formed biofilms and inhibition of biofilms formation.

Some of the recent studies have reported that the reactive oxygen species were causing antibiotic tolerance in *S. aureus* during systemic infections [5]. Further, oxidative immune response of the host appeared to be switched on during bacterial infections, resulting in increased oxidative stress to the host [6]. Additionally, the antibiotics used to treat infections might also cause an increase in the level of oxidative stress [7,8]. It was reported that antioxidants might prevent oxidative stress-induced pathology [9]. Biofilm formation in *S. aureus* is also enhanced in the presence of oxidative stress [10]. Therefore, providing antioxidants may help in the inhibition of biofilms formation and thus in the prevention of concomitant resistance to antibiotics among the pathogens.

In view of the facts mentioned above, our research endeavors have focused on the discovery of novel anti-infective therapeutics for the treatment and cure of drug resistant infections. Characterization of antioxidant properties of metabolites was also part of our study design due to its foreseeable application in therapy of infectious diseases. There are several approaches to develop drugs, for example, rational drug design (structure-based design of inhibitors against target), synthetic and combinatorial chemistry, high throughput screen of chemical libraries and mining of natural products [11,12]. However, from the stand point of discovery of novel pharmacophore or new classes of drug working on as yet unknown targets, mining of natural products is the obvious choice. Therefore, we chose natural products in the search for new antibiotics. Furthermore, we have considered actinobacteria from vast pool of natural product resources due to their versatility, ubiquity and ability to produce therapeutic compounds with extensive chemical diversity [13,14]. The genus *Streptomyces* of actinobacteria has been regarded as containing the most prolific producers of therapeutic compounds [15,16]. However, repeat discovery of known molecules remains a challenge while hunting the actinobacteria for drugs [17,18]. One of the possible approaches to avoid the repeat discovery of the drugs could be the sourcing of actinobacteria from niche habitats instead of common sources like soil. Accordingly, we have explored endophytic actinobacteria in an effort to enhance the chances of finding new compounds as potential drug candidates. Endophytic actinobacteria are the microorganisms that reside within the plant tissues without causing any adverse effect to plants [19]. Further, for increasing the prospect of strain novelty, we have selected a medicinally important plant, *Datura metel*, which had largely remained unexplored for endophytic population of actinobacteria.

In the present communication, identification of a novel strain, *Streptomyces californicus* strain ADR1 is reported from the plant *D. metel*. Secondary metabolites produced by the isolate ADR1 were characterized for their antibacterial, antibiofilm and antioxidant properties. Further, the metabolite preparations were analyzed for different class of therapeutically significant compounds produced by the isolate.

2. Materials and Methods

2.1. Isolation of Endophytic Actinobacteria

The endophytic actinobacteria were isolated from the plant, *Datura metel*. The ex-plants were surface sterilized by following the method reported elsewhere [20]. The sterilized plant parts were aseptically grinded by using autoclaved mortar pestle in phosphate-buffered saline (PBS) pH 7.0. The ground paste was spread over the following isolation media: nutrient agar, asparagine glycerol (AGS) agar) [21], humic acid—vitamin agar [22] and starch casein nitrate (SCN) agar [23]. The media were supplemented with cycloheximide (50 µg/mL). The plates were incubated at 28 °C for up to four weeks with regular observations for potential actinobacterial colonies.

The putative actinobacterial colonies were transferred and maintained on AGS medium. Purification of the isolates was achieved by repeated cycles of streaking on fresh plate. The purified cultures were screened for anti-bacterial against *S. aureus* ATCC 29213 by well diffusion method [24].

2.2. Molecular Identification and Characterization of the Isolate ADR1

Molecular identification of the strain ADR1 was based on 16S rRNA gene sequence analysis. The genomic DNA of the strain ADR1 was isolated using the method developed for Gram-positive bacteria [20] with a few modifications. Briefly, amplification of 16S rRNA gene was carried out using universal primers: V1f (50-AGAGTTTGATCMTGGCTCAG-30), V9r (50-AAGGAGGTGATCCANCCRCA-30), V3f (50-CCAGACTCCTACGGGAGGCAG-30) and V6r (50-ACGAGCTGACGACARCCATG-30) in a PCR machine (Mastercycler® nexus, Eppendorf International, Germany) by using the programme described elsewhere [25]. The amplified product was sequenced by Sanger's method using a 3130XL sequencer (Applied Biosystems, California, USA) for the 16S rRNA gene using universal primers as described above. The sequences were aligned in MEGA 6.0 to generate single consensus sequence. Homology search was performed using the standard Basic Local Alignment Search Tool (BLAST) sequence similarity search tool of the NCBI database to establish the identity of the isolate ADR1. Nucleotide sequences producing significant alignments after BLAST analysis with the 16S rRNA gene sequence of ADR1 were retrieved in FASTA format. These sequences were used to generate phylogenetic relationship of ADR1 with them by using the software, Phylogeny.fr [26,27]. The analysis was done by using advanced mode of this tool, which is an automated programme that performs step-by-step analysis starting from the multiple alignment of the sequences (MUSCLE 3.8.31) [28], alignment curation (Gblocks 0.91b), construction of phylogenetic tree (PhyML 3.1/3.0 aLRT) [29,30] to the visualization of phylogenetic tree (TreeDyn 198.3) [31]. The culture was characterized for its morphological features on different international *Streptomyces* protocol (ISP) media [32]. Isolate ADR1 was streaked on ISP1, ISP2, ISP3, ISP4, ISP5, ISP6 and ISP7 media plates and was incubated for 7 days at 28 °C for phenotypic and morphological observations. Single colony morphology of the culture was observed under Nikon stereo zoom microscope SMZ1270 at zooming ratio of 12.7:1 and resolution of 8×. Mycelial structure was observed under Nikon E600 microscope (Nikon, Tokyo, Japan) at a resolution of 100×.

2.3. Production of Secondary Metabolites

A single colony from freshly grown culture plate (72 h) was inoculated in 50 mL SCN broth (pH 7.4), which was incubated at 28 °C for 72 h to develop the pre-seed culture. The production medium (SCN broth, pH 7.2) was inoculated with the pre-seed culture (1%; *v/v*) to commence production of the secondary metabolites, which was carried out for 7 days at 28 °C in an incubator shaker (Adolf Kuhner AG, Birsfelden, Basel, Switzerland) run at 200 rpm. The cell-free broth was recovered by centrifugation at 5000× *g* for 20 min in Sorvall RC 5C plus centrifuge (Kendro Laboratory Products, Newtown, Connecticut, USA). The metabolites were recovered from the supernatant by using liquid-liquid extraction with equal volume of ethyl acetate. The extracted metabolites were dried by using rotary evaporator (50 °C) and vacuum oven (35 °C). The dried metabolite preparations were stored at 25 °C ± 2 °C till further use.

2.4. Antimicrobial Susceptibility Testing

The reference strains of bacterial pathogens used in this study were: *S. aureus* ATCC 29213, *S. aureus* ATCC 25923, *S. aureus* ATCC 13709, MRSA ATCC 43300, MRSA 562, *S. epidermis* ATCC 12228, *Enterococcus faecalis* ATCC 29212, *E. faecium* ATCC 49224 and *E. faecium* AIIMS. In-vitro antibacterial activity of the metabolite extract was determined on cation adjusted Muller Hinton agar (MHA) (Himedia, Mumbai, India) plates using well diffusion method [24]. The minimum inhibitory concentration (MIC_{90}) values were measured in a 96-well microtiter plates by the broth microdilution method as per the guidelines of Clinical and Laboratory Standards Institute (CLSI) [33]. Briefly, a stock solution of the metabolite extract (1mg/mL) was prepared in 0.2% DMSO and cation adjusted Muller Hinton broth. Bacterial pathogens (100 µL; 2×10^8 CFU/mL) and metabolite extract (100 µL) at concentrations varying from 125 to 0.122 µg/mL were added to the individual well in the microtitre plate. A sample control (ADR1 extract alone) and blank (media only) were included in each assay. After incubation for 24 h at 37 °C, iodonitrotetrazolium chloride (INT) (Sisco Research Laboratories, Mumbai, India) was added to the wells and the plates were incubated further for 30 min. The absorbance was measured on a multimode reader (Biotek Instruments, Winooski, Vermont, USA) at 490 nm. The value of MIC_{90} was considered to be the minimum concentration at which no visible growth could be observed. The following equation was used to compute the percent inhibition [34].

$$\text{Growth inhibition of pathogen (\%)} = [(\text{control OD}_{490\,nm} - \text{test OD}_{490\,nm})/\text{control OD}_{490\,nm}] \times 100 \qquad (1)$$

2.5. Antibiofilm Assay

Biofilms of *S. aureus* ATCC 25923, *S. aureus* ATCC 29213, MRSA ATCC 43300 and MRSA 562 were produced by using the method published elsewhere [35] in accordance with the CLSI guidelines [33]. Briefly, overnight grown reference cultures were suspended in tryptic soy broth (Himedia, Mumbai, India) supplemented with 2% glucose to attain turbidity equivalent to 0.5 McFarland standard (2×10^8 CFU/mL). A total of 100 µL of the cell suspension was transferred to the wells on the microtiter plate and was incubated at 37 °C for 24 h under static condition. Non-adherent cells were aspirated out along with the medium. The wells were rinsed with 100 µL of phosphate-buffered-saline (PBS). Fresh medium containing desired concentrations of ADR1 metabolites (from 250 to 0.49 µg/mL) were added to the wells on the microtiter plate, which was then incubated for the next 24 h at 37 °C under static condition. Viability of the biofilms was quantified by INT-calorimetric assay as described above. The following equation was used to find % inhibition of biofilm [35].

$$\text{Biofilm inhibition (\%)} = [(\text{control OD}_{490\,nm} - \text{test OD}_{490\,nm})/\text{control OD}_{490\,nm}] \times 100 \qquad (2)$$

2.6. Antioxidant Activity

The antioxidant potential of ADR1 metabolites was assessed by measuring reduction of DPPH (1, 1-diphenyl-2-picrylhydrazyl) free radicals as reported earlier [36]. Briefly, DPPH solution (0.1 mM) was prepared in methanol; 100 µL of this solution was added to 100 µL of the ADR1 metabolite preparations at different concentrations varying from 1000 to 7.81 µg/mL in 96-well microtitre plate. The plate was then incubated at 25 °C for 20 min in dark and the absorbance was measured at 517 nm. The scavenging strength was calculated using the following formula [36].

$$\text{\% scavenging activity} = [(\text{absorbance of DPPH control} - \text{absorbance of DPPH in the presence of metabolite})/\text{absorbance of DPPH control}] \times 100 \qquad (3)$$

2.7. Hemolytic Activities

Hemolytic activity was determined by disc diffusion assay using sheep blood agar (SBA) plates (Himedia, Mumbai, India). A total of 10 µL solution containing varying concentrations of the ADR1 metabolites (1000 to 7.8125 µg/mL) were dispensed on discs placed aseptically on the SBA plates and were incubated for 24 h at 37 °C. The type of hemolysis was observed as alpha, beta and gamma [37].

2.8. Secondary Metabolite Profiling and GC-MS Analysis

The ADR1 metabolite extract was used at a concentration of 5 mg/mL for chemical profiling of the classes of metabolites present in the extract. The tests for the different class of metabolites, for example, anthraquinones, glycosides, terpenoids, flavonoids, tannins, alkaloids, saponins, sterols, anthocyanins, coumarins, tannins, lactones, terpenes, fatty acids, proteins/amino acids and carbohydrates were carried out by using standard methods reported earlier [38–40].

Further analysis of the metabolites was carried out by employing GC-MS (GC-MS-QP2010 plus; Shimadzu, Kyoto, Japan) as outlined below. A constant column flow rate of 1.21 mL/min with helium gas was maintained in RESTEK capillary column (30 m × 0.25 mm I.D. × 0.25 µm film thickness).

Initial oven temperature was 100 °C for 3 min, which was increased to 250 °C for a hold time of 5 min, was further increased gradually to 280 °C where it was kept constant for 15 min. A total of 3 μL of sample (3 mg/mL) was injected in split mode (split ratio of 10.0) and linear velocity of the column was maintained at 40.9 cm/s. The mass fragmentation patterns (spectra) of the metabolites were obtained at electron ionization (EI) of 70 eV scanned over a m/z range of 40–650. The compounds detected were identified on the basis of comparison of the mass spectra with those available in the NIST14 and Wiley8 spectral library. The spectra having a match limit value lower than 700 were not considered.

2.9. Statistical Analysis

All the experiments were performed in triplicates. The data were expressed as mean ± standard deviation. The statistical analysis and significance of the test was performed by analysis of variance (ANOVA), using the software, graph pad prism 5.01. The graphs were also generated using grouped analysis in the graph pad prism and represented with SEM in the form of error bar.

3. Results and Discussion

Eight putative actinobacterial endophytes, designated as ADR1 to ADR8, were isolated from the plant, *Datura metel*. While no isolate could be found from the stem part of the plant, six were obtained from the root tissues and two were from the leaves. The antibacterial potency of these isolates was examined against the reference strain, *S. aureus* ATCC 25923. Based on the size of the zone of inhibition, the isolate ADR1 was chosen for further studies as it produced largest zone (22.5 ± 0.58 mm). Production of the metabolites by ADR1 was carried out under the conditions as described under Section 2.3. The metabolite extract was recovered as red colour hygroscopic sticky mass (approximately 120 mg/L).

3.1. Identification and Characterization of the Isolate ADR1

Amplification of the 16S rRNA gene from the genome of ADR1 produced a sequence of 1452 nucleotides. Blast analysis revealed 99.17% sequence identity of the ADR1 sequence with *S. californicus* strains with a query coverage of 99%. The phylogenetic relationship of the strain ADR1 can be seen in Figure 1, where it showed closest relationship with *S. californicus* strains. The 16S rRNA gene sequence obtained in this study was submitted as '*Streptomyces californicus* strain ADR1' to NCBI GenBank with accession no. KU299789.1.

A few strains of *S. californicus* had been reported earlier from the soil [41–44]. However, there are no reports of any endophytic strain of *S. californicus* till date to the best of our knowledge, making the present isolate as a new endophytic actinobacterial strain, designated as *S. californicus* strain ADR1.

The strain ADR1 was characterized for its cultural attributes on ISP media 1 to 7. The results (Table 1) suggested that the extent of growth of the culture varied from scanty to abundant on different ISP media. Further, differences with respect to the colour of substrate and aerial mycelia, and production of diffusible pigments were also noted as described in the Table 1. When compared with the cultural characteristics of non-endophytic *S. californicus* strains JCM 6910, MNM-1400 and G16, it was observed that ADR1 shared a few similarities, for example, colour of aerial mycelium on ISP 2, 3 and diffusible pigments on 4, with *S. californicus* strain JCM 6910, a soil isolate from Japan [41,44,45]. However, the growth of ADR1 was abundant on ISP3, while that of JCM 6910 was poor. No diffusible pigment was produced by ADR1 in ISP5, while violet pigment was produced by JCM 6910. Other soil isolates of *S. californicus*, strain MNM-1400 and strain G16, were morphologically very different from the strain ADR1 [44,45] Thus, the endophytic *S. californicus* strain ADR1 was evidently distinct from the soil isolates.

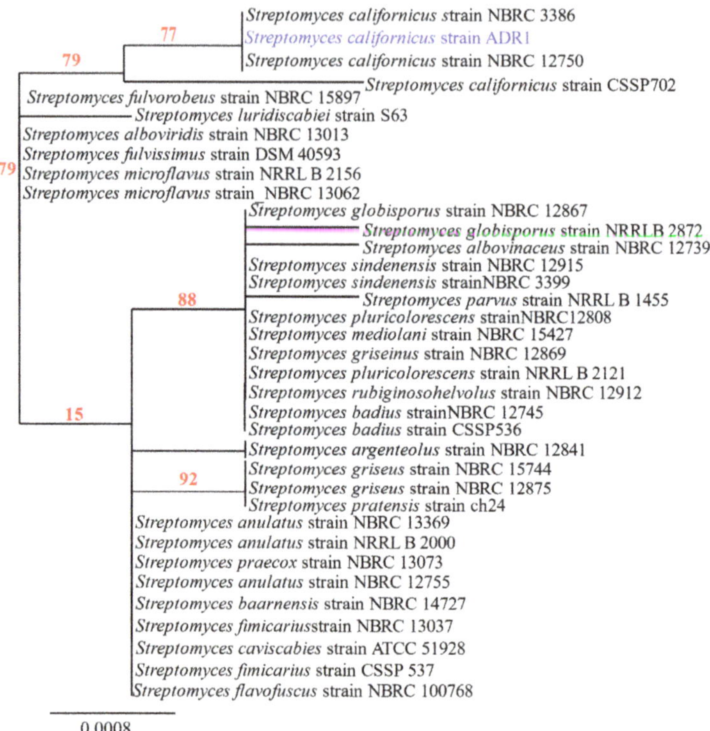

Figure 1. Phylogenetic analysis of isolate ADR1. Neighbour-joining phylogenetic tree showed maximum likelihood model showing the phylogenetic relationship of selected isolate (highlighted in blue) based on 16S rRNA gene sequence alignments. The numbers at the branching points are the percentages of occurrence in 500 bootstrapped trees. Bar indicated 0.0008 substitutions per nucleotide position.

Table 1. Cultural characteristics of the isolate ADR1 on international *Streptomyces* project (ISP) media.

ISP Media	Growth	Substrate Mycelium	Aerial Mycelium	Diffusible Pigments	Appearance
ISP-1 (Tryptone-Yeast Extract Broth)	Abundant	Crimson red	White	No	Shrinked and depressed with irregular edges
ISP-2 (Yeast extract- Malt extract Agar)	moderate	Wine Red	Pale green	Yellow	Elevated, smooth, regular edges
ISP-3 (Oatmeal agar)	Abundant	Wine Red	Dusty green	Light violet	Shrinked, pits formation, regular edges
ISP-4 (Inorganic salt starch agar)	Moderate	Dark pink	Light pink	Light pink	Flat, wavy edges, pointed centre
ISP-5 (Glycerol asparagine agar base)	Abundant	Pink	Dusty green	No	Elevated, Round, smooth edges
ISP-6 (Peptone yeast extract iron agar)	Moderate	Rusty red	White	Light pink	Elevated at centre, Round, smooth edges
ISP-7 (Tyrosine agar)	Scanty	Light pink	Whitish Pink	No	Pin pointed at centre, flat, round and smooth edges

A detailed view of the morphology was obtained through the study of single colonies (Figure 2; Panel A and B). Growth on different ISP media produced differences in colour and appearance of the colonies, which appeared as dense, depressed and rocky on ISP1, while on ISP2, 3 and 5 the aerial mycelia appeared to be fluffy and dusty. Clear exudates can be seen over the colony on ISP6. The colonies on ISP7 appeared scantly grown lacking distinct structures that were observed on other media. Prominent differences in the extent of growth, structure and pigmentation of the colonies on different ISP media were consistent with the earlier reports [46]. Such media-dependent phenotypic variations suggested that the primary and secondary metabolism of the culture varied significantly with changes in composition of the medium, which is in agreement with the current understanding of the physiology of the genus *Streptomyces* [47,48]. The microscopic observations showed highly branched flexuous mycelium and the arrangement of spores in a chain inside mycelium as shown in (Figure 2, panel C and D). These are some explicit features of *Streptomyces* species [32].

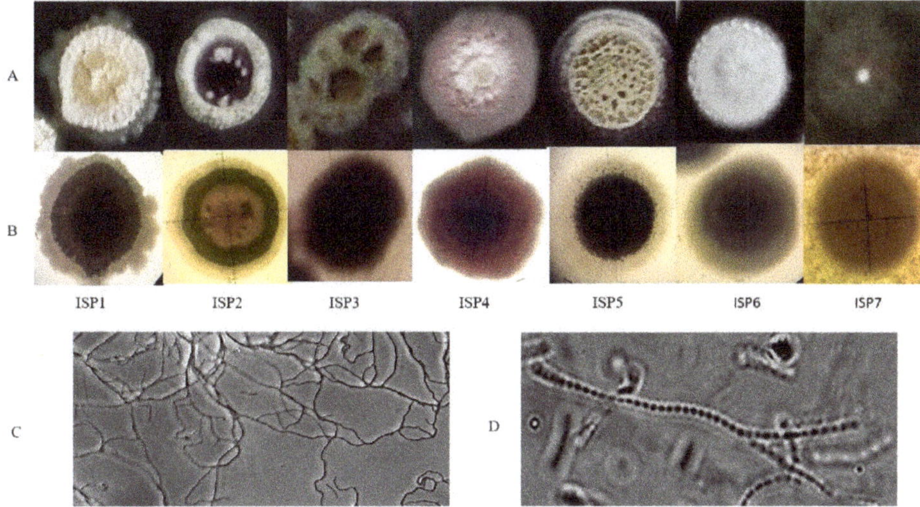

Figure 2. Morphological characterization of *S. californicus* ADR1. Colony observations were made on different ISP media. Panel (**A**) represented the magnified view showing exterior structures of the single colony while panel (**B**) showed interior view of the colony as observed under a Nikon Stereo Zoom Microscope SMZ1270; the mycelial network of branched hyphae could be observed in panel (**C**) and arrangement of spores in chains were viewed in panel (**D**).

3.2. Antibacterial Spectrum of ADR1 Metabolites Against Significant Gram-Positive Pathogens

The spectrum of activity of the ADR1 metabolites against different Gram-positive pathogens was determined as described under Section 2. The results presented in Table 2 showed that the metabolite extract possessed broad spectrum activity against Gram-positive pathogens as it effectively inhibited growth of all the reference strains. It may be noted that the activities of metabolite extract against MRSA strains were similar to those against *S. aureus* strains.

However, the above data provided more of a qualitative rather than quantitative estimate of the antibacterial activity. An accurate view of the potency of the metabolites was achieved by assessment of MIC$_{90}$ values against the pathogenic strains used in this study. As presented in Figure 3a, the MIC$_{90}$ values of the ADR1 metabolites against different strains of *S. aureus* were between $0.44 \pm 0.07 - 0.84 \pm 0.03$ µg/mL, while against *S. epidermis* ATCC1222 it was even lower (0.23 ± 0.01 µg/mL). It may be noted that the potency against MRSA strains (Figure 3b) was as good as it was against *S. aureus* strains. These results can be considered as significant since the MIC$_{90}$ values of the ADR1 metabolites stood better than the standard drug vancomycin (0.5 to 2 µg/mL) for MRSA strains [49]. Other than *S. aureus* strains, the MIC_{90} values of 1.92 ± 0.03 and 3.35 ± 0.18 µg/mL were observed against *E. faecium* strains AIIMS and ATCC 49214, respectively, which demonstrated good sensitivity of these strains to ADR1 metabolites. *E. faecalis* ATCC 29212 was the least sensitive to ADR1 metabolites among the reference pathogens since its MIC_{90} value (5.68 ± 0.20 µg/mL) stood at the maximum, but it was still better than some of the previously reported activities [50,51]. The MIC_{90} values of the metabolites from some of the recently reported *Streptomyces* spp. against the strains of *S. aureus* and MRSA were in the ranges of 2–125 µg/mL [52–54], which indicated wide variation in the anti-bacterial potency of the metabolites. The results of the present study clearly demonstrated that the anti-bacterial potency of ADR1 metabolites appeared at the bottom of the range or even lower, validating its promising potential for discovery of drugs against priority pathogens.

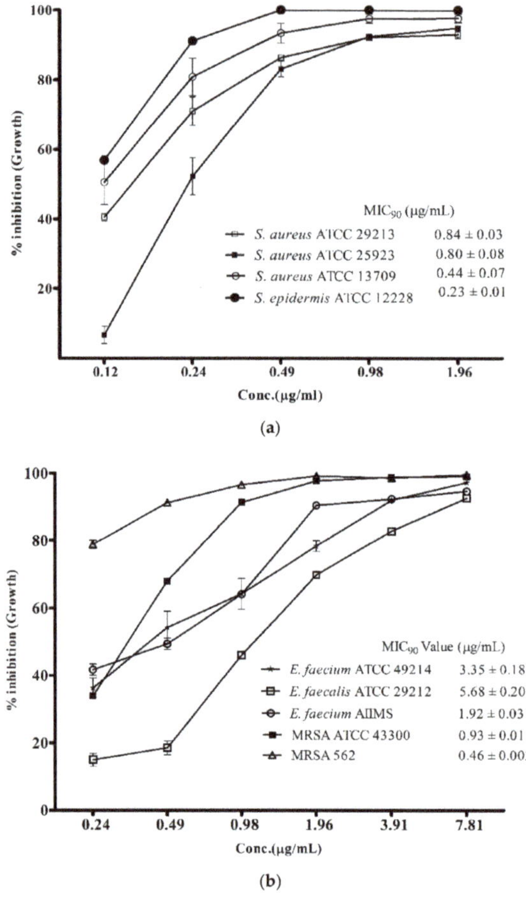

Figure 3. Potency of ADR1 metabolites for Gram-positive pathogens. Minimum inhibitory concentration (MIC_{90}) values of the metabolites against the target pathogens were corelated with its antibacterial potency. Percent inhibition in growth at different concentration of the metabolites was measured against the reference strains as shown in (**a**) and in (**b**). The data represented mean ± SEM values of experiments done in triplicate (*p* value < 0.0001).

Table 2. Spectrum of antibacterial activity of the metabolites produced by *S. californicus* ADR1.

S. No.	Reference Strains of Gram-Positive Pathogens	Zone of Inhibition (mm)
1	*Staphylococcus aureus* ATCC 29213	22.5 ± 0.58
2	*S. aureus* ATCC 25923	19 ± 0.42
3	*S. aureus* ATCC 13709	20 ± 0.5
4	*S. epidermis* ATCC 12228	18 ± 0.45
5	Methicillin-resistant *S. aureus* (MRSA) ATCC 43300	21.3 ± 0.27
6	MRSA 562 (clinical strain)	19 ± 0.25
7	*Enterococcus faecium* ATCC 49224	16.5 ± 0.4
8	*E. faecium* AIIMS	19.4 ± 0.47
9	*E. faecalis* ATCC 29212	17 ± 0.52

3.3. Anti-Biofilm Activity of ADR1 Metabolites Against S. aureus and MRSA

Biofilm-associated infections posed a greater challenge in treatment of infectious diseases as it is one of the major contributing factors in enhancing antibiotic-resistance among *S. aureus* and its methicillin-resistant strains [3]. Therefore, the discovery of drugs with potent anti-biofilm activity is needed more than ever before to combat the ever-growing global challenge of antibiotic resistance against the notorious pathogens like *S. aureus* and MRSA. In view of this, the biofilm inhibitory potential of the ADR1 metabolites was investigated. The results (Figure 4a) suggested that the ADR1 metabolites were able to effectively inhibit formation of biofilm by the *S. aureus* and the MRSA strains. Up to 90% reduction in the formation of biofilm could be achieved at significantly lower concentration of the metabolites; the BIC_{90} values were noted to be in the range of 0.74 ± 0.08 to 4.59 ± 0.71 µg/mL. The effectiveness of the ADR1 metabolites in the inhibition of the biofilm was found to be better than the previously reported activity of biofilm inhibition by actinobacterial metabolites where maximum inhibition of 83% was recorded at 265 µg/mL [35]. However, when the activity against pre-formed biofilms (24 h) was examined, the BIC_{90} values for *S. aureus* strains increased by many folds, for example, up to 45.69 ± 3.32 and 89.54 ± 0.40 µg/mL for *S. aureus* ATCC 25923 and *S. aureus* ATCC 29213, respectively. The biofilms produced by MRSA proved even more resistant (Figure 4b). It was reported that some of the well-known antibiotics like pyrrolomycin and related compounds could inhibit the biofilm only up to 67%–87% [55]. Similarly, in a study on the effect of antibiotics like rifampicin, polymyxin B, kanamycin and doxycyclin on reduction of *S. aureus* biofilm formation, only rifampicin was found to inhibit the biofilm by about 50% [56]. The effect of ADR1 metabolites on inhibition of biofilm formation as well as on the preformed biofilms was better than some previously reported metabolite extracts [57–59]. Inhibition of biofilm formation strongly suggested that the metabolites prevented adherence of *S. aureus* and MRSA cell to the polystyrene surface. Further, their ability to disrupt pre-formed biofilms might limit the biofilm-associated drug resistance among the pathogens.

3.4. Antioxidant Activity of the ADR1 Metabolites

Antioxidants assume significance for therapeutic applications in view of their role in neutralizing reactive oxygen species in patients fighting diseases involving infectious agents or metabolic disorders [60–62]. Hence, it was prudent to examine if the ADR1 metabolites possessed any such activity. The standard assay, which measured the reduction of DPPH free radical, revealed the antioxidant properties of ADR1 metabolites (Figure 5). The free radical scavenging activity of the metabolite extract and of ascorbic acid (a well-known antioxidant agent), followed a different pattern, where a sharp increase in DPPH scavenging activity (from 40% to 80%) was observed when concentration of the metabolite was increased from 62.5 to 125 µg/mL. However, in the case of ascorbic acid, there was no significant change in the activity over the above concentration range. The IC_{90} value for DPPH scavenging by ADR1 metabolite was achieved at the concentration of 217.24 ± 6.77 µg/mL, while that for the ascorbic acid it was 904.32 ± 12.93 µg/mL (Figure 5), which was approximately 4-fold higher compared to ADR1 metabolites. However, interestingly IC_{50} of the ascorbic acid was 4.617 ± 0.89 µg/mL while that of ADR1 extract was 77.41 ± 1.02 µg/mL (not plotted). It was recently reported that the endophytic actinobacterial strains BPSAC77, 101, 121 and 147 showed DPPH scavenging, with an IC_{50} value of 43.2 µg/mL, but the IC_{90} value was not reported in this study [63]. In another study, *Streptomyces* sp. had been found to scavenge DPPH free radicals at a much higher concentration, with an IC_{50} value at 435.31 ± 1.79 µg/mL [64]. In one of the recent studies, it was reported that IC_{50} values for DPPH scavenging activities of several isolates of actinobacteria varied from 12 ± 1.8 to 65 ± 3.2 µg/mL [65]. Such variations were also noted by others [54]. These reports indicated that there was a wide variation in the antioxidant activities of the metabolites produced by different actinobacterial strains. An in-depth characterization of relevant compounds in pure form may offer further insight into such differences in the activities.

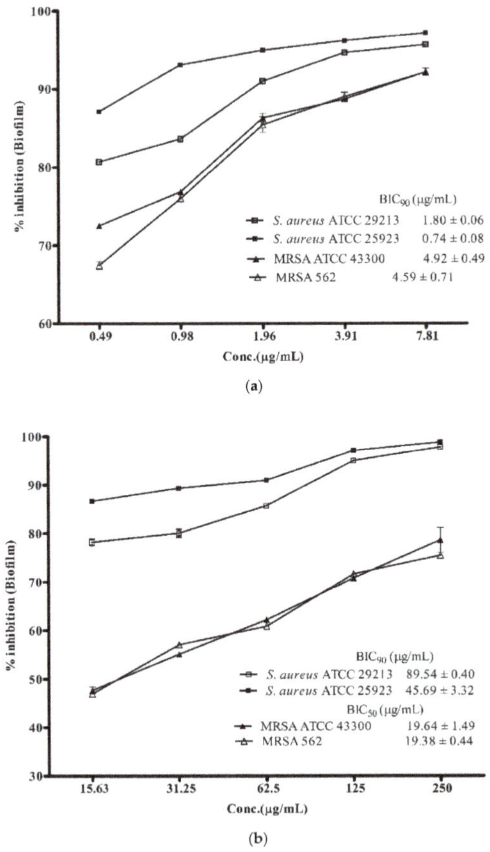

Figure 4. Anti-biofilm activity of ADR1 metabolites. (**a**). Inhibition of biofilm formation and (**b**) inhibition of pre-formed biofilm of *S. aureus* ATCC 29213, 25923, MRSA 562 and ATCC 43300 are shown at various concentration of the ADR1 metabolites. The data represented mean ± SEM values of experiments done in triplicate (*p* value < 0.0001).

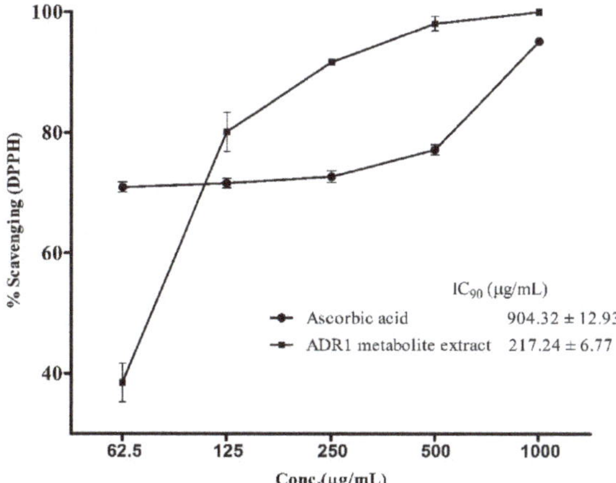

Figure 5. DPPH (1, 1-diphenyl-2-picrylhydrazyl) radical scavenging by ADR1 metabolites. Oxidation of DPPH free radicals were measured at the different concentrations of the metabolite. The data represented mean ± SEM values of the experiments done in triplicate (*p* value < 0.0001).

3.5. Haemolytic Activity

Drug-induced haemolytic anaemia (DIHA) and thrombocytopenia (DIT) are common adverse effects associated with antibiotics [66]. Reports suggested that the haemolytic effects of antimicrobial peptides tyrocidine A and gramicidin S, limited their use as topical agents [67,68]. Ceftriaxone, a third-generation cephalosporin was also reported to cause haemolysis, attracting advise for restricted administration of the drug [69]. So, it is very important for a potential drug to be tested for its haemolytic activity to secure critical information on its clinical suitability. In this context, the haemolytic effect of the ADR1 metabolites was tested. It was noted that the ADR1 metabolites did not show any sign of haemolysis in the concentration range from 7.8125 to 1000 µg/mL (Figure 6), which was far greater than the MIC$_{90}$ values against test pathogens.

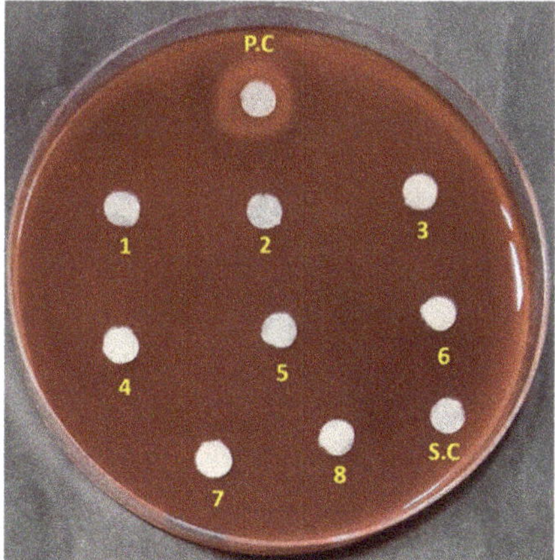

Figure 6. Haemolytic activity of the ADR1 metabolites. Spot No. 1 to 8 showed metabolite dilutions from higher (1000 µg/mL) to lower concentration (7.8125 µg/mL). 'PC' was positive control with 0. 1% SDS. SC represented solvent control.

In some previous studies, the extracts with low or no haemolytic activity were considered suitable for further characterization of their antimicrobial properties [70]. This implied that the metabolite extract of strain ADR1 could be safe for further investigation into specific metabolites as probable drug candidates.

3.6. Secondary Metabolite Profiling and GC-MS Analysis of Metabolite Extract

Characterization of bioactivities of the metabolites had demonstrated that the present endophytic strain, *S. californicus* ADR1 was an unexplored source of antimicrobial, antibiofilm and antioxidant agents, which might have potential clinical applications. It was therefore prudent to analyze the composition of the metabolite extract to unravel the class of compounds produced. Therefore, the ADR1 metabolite extract was subjected to chemical profiling using specific reagents. The tests revealed the classes of compounds present in the extract of metabolites produced by the strain ADR1, which included anthraquinones, anthocyanins, terpenoids, terpenes, flavonoids, phenols, alkaloids and glycosides (Table 3). Various compounds, which belonged to these classes have therapeutic significance and are routinely reported from phytochemical screening of plant extracts. But recent literature suggested that actinobacteria too are proving to be excellent sources of such class of compounds [14,71].

Table 3. Chemical profiling of ethyl acetate extract of secondary metabolites produced by *S. californicus* strain ADP4.

Chemical Class of Metabolites	Testes/Reagents Used	Observations	Results
Terpenoids	Salkowski Test	Reddish brown coloration at the interface	+
Phenols	Folin–Ciocalteu Test	Blue coloration was appeared	+
Flavonoids	Ferric chloride Test	Formation of greenish colour	+
	NaOH, HCl	Intense yellow coloration after adding HCl	
Terpenes	Salkowski Regent	Appearance of golden colour in the chloroform layer	+
Alkaloids	Wagner's Test	Formation of reddish-brown precipitate	+
Anthocyanins	HCl, Ammonia	Appearance of pink-red, turns blue	+
Anthraquinones	H_2SO_4, Chloroform, Ammonia	Light pink coloured layer of ammonia	+
Glycosides	Keller–Killiani Test	A reddish-brown colour ring at the junction of the two layers	+
Tannins	Lead Acetate Test	No precipitation was observed	−
Saponins	Foam Test	No frothing was observed	−
Lactones	Pyridine, sodium nitroprusside, NaOH	No change in coloration was observed	−
Coumarins	Alcoholic NaOH	Yellow fluorescence was not appeared on the paper soaked in NaOH	−
Sterols	Salkowski Test	No red colour was appeared in the lower layer (two layers formed)	−
Lignins	Gallic acid	No appearance of Olive-green coloration	−
Carbohydrates	Fehling's Test	No reddish violet ring appeared	−
Fatty acids	Ether	No appearance of transparence on filter paper	−
Proteins	Biuret Test	No violet coloration was observed	−

Note: '+': Present; '−': Absent.

While the results of chemical profiling were primarily qualitative in nature suggesting the presence or the absence of a specific chemical class, it did not offer any clue about the identity of the compounds. But it is important to have a deeper insight into the chemical class and identity of the specific compounds produced by the culture if these are to be considered for drug development. In pursuit of this goal,

GC-MS analysis of the ADR1 metabolites was undertaken to probe the identity of metabolites based on their mass spectrum.

The GC-MS analysis revealed the presence of alkaloids, terpene, terpenoids and glycosides in the ADR1 metabolite extract, thus confirming the findings of chemical profiling. The compounds were assigned probable identities based on similarity index (SI) value, generated from the mass spectral analyses of the test compounds and their comparison with the reference compounds listed in the NIST and Wiley library. An SI of more than 85 was considered for assignment of probable identity to the compounds [72]. Out of the total 30 peaks detected in the chromatogram, 20 peaks were specific to the culture *S. californicus* ADR1 while rest of the peaks were related to the production media and the solvent control (Figure S1). Only six of ADR1 metabolite peaks were found to have SI value greater than 85. Thus, identities could be assigned to only these compounds (Table 4). Majority of the compounds detected in the GC-MS chromatogram showed lower degree of similarity with SI values between 60–80 (Table S1), suggesting a high probability of occurrence of novel compounds among the secondary metabolites produced by *S. californicus* strain ADR1.

The first compound, detected at RT 11.945 min, showed SI of 88 with Methanoazulen-9-ol, decahydro-2, 2, 4, 8-tetramethyl-stereoisomer, which belonged to the class sesquiterpene and possessed antibacterial as well as antioxidant properties [73]. This compound was earlier reported as a major component of plant essential oils [74,75] but there are no reports on this from microbial sources as yet. Another compound in ADR1 metabolite extract was detected at RT 14.519 min, which showed SI of 90 with 5-z-methyl-2-z-hydroxycarbonyl-5-e-ethenyl-4-z-propen-2-ylcyclohexanone, was a terpenoid having structure similar to Asperaculane B, a known GABA-transaminase inhibitor having significance in epilepsy treatment [76]. Terpenes and terpenoids have wide therapeutic potential like antimalarial, antibacterial, antiinflammatory, antioxidant as well as wound healing properties [77]. A flavonoid that was found most abundant among the ADR1 metabolites, showed SI of 85 with 4′-Methoxy-2′-(trimethylsiloxy) acetophenone. Flavonoids are well regarded molecules with broad-spectrum biological activities as well as applications in nutraceuticals and cosmetic industry [78]. However, there are no reports on the activity of this molecule till date. An alkaloid with SI of 88 in comparison to pyridineethanamine, n-methyl-n-[2-(4-pyridinyl) ethyl] was detected among the ADR1 metabolites at RT 16.068 min. The compound is known for its importance in the treatment of vertigo [79]. Alkaloids are well regarded for their wide spectrum of therapeutic activities including antibacterial, antioxidant, anti-inflammatory, anti-diabetic, antimalarial and anticancer properties [80]. The presence of therapeutically important chemical classes of compounds in ADR1 metabolite extract underscored its importance in the discovery of novel compounds with prominent therapeutic potential.

Table 4. Similarity Index-based analysis of the compounds in the secondary metabolite extract of *S. californicus* ADR1. The data were obtained from GC-MS chromatogram. The reference compounds were from NIST and Wiley Library.

S. No.	RT (min)	Similarity Index (SI)	Reference Compounds	Chemical Class	Therapeutic Properties
1	11.945	88	Methanoazulen-9-ol, decahydro-2,2,4,8-tetramethyl-stereoisomer	Sesquiterpene (alpha-Caryophyllene alcohol)	Antibacterial, antioxidant, antiinflammatory [73]
2	13.089	96	Naphtho[2,3-g]-1,6,2,5-dioxasilaborocin	Organoboranic acid	No activity reported
3	13.231	85	2-[(trimethylsilyl)oxy]-4-methoxyacetophenone	Flavonoid	No activity reported
4	14.519	90	5-z-methyl-2-z-hydroxycarbonyl-5-e-ethenyl-4-z-propen- 2-ylcyclohexanone	Terpenoids (Asperaculane B type)	Anticonvulsant drug design pharmacophore, sodium channel blocker, GABA-transaminase inhibitors [76]
5	15.102	92	1,2-benzenedicarboxylic acid	Diisobutyl phthalate	No activity reported
6	16.068	88	2-pyridineethanamine, n-methyl-n-[2-(4-pyridinyl)ethyl]	Alkaloid (betahistine types)	Vasodilation and reduction of endolymphatic pressure [79]

4. Conclusions

The metabolites produced by *S. californicus* ADR1 had shown very low MIC_{90} values against a range of Gram-positive pathogens including those notified by the WHO as high priority pathogens. This suggested that it could be an excellent source for effective anti-infective compounds against some of the most challenging pathogens like *S. aureus*, MRSA, *S. epidermis, E. faecalis* and *E. faecium*. The ADR1 metabolites came across as potent inhibitors of biofilms of both methicillin-sensitive and methicillin-resistant strains of *S. aureus*. Along with the antimicrobial activity, the metabolites exhibited good antioxidant activity. Antioxidant property with no haemolytic activity indicated the potential of ADR1 metabolites for use as antioxidants. The above bioactivities could be attributed to the presence of several therapeutically significant class of compounds as revealed by the biochemical profiling and the GC-MS analysis. These constituents, individually or in combination, may account for the pharmacological actions of the extract. Low SI for several compounds in the ADR1 metabolite preparations was noteworthy as it indicated greater possibility of finding novel molecules.

Supplementary Materials: The following are available online at http://www.mdpi.com/2076-2607/8/6/929/s1, Figure S1: GC-MS Chromatogram of ADR1 metabolites; Table S1: Compounds detected in the ADR1 metabolites extract by GC-MS analysis using NIST & Wiley library as reference.

Author Contributions: A.K.D. has conceptualized the research project and guided the experimental work. R.S. has carried out the experiments and prepared the initial draft of the manuscript, which was corrected and finalized by A.K.D. All authors have read and agreed to the published version of the manuscript.

Funding: This research received no external funding.

Acknowledgments: The authors acknowledge the administrative and material support given by the Netaji Subhas University of Technology, New Delhi to carry out this work. Research fellowship to RS by NSUT is duly acknowledged. Services of Advanced Instrumentation Centre, Jawahar Lal Nehru University, New Delhi for GC-MS analysis, is duly acknowledged.

Conflicts of Interest: The authors declare no conflict of interest.

References

1. Chernov, V.M.; Chernova, O.A.; Mouzykantov, A.A.; Lopukhov, L.L.; Aminov, R.I. Omics of antimicrobials and antimicrobial resistance. *Expert Opin. Drug Discov.* **2019**, *14*, 455–468. [CrossRef] [PubMed]
2. *Prioritization of Pathogens to Guide Discovery, Research and Development of New Antibiotics for Drug-Resistant Bacterial Infections, Including Tuberculosis*; World Health Organization: Geneva, Switzerland, 2017.
3. Ch'ng, J.H.; Chong, K.; Lam, L.N.; Wong, J.J.; Kline, K.A. Biofilm-associated infection by enterococci. *Nat. Rev. Microbiol.* **2019**, *17*, 82–94. [CrossRef] [PubMed]
4. Xu, K.D.; McFeters, G.A.; Stewart, P.S. Biofilm resistance to antimicrobial agents. *Microbiol* **2000**, *146*, 547–549. [CrossRef] [PubMed]
5. Rowe, S.; Wagner, N.J.; Li, L.; Beam, J.E.; Wilkinson, A.D.; Radlinski, L.C.; Zhang, Q.; Miao, E.A.; Conlon, B.P. Reactive oxygen species induce antibiotic tolerance during systemic *Staphylococcus aureus* infection. *Nat. Microbiol.* **2020**, *5*, 282–290. [CrossRef] [PubMed]
6. Ivanov, A.V.; Bartosch, B.; Isaguliants, M.G. Oxidative stress in infection and consequent disease. *Oxidative Med. Cell. Longev.* **2017**, *2017*, 3496043. [CrossRef]
7. Kohanski, M.A.; Dwyer, D.J.; Hayete, B.; Lawrence, C.A.; Collins, J.J. A common mechanism of cellular death induced by bactericidal antibiotics. *Cell* **2007**, *130*, 797–810. [CrossRef]
8. Keren, I.; Wu, Y.; Inocencio, J.; Mulcahy, L.R.; Lewis, K. Killing by bactericidal antibiotics does not depend on reactive oxygen species. *Science* **2013**, *339*, 1213–1216. [CrossRef]
9. Liguori, I.; Russo, G.; Curcio, F.; Bulli, G.; Aran, L.; Della-Morte, D.; Gargiulo, G.; Testa, G.; Cacciatore, F.; Bonaduce, D.; et al. Oxidative stress, aging, and diseases. *Clin. Interv. Aging* **2018**, *13*, 757–772. [CrossRef]
10. Kulkarni, R.; Antala, S.; Wang, A.; Amaral, F.E.; Rampersaud, R.; Larussa, S.J.; Planet, P.J.; Ratner, A.J. Cigarette smoke increases *Staphylococcus aureus* biofilm formation via oxidative stress. *Infect. Immun.* **2012**, *80*, 3804–3811. [CrossRef]
11. Mak, K.K.; Pichika, M.R. Artificial intelligence in drug development: Present status and future prospects. *Drug Discov. Today* **2019**, *24*, 773–780. [CrossRef]

12. Ferreira, L.L.G.; Andricopulo, A.D. ADMET modeling approaches in drug discovery. *Drug Discov. Today* **2019**, *24*, 1157–1165. [CrossRef] [PubMed]

13. Patridge, E.; Gareiss, P.; Kinch, M.S.; Hoyer, D. An analysis of FDA-approved drugs: Natural products and their derivatives. *Drug Discov. Today* **2016**, *21*, 204–207. [CrossRef] [PubMed]

14. Singh, R.; Dubey, A.K. Diversity and applications of endophytic actinobacteria of plants in special and other ecological niches. *Front. Microbiol.* **2018**, *8*, 1767. [CrossRef] [PubMed]

15. Berdy, J. Thoughts and facts about antibiotics: Where we are now and where we are heading. *J. Antibiot.* **2012**, *65*, 385–395. [CrossRef] [PubMed]

16. Singh, R.; Dubey, A.K. Endophytic actinomycetes as emerging source for therapeutic compounds. *Indo Glob. J. Pharm. Sci.* **2015**, *5*, 106–116.

17. Silver, L.L. Challenges of antibacterial discovery. *Clin. Microbiol. Rev.* **2011**, *24*, 71–109. [CrossRef]

18. Guo, X.; Liu, N.; Li, X.; Ding, Y.; Shang, F.; Gao, Y.; Ruan, J.; Huang, Y. Red soils harbor diverse culturable actinomycetes that are promising sources of novel secondary metabolites. *Appl. Environ. Microbiol.* **2015**, *81*, 3086–3103. [CrossRef]

19. Nair, D.N.; Padmavathy, S. Impact of endophytic microorganisms on plants, environment and humans. *Sci. World J.* **2014**, *2014*, 250693. [CrossRef]

20. Coombs, J.T.; Franco, C.M. Isolation and identification of actinobacteria from surface-sterilized wheat roots. *Appl. Environ. Microbiol.* **2003**, *69*, 5603–5608. [CrossRef]

21. El-Nakeeb, M.A.; Lechevalier, H.A. Selective isolation of aerobic actinomycetes. *Appl. Microbiol.* **1963**, *11*, 75–77. [CrossRef]

22. Hayakawa, M.; Nomura, S. Humic acid-vitamin agar. A new medium for the selective isolation of soil actinomycetes. *J. Ferment. Technol.* **1987**, *65*, 501–509. [CrossRef]

23. Mohseni, M.; Norouzi, H.; Hamedi, J.; Roohi, A. Screening of antibacterial producing actinomycetes from sediments of the Caspian Sea. *Int. J. Mol. Cell. Med.* **2013**, *2*, 64–71. [PubMed]

24. Du Toit, E.A.; Rautenbach, M. A sensitive standardised micro-gel well diffusion assay for the determination of antimicrobial activity. *J. Microbiol. Methods* **2000**, *42*, 159–165. [CrossRef]

25. Farris, M.H.; Olson, J.B. Detection of Actinobacteria cultivated from environmental samples reveals bias in universal primers. *Lett. Appl. Microbiol.* **2007**, *45*, 376–831. [CrossRef]

26. Dereeper, A.; Guignon, V.; Blanc, G.; Audic, S.; Buffet, S.; Chevenet, F.; Dufayard, J.F.; Guindon, S.; Lefort, V.; Lescot, M.; et al. Phylogeny.fr: Robust phylogenetic analysis for the non-specialist. *Nucleic Acids Res.* **2008**, *1*, W465–W469. [CrossRef] [PubMed]

27. Dereeper, A.; Audic, S.; Claverie, J.M.; Blanc, G. BLAST-EXPLORER helps you building datasets for phylogenetic analysis. *BMC Evol. Biol.* **2010**, *10*, 1–6. [CrossRef] [PubMed]

28. Edgar, R.C. MUSCLE: Multiple sequence alignment with high accuracy and high throughput. *Nucleic Acids Res.* **2004**, *32*, 1792–1797. [CrossRef]

29. Guindon, S.; Gascuel, O. A simple, fast, and accurate algorithm to estimate large phylogenies by maximum likelihood. *Syst. Biol.* **2003**, *52*, 696–704. [CrossRef]

30. Anisimova, M.; Gascuel, O. Approximate likelihood ratio test for branchs: A fast, accurate and powerful alternative. *Syst. Biol.* **2006**, *55*, 539–552. [CrossRef]

31. Chevenet, F.; Brun, C.; Banuls, A.L.; Jacq, B.; Chisten, R. TreeDyn: Towards dynamic graphics and annotations for analyses of trees. *BMC Bioinform.* **2006**, *7*, 439. [CrossRef]

32. Shirling, E.B.; Gottlieb, D. Methods for characterization of *Streptomyces* species. *J. Syst. Evol. Microbiol.* **1966**, *16*, 313–340. [CrossRef]

33. CLSI. Methods for dilution antimicrobial susceptibility tests for Bacteria that grow aerobically, approved standard. 9th ed. CLSI document M07-A9. *Clin. Lab. Stand. Inst.* **2012**, *32*, 1–68.

34. Sivaranjani, M.; Gowrishankar, S.; Kamaladevi, A.; Pandian, S.K.; Balamurugan, K.; Ravi, A.V. Morin inhibits biofilm production and reduces the virulence of *Listeria monocytogenes*—An in vitro and in vivo approach. *Int. J. Food Microbiol.* **2016**, *237*, 73–82. [CrossRef]

35. Bakkiyaraj, D.; Pandian, K.S.T. In vitro and in vivo antibiofilm activity of a coral associated actinomycete against drug resistant *Staphylococcus aureus* biofilms. *Biofouling* **2010**, *26*, 711–717. [CrossRef]

36. Rahman, M.M.; Islam, M.B.; Biswas, M.; Khurshid Alam, A.H.M. In vitro antioxidant and free radical scavenging activity of different parts of *Tabebuia pallida* growing in Bangladesh. *BMC Res. Notes* **2015**, *8*, 621. [CrossRef]

37. Russell, F.M.; Biribo, S.S.N.; Selvaraj, G.; Oppedisano, F.; Warren, S.; Seduadua, A.; Mulholland, E.K.; Carapetis, J.R. As a bacterial culture medium, citrated sheep blood agar is a practical alternative to citrated human blood agar in laboratories of developing countries. *J. Clin. Microbiol.* **2006**, *44*, 3346–3351. [CrossRef]

38. Ayoola, G.A.; Coker, H.A.B.; Adesegun, S.A.; Bello, A.A.A.; Obaweya, K.; Ezennia, E.C.; Atangbayila, T.O. Phytochemical screening and antioxidant activities of some selected medicinal plants used for malaria therapy in southwestern Nigeria. *Trop. J. Pharm. Res.* **2008**, *7*, 1019–1024.

39. Subramaniam, D.; Menon, T.; Elizabeth, H.L.; Swaminathan, S. Anti-HIV-1 activity of *Sargassum swartzii* a marine brown alga. *BMC Infect. Dis.* **2014**, *14*, E43. [CrossRef]

40. Savithramma, N.; Rao, M.L.; Suhrulatha, D. Screening of medicinal plants for secondary metabolites. *Middle-East J. Sci. Res.* **2011**, *8*, 579–584.

41. Suetsuna, K.; Seino, A.; Kudo, T.; Osajima, Y. Production, and biological characterization, of Dideoxygriseorhodin C by a *Streptomyces* sp. and its taxonomy. *Agric. Biol. Chem.* **1989**, *53*, 581–583. [CrossRef]

42. Panzone, G.; Trani, A.; Ferrari, P.; Gastaldo, L.; Colombo, L. Isolation and structure elucidation of 7,8-dideoxy-6-oxo-griseorhodin C produced by *Actinoplanes ianthinogenes*. *J. Antibiot.* **1997**, *50*, 665–670. [CrossRef] [PubMed]

43. Tsuge, N.; Furihata, K.; Shin-Ya, K.; Hayakawa, Y.; Seto, H. Novel antibiotics pyrisulfoxin A and B produced by *Streptomyces californicus*. *J. Antibiot.* **1999**, *52*, 505–507. [CrossRef] [PubMed]

44. Srinivasan, S.; Bhikshapathi, D.V.R.N.; Krishnaveni, J.; Veerabrahma, K. Isolation of borrelidin from *Streptomyces californicus*—An Indian soil isolate. *Indian J. Biotechnol.* **2008**, *7*, 349–355.

45. Gozari, M.; Mortazavi, M.S.; Bahador, N.; Tamadoni Jahromi, S.; Rabbaniha, M. Isolation and screening of antibacterial and enzyme producing marine actinobacteria to approach probiotics against some pathogenic vibrios in shrimp *Litopenaeus vannamei*. *Iran J. Fish Sci.* **2016**, *15*, 630–644.

46. Sharma, P.; Singh, T.A.; Bharat, B.; Bhasin, S.; Modi, H.A. Approach towards different fermentative techniques for the production of bioactive actinobacterial melanin. *Beni-Suef Univ. J. Basic Appl. Sci.* **2018**, *7*, 695–700. [CrossRef]

47. Rao, M.P.N.; Xiao, M.; Li, W.J. Fungal and bacterial pigments: Secondary metabolites with wide applications. *Front. Microbiol.* **2017**, *8*, 1113.

48. Ser, H.L.; Yin, W.F.; Chan, K.G.; Khan, T.M.; Goh, B.H.; Lee, L.H. Antioxidant and cytotoxic potentials of *Streptomyces gilvigriseus* MUSC 26T isolated from mangrove soil in Malaysia. *Prog. Microbes Mol. Biol.* **2018**, *1*, a0000002. [CrossRef]

49. Chaudhari, C.N.; Tandel, K.; Grover, N.; Bhatt, P.; Sahni, A.K.; Sen, S.; Prahraj, A.K. In vitro vancomycin susceptibility amongst methicillin resistant *Staphylococcus aureus*. *Med. J. Armed Forces India* **2014**, *70*, 215–219. [CrossRef]

50. Horiuchi, K.; Shiota, S.; Hatano, T.; Yoshida, T.; Kuroda, T.; Tsuchiya, T. Antimicrobial activity of oleanolic acid from *Salvia officinalis* and related compounds on vancomycin-resistant *Enterococci* (VRE). *Biol. Pharm. Bull.* **2007**, *30*, 1147–1149. [CrossRef]

51. Maasjost, J.; Mühldorfer, K.; Cortez de Jäckel, S.; Hafez, H.M. Antimicrobial susceptibility patterns of *Enterococcus faecalis* and *Enterococcus faecium* isolated from poultry flocks in Germany. *Avian Dis.* **2015**, *59*, 143–148. [CrossRef]

52. Gos, F.M.W.R.; Savi, D.C.; Shaaban, K.A.; Thorson, J.S.; Aluizio, R.; Possiede, Y.M.; Rohr, J.; Glienke, C. Antibacterial activity of endophytic actinomycetes isolated from the medicinal plant *Vochysia divergens* (Pantanal, Brazil). *Front. Microbiol.* **2017**, *8*, 1642. [CrossRef] [PubMed]

53. Girão, M.; Ribeiro, I.; Ribeiro, T.; Azevedo, I.C.; Pereira, F.; Urbatzka, R.; Leão, P.N.; Carvalho, M.F. Actinobacteria isolated from *Laminaria ochroleuca*: A source of new bioactive compounds. *Front. Microbiol.* **2019**, *10*, 683. [CrossRef] [PubMed]

54. Siddharth, S.; Vittal, R.R.; Wink, J.; Steinert, M. Diversity and bioactive potential of actinobacteria from unexplored regions of Western Ghats, India. *Microorganisms* **2020**, *7*, 225. [CrossRef]

55. Schillaci, D.; Petruso, S.; Raimondi, M.V.; Cusimano, M.G.; Cascioferro, S.; Scalisi, M.; La Giglia, M.A.; Vitale, M. Pyrrolomycins as potential anti-Staphylococcal biofilms agents. *Biofouling* **2010**, *26*, 433–438. [CrossRef]

56. Tote, K.; Berghe, D.V.; Deschacht, M.; de Wit, K.; Maes, L.; Cos, P. Inhibitory efficacy of various antibiotics on matrix and viable mass of *Staphylococcus aureus* and *Pseudomonas aeruginosa* biofilms. *Int. J. Antimicrob. Agents* **2009**, *33*, 525–531. [CrossRef] [PubMed]

57. Melo, T.A.; Dos Santos, T.F.; de Almeida, M.E.; Junior, L.A.; Andrade, E.F.; Rezende, R.P.; Marques, L.M.; Romano, C.C. Inhibition of *Staphylococcus aureus* biofilm by Lactobacillus isolated from fine cocoa. *BMC Microbiol.* **2016**, *16*, 250. [CrossRef]

58. Brambilla, L.Z.S.; Endo, E.H.; Cortez, D.A.G.; Filhoa, B.P.D. Anti-biofilm activity against *Staphylococcus aureus* MRSA and MSSA of neolignans and extract of *Piper regnellii*. *Rev. Bras. Farmacogn.* **2017**, *27*, 112–117. [CrossRef]

59. Yu, H.; Liu, M.; Liu, Y.; Qin, L.; Jin, M.; Wang, Z. Antimicrobial Activity and mechanism of action of *Dracocephalum moldavica* L. extracts against clinical isolates of *Staphylococcus aureus*. *Front. Microbiol.* **2019**, *10*, 1249. [CrossRef]

60. Ouyang, Y.; Li, J.; Peng, Y.; Huang, Z.; Ren, Q.; Lu, J. The role and mechanism of thiol-dependent antioxidant system in bacterial drug susceptibility and resistance. *Curr. Med. Chem.* **2020**, *27*, 1940–1954. [CrossRef]

61. Elswaifi, S.F.; Palmieri, J.R.; Hockey, K.S.; Rziqalinski, B.A. Antioxidant nanoparticles for control of infectious disease. *Infect. Disord. Drug Targets (Former. Curr. Drug Targets-Infect. Disord.)* **2009**, *9*, 445–452. [CrossRef]

62. Pellegrino, D. Antioxidants and cardiovascular risk factors. *Diseases* **2016**, *4*, 11. [CrossRef]

63. Passari, A.K.; Mishra, V.K.; Singh, G.; Singh, P.; Kumar, B.; Gupta, V.K.; Sarma, R.K.; Saikia, R.; Donovan, A.O.; Singh, B.P. Insights into the functionality of endophytic actinobacteria with a focus on their biosynthetic potential and secondary metabolites production. *Sci. Rep.* **2018**, *7*, 11809. [CrossRef] [PubMed]

64. Nimal, C.I.V.; Kumar, P.P.; Agastian, P. In vitro α-glucosidase inhibition and antioxidative potential of an endophyte species (*Streptomyces sp. loyola* UGC) isolated from *Datura stramonium* L. *Curr. Microbiol.* **2013**, *67*, 69–76. [CrossRef] [PubMed]

65. Chaudhuri, M.; Paul, A.K.; Pal, A. Isolation and assessment of metabolic potentials of bacteria endophytic to carnivorous plants *Drosera Burmannii* and *Utricularia* Spp. *Biosci. Biotechnol. Res. Asia* **2019**, *16*, 731–741. [CrossRef]

66. Garratty, G.; Arndt, P. Drugs that have been shown to cause drug-induced immune hemolytic anemia or positive direct antiglobulin tests: Some interesting findings since 2007. *Immunohematology* **2014**, *30*, 66–79.

67. Marques, M.A.; Citron, D.M.; Wang, C.C. Development of Tyrocidine A analogues with improved antibacterial activity. *Bioorganic Med. Chem.* **2007**, *15*, 6667–6677. [CrossRef]

68. Mosges, R.; Baues, C.M.; Schroder, T.; Sahin, K. Acute bacterial otitis externa: Efficacy and safety of topical treatment with an antibiotic ear drop formulation in comparison to glycerol treatment. *Curr. Med. Res. Opin.* **2011**, *27*, 871–878. [CrossRef]

69. Guleria, V.S.; Sharma, N.; Amitabh, S.; Nair, V. Ceftriaxone-induced hemolysis. *Indian J. Pharmacol.* **2013**, *45*, 530–531. [CrossRef]

70. Ghosh, T.; Biswas, M.K.; Chatterjee, S.; Roy, P. In-vitro study on the hemolytic activity of different extracts of Indian medicinal plant *Croton bonplandianum* with phytochemical estimation: A new era in drug development. *J. Drug Deliv. Ther.* **2018**, *8*, 155–160. [CrossRef]

71. Srivastava, V.; Singla, R.K.; Dubey, A.K. Inhibition of biofilm and virulence factors of *Candida albicans* by partially purified secondary metabolites of *Streptomyces chrestomyceticus* strain ADP4. *Curr. Top. Med. Chem.* **2018**, *11*, 925–945. [CrossRef]

72. Gumbi, B.P.; Moodley, B.; Birungi, G.; Ndungu, P.G. Target, suspect and non-target screening of silylated derivatives of polar compounds based on single Ion monitoring GC-MS. *Int. J. Environ. Res. Public Health* **2019**, *16*, 4022. [CrossRef] [PubMed]

73. Salem, M.Z.M.; Zayed, M.Z.; Ali, H.M.; Abd El-Kareem, M.S.M. Chemical composition, antioxidant and antibacterial activities of extracts from *Schinus molle* wood branch growing in Egypt. *J. Wood Sci.* **2016**, *62*, 548–561. [CrossRef]

74. Ghaffari, T.; Kafil, H.S.; Asnaashari, S.; Farajnia, S.; Delazar, A.; Baek, S.C.; Hamishehkar, H.; Kim, K.H. Chemical composition and antimicrobial activity of essential oils from the aerial parts of *Pinus eldarica* grown in Northwestern Iran. *Molecules* **2019**, *24*, E3203. [CrossRef]

75. Xiong, L.; Peng, C.; Zhou, Q.M.; Wan, F.; Xie, X.F.; Guo, L.; Li, X.H.; He, C.J.; Dai, O. Chemical composition and antibacterial activity of essential oils from different parts of *Leonurus japonicus* Houtt. *Molecules* **2013**, *18*, 963–973. [CrossRef]

76. Gao, Y.Q.; Guo, C.J.; Zhang, Q.; Zhou, W.M.; Wang, C.C.; Gao, J.M. Asperaculanes A and B, two sesquiterpenoids from the fungus *Aspergillus aculeatus*. *Molecules* **2014**, *20*, 325–334. [CrossRef] [PubMed]
77. Hussein, R.A.; El-Anssary, A.A. *Plants Secondary Metabolites: The Key Drivers of the Pharmacological Actions of Medicinal Plants in Herbal Medicine*; Philip, F., Ed.; IntechOpen: London, UK, 2019; pp. 11–30.
78. Karak, P. Biological activities of flavonoids: An overview. *Int. J. Pharm. Sci.* **2019**, *10*, 1567–1574.
79. Gbahou, F.; Davenas, E.; Morisset, S.; Arrang, J.M. Effects of betahistine at histamine H3 receptors: Mixed inverse agonism/agonism in vitro and partial inverse agonism in vivo. *J. Pharmacol. Exp. Ther.* **2010**, *334*, 945–954. [CrossRef]
80. Debnath, B.; Uddina, M.D.J.; Pataric, P.; Das, M., Maitia, D., Manna, K. Estimation of alkaloids and phenolics of five edible cucurbitaceous plants and their antibacterial activity. *Int. J. Pharm. Pharm. Sci.* **2018**, *7*, 223–227.

Article

Taxonomic Characterization and Secondary Metabolite Analysis of NEAU-wh3-1: An *Embleya* Strain with Antitumor and Antibacterial Activity

Han Wang [1,†], Tianyu Sun [1,†], Wenshuai Song [1], Xiaowei Guo [1], Peng Cao [1], Xi Xu [1], Yue Shen [1,2,*] and Junwei Zhao [1,*]

1   Key Laboratory of Agricultural Microbiology of Heilongjiang Province, Northeast Agricultural University, No. 600 Changjiang Road, Xiangfang District, Harbin 150030, China; wanghan507555536@gmail.com (H.W.); sty1561214024@163.com (T.S.); wenshuaisong@163.com (W.S.); guoweizi@hotmail.com (X.G.); cp511@126.com (P.C.); xuxi1758899581@126.com (X.X.)
2   College of Science, Northeast Agricultural University, No. 600 Changjiang Road, Xiangfang District, Harbin 150030, China
*   Correspondence: shenyuelele@163.com (Y.S.); guyan2080@126.com (J.Z.)
†   These authors contributed equally to this work.

Received: 28 February 2020; Accepted: 18 March 2020; Published: 20 March 2020

Abstract: Cancer is a serious threat to human health. With the increasing resistance to known drugs, it is still urgent to find new drugs or pro-drugs with anti-tumor effects. Natural products produced by microorganisms have played an important role in the history of drug discovery, particularly in the anticancer and anti-infective areas. The plant rhizosphere ecosystem is a rich resource for the discovery of actinomycetes with potential applications in pharmaceutical science, especially *Streptomyces*. We screened *Streptomyces*-like strains from the rhizosphere soil of wheat (*Triticum aestivum* L.) in Hebei province, China, and thirty-nine strains were obtained. Among them, the extracts of 14 isolates inhibited the growth of colon tumor cell line HCT-116. Strain NEAU-wh-3-1 exhibited better inhibitory activity, and its active ingredients were further studied. Then, 16S rRNA gene sequence similarity studies showed that strain NEAU-wh3-1 with high sequence similarities to *Embleya scabrispora* DSM 41855[T] (99.65%), *Embleya hyalina* MB891-A1[T] (99.45%), and *Streptomyces lasii* 5H-CA11[T] (98.62%). Moreover, multilocus sequence analysis based on the five other house-keeping genes (*atp*D, *gyr*B, *rpo*B, *rec*A, and *trp*B) and polyphasic taxonomic approach comprising chemotaxonomic, phylogenetic, morphological, and physiological characterization indicated that the isolate should be assigned to the genus *Embleya* and was different from its closely related strains, therefore, it is proposed that strain NEAU-wh3-1 may be classified as representatives of a novel species of the genus *Embleya*. Furthermore, active substances in the fermentation broth of strain NEAU-wh-3-1 were isolated by bioassay-guided analysis and identified by nuclear magnetic resonance (NMR) and mass spectrometry (MS) analyses. Consequently, one new Zincophorin analogue together with seven known compounds was detected. The new compound showed highest antitumor activity against three human cell lines with the 50% inhibition (IC_{50}) values of 8.8–11.6 μg/mL and good antibacterial activity against four pathogenic bacteria, the other known compounds also exhibit certain biological activity.

Keywords: Rhizosphere soil; *Embleya*; NEAU-wh-3-1; compound; antitumor activity; antibacterial activity

---

1. Introduction

Tumor, especially malignant tumor, has become one of the major diseases, which is a serious threat to the health of people around the world [1,2]. According to records of the World Health Organization

(WHO) in 2018, more than 9 million people died of cancer, which was the second leading cause of death worldwide [3]. This figure will further rise because of aging, intensification of industrialization and urbanization, lifestyle modifications, etc. [2]. Thus, the burden of cancer cannot be ignored and the search for effective anticancer drugs is urgent [4]. On the other hand, the severe cancer incidence is also an invisible spur to the development of anti-tumor drugs throughout the world. So far, chemotherapy is still an important method for cancer treatment. Among the chemotherapeutics used, antitumor antibiotics derived from natural products account for a large proportion [5–7]. Natural product antibiotics were derived from various source materials including terrestrial plants, terrestrial microorganisms, marine organisms, and some invertebrates [8].

Microbial natural products have, in fact, been an excellent resource for drug discovery, particularly in the anticancer and anti-infective areas [9–11]. The phylum Actinobacteria accounts for a high proportion of soil microbial biomass and contains the most economically significant prokaryotes, producing more than half of the bioactive compounds in a literature survey, including antibiotics, antitumor agents, and enzymes [8,12–14]. Many famous antibiotics, such as bleomycin (BLM), mitomycin, anthracyclines [15], actinomycin D (ActD), polyether ionophore antibiotics, tetracyclines, quinolones, and so on, are derived from actinomycetes, which played an important role in the drug market [16]. Zincophorin, also referred to as M144255 or griseochellin, is a polyoxygenated ionophoric antibiotic [17], and has been reported to possess strong activity against Gram-positive bacteria and have strong cytotoxicity against human lung carcinoma cells A549 and Madin-Darby canine kidney cells MDCK [17,18], which was also isolated from actinomycetes. As the main genus of Actinobacteria, *Streptomyces* is the largest antibiotic producer. More than 70% of nearly 10,000 microbial origin compounds are produced by *Streptomyces* while some rare actinobacterial genera only accounted for less than 30% [19–22]. As abundant resources of larger number and wider variety of new antibiotics, *Streptomyces* strains have been continuously noted rather than any other actinomycete genera [19,23]. *Streptomyces* are widely distributed in terrestrial ecosystems, especially in the soil [24,25]. However, as time goes on, the possibility of finding novel compounds from *Streptomyces* in conventional soil has decreased and the rediscovery rate is high [22,26]. In recent years, studies on actinomycetes from diverse habitats have suggested new chemical structures and bioactive compounds [27,28]. Rhizosphere soil, the thin layer of soil around the roots of plants, has been a potential region for the discovery of functional microbes due to its special ecological environment. As early as the beginning of the last century, Hiltner proposed that there are more microorganisms in rhizosphere soil than surrounding soil [29–31]. There is a close relationship between rhizosphere microorganisms and plants. Plants can release organic compounds and signal molecules through root secretions to recruit microbial flora that are beneficial to their own growth. Microbes can control plant pathogens and pests by synthesizing multiple antibiotics, thereby indirectly promoting plant growth [32–34]. In recent years, many biologically active microorganisms and active substances produced by their secondary metabolism have been isolated from plant rhizosphere soil [35–38].

The genus *Embleya*, was very recently transferred from genus *Streptomyces* and established by Nouioui et al [39] and is a new member of the family *Streptomycetaceae* in the order *Streptomycetales* [39,40]. *Embleya* forms well-branched substrate mycelia with long aerial hyphae in open spirals and contains LL-diaminopimelic acid in the cell wall peptidoglycan, MK-9(H$_4$) or MK-9(H$_6$) as the major isoprenoid quinone and phosphatidylethanolamine (PE) as the predominant phospholipid [41], which is very similar to that of *Streptomyces* [42]. At present, the genus comprises only two species: *Embleya scabrispora* and *Embleya hyaline*. *Embleya scabrispora* was originally proposed as *Streptomyces scabrisporus* sp. nov. [43], and it has been reclassified to the genus *Embleya* as the type species [39,40], it could produce hitachimycin with antitumor, antibacterial, and antiprotozoal activities [44–46]; and *Embleya hyaline* was first described as *Streptomyces hyalinum* [41,47], and it has been reported to produce nybomycin which is an effective agent against antibiotic-resistant *Staphylococcus aureus* and it was called a reverse antibiotic [48].

In this study, an *Embleya* strain, NEAU-wh-3-1, with better antitumor activity was isolated from the wheat rhizosphere soil. The taxonomic identity of strain NEAU-wh3-1 was determined by a combination of 16S rRNA gene sequence and five other house-keeping genes (*atp*D, *gyr*B, *rpo*B, *rec*A, and *trp*B) analysis with morphological and physiological characteristics. The active substances of strain NEAU-wh-3-1 were also isolated, identified, and determined. Furthermore, the cytotoxicity and antimicrobial activity of the isolated compounds were tested.

2. Materials and Methods

2.1. Isolation of Streptomyces-Like Strains

Rhizosphere soil of wheat (*Triticum aestivum* L.) was collected from Langfang, Hebei Province, Central China (39°32′ N, 116°40′ E). The soil sample should be protected from light and air-dried at room temperature for 14 days before isolation for *Streptomyces*-like strains. After drying, the soil sample was ground into powder and then suspended in sterile distilled water followed by a standard serial dilution technique. The diluted soil suspension was spread on humic acid-vitamin agar (HV) [49] supplemented with cycloheximide (50 mg L^{-1}) and nalidixic acid (20 mg L^{-1}). After 28 days of aerobic incubation at 28 °C, colonies were transferred and purified on the International *Streptomyces* Project (ISP) medium 3 [50], and maintained as glycerol suspensions (20%, *v/v*) at −80 °C for long-period preservation.

2.2. Screening of Strains with Antitumor Activity

All the isolated were cultured on ISP medium 2 (yeast extract-malt extract agar) and incubated at 28 °C for 7 days. The spores of the strains were transferred into 250 mL Erlenmeyer flasks containing 30 mL of the production broth containing maltodextrin 4%, lactose 4%, yeast extract 0.5%, and MOPS 2%, at pH 7.2–7.4. on a rotary shaker at 250 r.p.m at 28 °C. After seven days, the production broth was extracted with an equal volume of methanol for approximately 24 h. After filtration, the filtrate substances were evaporated under reduced pressure at 50 °C to yield the crude extract and then dissolved in DMSO (dimethyl sulfoxide) at concentrations of 20 µg/mL and 100 µg/mL. The HCT-116 (human colorectal carcinoma) cell lines were maintained in Dulbecco's modified Eagle's medium supplemented with 10% (*w/v*) fetal bovine serum in a humidified incubator at 37 °C of 5% CO$_2$ incubator. The antitumor activities of extracts with two concentrations were investigated by the SRB (Sulforhodamine B) colorimetric method. Briefly, treated cells were harvested and seeded at a density of 5×10^4 cells/well into a sterile flat bottom 96-well plate for 24 h, the cells were treated with different concentrations of the extracts for 48 h and growth inhibition was measured by determining the optical density at 510 nm, and the assay was performed basing on an established method [51].

2.3. Morphological and Physiological and Biochemical Characteristics of NEAU-wh3-1

Gram staining was carried out by using the standard method and morphological characteristics were observed by light microscopy (Nikon ECLIPSE E200, Nikon Corporation, Tokyo, Japan) and scanning electron microscopy (Hitachi SU8010, Hitachi Co., Tokyo, Japan) using cultures grown on ISP 3 agar at 28 °C for 2 weeks. Samples for scanning electron microscopy were prepared as described by Jin et al. [52]. Growth at different temperatures (4, 10, 15, 20, 25, 28, 32, 37, 40, and 45 °C) was determined on ISP 3 medium after incubation for 14 days. Growth tests for pH range (pH 4.0–12.0, at intervals of 1.0 pH unit) using the buffer system described by Zhao et al. [53] and tolerance of various NaCl concentrations (0–10%, with an interval of 1%, *w/v*) were tested in GY (Glucose-yeast extract powder) medium (glucose 1%, yeast extract 1%, K$_2$HPO$_4$ 3H$_2$O 0.05%, MgSO$_4$ 7H$_2$O 0.05%, *w/v*, pH 7.2) at 28 °C for 14 days on a rotary shaker. Hydrolysis of Tweens (20, 40, and 80) and production of urease were tested as described by Smibert and Krieg [54]. The utilization of sole carbon and nitrogen sources, decomposition of cellulose, hydrolysis of starch and aesculin, reduction of nitrate, coagulation and peptonization of milk, liquefaction of gelatin, and production of H_2S were examined as described previously [55,56].

2.4. Chemotaxonomic Analysis of NEAU-wh3-1

Biomass for chemotaxonomic characterization was prepared by growing strain NEAU-wh3-1 in ISP 2 broth in shake flasks at 28 °C for 7 days. Cells were harvested by centrifugation, washed twice with distilled water, and freeze-dried. The whole-cell sugars were analyzed according to the procedures developed by Lechevalier and Lechevalier [57]. The polar lipids were examined by two-dimensional TLC (thin layer chromatography) and identified using the method of Minnikin et al. [58]. Menaquinones were extracted from freeze-dried biomass and purified according to Collins [59]. *Streptomyces lutosisoli* DSM 42165^T [60] was used as the reference strain for identification of menaquinones. Extracts were analyzed by a HPLC-UV method [61] using an Agilent Extend-C18 Column (150 × 4.6 mm, i.d. 5 μm) (Agilent Corp., Santa Clara, CA, USA) at 270 nm.

2.5. Phylogenetic Analysis of NEAU-wh3-1

Extraction of genomic DNA, PCR amplification of the 16S rRNA gene sequence and sequencing of PCR products were carried out using a standard procedure [62]. The PCR product was purified and cloned into the vector pMD19-T (Takara Bio Inc., Dalian, China) and sequenced using an Applied Biosystems DNA sequencer (model 3730XL, Applied Biosystems Inc., Foster City, California, USA). The almost complete 16S rRNA gene sequence of strain NEAU-wh3-1, comprising 1487 bp, was obtained and compared with type strains available in the EzBioCloud server [63] and retrieved using NCBI BLAST (https://blast.ncbi.nlm.nih.gov/Blast.cgi;), and then submitted to the GenBank database. The phylogenetic tree was constructed based on the 16S rRNA gene sequences of strain NEAU-wh3-1 and related reference species. Sequences were multiply aligned in Molecular Evolutionary Genetics Analysis (MEGA) using the Clustal W algorithm and trimmed manually where necessary. Phylogenetic trees were generated with the neighbor-joining [64] and maximum-likelihood [65] algorithms using MEGA software version MEGA 7.0 [66]. The stability of the topology of the phylogenetic tree was assessed using the bootstrap method with 1000 replicates [67]. A distance matrix was generated using Kimura's two-parameter model [68]. All positions containing gaps and missing data were eliminated from the dataset (complete deletion option). The *gyr*B gene was amplified with primers PF-1 and PR-2 [69] under the PCR program for 16S rRNA gene. PCR of the *atp*D, *rec*A, *rpo*B, and *trp*B genes were performed using primers and amplification conditions described by Guo et al. [70]. The sequence data were exported as single gene alignments or a concatenated five-gene alignment for subsequent analysis as described above. Trimmed sequences of the five housekeeping genes were concatenated head-to-tail in-frame in the order *atp*D (430 bp)-*gyr*B (354 bp)-*rec*A (431 bp)-*rpo*B (208 bp)-*trp*B (556 bp). Phylogenetic analysis was performed as described above.

2.6. Production

The strain *Embleya* sp. NEAU-wh3-1 was grown on the ISP medium 2 (yeast extract-malt extract agar) and incubated for 6–7 days at 28 °C. The spores of the strain were transferred into two 1.0 L Erlenmeyer flasks containing 250 mL of the seed medium and incubated at 28 °C for 48 h on a rotary shaker at 250 r.p.m. All of the media were sterilized at 121 °C for 30 min. The seed culture (8%) was transferred into 60 flasks (1.0 L) containing 250 mL of production broth. The production broth was composed of maltodextrin 4%, lactose 4%, yeast extract 0.5%, MOPS 2%, at pH 7.2–7.4. The flasks were incubated at 28 °C for 7 days, shaken at 250 r.p.m.

2.7. Extraction and Isolation

The final 15.0 L production broth was filtered to separate supernatant and mycelial cake. The supernatant was subjected to a Diaion HP-20 resin column and eluted with 95% EtOH. The mycelial cake was washed with water (3 L) and subsequently extracted with MeOH (3 L) to obtain soluble material. The MeOH extract and the EtOH eluents were evaporated under reduced pressure at 50 °C to yield the crude extract (24 g). The crude extract was chromatographed on a silica gel column and eluted with a stepwise gradient of CH_2Cl_2/MeOH (95:5/90:10/85:15/80:20/70:30/65:35, *v/v*) and giving three fractions (Fr.1–Fr.3) based on the TLC profiles, which was performed on silica-gel plates with solvent system of $CHCl_3$/MeOH (9:1, *v/v*). The Fr.1 was subjected to a Sephadex LH-20 column eluted with CH_2Cl_2/MeOH (1:1, *v/v*) and detected by TLC to give two subfractions (Fr.1-1-Fr.1-2). The Fr.1-1 was further isolated by semi-preparative HPLC (Agilent 1100, Zorbax SB-C18, 5 μm, 250 × 9.4 mm inner diameter; 1.5 mL min^{-1}; 254 nm; Agilent, PaloAlto, CA, USA) eluting with CH_3CN/H_2O (90:10, *v/v*) to give compound **1** (t_R 25.06 min, 10.5 mg), the Fr.1-2 was further isolated by preparative HPLC (Shimadzu LC-8 A, Shimadzu-C18, 5 μm, 250 × 20 mm inner diameter; 20 mL min^{-1}; 220 /254 nm; Shimadzu, Kyoto, Japan) eluting with a stepwise gradient MeOH/H_2O (30–80%, *v/v* 30 min), and giving compound **2** (t_R 12.7 min, 7.5 mg), compound **3** (t_R 17.5 min, 12.7 mg) and compound **4** (t_R 22.6 min, 16.3 mg). The Fr.2 was subjected to another silica gel column eluted with n-hexane/acetone (95:5-60:40, *v/v*) and further purified by semi-preparative HPLC (Agilent 1100, Zorbax SB-C18, 5 μm, 250 × 9.4 mm inner diameter; 1.5 mLmin^{-1}; 254 nm; Agilent, PaloAlto, CA, USA) eluting with CH_3CN/H_2O (75:25, *v/v*) to give compound **5** (t_R 15.1 min, 13.5 mg) and compound **6** (t_R 24.3 min, 18.5 mg). Fr.3 was treated by an another silica gel column and eluted with a stepwise gradient of n-hexane/acetone (100:0-40:60, *v/v*) to give three fractions Fr.3-1–Fr.3-3 according to their TLC profiles, which was observed on silica-gel plates with solvent system of n-hexane/acetone (1:3, *v/v*). The Fr.3-3 was further purified by semi-preparative HPLC (Agilent 1260, Zorbax SB-C18, 5 μm, 250 × 9.4 mm inner diameter; 1.5 mL min^{-1}; 220nm; 254 nm; Agilent, PaloAlto, CA, USA) eluting with CH_3CN/H_2O (45:55, *v/v*) to obtain compounds **7** (t_R 25.1 min, 13.0 mg) and **8** (t_R 30.1 min, 7.5 mg).

2.8. General Experimental Procedures

IR spectra were recorded on a Thermo Nicolet Avatar FT-IR-750 spectrophotometer (Thermo, Tokyo, Japan) using KBr disks. Optical rotations were measured on a Perkin-Elmer 341 polarimeter (PerkinElmer, Inc. Suzhou, China). UV spectra were recorded on a Varian CARY 300 BIO spectrophotometer (Varian, Cary, NC, USA). The HR-ESI-MS and ESI-MS were taken on a Q-TOF Micro LC-MS-MS mass spectrometer (Waters Co, Milford, MA, U.S.A.). Nuclear magnetic resonance (NMR) spectra (400 MHz for ^{1}H and 100 MHz for ^{13}C) were measured with a Bruker DRX-400 spectrometer (Bruker, Rheinstetten, Germany). HPLC analysis was performed on a preparative HPLC (Shimadzu LC-8 A, Shimadzu-C18, 5 μm, 250 × 20 mm inner diameter; 20 mL min^{-1}; 220/254 nm; Shimadzu, Kyoto, Japan) as well as a semipreparative HPLC (Agilent 1100, Zorbax SB-C18, 5 μm, 250 × 9.4 mm inner diameter; 1.5 mL/min; 220/254 nm; Agilent, Palo Alto, CA, USA). Column chromatography were consisted of silica gel (100–200 mesh, Qingdao Haiyang Chemical Group Co., Qingdao, China) as well as Sephadex LH-20 gel (GE Healthcare, Glies, UK), which were analyzed by thin-layer chromatography (TLC). TLC was performed on silica-gel plates (HSGF254, Yantai Chemical Industry Research Institute, Yantai, China) and the developed plates were observed under a UV lamp at 254 nm or by heating after spraying with sulfuric acid-ethanol, 5:95 (*v/v*).

2.9. Biological Assays

The cytotoxicity of the eight compounds was assayed by cell counting kit-8 (CCK-8) colorimetric method [71] in vitro against the human leukemia cells K562, hepatocellular liver carcinoma cell line HepG2, and the human colon tumor cell line HCT-116. The cell lines were routinely in Dulbecco's Modified Eagle's Medium (DMEM) containing 10% calf serum at 37 °C for 4 h in a humidified atmosphere of 5% CO_2 incubator. The adherent cells at logarithmic phase were digested by pancreatic enzymes and inoculated onto 96-well culture plate at a density of 1.0×10^4 cells per/well. Test samples and control were dissolved in DMSO (dimethyl sulfoxide) and then added to the medium, incubated for 72 h. Then, the cell counting kit-8 (CCK-8, Dojindo, Kumamoto, Japan) reagent was added to the medium followed by further incubation for 3 h. Absorbance at 450 nm with a 600 nm reference was measured thereafter using a SpectraMax M5 microplate reader (Molecular Devices Inc., Sunnyvale, CA, USA). The inhibitory rate of cell proliferation was expressed as IC_{50} values and calculated by the following formula:

$$\text{Growth inhibition (\%)} = [\text{ODcontrol-ODtreated}]/\text{ODcontrol} \times 100$$

Doxorubicin was tested as a positive control, and cell solutions containing 0.5% DMSO were tested as a negative control.

The antibacterial activities of the isolated compounds were tested against Gram-positive bacteria *Staphylococcus aureus*, *Bacillus subtilis*, and *Sarcina lutea* and Gram-negative bacteria *Klebsiella pneumoniae* and *Escherichia coli* with the minimum inhibitory concentration (MIC) method recommended by the Clinical and Laboratory Standards Institute [72].

3. Results

3.1. Isolation and Screening of an Antitumor Compound Producing Strains

Thirty-nine strains belonging to actinomycetes were isolated from the soil samples. The crude extracts of these isolates were examined for their cytotoxic activity at dilution concentrations of 100 µg/mL and 20 µg/mL. As a result of primary screening, fourteen strains showed cytotoxic activity to human colon tumor cell line HCT-116 (Figure 1). Due to the superior cytotoxic activity of strain NEAU-wh3-1, which inhibition rate was greater than 80% at both concentrations, further chemical investigations were performed on this strain.

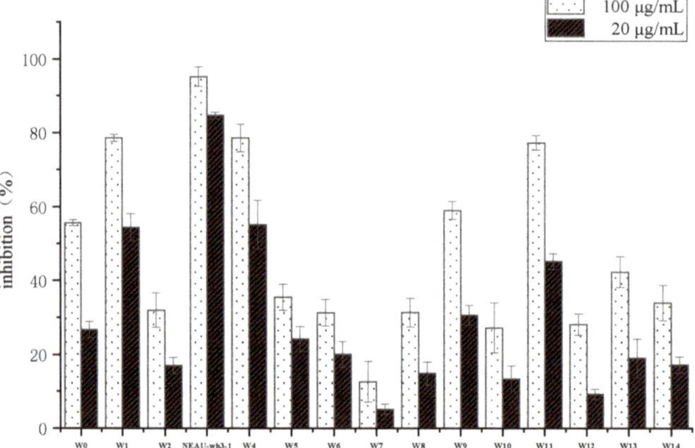

Figure 1. Antitumor activities of extracts obtained from fourteen isolates against human colon tumor cell line HCT-116.

3.2. Polyphasic Taxonomic Characterization of NEAU-wh3-1

Morphological observation of 2-week-old cultures of strain NEAU-wh3-1 grown on ISP 3 medium revealed that the strain has the typical characteristics of genus *Embleya* and formed well-developed, branched substrate hyphae and aerial mycelium that differentiated into spiral spore chains consisted of cylindrical spores (0.6–0.8µm × 0.9–1.3µm), the spores were rough-surfaced and non-motile (Figure 2). Strain NEAU-wh3-1 was found to grow at a temperature range of 4 to 37 °C (optimum temperature 28 °C), pH 5 to 12 (optimum pH 7), and NaCl tolerance of 0% to 3% (optimum NaCl of 1%). The physiological and biochemical properties of strain NEAU-wh3-1, *Embleya scabrispora* DSM 41855[T], *Embleya hyalina* MB891-A1[T], and *Streptomyces lasii* 5H-CA11[T] are given in Table 1.

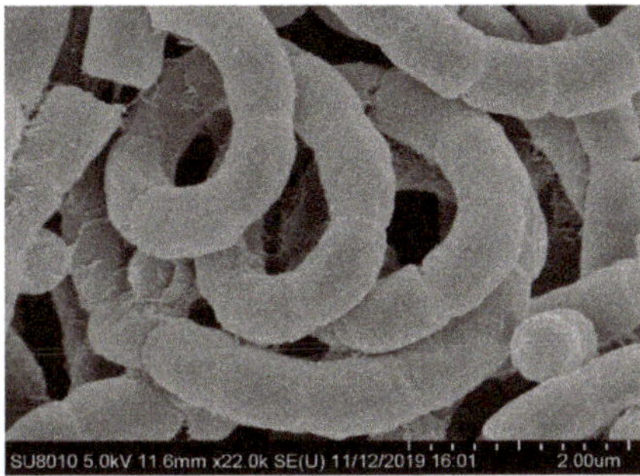

Figure 2. Scanning electron micrograph of spore chains of strain NEAU-wh3-1 grown on ISP 3 agar for 2 weeks at 28 °C.

Table 1. The physiological and biochemical properties of strain NEAU-wh3-1, *Embleya scabrispora* DSM 41855[T], *Embleya hyalina* MB891-A1[T], *Streptomyces lasii* 5H-CA11[T].

Characteristic	1	2[a,c]	3[b]	4[c]
Decomposition of				
Cellulose	−	−	ND	+
Tween 20	−	+	ND	+
Tween 40	+	+	ND	+
Tween 80	−	+	ND	+
Liquefaction of gelatin	−	−	+	+
Growth temperature (°C)	4–37	18–36	10–28	15–38
pH range for growth	5–12	4–10	6–11	5–11
NaCl tolerance range (*w/v*, %)	0–3	0–3	0–1	0–2.5
Milk coagulation	−	w	−	+
Nitrate reduction	−	+	−	−
Starch hydrolysis	−	w	−	−
Carbon source utilization				
L-arabinose	+	±	−	−
Dulcitol	+	w	−	W
D-Fructose	+	−	+	−
D-Galactose	+	−	+	−
D-Glucose	+	+	+	+
Inositol	−	+	+	−
Lactose	+	+	−	+
D-Maltose	+	−	±	−
D-Mannitol	+	−	−	−
D-Mannose	+	−	+	+
D-Raffinose	+	−	−	+
D-Ribose	−	+	ND	−
D-Sorbitol	+	−	−	−
D-Sucrose	+	±	W	+
D-Xylose	+	+	−	−
L-Rhamnose	−	+	+	−
Nitrogen source utilization				
L-Alanine	+	W	ND	+
L-Arginine	+	W	ND	+
L-Asparagine	+	W	ND	−
L-Aspartic acid	+	+	ND	+
Creatine	+	w	ND	−
L-Glutamic acid	+	w	ND	+
L-Glutamine	+	w	ND	+
Glycine	+	w	ND	+
L-Proline	−	+	ND	+
L-Serine	+	−	ND	−
L-Threonine	+	w	ND	+
L-Tyrosine	+	+	ND	+
Phospholipids	DPG, PE, PI, UL	PE, PGL	DPG, PE, PI	DPG, PME, PI, PIM, GL
Menaquinones	MK-9(H$_4$), MK-9(H$_6$), MK-9(H$_8$)	MK-9(H$_2$), MK-9(H$_4$), MK-9(H$_6$)	MK-9(H$_4$), MK-9(H$_6$), MK-9(H$_8$)	MK-9(H$_4$), MK-9(H$_6$), MK-9(H$_8$)
Whole cell-wall sugars	Arabinose, glucose, ribose	Arabinose	Arabinose, glucose	Glucose, ribose

Strains: 1, NEAU-wh3-1; 2, *Embleya scabrispora* DSM 41855[T]; 3, *Embleya hyalina* MB891-A1[T]; 4, *Streptomyces lasii* 5H-CA11[T]. Abbreviation: +, positive; −, negative; ±, ambiguous; ND, not determined; w, weak; DPG, diphosphatidylglycerol; PME, phosphatidylmonomethylethanolamine; PE, phosphatidylethanolamine; PI, phosphatidylinositol; PIM, phosphatidylinositolmannoside; UL, unidentified lipid; GL, glucosamine-containing lipid; PGL, phospholipid containing glucosamine. All data are from this study except where marked. [a] Data from Ping et al. [43]; [b] Data from Komaki et al. [41]; [c] Data from Liu et al. [73].

Chemotaxonomic analyses revealed that strain NEAU-wh3-1 contained LL-diaminopimelic acid as cell wall diamino acid. The whole-cell sugar was found to contain arabinose, glucose, and ribose. The phospholipid profile consisted of diphosphatidylglycerol (DPG), phosphatidylethanolamine (PE), phosphatidylinositol (PI), and two unidentified lipids (ULs) (Supplementary Figure S1). The menaquinones detected were MK-9(H$_4$) (46.5%), MK-9(H$_6$) (45.8%), and MK-9(H$_8$) (7.7%).

The almost complete 16S rRNA gene sequence of strain NEAU-wh3-1 (1487 bp) was determined and deposited with the accession number MN928616 in the GenBank/EMBL (European Molecular Biology Laboratory)/DDBJ (DNA Data Bank of Japan) databases. EzBioCloud analysis suggests that strain NEAU-wh3-1 shared the highest 16S rRNA gene sequence similarities with *Embleya scabrispora* DSM 41855^T (99.65%), *Embleya hyalina* MB891-A1^T (99.45%), and *Streptomyces lasii* 5H-CA11^T (98.62%). Phylogenetic analysis based on the 16S rRNA gene sequences indicated that the strain formed a stable cluster with *E. scabrispora* DSM 41855^T, *E. hyalina* MB891-A1^T, and *S. lasii* 5H-CA11^T based on neighbor-joining algorithm (Figure 3) and also supported by the maximum-likelihood algorithm (Supplementary Figure S2). To further clarify the affiliation of strain NEAU-wh3-1 to its closely related strains, partial sequences of housekeeping genes including *atp*D, *gyr*B, *rec*A, *rpo*B, and *trp*B were obtained. GenBank accession numbers of the sequences are displayed in Table S1. The phylogenetic tree based on the neighbor-joining tree constructed from the concatenated sequence alignment (1979 bp) of five housekeeping genes (Figure 4) suggested that the isolate clustered with *E. scabrispora* DSM 41855^T and *E. hyalina* MB891-A1^T, and also supported by the maximum-likelihood algorithm (*Streptomyces lasii* 5H-CA11^T lacks housekeeping genes; Supplementary Figure S3). Moreover, pairwise distances calculated for NEAU-wh3-1 and the related species using concatenated sequences of *atp*D-*gyr*B-*rec*A-*rpo*B-*trp*B were well above 0.007 (Table S2) for the related species, which was considered to be the threshold for species determination [74].

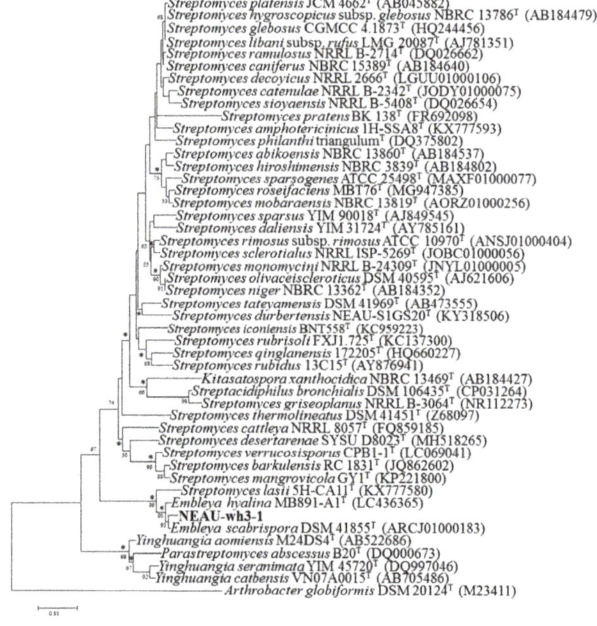

Figure 3. Neighbor-joining tree showing the phylogenetic position of strain NEAU-wh3-1 (1487 bp) and related taxa based on 16S rRNA gene sequences. Bootstrap values > 50% (based on 1000 replications) are shown at branch points. *Arthrobacter globiformis* DSM 20124^T (M23411) was used as an outgroup. *Asterisks* indicate branches also recovered in the maximum-likelihood tree; Bar, 0.01 substitutions per nucleotide position.

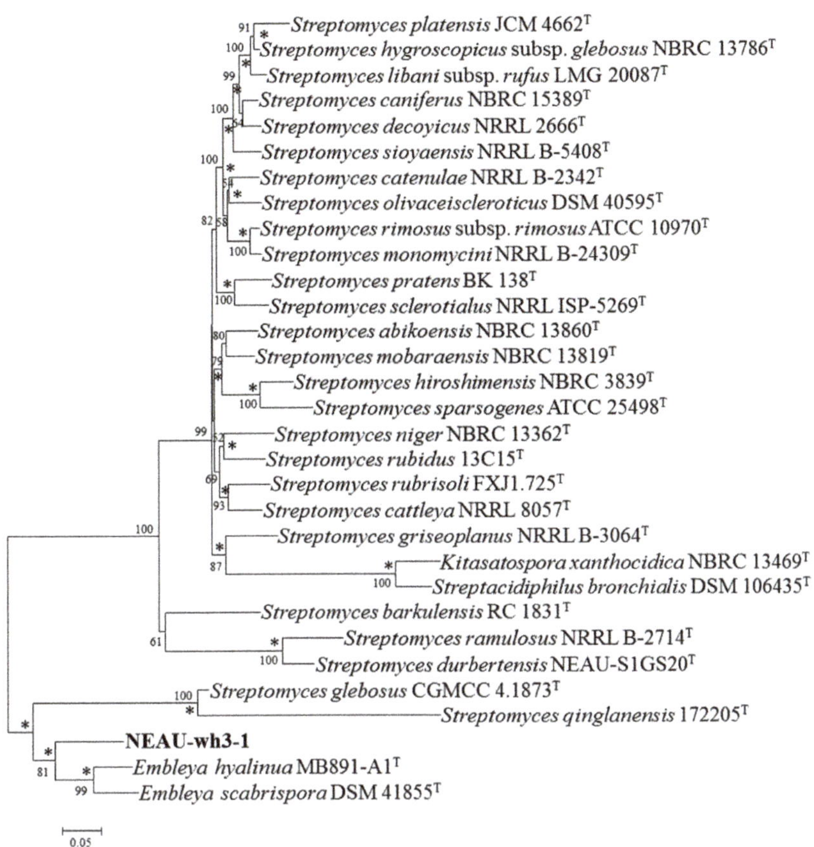

Figure 4. Neighbor-joining tree based on multilocus sequence analysis (MLSA) analysis of the concatenated partial sequences (1979 bp) from five housekeeping genes (*atp*D, *gyr*B, *rec*A, *rpo*B, and *trp*B) of strain NEAU-wh3-1 (in bold) with related taxa. Only bootstrap values above 50% (percentages of 1000 replications) are indicated. Asterisks indicate branches also recovered in the maximum-likelihood tree; Bar, 0.05 substitutions per nucleotide position.

3.3. Structural Elucidation

The strain NEAU-wh3-1 was grown preparative scale in 15.0 L of production broth for 7 days. Bioassay-guided isolation of the active components of the strain yielded eight main bioactive compounds. Compounds **2–8** are known compounds, which structures were elucidated as conglobatin (**2**) [75], piericidin C1 (**3**) [76], piericidin C5 (**4**) [77], piericidin A1 (**5**) [78], piericidin A3 (**6**) [76], Mer-A 2026 A (**7**) [79], and BE-52211 D (**8**) [80] by analysis of their spectroscopic data and comparison with literature values (Figure 5, Figures S12–S27). Compound **1** is a new zincophorin analogue (Figure 6, Figures S4–S11) [17].

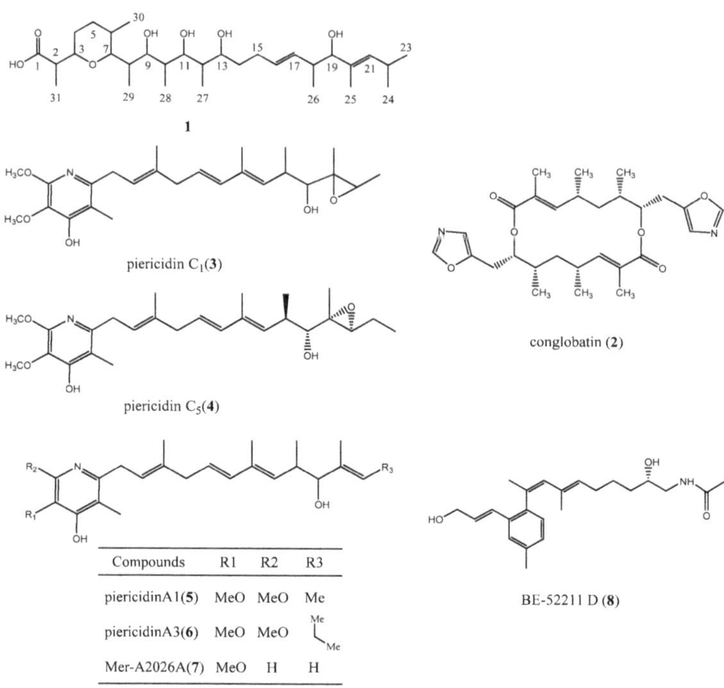

Figure 5. The structures of compounds **1–8**. Compound **1** was isolated as white solid with $[\alpha]_D^{25} + 15$ (c 0.043, EtOH) and UV (EtOH) λmax nm (log ε): 202 (4.53). Its molecular formula was established as $C_{31}H_{56}O_7$ by HR-ESI-MS at *m/z* 539.3942 [M-H]⁻ (calcd 539.3953 as $C_{31}H_{55}O_7$). The IR spectrum revealed hydroxyl absorption at 3320 cm⁻¹ and carbonyl absorption at 1735 cm⁻¹, as well as methyl and methylene absorptions at 2953 cm⁻¹ and 2924 cm⁻¹.

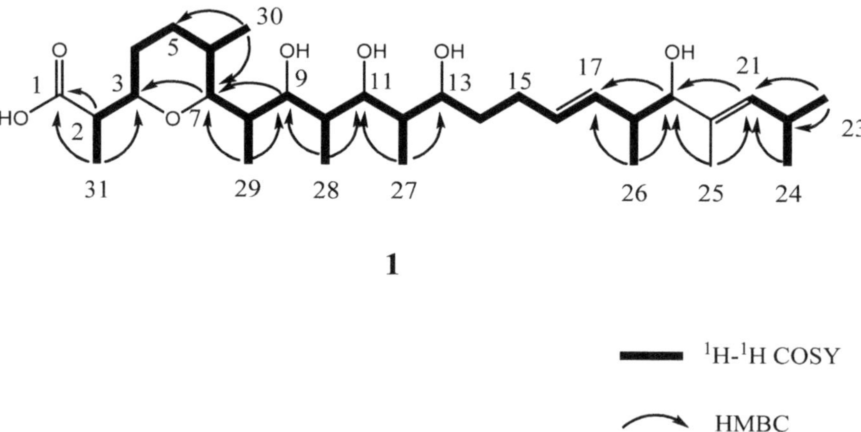

Figure 6. 2D nuclear magnetic resonance (NMR) correlations of compound **1**.

Analysis of ¹H NMR spectrum of **1** revealed the presence of three olefinic protons at δ_H 5.52 (1H, m), 5.36 (1H, m), 5.20 (1H, d, *J* = 8.9 Hz), seven aliphatic methine protons at δ_H 4.07 (1H, m), 4.06 (1H, m), 3.77 (1H, d, *J* = 8.9 Hz), 3.72 (1H, dd, *J* = 9.6, 2.1 Hz), 3.58 (1H, d, *J* = 9.2 Hz), 3.49 (1H, m), 3.28 (1H, m), seven methylene protons at δ_H 2.18 (1H, m), 2.13 (1H, m), 1.77 (1H, m), 1.67 (2H, m), 1.40 (1H, m),

1.35 (1H, m), 1.28 (1H, m), one singlet methyl at δ_H 1.63 s, in addition to eight doublet methyl protons at δ_H 1.18 (3H, d, J = 7.1 Hz), 1.15 (3H, d, J = 7.2 Hz), 1.11 (3H, d, J = 7.0 Hz), 0.98 (3H, d, J = 6.5 Hz), 0.97 (3H, d, J = 6.6 Hz), 0.86 (3H, d, J = 6.7 Hz), 0.81 (3H, d, J = 6.5 Hz), 0.70 (3H, d, J = 6.7 Hz). The ^{13}C NMR and DEPT135 spectra (Table 2) of 1 showed 31 resonances attributable to a carbonyl carbon at δ_C 175.6, one sp^2 quaternary carbon at δ_C 132.3, three sp^2 methines at δ_C 136.6, 134.5, 132.4. In the sp^3-carbon region, the spectrum showed six oxygenated methines at δ_C 84.3, 83.3, 81.7, 76.1, 74.0, 69.4, six methines at δ_C 42.2, 37.1, 36.5, 36.0, 33.0, 26.1, four methylenes at δ_C 34.2, 28.6, 27.1, 24.9 and nine methyl carbons at δ_C 22.9, 22.9, 17.1, 16.7, 15.4, 12.3, 11.3, 10.8, 10.1. The ^{1}H–^{1}H COSY correlations (Figure 6) of   H-2/H-3/H$_2$-4/H$_2$-5/H-6/H-7/H-8/H-9/H-10/H-11/H-12/H-13/H$_2$-14/H$_2$-15/H-16/H-17/H-18/H-19 established connectivity from H-2 atom along the chain through to C-19 atom. The correlations between H-21/H-22/H$_3$-23/H$_3$-24, H-12/H$_3$-27, H-10/H$_3$-28, H-18/H$_3$-26, H-8/H$_3$-29 protons in the ^{1}H–^{1}H COSY spectrum (Figure 6) indicated the five structural units of C-21–C-24, C-18–C-26, C-12–C-27, C-10-C-28, C-8-C-29. The observed HMBC (heteronuclear multiple bond correlation) correlations (Figure 6) from H$_3$-23, H$_3$-24 to C-21, C-22, from H$_3$-25 to C-19, C-21, from H$_3$-26 to C-17, C-18, and C-19, from H$_3$-27 to C-11, C-12, and C-13, from H$_3$-28 to C-9, C-10, C-11, from H$_3$-29 to C-7, C-8, C-9, from H$_3$-30 to C-5, C-6, C-7, from H$_3$-31 to C-2 and C-3, from H-21 to C-19, H-20 to C-18, from H-19 to C-17, from H$_3$-23 to C-21 established the linkage of C-2–C-22. The carbonyl group was connected with C-2 by the HMBC corrections from H-2 and H$_3$-31 to C-1 (δ_C 175.6). The correlations from H-7 (δ_H 3.77 d, J = 8.9 Hz) to C-3 (δ_C 74.0) indicated the linkage of C-3 and C-7 through an oxygen atom to form a tetrahydropyran ring. Taking the molecular formula of $C_{31}H_{56}O_7$ into account, four hydroxyl groups were situated at C-9, C-11, C-13, C-19, respectively, and a carboxyl group was situated at C-1. Comparison the NMR data of 1 with Zincophorin [17], a mocarboxylic acid ionophore contains one single tetrahydropyran ring, which was isolated from a strain of *Streptomyces griseus*, implied that 1 was identified to be an analogue of Zincophorin, the difference between two compounds was that the terminal ethyl group in Zincophorin was replaced by a H proton in compound 1. On the basis of the above spectroscopic data, a gross structure of 1 was established and named Zincophorin B, and the ^{1}H and ^{13}C resonances in 1 were assigned (Table 2).

Table 2. ^{1}H and ^{13}C NMR data of compound 1 in CDCl$_3$.

No.	δ_H (J in Hz)	δ_C (p.p.m)	No.	δ_H (J in Hz)	δ_C (p.p.m)
1		175.6	16	5.52 m	132.4
2	3.28 m	37.1	17	5.36 m	134.5
3	4.07 m	74.0			
4	1.67 m	24.9	18	2.27 m	42.2
			19	3.58 d (9.2)	81.7
5a	1.28 m	27.1	20		132.3
5b	1.40 m		21	5.20 d (8.9)	136.6
6	1.52 m	26.1	22	2.59 m	26.1
7	3.77 d (8.9)	76.1	23	0.97 d (6.6)	22.9
8	2.04 m	33.0	24	0.98 d (6.5)	22.9
9	3.72 dd (9.6, 2.1)	84.3	25	1.63 s	10.1
10	2.01 m	36.5	26	0.86 d (6.7)	16.7
11	3.49 m	83.3	27	1.15 d (7.2)	10.8
12	1.75 m	36.0	28	0.70 d (6.7)	12.3
			29	1.11 d (7.0)	11.3
13	4.06 m	69.4	30	0.81 d (6.5)	17.1
14a	1.35 m	34.2	31	1.18 d (7.1)	15.4
14b	1.77 m				
15a	2.13 m	28.6			
15b	2.18 m				

3.4. Biological Activity

The cytotoxic activities of compounds 1–8 against K562, HCT-116, and HepG2 cancer cell lines are showed in Table 3. Eight compounds restrained proliferation of the tested cells and compound **1** showed the highest cytotoxic activity, and the average IC_{50} values were lower than 10.0 µg/mL.

Table 3. The cytotoxicity of compounds 1–8.

Compound	IC_{50} (µg/mL)		
	K562	HCT-116	HepG2
1	8.8 ± 1.5	9.5 ± 0.8	9.6 ± 5.6
2	57.1 ± 7.3	75.42 ± 2.1	—
3	—	68.39 ± 3.3	53.78 ± 6.7
4	—	36.8 ± 5.6	17.5 ± 1.9
5	28.3 ± 1.1	14.3 ± 1.6	27.3 ± 5.8
6	36.6 ± 2.4	21.6 ± 4.1	79.7 ± 5.9
7	—	112.3 ± 5.7	—
8	11.42 ± 3.05	15.13 ± 1.76	10.83 ± 3.47
Doxorubicin	1.1 ± 0.1	0.9 ± 0.3	2.1 ± 0.2

The result of minimum inhibitory concentrations (MICs) showed that compound **1** showed good activities against Gram-positive bacteria *Staphylococcus aureus*, *Sarcina lutea*, and *Bacillus subtilis*, and the Gram-negative bacteria *Klebsiella pneumoniae* in vitro (Table 4). Compound **8** showed weak antibacterial activity against two Gram-positive bacterium and the minimum inhibitory concentrations (MICs) of compounds **2–7** were determined to be >10 mg/mL, so they had no activity against these tested pathogens.

Table 4. The antibacterial activity of compounds 1-8.

Compounds	MIC (µg/mL)				
	Gram-Positive Bacteria			Gram-Negative Bacteria	
	Staphylococcus aureus	*Sarcina lutea*	*Bacillus subtilis*	*Klebsiella pneumonie*	*Escherichia coli*
1	31.0 ± 2.5	44.0 ± 5.8	3.5 ± 0.5	25.0 ± 1.5	—
2–7	—	—	—	—	—
8	210.0 ± 20.0	190.0 ± 15.0	—	—	—

4. Discussion

In this research, the results of morphological, physiological, and biochemical tests showed that strain NEAU-wh3-1 has typical characteristics of the genus *Embleya* [41]. Such as containing LL-diaminopimelic acid as the cell wall peptidoglycan, MK-9(H_4), and MK-9(H_6) as the major menaquinones, phosphatidylethanolamine (PE) as the predominant phospholipid and arabinose in the whole sugars. Moreover, strain NEAU-wh3-1 formed spiral spore chains and the spore surface was rough, which are consistent with *E. scabrispora* DSM 41855[T] and *E. hyalina* MB891-A1[T] [39,41]. In addition, the phylogenetic trees constructed from the 16S rRNA gene sequences and the concatenated sequences alignment (1979 bp) of five housekeeping genes all suggested that the isolate should be assigned to the genus *Embleya*.

However, some obvious differences could also be found between strain NEAU-wh3-1 and its closely related strains regarding several phenotypic and chemotaxonomic characteristics (Table 1). The isolate was able to grow at 4 °C, in contrast to its closely related strains, which were not. The composition of phospholipids and menaquinones of strain NEAU-wh3-1 was also different from its related species, *E. scabrispora* DSM 41855[T], *E. hyalina* MB891-A1[T] and *S. lasii* 5H-CA11[T]. Most notably, the whole-cell sugars of strain NEAU-wh3-1 was found to contain arabinose, glucose, and ribose, while *E. scabrispora* DSM 41855^T only contains arabinose, *E. hyalina* MB891-A1^T contains arabinose and glucose and *S. lasii* 5H-CA11^T contains glucose and ribose, which also could distinguish the strain from its closely related strains. Other phenotypic differences include the temperature and pH range of growth, patterns of carbon and nitrogen utilization, hydrolysis of cellulose, starch, and Tweens (20, 40, and 80), liquefaction of gelatin, peptonization of milk, production of H$_2$S, and urease and reduction of nitrate. Therefore, it is evident from the phenotypic, genotypic, and chemotaxonomic data that strain NEAU-wh3-1 may represent a novel species of the genus *Embleya*.

The genus *Embleya*, recently transferred from genus *Streptomyces*, is a new member of the family *Streptomycetaceae* [39,40]. At present, it contains only two species: *Embleya scabrispora*, could produce hitachimycin with antitumor, antibacterial, and antiprotozoal activities [44–46]; and *Embleya hyaline*, could produce nybomycin which is an effective agent against antibiotic-resistant *Staphylococcus aureus* and is called a reverse antibiotic [48]. During the study of the chemical properties of the active ingredients of strain NEAU-wh3-1, eight active compounds were obtained, including one macrolide dilactone, five piericidins, one β-hydroxy acetamides, an analogue of monocarboxylic acid ionophore, which were observed to fit into at least three types based on their molecular skeletons. This shows to some extent that this strain has the ability to produce metabolites with a wide variety of different skeletal structures. Out of these compounds, Piericidins (**3-7**) were a class of polyene alpha-pyridone heterocyclic antibiotics, among them, Piericidin A1(**5**) was first reported [81], which was isolated from *Streptomyces mobaraensis*. Piericidins exhibit interesting biological activities, in particular antitrypanosomal [82]. In our research, compounds **3-7** exhibited different degrees of cytotoxicity on three types of tumor cells, but they did not show any antibacterial activity, which was consistent with previous reports [78,81]. As a Piericidins-producing strain, NEAU-wh3-1 has certain application potential in pest control. Conglobatin (**2**) is an unusual 16-membered macrocyclic diolide originally isolated from a polyether-producing strain of *Streptomyces conglobatus* ATCC 31005^T and was reported to be essentially devoid of antifungal, antibacterial, antitumor, and antiprotozoal activity at that time [75]. However, in recent research, FW-04-806 is identical in structure to conglobatin, and it has been reported to inhibit the growth of a human chronic myelocytic leukemia K562 cell line with an IC$_{50}$ of 6.66 μg/mL, further study also investigated the effects of FW-04-806 on SKBR3 and MCF-7, respectively [83]. Its mechanism of action appears to be novel, via direct binding to the N-terminal domain of Hsp90 and disruption of its interaction with co-chaperone Cdc37 [84]. In our antitumor activity test, Conglobatin (**2**) showed good bioactivity against two tumor cell lines, supporting it at least partially accounted for the cytotoxic activity of the strain NEAU-wh3-1 extract. BE-52211 D (**8**) was a cytotoxic metabolite from a strain of *Streptomyces* and had been reported to have moderate cytotoxicity against human hepatocellular liver carcinoma cells HepG2, human leukemia cells K562, and human colon carcinoma cells HCT-116 with the IC$_{50}$ values of >10 μg/mL [80], which is consistent with the result in the present study. Compound **1** structurally related to zincophorin, which was also referred to as M144255 or griseochellin and is a polyoxygenated ionophoric antibiotic isolating from *Streptomyces griseus* in 1984 [17]. It has been reported to possess strong in vivo activity against Gram-positive bacteria and have strong cytotoxicity against human lung carcinoma cells A549 and Madin-Darby canine kidney cells MDCK [18]. No biological activity has been reported against Gram-negative bacteria, yeasts, and fungi [85]. The second member in the zincophorin family named CP-78545, was found in the culture broth of *Streptomyces* sp. N731-45. The structural difference between them is that CP-78545 has an extra terminal double bond; but they have similar spectrum and potency on biological properties except for the antitumor activity (no reports) [86]. In our antitumor and antimicrobial assays, compound **1** showed the highest antitumor activity against three human cell lines and good antibacterial activity against Gram-negative bacteria. To our knowledge, this is the first report of this kind of compound with antibacterial activity against Gram-negative bacteria. This study has enriched the activity spectrum of Zincophorins.

5. Conclusions

Strain producing a new compound with strong antitumor activity, isolated from the rhizosphere soil of wheat (*Triticum aestivum* L.) in HeBei province, China. Morphological and chemotaxonomic features together with phylogenetic analysis suggested that strain NEAU-wh3-1 belonged to the genus *Embleya*. Cultural and biochemical characteristics combined with multilocus sequence analysis clearly revealed that strain NEAU-wh3-1 may represent a novel species of the genus *Embleya*. Moreover, eight compounds, including one new compound with higher antitumor activities against three human cell lines, were isolated from the strain.

Supplementary Materials: The following are available online at http://www.mdpi.com/2076-2607/8/3/441/s1, Table S1: GenBank accession numbers of the sequences used in MLSA; Table S2: MLSA distance values for selected strains in this study; Figure S1: The polar lipids of strain NEAU-wh3-1; Figure S2: Maximum-likelihood tree based on 16S rRNA gene sequences showing relationship between strain NEAU-wh3-1 and related taxa; Figure S3: Maximum-likelihood tree based on multilocus sequence analysis (MLSA)analysis of the concatenated partial sequences (1979 bp) from five housekeeping genes (*atpD*, *gyrB*, *recA*, *rpoB*, and *trpB*) of strain NEAU-wh3-1 (in bold) with related taxa; Figure S4: ^{1}H NMR (400 MHz) spectrum of compound **1** in CDCl$_3$; Figure S5: ^{13}C NMR (150 MHz) spectrum of compound **1** (in CDCl$_3$); Figure S: ^{1}H-^{1}H COSY spectrum (400 MHz) of compound **1** (in CDCl$_3$); Figure S7: HSQC spectrum (400 MHz) of compound **1** (in CDCl$_3$); Figure S8: HMBC spectrum (400 MHz) of compound **1** (in CDCl$_3$); Figure S9: IR spectrum of compound **1**(in EtOH); Figure S10: UV spectrum of compound **1** (in EtOH); Figure S11: The HRESIMS spectrum of compound **1**; Figure S12: ^{1}H NMR (400 MHz) spectrum of compound **2** in CDCl$_3$; Figure S13: ^{13}C NMR (150 MHz) spectrum of compound **2** in CDCl$_3$; Figure S14: The ESI-MS spectrum of compound **2**; Figure S15: ^{1}H NMR (400 MHz) spectrum of compound **3** in CDCl$_3$; Figure S16: The ESI-MS spectrum of compound **3**; Figure S17: ^{1}H NMR (400 MHz) spectrum of compound **4** in CDCl$_3$; Figure S18: The ESI-MS spectrum of compound **4**; Figure S19: ^{1}H NMR (400 MHz) spectrum of compound **5** in CDCl$_3$; Figure S20: The ESI-MS spectrum of compound **5**; Figure S21: ^{1}H NMR (400 MHz) spectrum of compound **6** in CDCl$_3$; Figure S22: The ESI-MS spectrum of compound **6**; Figure S23: ^{1}H NMR (400 MHz) spectrum of compound **7** in CDCl$_3$; Figure S24: ESI-MS spectrum of compound of compound **7**; Figure S25: ^{1}H NMR (400 MHz) spectrum of compound **8** in CDCl$_3$; Figure S26: ^{13}C NMR (150 MHz) spectrum of compound **8** in CDCl$_3$; Figure S27: The HRESIMS spectrum of compound **8**.

Author Contributions: H.W., T.S., and W.S. performed the experiments. X.G. prepared the figures and tables. P.C. and X.X. analyzed the data. Y.S. and J.Z. designed the experiments and reviewed the manuscript. All authors have read and agreed to the published version of the manuscript.

Funding: This work was supported in part by grants from the Heilongjiang Postdoctoral Fund (LBH-Z17015) and the Scientific Research Foundation for Settled Postdoctoral of Heilongjiang Province (LBH-Q19082).

Conflicts of Interest: The authors declare that there are no conflict of interest.

References

1. Torre, L.A.; Bray, F.; Siegel, R.L.; Ferlay, J.; Lortet-Tieulent, J.; Jemal, A. Global cancer statistics, 2012. *CA. Cancer J. Clin.* **2015**, *65*, 87–108. [CrossRef]

2. Commander, H.; Whiteside, G.; Perry, C. Vandetanib: First global approval. *Drugs* **2011**, *71*, 1355–1365. [CrossRef]

3. Zhu, F.; Zhao, X.; Li, J.; Guo, L.; Bai, L.; Qi, X. A new compound Trichomicin exerts antitumor activity through STAT3 signaling inhibition. *Biomed. Pharmacother.* **2020**, *121*, 109608. [CrossRef]

4. Zhang, Z.; Yu, X.; Wang, Z.; Wu, P.; Huang, J. Anthracyclines potentiate anti-tumor immunity: A new opportunity for chemoimmunotherapy. *Cancer Lett.* **2015**, *369*, 331–335. [CrossRef]

5. Sznarkowska, A.; Kostecka, A.; Meller, K.; Bielawski, K.P. Inhibition of cancer antioxidant defense by natural compounds. *Oncotarget* **2017**, *8*, 15996–16016. [CrossRef]

6. Newman, D.J.; Cragg, G.M. Natural products as sources of new drugs over the last 25 years. *J. Nat. Prod.* **2007**, *70*, 461–477. [CrossRef]

7. Cragg, G.M.; Newman, D.J.; Snader, K.M. Natural products in drug discovery and development. *J. Nat. Prod.* **1997**, *60*, 52–60. [CrossRef]

8. Chin, Y.W.; Balunas, M.J.; Chai, H.B.; Kinghorn, A.D. Drug discovery from natural sources. *Drug Addict. From Basic Res. Ther.* **2008**, *8*, 17–39.

9.   Pettit, R.K. Culturability and Secondary Metabolite Diversity of Extreme Microbes: Expanding Contribution of Deep Sea and Deep-Sea Vent Microbes to Natural Product Discovery. *Mar. Biotechnol.* **2011**, *13*, 1–11. [CrossRef]

10.   Qin, S.; Li, J.; Chen, H.H.; Zhao, G.Z.; Zhu, W.Y.; Jiang, C.L.; Xu, L.H.; Li, W.J. Isolation, diversity, and antimicrobial activity of rare actinobacteria from medicinal plants of tropical rain forests in Xishuangbanna China. *Appl. Environ. Microbiol.* **2009**, *75*, 6176–6186. [CrossRef]

11.   Parkinson, D.R.; Arbuck, S.G.; Moore, T.; Pluda, J.M.; Christian, M.C. Clinical development of anticancer agents from natural products. *Stem Cells* **1994**, *12*, 30–43. [CrossRef]

12.   Berdy, J. Bioactive microbial metabolites. *J Antibiot.* **2005**, *58*, 1–26. [CrossRef]

13.   Ventura, M.; Canchaya, C.; Tauch, A.; Chandra, G.; Fitzgerald, G.F.; Chater, K.F.; van Sinderen, D. Genomics of Actinobacteria: Tracing the Evolutionary History of an Ancient Phylum. *Microbiol. Mol. Biol. Rev.* **2007**, *71*, 495–548. [CrossRef]

14.   Goodfellow, M.; Williams, S.T. Ecology of *actinomycetes*. *Annu. Rev. Microbiol.* **1983**, *37*, 189–216. [CrossRef]

15.   Rabbani, A.; Finn, R.M.; Ausió, J. The anthracycline antibiotics: Antitumor drugs that alter chromatin structure. *BioEssays* **2005**, *27*, 50–56. [CrossRef]

16.   Demain, A.L.; Sanchez, S. Microbial drug discovery: 80 Years of progress. *J. Antibiot.* **2009**, *62*, 5–16. [CrossRef]

17.   Brooks, H.A.; Gardner, D.; Poyser, J.P.; King, T.J. The structure and absolute stereochemistry of zincophorin (antibiotic m144255): a monobasic carboxylic acid ionophore having a remarkable specificity for divalent cations. *J. Antibiot.* **1984**, *37*(11), 1501–1504. [CrossRef]

18.   Walther, E.; Boldt, S.; Kage, H.; Lauterbach, T.; Martin, K.; Roth, M.; Hertweck, C.; Sauerbrei, A.; Schmidtke, M.; Nett, M. Zincophorin-biosynthesis in *Streptomyces griseus* and antibiotic properties. *GMS Infect. Dis.* **2016**, *4*, Doc08.

19.   Yang, S.X.; Gao, J.M.; Zhang, A.L.; Laatsch, H.S. A novel toxic macrolactam polyketide glycoside produced by actinomycete *Streptomyces sannanensis*. *Bioorg. Med. Chem. Lett.* **2011**, *21*, 3905–3908. [CrossRef]

20.   Subramani, R.; Aalbersberg, W. Culturable rare Actinomycetes: Diversity, isolation and marine natural product discovery. *Appl. Microbiol. Biotechnol.* **2013**, *97*, 9291–9321. [CrossRef]

21.   Heidari, B.; Mohammadipanah, F. Isolation and identification of two alkaloid structures with radical scavenging activity from *Actinokineospora* sp. UTMC 968, a new promising source of alkaloid compounds. *Mol. Biol. Rep.* **2018**, *45*, 2325–2332. [CrossRef] [PubMed]

22.   Gao, M.Y.; Qi, H.; Li, J.S.; Zhang, H.; Zhang, J.; Wang, J.D.; Xiang, W.S. A new polysubstituted cyclopentene derivative from *Streptomyces* sp. HS-NF-1046. *J. Antibiot.* **2017**, *70*, 216–218. [CrossRef] [PubMed]

23.   Peláez, F. The historical delivery of antibiotics from microbial natural products - Can history repeat? *Biochem. Pharmacol.* **2006**, *71*, 981–990. [CrossRef]

24.   Kekuda, P.; Onkarappa, R.; Jayanna, N. Characterization and Antibacterial Activity of a Glycoside Antibiotic from *Streptomyces variabilis* PO-178. *Sci. Technol. Arts. Res. J.* **2015**, *3*, 116. [CrossRef]

25.   Miao, V.; Davies, J. Actinobacteria: The good, the bad, and the ugly. *Antonie van Leeuwenhoek* **2010**, *98*, 143–150. [CrossRef]

26.   Subramani, R.; Aalbersberg, W. Marine actinomycetes: An ongoing source of novel bioactive metabolites. *Microbiol. Res.* **2012**, *167*, 571–580. [CrossRef]

27.   Clardy, J.; Fischbach, M.A.; Walsh, C.T. New antibiotics from bacterial natural products. *Nat. Biotechnol.* **2006**, *24*, 1541–1550. [CrossRef]

28.   Ayed, A.; Slama, N.; Mankai, H.; Bachkouel, S.; ElKahoui, S.; Tabbene, O.; Limam, F. *Streptomyces tunisialbus* sp. nov., a novel *Streptomyces* species with antimicrobial activity. *Antonie van Leeuwenhoek* **2018**, *111*, 1571–1581. [CrossRef]

29.   Mccully, M.; Harper, J.D.I.; An, M.; Kent, J.H. The rhizosphere: the key functional unit in plant soil microbial interactions in the field implications for the understanding of allelopathic effects. *Pol. J. Vet. Sci.* **2005**, *15*, 493–498.

30.   Hiltner, L. Uber neuer erfahrungen und probleme auf dem gebiet der berücksichtigung unter besonderer berücksichtigung der gründüngung und brache. *Arbeiten der Deustchen Landwirtschaftsgesellesschaft* **1904**, *32*, 1405–1417.

31.   Berendsen, R.L.; Pieterse, C.M.J.; Bakker, P.A.H.M. The rhizosphere microbiome and plant health. *Trends. Plant Sci.* **2012**, *17*, 478–486. [CrossRef]

32. Loper, J.E.; Gross, H. Genomic analysis of antifungal metabolite production by *Pseudomonas fluorescens* Pf-5. *Eur. J. Plant Pathol.* **2007**, *119*, 265–278. [CrossRef]

33. Reiter, B.; Sessitsch, A.; Nowak, J.; Cle, C. Endophytic colonization of *Vitis vinifera* L. by plant growth-promoting bacterium *Burkholderia* sp. strain PsJN. *Society* **2005**, *71*, 1685–1693.

34. Lanteigne, C.; Gadkar, V.J.; Wallon, T.; Novinscak, A.; Filion, M. Production of DAPG and HCN by Pseudomonas sp. LBUM300 contributes to the biological control of bacterial canker of tomato. *Phytopathology* **2012**, *102*, 967–973. [CrossRef]

35. Osei, E.; Kwain, S.; Mawuli, G.T.; Anang, A.K.; Owusu, K.B.A.; Camas, M.; Camas, A.S.; Ohashi, M.; Alexandru-Crivac, C.N.; Deng, H. Paenidigyamycin A, potent antiparasitic imidazole alkaloid from the ghanaian *Paenibacillus* sp. De2Sh. *Mar. Drugs* **2019**, *17*, 9. [CrossRef]

36. Mukhtar, S.; Mehnaz, S.; Mirza, M.S.; Malik, K.A. Isolation and characterization of bacteria associated with the rhizosphere of halophytes (Salsola stocksii and Atriplex amnicola) for production of hydrolytic enzymes. *Brazilian J. Microbiol.* **2019**, *50*, 85–97. [CrossRef]

37. Orfali, R.; Perveen, S. Secondary metabolites from the *Aspergillus* sp. in the rhizosphere soil of *Phoenix dactylifera* (Palm tree). *BMC Chem.* **2019**, *13*, 1–6. [CrossRef]

38. Wang, R.J.; Zhang, S.Y.; Ye, Y.H.; Yu, Z.; Qi, H.; Zhang, H.; Xue, Z.L.; Wang, J.D.; Wu, M. Three new isoflavonoid glycosides from the mangrove-derived actinomycete micromonospora aurantiaca 110B. *Mar. Drugs* **2019**, *17*, 294. [CrossRef]

39. Nouioui, I.; Carro, L.; García-López, M.; Meier-Kolthoff, J.P.; Woyke, T.; Kyrpides, N.C.; Pukall, R.; Klenk, H.-P.; Goodfellow, M.; Göker, M. Genome-Based Taxonomic Classification of the Phylum Actinobacteria. *Front. Microbiol.* **2018**, *9*, 2007. [CrossRef]

40. Oren, A.; Garrity, G.M. List of new names and new combinations previously effectively, but not validly, published. *Int. J. Syst. Evol. Microbiol.* **2018**, *68*, 2707–2709. [CrossRef]

41. Komaki, H.; Hosoyama, A.; Kimura, A.; Ichikawa, N.; Igarashi, Y.; Tamura, T. Classification of '*Streptomyces hyalinum*' Hamada and Yokoyama as *Embleya hyalina* sp. nov., the second species in the genus *Embleya*, and emendation of the genus *Embleya*. *Int. J. Syst. Evol. Microbiol.* **2020**, 1–5. [CrossRef]

42. Kämpfer, P.; Genus, I. Streptomyces Waksman and Henrici 1943, 339 [AL]. In *Bergey's Manual of Systematic Bacteriology*, 2nd ed.; Springer: New York, NY, USA, 2012; pp. 1679–1680.

43. Ping, X.; Takahashi, Y.; Seino, A.; Iwai, Y.; Omura, S. *Streptomyces scarbrisporus* sp. nov. *Int. J. Syst. Evol. Microbiol.* **2004**, *54*, 577–581. [CrossRef]

44. Omura, S.; Nakagawa, A.; Shibata, K.; Sano, H. The structure of hitachimycin, a novel macrocyclic lactam involving β-phenylalanine. *Tetrahedron Lett.* **1982**, *23*, 4713–4716. [CrossRef]

45. Komiyama, K.; Edanami, K.I.; Yamamoto, H.; Umezawa, I. Antitumor activity of a new antitumor antibiotic, stubomycin. *J. Antibiot.* **1982**, *35*, 703–706. [CrossRef]

46. Shibata, K.; Satsumabayashi, S.; Sano, H.; Komiyama, K.; Nakagawa, A.; Omura, S. Chemical modification of hitachimycin. Synthesis, antibacterial, cytocidal and in vivo antitumor activities of hitachimycin derivatives. *J. Antibiot.* **1988**, *41*, 614–623. [CrossRef]

47. Naganawa, H.; Wakashiro, T.; Yagi, A.; Kondo, S.; Takita, T.; Hamada, M.; Maeda, K.; Umezawa, H. Deoxynybomycin from a *streptomyces*. *J. Antibiot.* **1970**, *23*, 365–368. [CrossRef]

48. Hiramatsu, K.; Igarashi, M.; Morimoto, Y.; Baba, T.; Umekita, M.; Akamatsu, Y. Curing bacteria of antibiotic resistance: Reverse antibiotics, a novel class of antibiotics in nature. *Int. J. Antimicrob. Agents* **2012**, *39*, 478–485. [CrossRef]

49. Hayakawa, M.; Nonomura, H. Humic acid-vitamin agar, a new medium for the selective isolation of soil actinomycetes. *J. Ferment. Technol.* **1987**, *65*, 501–509. [CrossRef]

50. Shirling, E.B.; Gottlieb, D. Methods for characterization of *Streptomyces* species. *Int. J. Syst. Bacteriol.* **1966**, *16*, 313–340. [CrossRef]

51. Vichai, V.; Kirtikara, K. Sulforhodamine B colorimetric assay for cytotoxicity screening. *Nat. Protoc.* **2006**, *1*, 1112–1116. [CrossRef]

52. Jin, L.; Zhao, Y.; Song, W.; Duan, L.; Jiang, S.; Wang, X.; Zhao, J.; Xiang, W. *Streptomyces inhibens* sp. Nov., a novel actinomycete isolated from rhizosphere soil of wheat (*Triticum aestivum* L.). *Int. J. Syst. Evol. Microbiol.* **2019**, *69*, 688–695. [CrossRef] [PubMed]

53. Zhao, J.W.; Han, L.Y.; Yu, M.Y.; Cao, P.; Li, D.M.; Guo, X.W.; Liu, Y.Q.; Wang, X.J.; Xiang, W.S. Characterization of *Streptomyces sporangiiformans* sp. nov., a Novel Soil Actinomycete with Antibacterial Activity against *Ralstonia solanacearum*. *Microorganisms* **2019**, *7*, 360. [CrossRef] [PubMed]

54. Smibert, R.M.; Krieg, N.R. Phenotypic characterization. In *Methods for General and Molecular Bacteriology*, 2nd ed.; Gerhardt, P., Murray, R.G.E., Wood, W.A., Krieg, N.R., Eds.; American Society for Microbiology: Washington, DC, USA, 1994; pp. 607–654.

55. Gordon, R.E.; Barnett, D.A.; Handerhan, J.E.; Pang, C. Nocardia coeliaca, Nocardia autotrophica, and the nocardin strain. *Int. J. Syst. Bacteriol.* **1974**, *24*, 54–63. [CrossRef]

56. Yokota, A.; Tamura, T.; Hasegawa; Huang, L.H. *Catenuloplanes japonicas* gen. nov., sp. nov., nom. rev., a new genus of the order *Actinomycetales*. *Int. J. Syst. Bacteriol.* **1993**, *43*, 805–812. [CrossRef]

57. Lechevalier, M.P.; Lechevalier, H.A. The chemotaxonomy of actinomycetes. In *Actinomycete taxonomy*, 2nd ed.; Dietz, A., Thayer, D.W., Eds.; special publication for Society of Industrial Microbiology: Arlington, TX, USA, 1980; pp. 227–291.

58. Minnikin, D.E.; O'Donnell, A.G.; Goodfellow, M.; Alderson, G.; Athalye, M.; Schaal, A.; Parlett, J.H. An integrated procedure for the extraction of bacterial isoprenoid quinones and polar lipids. *J. Microbiol. Methods* **1984**, *2*, 233–241. [CrossRef]

59. Collins, M.D. Isoprenoid quinone analyses in bacterial classification and identification. In *Chemical methods in bacterial systematics*; Goodfellow, M., Minnikin, D.E., Eds.; Academic Press: London, UK, 1985; pp. 267–284.

60. Shen, Y.; Sun, T.; Jiang, S.; Mu, S.; Li, D.; Guo, X.; Zhang, J.; Zhao, J.; Xiang, W. *Streptomyces lutosisoli* sp. nov., a novel actinomycete isolated from muddy soil. *Antonie Van Leeuwenhoek, Int. J. Gen. Mol. Microbiol.* **2018**, *111*, 2403–2412. [CrossRef]

61. Wu, C.; Lu, X.; Qin, M.; Wang, Y.; Ruan, J. Analysis of menaquinone compound in microbial cells by HPLC. *Microbiology* **1989**, *16*, 176–178.

62. Kim, S.B.; Brown, R.; Oldfield, C.; Gilbert, S.C.; Iliarionov, S.; Goodfellow, M. Gordonia amicalis sp. nov., a novel dibenzothiophene-desulphurizing actinomycete. *Int. J. Syst. Evol. Microbiol.* **2000**, *50*, 2031–2036. [CrossRef]

63. Yoon, S.H.; Ha, S.M.; Kwon, S.; Lim, J.; Kim, Y.; Seo, H.; Chun, J. Introducing EzBioCloud: A taxonomically united database of 16S rRNA gene sequences and whole-genome assemblies. *Int. J. Syst. Evol. Microbiol.* **2017**, *67*, 1613–1617. [CrossRef]

64. Saitou, N.; Nei, M. The neighbor-joining method: a new method for reconstructing phylogenetic trees. *Mol. Biol. Evol.* **1987**, *4*, 406–425.

65. Felsenstein, J. Evolutionary trees from DNA sequences: A maximum likelihood approach. *J. Mol. Evol.* **1981**, *17*, 368–376. [CrossRef] [PubMed]

66. Kumar, S.; Stecher, G.; Tamura, K. Mega7: molecular evolutionary genetics analysis version 7.0 for bigger datasets. *Mol. Biol. Evol.* **2016**, *33*, 1870–1874. [CrossRef] [PubMed]

67. Felsenstein, J. Confidence Limits on Phylogenies: an Approach Using the Bootstrap. *Evolution (N. Y).* **1985**, *39*, 783–791.

68. Kimura, M. A simple method for estimating evolutionary rates of base substitutions through comparative studies of nucleotide sequences. *J. Mol. Evol.* **1980**, *16*, 111–120. [CrossRef]

69. Hatano, K.; Nishii, T.; Kasai, H. Taxonomic re-evaluation of whorl-forming *Streptomyces* (formerly *Streptoverticillium*) species by using phenotypes, DNA-DNA hybridization and sequences of gyrB, and proposal of *Streptomyces luteireticuli* (ex Katoh and Arai 1957) corrig., sp. nov., nom. rev. *Int. J. Syst. Evol. Microbiol.* **2003**, *53*, 1519–1529. [CrossRef]

70. Guo, Y.P.; Zheng, W.; Rong, X.Y.; Huang, Y. A multilocus phylogeny of the *Streptomyces griseus* 16S rRNA gene clade: Use of multilocus sequence analysis for streptomycete systematics. *Int. J. Syst. Evol. Microbiol.* **2008**, *58*, 149–159. [CrossRef]

71. Wang, J.; Zhang, H.; Ying, L.; Wang, C.; Jiang, N.; Zhou, Y.; Wang, H.; Bai, H. Five new epothilone metabolites from *Sorangium cellulosum* strain So0157-2. *J. Antibiot.* **2009**, *62*, 483–487. [CrossRef]

72. Wayne, P.A. Clinical and Laboratory Standards Institute. Methods for dilution antimicrobial susceptibility tests for bacteria that grow aerobically; approved standard—ninth edition. *Clin. Lab. Standards Institute* **2012**, *32*, 2.

73. Liu, C.; Han, C.; Jiang, S.; Zhao, X.; Tian, Y.; Yan, K.; Wang, X.; Xiang, W. *Streptomyces lasii* sp. nov., a Novel Actinomycete with Antifungal Activity Isolated from the Head of an Ant (*Lasius flavus*). *Curr. Microbiol.* **2018**, *75*, 353–358. [CrossRef]

74. Rong, X.; Huang, Y. Taxonomic evaluation of the *Streptomyces hygroscopicus* clade using multilocus sequence analysis and DNA-DNA hybridization, validating the MLSA scheme for systematics of the whole genus. *Syst. Appl. Microbiol.* **2012**, *35*, 7–18. [CrossRef]

75. Westley, J.W.; Liu, C.M.; Evans, R.H.; Blount, J.F. Conglobatin, a novel macrolide dilactone from *Streptomyces conglobatus* ATCC 31005. *J. Antibiot.* **1979**, *32*, 874–877. [CrossRef]

76. Yoshida, S.; Yoneyama, K.; Shiraishi, S.; Watanabe, A.; Takahashi, N. Chemical structures of new piericidins produced by *Streptomyces pactum*. *Agric. Biol. Chem.* **1977**, *41*, 855–862. [CrossRef]

77. Kubota, N.K.; Ohta, E.; Ohta, S.; Koizumi, F.; Suzuki, M.; Ichimura, M.; Ikegami, S. Piericidins C5 and C6: New 4-pyridinol compounds produced by *Streptomyces* sp. and *Nocardioides* sp. *Bioorganic Med. Chem.* **2003**, *11*, 4569–4575. [CrossRef]

78. Tamura, S.; Takahashi, N.; Miyamoto, S.; Mori, R.; Suzuki, S.; Nagatsu, J. Isolation and physiological activities of piericidin A, a natural insecticide produced by *Streptomyces*. *Agr. Biol. Chem.* **1963**, *27*, 576–582. [CrossRef]

79. Kominato, K.; Watanabe, Y.; Hirano, S.I.; Kioka, T.; Tone, H. Mer-a2026a and b, novel piericidins with vasodilating effect. ii. physico-chemical properties and chemical structures. *J. Antibiot.* **1994**, *48*, 103–105. [CrossRef]

80. Wang, H.; Zhao, X.L.; Gao, Y.H.; Qi, H.; Zhang, H.; Xiang, W.S.; Wang, J.D.; Wang, X.J. Two new cytotoxic metabolites from *Streptomyces* sp. HS-NF-813. *J. Asian. Nat. Prod. Res.* **2018**, *22*, 249–256. [CrossRef]

81. Hall, C.; Wu, M.; Crane, F.L.; Takahashi, H.; Tamura, S.; Folkers, K. Piericidin A: A new inhibitor of mitochondrial electron transport. *Biochem. Biophys. Res. Commun.* **1966**, *25*, 373–377. [CrossRef]

82. Shaaban, K.A.; Helmke, E.; Kelter, G.; Fiebig, H.H.; Laatsch, H. Glucopiericidin C: A cytotoxic piericidin glucoside antibiotic produced by a marine *Streptomyces* isolate. *J. Antibiot.* **2011**, *64*, 205–209. [CrossRef]

83. Huang, W.; Ye, M.; Zhang, L.; Wu, Q.; Zhang, M.; Xu, J.; Zheng, W. FW-04-806 inhibits proliferation and induces apoptosis in human breast cancer cells by binding to N-terminus of Hsp90 and disrupting Hsp90-Cdc37 complex formation. *Mol. Cancer* **2014**, *13*, 1–13. [CrossRef]

84. Huang, W.; Wu, Q.D.; Zhang, M.; Kong, Y.L.; Cao, P.R.; Zheng, W.; Xu, J.H.; Ye, M. Novel Hsp90 inhibitor FW-04-806 displays potent antitumor effects in HER2-positive breast cancer cells as a single agent or in combination with lapatinib. *Cancer Lett.* **2015**, *356*, 862–871. [CrossRef]

85. Song, Z.; Lohse, A.G.; Hsung, R.P. Challenges in the synthesis of a unique mono-carboxylic acid antibiotic, (+)-zincophorin. *Cheminform* **2009**, *26*, 560–571. [CrossRef] [PubMed]

86. Dirlam, J.P.; Belton, A.M.; Chang, S.P.; Cullen, W.P.; Huang, L.H.; Kojima, Y.; Maeda, H.; Nishiyama, S.; Oscarson, J.R.; Sakakibara, T.; et al. Cp-78,545, a new monocarboxylic acid ionophore antibiotic related to zincophorin and produced by a *streptomyces*. *J. Antibiot.* **1989**, *42*, 1213–1220. [CrossRef] [PubMed]

 *microorganisms*

Article

A *Streptomyces* sp. NEAU-HV9: Isolation, Identification, and Potential as a Biocontrol Agent against *Ralstonia solanacearum* of Tomato Plants

Ling Ling [1], Xiaoyang Han [1], Xiao Li [1], Xue Zhang [1], Han Wang [1], Lida Zhang [1], Peng Cao [1], Yutong Wu [1], Xiangjing Wang [1], Junwei Zhao [1,*] and Wensheng Xiang [1,2,*]

[1]   Key Laboratory of Agricultural Microbiology of Heilongjiang Province, Northeast Agricultural University, No. 59 Mucai Street, Xiangfang District, Harbin 150030, China; LLYNL2621161093@163.com (L.L.); hanxy139251@163.com (X.H.); Lx1244070003@126.com (X.L.); zhangxue_968425@163.com (X.Z.); wanghan507555536@gmail.com (H.W.); yone910310@163.com (L.Z.); cp511@126.com (P.C.); 18103699151@163.com (Y.W.); wangneau2013@163.com (X.W.)

[2]   State Key Laboratory for Biology of Plant Diseases and Insect Pests, Institute of Plant Protection, Chinese Academy of Agricultural Sciences, Beijing 100193, China

*   Correspondence: guyan2080@126.com (J.Z.); xiangwensheng@neau.edu.cn (W.X.)

Received: 25 October 2019; Accepted: 12 February 2020; Published: 1 March 2020

Abstract: *Ralstonia solanacearum* is an important soil-borne bacterial plant pathogen. In this study, an actinomycete strain named NEAU-HV9 that showed strong antibacterial activity against *Ralstonia solanacearum* was isolated from soil using an in vitro screening technique. Based on physiological and morphological characteristics and 98.90% of 16S rRNA gene sequence similarity with *Streptomyces panaciradicis* 1MR-8^T, the strain was identified as a member of the genus *Streptomyces*. Tomato seedling and pot culture experiments showed that after pre-inoculation with the strain NEAU-HV9, the disease occurrence of tomato seedlings was effectively prevented for *R. solanacearum*. Then, a bioactivity-guided approach was employed to isolate and determine the chemical identity of bioactive constituents with antibacterial activity from strain NEAU-HV9. The structure of the antibacterial metabolite was determined as actinomycin D on the basis of extensive spectroscopic analysis. To our knowledge, this is the first report that actinomycin D has strong antibacterial activity against *R. solanacearum* with a MIC (minimum inhibitory concentration) of 0.6 mg L^{-1} (0.48 μmol L^{-1}). The in vivo antibacterial activity experiment showed that actinomycin D possessed significant preventive efficacy against *R. solanacearum* in tomato seedlings. Thus, strain NEAU-HV9 could be used as BCA (biological control agent) against *R. solanacearum*, and actinomycin D might be a promising candidate for a new antibacterial agent against *R. solanacearum*.

Keywords: antibacterial activity; *Ralstonia solanacearum*; *Streptomyces* sp. NEAU-HV9; actinomycin D

---

1. Introduction

Tomato is one of the world's most important vegetable crops, with a global annual yield of approximately 160 million tons [1,2]. In China, long term continuous cropping is the main planting practice for tomato, which has led to serious soilborne diseases [3]. *Ralstonia solanacearum* [4] is an important soilborne bacterial plant pathogen [5]. Bacterial wilt caused by *R. solanacearum* is a serious and common disease, which reduces the yield of tomato and many other crops in tropical, subtropical, and warm-temperature regions of the world [6]. Because of worldwide distribution and a large host range of more than 200 plant species in 50 families, including pepper, tomato, tobacco, potato, peanut, and banana, this soil bacterium has been recognized as one of the causative agents of bacterial wilt disease and is one of the leading models in pathogenicity [5]. In the absence of host plants, this bacterium can be free-living as a saprophyte in the soil or in water [7]. Plant breeding, field sanitation, crop rotation, and use of bactericides have met with only limited success for *R. solanacearum* [8]. Furthermore, pathogenic microbial multi-drug resistance is also increasing. Therefore, new natural resources and antibiotics for suppressing this soilborne disease are needed.

Various recent studies have showed that biological control of bacterial wilt disease could be achieved using antagonistic bacteria [8,9]. The suppressive effect of some antagonistic bacteria on *R. solanacearum* was reported by Toyota and Kimura [10]. Moreover, the use of antagonistic bacteria to be effective in control of *R. solanacearum* has been proved by Ciampi-Panno et al. under field conditions [8]. *Streptomycetes* are gaining interest in agriculture as plant growth promoting (PGP) bacteria and/or biological control agents (BCAs) [11,12]. The *Streptomyces* genus comprises Gram-positive bacteria which show a filamentous form; they can grow in various environments. Several *Streptomyces* species such as *S. aureofaciens*, *S. avermitilis*, *S. lividans*, *S. humidus*, *S. hygroscopicus*, *S. lydicus*, *S. plicatus*, *S. olivaceoviridis*, *S. roseoflavus*, *S. scabies* and *S. violaceusniger* have been used to control soilborne diseases due to their greatly antagonistic activities by production of various antimicrobial substances [13–15].

Actinobacteria are famous for producing a variety of natural bioactive metabolites. *Streptomyces* is an important source of bioactive compounds among all members of antibiotic production, accounting for two-thirds of commercially available antibiotics [16]. Actinomycins belonging to a family of chromopeptide lactones are produced by various *Streptomyces*. Among several antibiotics produced by this genus, actinomycins are prominent. More than 20 naturally-occurring actinomycins were isolated and observed to have commonality of two pentapeptidolactone moieties with an actnoyl chromophore [17]; however, they differ in functional and/or positional group. Among actinomycins, actinomycin D has been widely studied and used clinically as an anticancer drug, especially in the treatment of childhood rhabdomyosarcoma, infantile kidney tumors and several other malignant tumors [18,19]. However, no reports have been published on actinomycin D against phytopathogen *R. solanacearum*.

In the existing protocol for virulence assays, one-month old tomato plantlets are soil-inoculated with the bacterium and wilting symptoms, if any, are observed and recorded. In usual ground work, tomato seeds are sown to obtain seedlings that take 5–6 days to sprout. Seedlings are then transferred to pots containing soil and grown in a greenhouse for about one month. Following this, plants are shifted to a growth chamber where plants are inoculated with the pathogen by soil drench or the stem inoculation method [20,21]. Using this approach, it usually takes 40 days to perform a single virulence assay. The infection achieved in this way is generally not axenic as the soil conditions used are not devoid of other bacterial communities that can colonize the plant during its growth prior to the infection study. Singh et al. [22] described a simple assay to study the pathogenicity of *R. solanacearum* on freshly grown tomato seedlings instead of fully-grown tomato plants. From seed germination to completion of the infection process, the study takes around 15 to 20 days. Pathogenicity due to *R. solanacearum* was also demonstrated when there is no significant plant growth since no mineral/growth inducing factors have been added into the water [23]. Under this same condition, there are reports of the bacterium's survivability without any growth [24]. The death of tomato seedlings was actually occurring due to the presence of *R. solanacearum* in the water. On the basis of the previous study, we have discussed an approach to study biological assays in tomato seedlings.

In this study, a *Streptomyces* sp., NEAU-HV9, was isolated and showed strong antimicrobial activity against *R. solanacearum*. The taxonomic identity of NEAU-HV9 was determined by a combination of 16S rRNA gene sequence analysis with morphological and physiological characteristics. The potential control of actinomycin D produced by the strain NEAU-HV9 against *R. solanacearum* was also investigated.

2. Materials and Methods

2.1. Sample Collection

Soil samples were collected from a field situated in Bama yao Autonomous County, Hechi City, Guangxi zhuang Autonomous Region (24°15′ N, 107°26′ E). The collected soil samples were brought to the laboratory in sterile bags and kept at 4 °C until further analysis. Before isolation of actinomycetes, the soil samples were air-dried at room temperature.

2.2. Screening and Isolation of Actinomycetes

The soil sample (5 g) was mixed with 45 mL distilled water and followed by an ultrasonic treatment (160 W) for 3 min. The soil suspension was incubated at 28 °C and 250 rpm on a rotary shaker for 30 min. Subsequently, the supernatant was collected and subjected to serial dilutions from 10^{-2} to 10^{-5}. Each dilution (200 µL) was spread on a plate of humic acid-vitamin (HV) agar [25] supplemented with cycloheximide (50 mg L^{-1}) and nalidixic acid (20 mg L^{-1}). Colonies were transferred and purified on International *Streptomyces* Project (ISP) medium 3 [26] and stored for a long time in glycerol suspensions (20%, *v/v*) at −80 °C after 14 days of aerobic incubation at 28 °C.

2.3. Screening of Antagonistic Actinobacteria Strains

The isolates were screened using the agar well diffusion method, and *R. solanacearum* was used as the indicator bacterium [27]. To further investigate the antibacterial components produced by the isolated cultures, these strains were cultured in ISP 2 medium [26] and the inhibitory activities of the supernatant and cell precipitate were tested. Initially, the isolated cultures were grown in ISP 2 medium and incubated at 28 °C on a rotary shaker. After 7 days of incubation, the supernatants were obtained by centrifugation at 8000 rpm and 4 °C for 10 min and subsequently filtrated with a 0.2 µm membrane filter. The cell precipitates were extracted with an equal volume of methanol for approximately 24 h [28]. A cell suspension (1 mL at 1×10^8 cfu mL^{-1}) of *R. solanacearum* was aseptically plated onto Bactoagar-glucose (BG) media supplemented with 0.5% glucose [22]. Supernatant and methanol extracts were collected from each isolate and tested initially for antimicrobial activity against *R. solanacearum*; each well contained 200 µL of supernatant or methanol extract. The plates were incubated at 37 °C for 12 h to test antibacterial activity. The diameters of inhibition zones were measured by using vernier calipers [29]. The experiments were conducted twice. The isolates that showed activities against tested organisms were collected and maintained. Among the collected isolates, the potential isolate designated as NEAU-HV9 was selected for further studies.

2.4. Morphological and Biochemical Characteristics of NEAU-HV9

Morphological characteristics, using cultures grown on ISP 3 medium at 28 °C for 2 weeks, were observed by light microscopy (Nikon ECLIPSE E200, Nikon Corporation, Tokyo, Japan) and scanning electron microscopy (Hitachi SU8010, Hitachi Co., Tokyo, Japan). Scanning electron microscopy samples were prepared as described by Jin et al. [30]. Cultural characteristics were determined using 2-week cultures grown at 28 °C on Czapek's agar [31], Bennett's agar [32], Nutrient agar [33], ISP 1 agar and ISP 2-7 media [26]. The color designation of substrate mycelium and aerial mycelium was done with ISCC–NBS (Inter-Society Color Council-National Bureau of Standards) Color Charts Standard Sample No. 2106 [34]. Growth at different temperatures (10 °C, 15 °C, 18 °C, 20 °C, 25 °C, 28 °C, 32 °C, 35 °C, 37 °C and 40 °C) was determined on ISP 3 medium after incubation for 14 days. Growth tests for pH range (pH 4.0–10.0, at intervals of 1.0 pH unit) using the buffer system described by Zhao et al. [35] and NaCl tolerance (0%, 1%, 2%, 3%, 4%, 5%, 6%, 7%, 8%, 9% and 10%, w/v) were tested in ISP 2 broth at 28 °C for 14 days on a rotary shaker. Biochemical testing (decomposition of adenine, casein, hypoxanthine, tyrosine, xanthine and cellulose, hydrolysis of starch, aesculin and gelatin, milk peptonization and coagulation, nitrate reduction and H_2S production), the utilization of sole carbon and nitrogen sources were examined as described previously [36,37].

2.5. Phylogenetic Analysis of NEAU-HV9

Strain NEAU-HV9 was cultured in ISP 2 medium for 3 days at 28 °C to harvest cells. The genomic DNA was isolated using a Bacteria DNA Kit (TIANGEN Biotech, Co. Ltd., Beijing, China). The universal bacterial primers 27F and 1541R were used to carry out PCR amplification of the 16S rRNA gene sequence [38,39]. The purified PCR product cloned into the vector pMD19-T (Takara) and sequenced by using an Applied Biosystems DNA sequencer (model 3730XL, Applied Biosystems Inc., Foster City, California, USA). The almost complete 16S rRNA gene sequence (1510 bp) was uploaded to the EzBioCloud server (Available online: https://www.ezbiocloud.net/) [40] to calculate pairwise 16S rRNA gene sequence similarity between strain NEAU-HV9 and related similar species. The phylogenetic tree was reconstructed with neighbor-joining trees [41] using MEGA 7.0 software [42]. The confidence value of branches of the neighbor-joining tree was assessed using bootstrap resampling with 1000 replication [43]. A distance matrix was calculated using Kimura's two-parameter model [44]. All positions containing gaps and missing data were eliminated from the dataset (complete deletion option).

2.6. Fermentation

Strain NEAU-HV9 was grown and maintained for 7 days at 28 °C on ISP 3 medium agar plates. Fermentation involved the generation of a seed culture. The stock culture was transferred into two 250 mL Erlenmeyer flasks containing 50 mL of the ISP2 medium and incubated at 28 °C for 72 h on a rotary shaker at 250 rpm. All of the media were sterilized at 121 °C for 20 min. The seed culture (5%) was transferred into 75 flasks (250 mL) containing 100 mL of production medium. The production medium was composed of maltodextrin 4%, lactose 4%, yeast extract 0.5%, Mops 2% at pH 7.2–7.4. The flasks were incubated at 28 °C for 7 days, shaken at 250 rpm. The final 7.5 L fermentation broth was filtered to separate the supernatant and the mycelial cake. The supernatant was extracted with ethyl acetate three times (3 × 2 L), and the mycelial cake was extracted with MeOH (3 L). The organic phase was evaporated under reduced pressure at 55 °C to yield the red crude extract (5.2 g).

2.7. Isolation and Purification of Antibacterial Compounds

Crude extract from the mycelium and supernatant was combined and subjected to silica gel column chromatography (Qingdao Haiyang Chemical Group, Qingdao, China; 100–200 mesh; 100 × 3 cm column) using a gradient of ethyl acetate−MeOH (100:0−90:10) to yield three fractions (Fr.1-Fr.3) based on the TLC (thin layer chromatography) profiles. TLC was performed on silica-gel plates with solvent of ethyl acetate/MeoH (4:1). All fractions (Fr.) were screened against *R. solanacearum*. The most active, Fr.1 and Fr.2, were applied to a Sephadex LH-20 column eluted with CH_2Cl_2/MeOH (1:1, *v/v*) and then further purified by semipreparative HPLC (Agilent 1260, Zorbax SB-C18, 5 μm, 250 × 9.4 mm inner diameter; 1.5 mL/min; 220 nm; 254 nm; Agilent, Palo Alto, CA, USA) MeOH/H2O (90:10, *v/v*) to obtain Compound 1 (*t*R 10.928 min, 9.3 mg) and Compound 2 (*t*R 12.367 min, 60.4 mg). We chose the main product, Compound 2, for further research. NMR spectra (1H and 13C) were measured with a Bruker DRX-400 (400 MHz for ^{1}H and 100 MHz for ^{13}C) spectrometer (Bruker, Rheinstetten, Germany). The ESI-MS (electrospray ionization mass spectra) spectra were taken on a Q-TOF Micro LC-MS-MS mass spectrometer (Waters Co, Milford, MA, USA).

2.8. Determination of Minimum Inhibitory Concentration (MIC)

The minimum inhibitory concentration (MIC) of the antibacterial compounds was determined as described by Rathod et al. [45]. *R. solanacearum* was grown in BG medium with 0.5% glucose in shake flasks at 28 °C for 24 h. Cells were harvested by centrifugation, washed with 0.85% saline twice, then the supernatant was discarded and 0.85% saline was added to the washed cells. The suspensions were standardized to an optical density (OD) of 0.2 at 540 nm. Antibacterial compounds were two-fold serially diluted to obtain concentrations ranging from 0.2 to 12.8 mg L^{-1} and one tube without drug served as a control. All of the tubes were inoculated with 1 mL of suspension of *R. solanacearum* above and incubated at 37 °C for 12 to 16 h. The turbidity of each tube with respect to the control tube was measured. The MIC value was defined as the lowest concentration of a compound that completely inhibits growth.

2.9. Biological Assays in Tomato Seedlings

Germination of tomato seedlings and preparation of bacterial inoculum were prepared as described by Singh et al. [22]. Freshly grown *R. solanacearum* was inoculated into 50 mL BG media broth with 0.5% glucose and incubated at 28 °C and 150 rpm for 24 h. The bacterial cultures were obtained by centrifugation at 4000 rpm and 4 °C for 15 min and were then resuspended in an equal volume of sterile distilled water to obtain a concentration of approximately 10^9 cfu mL^{-1}. Strain NEAU-HV9 was cultured in ISP 2 broth on rotary shaker for 3 days at 28 °C and centrifuged at 10,000 rpm. Subsequently, cell pellets were diluted in 0.85% (*w/v*) NaCl solution and adjusted to 10^7, 10^8 or 10^9 cfu mL^{-1}. Root inoculation of *R. solanacearum* in tomato seedlings was carried out as described by Singh et al. [22]. About 15 to 20 mL of *R. solancearum* inoculum was taken in a sterile container. Tomato seedlings (6 to 7 days old) were picked one at a time from the germinated seedling tray and then the roots of each seedling were dipped in the bacterial inoculum (up to the root-shoot junction). Four treatments were established as follows: TR 1 (tomato seedlings were pre-inoculated with suspension (10^7, 10^8 or 10^9 cfu mL^{-1}) of strain NEAU-HV9 and then inoculated with *R. solanacearum*); TR 2 (tomato seedlings were pre-inoculated with *R. solanacearum* and then transferred to microfuge tubes with the addition of 1 to 1.5 mL of sterile water and active fraction, where the final treatment concentrations were 1 × MIC and 2 × MIC, respectively); CK 1 (tomato seedlings were inoculated with sterile water); and CK 2 (tomato seedlings only were inoculated with *R. solanacearum*). For all of the treatments, the root-dip inoculated seedlings were transferred to an empty 1.5 mL sterile microfuge tube. After approximately 5 minutes, 1 to 1.5 mL of sterile water was added to the tube. All the inoculated seedlings, along with the controls, were transferred to a growth chamber maintained at 28 °C with 75% relative humidity (RH) and a 12-h photoperiod. Seedlings were analyzed for disease progression after 7 days. Sets of 4 seedlings were recruited in each dilution inoculation, and each assay was performed in triplicate.

2.10. Pot Culture Experiments

Prior to use, seed surfaces were disinfected with 2% sodium hypochlorite for 2 min [46]. Both germination and plant growth conditions followed 75–90% RH and a 12-h photoperiod at 28 °C. Four treatments were established as follows: TR 1 (one day before transplanting the test plants, strain NEAU-HV9 was added into the sterilized soil so that each gram of soil received about 1×10^9 cfu g^{-1} bacterial cells. Seven day old tomato seedlings were transferred to the soil; after three weeks, plants were inoculated with a suspension ($OD_{600} = 0.3$) of *R. solanacearum*); TR 2 (seven day old tomato seedlings were transferred to sterilized soil; after three weeks, tomato seedlings were irrigated with a solution of actinomycin D (0.6 mg L^{-1}). After one day, tomato plants were inoculated with a suspension ($OD_{600} = 0.3$) of *R. solanacearum* by pouring it onto the soil of unwounded plants at a final concentration of 1×10^7 cfu g^{-1} of soil [47]); CK 1 (seven day old tomato seedlings were transferred to sterilized soil; after three weeks, tomato seedlings were irrigated with sterilized water as positive control); and CK 2 (seven day old tomato seedlings were transferred to sterilized soil; after three weeks, tomato seedlings were inoculated with a suspension ($OD_{600} = 0.3$) of *R. solanacearum* as a negative control). All plants were kept in the greenhouse at 24–28 °C and 75–90% RH with a 12-h photoperiod. Treatments were replicated three times with five plants per replication. The disease incidence was rated using the 0–4 scale [48].

3. Results

3.1. Isolation and Identification of an Antimicrobial Compound Producing Strain

More than 20 isolates from the soil samples were isolated, purified, and screened for bioactivity against *R. solanacearum*. Among them, only four isolates showed bioactivity against *R. solanacearum*. Since the methanol extract of the the cell pellet and supernatant of one isolate, designated as NEAU-HV9, revealed a higher activity (30.5 mm and 32.8 mm) against the tested bacterial strain (Table 1, Figure S2), this strain was selected for further studies.

Table 1. Bioactivities of the supernatant and cell pellet of NEAU-HV9 against *R. solanacearum*.

	Methanol Extract of Cell Pellet	**Supernatant**
Inhibitory zone diameters (mm)	30.5	32.8

Data shown are the mean of two replications.

Strain NEAU-HV9 was aerobic, Gram-stain positive and formed well-developed, branched substrate hyphae and aerial mycelium that differentiated into spiral spore chains with oval spores (Figure 1). The spore surface was wrinkled. It had good growth on ISP 1, ISP 2, ISP 3, ISP 4, ISP 5, ISP 6, ISP 7, Bennett's agar and Nutrient agar, and poor growth on Czapek's agar (Figure S1). The data on the growth characteristics of NEAU-HV9 in different media are given in Table S1.

Figure 1. Scanning electron micrograph of strain NEAU-HV9 grown on International *Streptomyces* Project (ISP) 3 agar for 2 weeks at 28 °C.

Further characterization of NEAU-HV9 was performed by evaluating various biochemical tests (Table S2). Growth at 15 °C to 37 °C (optimum: 28 °C) and in the range of pH 5 to 9 (optimum: pH 7.0). Tolerate up to 7% (*w/v*) NaCl in the culture medium. Positive for hydrolysis of starch, production of H_2S, hydrolysis of aesculin and decomposition of adenine, hypoxanthine, tyrosine and xanthine, negative for reduction of nitrate, coagulation and peptonization of milk, liquefaction of gelatin and decomposition of casein. D-Glucose, D-maltose, D-mannitol, D-galactose, inositol, D-mannose, L-rhamnose and D-sucrose are utilized as sole carbon sources, but not L-arabinose, dulcitol, D-fructose, lactose, D-ribose, D-sorbitol or D-xylose. L-Alanine, D-arginine, L-asparagine, L-aspartic acid, L-glutamic acid, L-glutamine, glycine, L-proline, L-serine, L-threonine and L-tyrosine are utilized as sole nitrogen sources, but not creatine. The above growth data of isolate NEAU-HV9 denote that the isolate has the typical characteristics of the genus *Streptomyces*.

Recently, it has been suggested that the 16S rRNA gene can be used as a reliable molecular clock due to 16S rRNA sequences from distantly related bacterial lineages having similar functionalities [49]. Basically, the 16S rRNA gene sequence, comprising of about 1500 bp with hyper variable and conserved regions, is universal in all bacteria. According to Woese's report [50], comparing a stable part of the genetic code could determine phylogenetic relationships of bacteria. The hyper variable regions of the 16S rRNA gene sequences provide species-specific signature sequences, so it is widely used in bacterial identification all over the world. Therefore, the almost-complete 16S rRNA gene sequence (1510 bp) of strain NEAU-HV9 was obtained and has been deposited as MN578143 in the GenBank, EMBL (European Molecular Biology Laboratory) and DDBJ (DNA Data Bank of Japan) databases. BLAST sequence analysis of the 16S rRNA gene sequence indicated that strain NEAU-HV9 was related to members of the genus *Streptomyces*. The EzBioCloud analysis showed that strain NEAU-HV9 was most closely related to *Streptomyces panaciradicis* 1MR-8[T] and *Streptomyces sasae* JR-39[T] with a gene sequence similarity of 98.90% and 98.89%, respectively. In conclusion, based on the 16S rRNA gene sequence and the genetic identity of isolate NEAU-HV9, the isolated strain was further identified by neighbor-joining tree (Figure 2), and was also found to belong to the genus *Streptomyces*.

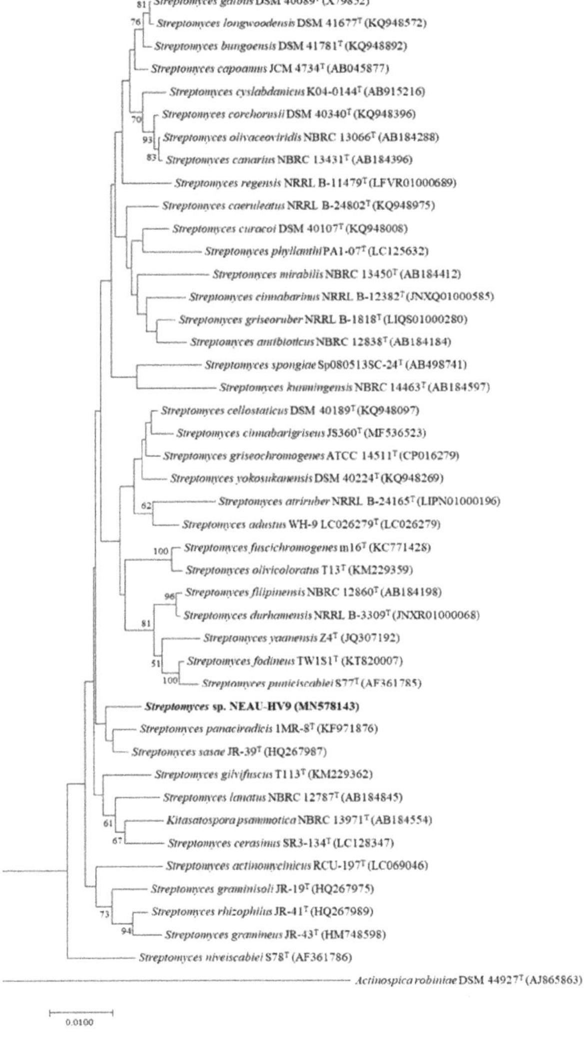

Figure 2. Neighbor-joining phylogenetic tree based on 16S rRNA gene sequences showing the relationships among strain NEAU-HV9 (bold) and members of the genus *Streptomyces*. Bootstrap percentages (≥50%) based on 1000 resamplings are listed at the nodes. *Actinospica robiniae* DSM 44927[T] was used as the out-group. Scale bar represents 0.01 nucleotide substitutions per site.

3.2. Structural Characterization of Compound

The active component was isolated from fermentation medium (7.5 L) and one bioactive compound was obtained as red, amorphous powder. The compound had UV visible spectra at 215 nm, 440 nm in methanol. The compound showed absorptions at 220 nm and 254 nm with a retention time of 12.367 min (Figure S3), similar to that of actinomycin class of compounds [51,52]. The structure of the compound was further elucidated by ^{1}H NMR, ^{13}C NMR, and MS analysis as well as comparison with previously reported data. The ESI-MS of the isolated compound revealed molecular ion peaks at m/z 1277.6 [M+Na]$^+$ (Figure S4), which was identical to that of actinomycin D [53]; ^{1}H and ^{13}C spectra of the isolated compound in CD$_4$O also showed great similarities to that of actinomycin D [52,53] (Figures S5 and S6). In addition, the retention time of commercial actinomycin D (Biotopped, purity: ≥98%) was 12.328 min (Figure S7), and the retention time of compound 2 was 12.367 min (Figure S3). Compound 2 and commercial actinomycin D have similar activity against *R. solanacearum* (Figure S8). The above results showed that the structure of the main active compound was confirmed to be actinomycin D (Figure 3).

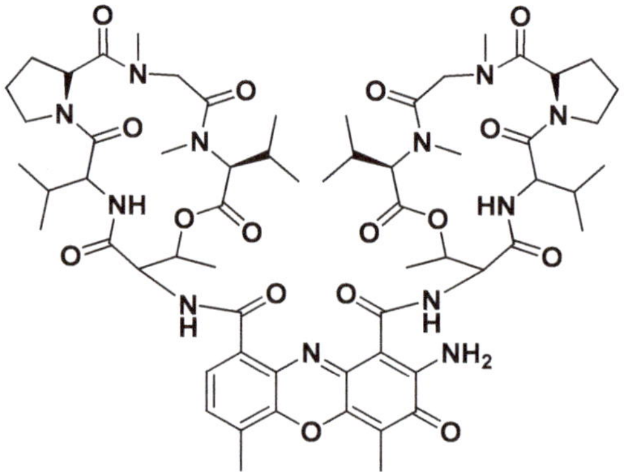

Figure 3. Chemical structure of actinomycin D.

3.3. Bioactivity of Isolated Compound

3.3.1. Minimum Inhibitory Concentration (MIC)

The minimum inhibitory concentration (MIC) of the antibacterial compound was determined as described by Rathod et al. [45]. The minimum inhibitory concentration of actinomycin D was determined as 0.6 mg L^{-1} (0.48 μmol L^{-1}) against *R. solanacearum* (Table 2).

Table 2. Minimum inhibitory concentration (MIC) values of actinomycin D against *R. solanacearum*.

Pathogen	MIC (mg/L)
R. solanacearum	0.6 ± 0.2

Data shown are the mean of three replications.

3.3.2. Biological Assays in Tomato Seedlings

The efficacy of the selected antagonist for the control of *R. solanacearum* was evaluated on tomato seedlings (Figure 4 and Figure S9, Table 3). The disease assessment was carried out using the method described in Kumar [23]. For the plants in the TR 1 group, the 10^9 cfu mL^{-1} suspension of NEAU-HV9 was effective against *R. solanacearum* when compared with the control (CK 2); all seedlings were as healthy as the CK 1 group. The 10^8 cfu mL^{-1} suspension of NEAU-HV9 showed very weak bioactivity against *R. solanacearum* compared with the control (CK 2); only one seedling was healthy and others were wilted. The 10^7 cfu mL^{-1} suspension of NEAU-HV9 exhibited no bioactivity against *R. solanacearum*; all seedlings were wilted, the same as the seedlings that were dried.

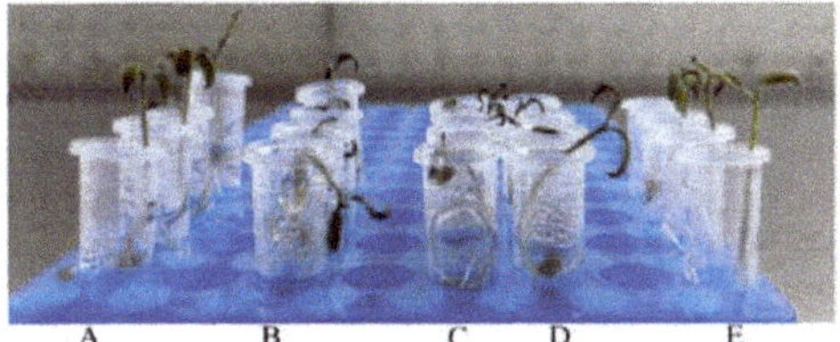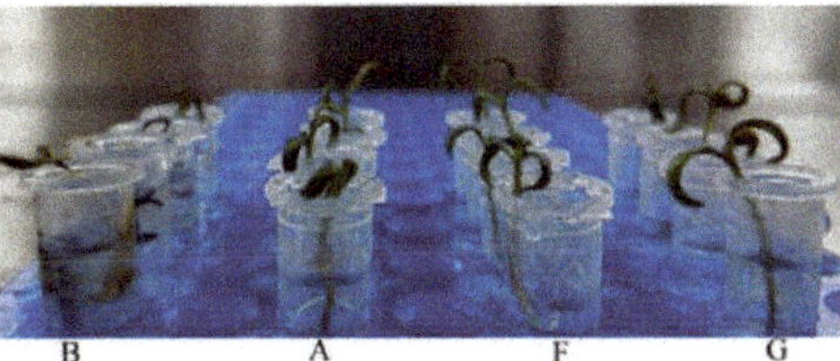

Figure 4. Control efficiency of strain NEAU-HV9 against *R. solanacearum*. A, tomato seedlings were inoculated with sterile water (CK 1); B, tomato seedlings only were inoculated with *R. solanacearum* (CK 2); C, tomato seedlings were pre-inoculated with suspension (10^7 cfu mL^{-1}) of NEAU-HV9 and then inoculated with *R. solanacearum* (TR 1); D, tomato seedlings were pre-inoculated with suspension (10^8 cfu mL^{-1}) of NEAU-HV9 and then inoculated with *R. solanacearum* (TR 1); E, tomato seedlings were pre-inoculated with suspension (10^9 cfu mL^{-1}) of NEAU-HV9 and then inoculated with *R. solanacearum* (TR 1); F, actinomycin D at the concentration 1 × MIC (TR 2); G, actinomycin D at the concentration 2 × MIC (TR 2).

Table 3. Effect of the strain NEAU-HV9 and actinomycin D on the incidence and control of tomato bacterial wilt in tomato seedlings.

Treatment	Wilt Incidence (%)	Control Efficacy (%)
NEAU-HV9 (10^7 cfu mL^{-1})	100 ± 0 [a]	0 ± 0 [a]
NEAU-HV9 (10^8 cfu mL^{-1})	93 ± 11.6 [b]	6.6 ± 11.5 [b]
NEAU-HV9 (10^9 cfu mL^{-1})	0 ± 0 [c]	100 ± 0 [c]
Actinomycin D (1 × MIC)	0 ± 0 [c]	100 ± 0 [c]
Actinomycin D (2 × MIC)	0 ± 0 [c]	100 ± 0 [c]
Control	100 ± 0 [a]	...

Data shown are the mean of three replications. Means within the same column followed by the same letter are not significantly different ($p = 0.05$) according to Fisher's least significant difference test.

Actinomycin D was highly effective against *R. solanacearum* in tomato seedlings (Figure 4 and Figure S9, Table 3). It was notable that all seedlings of the control (CK 2) were wilted, however after treatment with actinomycin D at the concentration 1 × MIC and 2 × MIC, none of the seedlings exhibited disease symptoms; the control efficacy of the formulation was 100%.

3.4. Pot Culture Experiments

In the pot culture experiments, NEAU-HV9 and actinomycin D effectively suppressed the development of bacterial wilt caused by *R. solanacearum* (Table 4, Figure S10). The negative control treatment had 73.9% relative disease incidence. For strain NEAU-HV9 and actinomycin D, the control efficacies of the formulations were 82% and 100%, respectively.

Table 4. Effect of the strain NEAU-HV9 and actinomycin D on the incidence and control of tomato bacterial wilt in pot culture experiments.

Treatment	Wilt Incidence (%)	Control Efficacy (%)
NEAU-HV9	13.3 ± 5.8 [b]	82 ± 6 [a]
Actinomycin D	0 ± 0 [b]	100 ± 0 [a]
Control	73.9 ± 6.6 [a]	

Wilt incidence (WI) was calculated as the percentage of leaves that were completely wilted. Control efficacy was calculated using the following formula: control efficacy (%) = 100 × (WI of control − WI of treatment)/WI of control. Data shown are the mean of three replications. Means within the same column followed by the same letter are not significantly different ($p = 0.05$) according to Fisher's least significant difference test.

4. Discussion

Soil-borne diseases have caused a significant decline in yield in the monoculture tomato field [3]. *Ralstonia solanacearum* is an important soil-borne bacterial plant pathogen which is distributed all over the world [5]. Recently, the biological control of soil-borne diseases has attracted more attention due to its environmental friendliness and high efficiency [54]. Therefore, isolation, screening and application of highly efficient antagonistic microorganisms is a key factor in biological control. With this outlook, a *Streptomyces* sp. strain NEAU-HV9 was isolated and found to exhibit antibacterial activities against *R. solanacearum* in the present study. By using 16S rRNA gene sequence analysis, combined with morphological, cultural and physiological characteristics, the results showed that strain NEAU-HV9 belongs to members of the genus *Streptomyces* and was most closely related to *Streptomyces panaciradicis* 1MR-8^T and *Streptomyces sasae* JR-39^T with gene sequence similarities of 98.90% and 98.89%, respectively.

Actinobacteria, particularly *Streptomyces*, are ubiquitous in the rhizosphere soil and can protect plant from pathogenic fungi/bacteria [55], so they have always been used in agriculture [56]. For instance, several *Streptomyces* species such as strains CAI-24, CAI-121, CAI-127, KAI-32 and KAI-90 have been used as BCAs against *Fusarium* wilt in chickpea plants [57]. The *Streptomyces* sp. CB-75, selected from banana rhizosphere soil, showed antifungal activity against 11 plant pathogenic fungi [54]. In this study, the *Streptomyces* sp. NEAU-HV9 exhibited strong antagonistic activity against *R. solanacearum*. According to the study of Singh et al., susceptibility of early stages of tomato seedlings toward the pathogen was confirmed by root-inoculation of *R. solanacearum* in early stages of tomato seedlings [22]. The antagonistic strains should reach a certain amount to demonstrate a significant biocontrol effect [58,59]. In this study, we inoculated very high numbers of *R. solanacearum* and very high levels of *Streptomyces* sp. NEAU-HV9 (10^9 cfu mL^{-1}) in small tubes in the TR 1 group. After culturing for seven days, all tomato seedlings were as healthy as the CK 1 control group (Table 3). There are only *R. solanacearum* and *Streptomyces* sp. NEAU-HV9 in this artificial system, which can better prove that a single NEAU-HV9 was able to be effective against *R. solanacearum*. On the seventh day, more than 90% of seedlings inoculated with *R. solanacearum* were found to be killed, but water-inoculated control seedlings were not wilted/dried [23]. Freshly grown tomato seedlings are too small to carry out detailed disease assessment, and can only be described as healthy, healthy wilted or dried. In the tests of this study, tomato seedlings inoculated with suspension (10^7 or 10^8 cfu mL^{-1}) of NEAU-HV9 and *R. solanacearum* showed healthy wilted and dried disease phenotypes at different levels, while all tomato seedlings inoculated with suspension (10^9 cfu mL^{-1}) of NEAU-HV9 and *R. solanacearum* were healthy (Figure 4). The results indicated that there are only *R. solanacearum* and *Streptomyces* sp. NEAU-HV9 in this artificial system, which can better prove that a single NEAU-HV9 was able to be effective against *R. solanacearum*. In addition, strain NEAU-HV9 effectively controlled *R. solanacearum* on larger plants in pot culture experiments (Table 4). Thus, the test presented in this study is viable for a preliminary screening of antagonistic actinobacterial strains against *R. solanacearum* and has important aspects with respect to reduced time, space consumption and economics. Meanwhile, the results showed the possibility of using *Streptomyces* sp. NEAU-HV9 as bioinoculant for *R. solanacearum*.

A wide range of bioactive secondary metabolites with anti-inflammatory, antibacterial, antifungal, antialgal, antimalarial and anticancer activities were produced by actinomycetes. Actinomycetes have produced about two-thirds of available antibiotics that have great practical value [60,61]. For example, *Streptomyces* TP-A0595 produced an antagonist that was determined as 6-prenylindole and effective against *Alternaria brassicicola* by inhibiting the formation of infection hyphae [62]. *Streptomyces griseus* H7602 produced a monomer compound that has suppressive effect on infection by *Phytophthora capsici* [63]. Some well-known antibiotics have been isolated from *Streptomyces* and used as fungicides. Many types of antibiotics with high antibacterial activity were produced from *Streptomyces spectabilis*, including streptovaricin [64], desertomycin [65] and spectinomycin [66], and they have high application value in the pharmaceutical industry [67]. Actinomycin D (or Dactinomycin) is a proverbial antitumor-antibiotic drug, which belongs to the actinomycin family and was isolated from *Streptomyces*. Actinomycin D has been demonstrated to have various biological activities.

Gram-negative bacteria were largely inhibited by using 10–100 mg per liter concentrations [68]. Actinomycin D produced by the bacterium *Streptomyces hydrogenans* IB310 was effective against both bacterial and fungal phytopathogens [51], and the authors proposed that actinomycin D might be developed as an antibacterial agent used in agriculture. However, there are no reports on antibacterial activities against *R. solanacearum* and it is not currently used in agriculture. In this study, *Streptomyces* sp. NEAU-HV9, which showed strong antibacterial activity against *R. solanacearum*, was isolated and identified. To learn more about the chemical nature of the antibacterial activity of the culture filtrate, the active compound actinomycin D was finally obtained. In this paper, we tested the in vitro antibacterial activity of actinomycin D against *R. solanacearum* and obtained a MIC value of 0.6 mg L^{-1} (0.48 µmol L^{-1}), which was many fold lower than other reported new natural antibacterial agents [69], synthesized antibacterial agents and those of commercial fungicides including gentamicin and streptomycin [70]. The antibacterial activity of actinomycin D against *R. solanacearum* tomato seedlings treated with 1 × MIC and 2 × MIC were determined. None of the seedlings inoculated with actinomycin D exhibited disease symptoms and the phytotoxic rating of actinomycin D was similar to that of a water control. Thus, actinomycin D was not phytotoxic at a concentration of 0.6 mg L^{-1} (0.48 µmol L^{-1}). The results suggest that actinomycin D might be useful as a candidate pesticide for the treatment of *Ralstonia solanacearum* in tomato.

5. Conclusions

In summary, this study found that *Streptomyces* sp. NEAU-HV9 exerted significant antibacterial activity against *R. solanacearum*, and actinomycin D, which was produced by *Streptomyces* sp. NEAU-HV9, exhibited a minimum inhibitory concentration (MIC) against *R. solanacearum* of 0.6 mg L^{-1} (0.48 µmol L^{-1}). In addition, *Streptomyces* sp. NEAU-HV9 and actinomycin D can effectively inhibit the occurrence of *R. solanacearum*. From the results, it is obvious that *Streptomyces* sp. NEAU-HV9 is an important microbial resource as a biological control against *R. solanacearum* and actinomycin D is a promising candidate for the development of potential antibacterial biocontrol agents.

Supplementary Materials: The following are available online at http://www.mdpi.com/2076-2607/8/3/351/s1: Table S1: Growth and cultural characteristics of strain NEAU-HV9 after 2 weeks at 28 °C; Table S2: Physiological and biochemical characteristics of strain NEAU-HV9; Figure S1: Cultural characteristics of strain NEAU-HV9 observed on ISP 1, ISP 2, ISP 3, ISP 4, ISP 5, ISP6, ISP7, Nutrient agar, Bennett's agar and Czapek's agar after being incubated at 28 °C for 2 weeks; Figure S2: Bioactivities of the supernatant and cell pellet of NEAU-HV9 against *R. solanacearum*; Figure S3: The HPLC profiles of crude extract produced by *Streptomyces* NEAU-HV9; Figure S4: Mass Spectrometry of actinomycin D (C$_{62}$H$_{86}$N$_{12}$O$_{16}$Na: 1277.6); Figure S5: ^{1}H NMR of actinomycin D; Figure S6: ^{13}C NMR of actinomycin D; Figure S7: The HPLC profiles of commercial actinomycin D; Figure S8: Bioactivities of commercial actinomycin D and the main product of NEAU-HV9 against *R. solanacearum*; Figure S9: Control efficiency of strain NEAU-HV9 against *R. solanacearum*; Figure S10: Control efficiency of the actinomycin D and strain NEAU-HV9 against *R. solanacearum*.

Author Contributions: L.L., X.H., X.L., H.W. and Y.W. performed the experiments. X.Z., L.Z. and P.C. analyzed the data. L.L. wrote the paper. X.W. and X.H. prepared the figures and tables. J.Z. and W.X. designed the experiments and reviewed the manuscript. All authors have read and agreed to the published version of the manuscript.

Funding: This work was supported in part by grants from the National Natural Youth Science Foundation of China (No. 31701858), the China Postdoctoral Science Foundation (2018M631907), the Heilongjiang Postdoctoral Fund (LBH-Z17015), the University Nursing Program for Young Scholars with Creative Talents in Heilongjiang Province (UNPYSCT-2017017) and the "Young Talents" Project of Northeast Agricultural University (17QC14).

Conflicts of Interest: The authors declare that there are no conflicts of interest.

References

1. Lee, C.G.; Lida, T.; Uwagaki, Y.; Otani, Y.; Nakaho, K.; Ohkuma, M. Comparison of prokaryotic and eukaryotic communities in soil samples with and without tomato bacterial wilt collected from different fields. *Microbes Eniviron.* **2017**, *32*, 376–385. [CrossRef] [PubMed]
2. Singh, V.K.; Singh, A.K.; Kumar, A. Disease management of tomato through PGPB: Current trends and future perspective. *3Biotech* **2017**, *7*, 255. [CrossRef] [PubMed]
3. Zhou, X.G.; Wu, F.Z. Dynamics of the diversity of fungal and Fusarium communities during continuous cropping of cucumber in the green house. *FEMS Microbiol. Ecol.* **2012**, *802*, 469–478. [CrossRef] [PubMed]
4. Yabuuchi, E.; Kosako, Y.; Yano, I.; Hota, H.; Nishiuchi, Y. Transfer of two *Burkholderia* and an *Alcaligenes species* to *Ralstonia* gen. nov., *Ralstonia solanacearum* (Smith, 1986) comb. nov. *Microbiol. Immun.* **1995**, *39*, 897–904. [CrossRef]
5. Hayward, A.C. Biology and epidemiology of bacterial wilt caused by *Pseudomonas solanacearum*. *Ann. Rev. Phytopathol.* **1991**, *29*, 65–87. [CrossRef]
6. Vu, T.T.; Kim, J.C.; Choi, Y.H.; Choi, G.J.; Jang, K.S.; Choi, T.H.; Yoon, T.M.; Lee, S.W. Effect of Gallotannins Derived from *Sedum takesimense* on Tomato Bacterial Wilt. *Plant Dis.* **2013**, *97*, 1593–1598. [CrossRef]
7. Genin, S.; Boucher, C. Lessons learned from the genome analysis of *Ralstonia solanacearum*. *Ann. Rev. Phytopathol.* **2004**, *42*, 107–134. [CrossRef]
8. Ciampi-Panno, L.; Fernandez, C.; Bustamante, P.; Andrade, N.; Ojeda, S.; Conteras, A. Biological control of bacterial wilt of potatoes caused by *Pseudomonas solanacearum*. *Am. Potato J.* **1989**, *66*, 315–332. [CrossRef]
9. McLaughlin, R.J.; Sequeira, L.; Weingartner, D.P. Biocontrol of bacterial wilt of potato with an avirulent strain of *Pseudomonas solanacearum*: Interactions with root-knot nematodes. *Am. Potato J.* **1990**, *67*, 93–107. [CrossRef]
10. Toyota, K.; Kimura, M. Suppresision of *Ralstonia solanacearum* in soil following colonization by other strains of *R. solanacearum*. *Soil Sci. Plant Nutr.* **2000**, *46*, 449–459.
11. Viaene, T.; Langendries, S.; Beirinckx, S.; Maes, M.; Goormachtig, S. *Streptomyces* as a plant's best friend? *FEMS Microbiol. Ecol.* **2016**, *92*. [CrossRef] [PubMed]
12. Dias, M.P.; Bastos, M.S.; Xavier, V.B.; Cassel, E.; Astarita, L.V.; Santarémm, E.R. Plant growth and resistance promoted by *Streptomyces* spp. in tomato. *Plant Physiol. Biochem.* **2017**, *118*, 479–493. [CrossRef] [PubMed]
13. Xio, K.; Kinkel, L.L.; Samac, D.A. Biological control of *Phytophthora* root rots on alfalfa and soybean with *Streptomyces*. *Biol. Control* **2002**, *23*, 285–295. [CrossRef]
14. El-Tarabily, K.A.; Sivasithamparam, K. Non-streptomycete Actinobacteria as biocontrol agents of soil-borne fungal plant pathogens and as plant growth promoters. *Soil Biol. Biochem.* **2006**, *38*, 1505–1520. [CrossRef]
15. Taechowisan, T.; Chuaychot, N.; Chanaphat, S.; Wanbanjob, A.; Tantiwachwutikul, P. Antagonistic effects of Streptomyces sp. SRM1 on colletotrichum musae. *Biotechnology* **2009**, *8*, 86–92. [CrossRef]
16. Bentley, S.D.; Chater, K.F.; Cerden, A.M.; Challis, G.L.; Thomson, N.R.; James, K.D.; Harris, D.E.; Quail, M.A.; Kieser, H.; Harper, D.; et al. Complete genome sequence of the model actinomycete *Streptomyces coelicolor* A3. *Nature* **2002**, *417*, 141–147. [CrossRef]
17. Brockmann, H. Die actinomycine. *Angew. Chem.* **1960**, *87*, 1767–1947.
18. Farber, S.; D'Angio, G.; Evans, A.; Mitus, A. Clinical studies of actinomycin D with special reference to Wilms' tumor in children. *J. Urol.* **2002**, *168*, 2560–2562. [CrossRef]
19. Womer, R.B. Soft tissue sarcomas. *Eur. J. Cancer* **1997**, *33*, 2230–2234. [CrossRef]
20. Monteiro, F.; Genin, S.; van, D.I.; Valls, M. A luminescent reporter evidences active expression of *Ralstonia solanacearum* type III secretion system genes throughout plant infection. *Microbiology* **2012**, *158*, 2107–2116. [CrossRef]
21. Guidot, A.; Jiang, W.; Ferdy, J.B.; Thébaud, C.; Barberis, P.; Gouzy, J.; Genin, S. Multihost experimental evolution of the pathogen *Ralstonia solanacearum* unveils genes involved in adaptation to plants. *Mol. Biol. Evol.* **2014**, *11*, 2913–2928. [CrossRef] [PubMed]
22. Singh, N.; Phukan, T.; Sharma, P.L.; Kabyashree, K.; Barman, A.; Kumar, R.; Sonti, R.V.; Genin, S.; Ray, S.K. An Innovative Root Inoculation Method to Study *Ralstonia solanacearum* Pathogenicity in Tomato Seedlings. *Phytopathology* **2018**, *108*, 436–442. [CrossRef]
23. Kumar, R. Studying Virulence Functions of *Ralstonia solanacearum*, the Causal Agent of Bacterial Wilt in Plants. Ph.D. Thesis, Tezpur University, Tezpur, India, 2014.

24. van Elsas, J.D.; Kastelein, P.; de Vries, P.M.; van Overbeek, L.S. Effects of ecological factors on the survival and physiology of *Ralstonia solanacearum* bv 2 in irrigation water. *Can. J. Microbiol.* **2001**, *47*, 842–854. [CrossRef] [PubMed]

25. Hayakawa, M.; Nonomura, H. Humic acid-vitamin agar, a new medium for selective isolation of soil actinomycetes. *J. Ferment. Technol.* **1987**, *65*, 501–509. [CrossRef]

26. Shirling, E.B.; Gottlieb, D. Methods for characterization of *Streptomyces* species. *Int. J. Syst. Bacteriol.* **1966**, *16*, 313–340. [CrossRef]

27. Zhang, D.C.; Brouchkov, A.; Griva, G.; Schinner, F.; Margesin, R. Isolation and characterization of bacteria from ancient siberian permafrost sediment. *Biology* **2013**, *2*, 85–106. [CrossRef]

28. Fu, Y.S.; Yan, R.; Liu, D.; Zhao, J.W.; Song, J.; Wang, X.J.; Cui, L.; Zhang, J.; Xiang, W.S. Characterization of *Sinomonas gamaensis* sp. nov., a Novel Soil Bacterium with Antifungal Activity against *Exserohilum turcicum*. *Microorganisms* **2019**, *7*, 170. [CrossRef] [PubMed]

29. Gao, F.; Wu, Y.; Wang, M. Identification and antifungal activity of an actinomycete strain against *Alternaria* spp. *Span. J. Agric. Res.* **2014**, *12*, 1158–1165. [CrossRef]

30. Jin, L.Y.; Zhao, Y.; Song, W.; Duan, L.P.; Jiang, S.W.; Wang, X.J.; Zhao, J.W.; Xiang, W.S. *Streptomyces inhibens* sp. nov., a novel actinomycete isolated from rhizosphere soil of wheat (*Triticum aestivum* L.). *Int. J. Syst. Evol. Microbiol.* **2019**, *69*, 688–695. [CrossRef]

31. Waksman, S.A. *The Actinomycetes. A Summary of Current Knowledge*; Ronald: New York, NY, USA, 1967.

32. Jones, K.L. Fresh isolates of actinomycetes in which the presence of sporogenous aerial mycelia is a fluctuating characteristic. *J. Bacteriol.* **1949**, *57*, 141–145. [CrossRef]

33. Waksman, S.A. Classification, identification and descriptions of genera and species. In *The Actinomycetes*; Williams and Wilkins: Baltimore, MD, USA, 1961; Volume 2.

34. Kelly, K.L. *Inter-Society Color Council-National Bureau of Standards Color-Name Charts Illustrated with Centroid Colors*; Government Printing Office: Washington, DC, USA, 1964.

35. Zhao, J.W.; Han, L.Y.; Yu, M.Y.; Cao, P.; Li, D.M.; Guo, X.W.; Liu, Y.Q.; Wang, X.J.; Xiang, W.S. Characterization of *Streptomyces sporangiiformans* sp. nov., a Novel Soil Actinomycete with Antibacterial Activity against *Ralstonia solanacearum*. *Microorganisms.* **2019**, *7*, 360. [CrossRef] [PubMed]

36. Gordon, R.E.; Barnett, D.A.; Handerhan, J.E.; Pang, C. *Nocardia coeliaca, Nocardia autotrophica*, and the nocardin strain. *Int. J. Syst. Bacteriol.* **1974**, *24*, 54–63. [CrossRef]

37. Yokota, A.; Tamura, T.; Hasegawa, T.; Huang, L.H. *Catenuloplanes japonicas* gen. nov., sp. nov., nom. rev., a new genus of the order *Actinomycetales*. *Int. J. Syst. Bacteriol.* **1993**, *43*, 805–812. [CrossRef]

38. Woese, C.R.; Gutell, R.; Gupta, R.; Noller, H.F. Detailed analysis of the higher-order structure of 16S-like ribosomal ribonucleic acids. *Microbiol. Rev.* **1983**, *47*, 621–669. [CrossRef] [PubMed]

39. Springer, N.; Ludwig, W.; Amann, R.; Schmidt, H.J.; Görtz, H.D.; Schleifer, K.H. Occurrence of fragmented 16S rRNA in an obligate bacterial endosymbiont of *Paramecium caudatum*. *Proc. Natl. Acad. Sci. USA* **1993**, *90*, 9892–9895. [CrossRef]

40. Yoon, S.H.; Ha, S.M.; Kwon, S.; Lim, J.; Kim, Y.; Seo, H.; Chun, J. Introducing EzBioCloud: A taxonomically united database of 16S rRNA and whole genome assemblies. *Int. J. Syst. Evol. Microbiol.* **2017**, *67*, 1613–1617. [CrossRef]

41. Saitou, N.; Nei, M. The neighbor-joining method: A new method for reconstructing phylogenetic trees. *Mol. Biol. Evol.* **1987**, *4*, 406–425.

42. Kumar, S.; Stecher, G.; Tamura, K. Mega7: Molecular evolutionary genetics analysis version 7.0 for bigger datasets. *Mol. Biol. Evol.* **2016**, *33*, 1870–1874. [CrossRef]

43. Felsenstein, J. Confidence limits on phylogenies: An approach using the bootstrap. *Evolution* **1985**, *39*, 783–791. [CrossRef]

44. Kimura, M. A simple method for estimating evolutionary rates of base substitutions through comparative studies of nucleotide sequences. *J. Mol. Evol.* **1980**, *16*, 111–120. [CrossRef] [PubMed]

45. Rathod, B.B.; Korasapati, R.; Sripadi, P.; Reddy, S.P. Novel actinomycin group compound from newly isolated *Streptomyces* sp. RAB12: Isolation, characterization, and evaluation of antimicrobial potential. *Appl. Microbiol. Biotechnol.* **2018**, *102*, 1241–1250. [CrossRef]

46. Guo, J.H.; Qi, H.Y.; Guo, Y.H.; Ge, H.; Gong, L.Y.; Zhang, L.X.; Sun, P.H. Biocontrol of tomato wilt by growth-promoting rhizobacteria. *Biol. Control* **2004**, *29*, 66–72. [CrossRef]

47. Roy, N.; Choi, K.; Khan, R.; Lee, S.W. Culturing Simpler and Bacterial Wilt Suppressive Microbial Communities from Tomato Rhizosphere. *Plant Pathol. J.* **2019**, *35*, 362–371. [PubMed]
48. Roberts, D.P.; Denny, T.P.; Schell, M.A. Cloning of the egl gene of *Pseudomonas solanacearum* and analysis of its role in phytopathogenicity. *J. Bacteriol.* **1988**, *170*, 1445–1451. [CrossRef] [PubMed]
49. Tsukuda, M.; Kitahara, K.; Miyazaki, K. Comparative RNA function analysis reveals high functional similarity between distantly related bacterial 16 S rRNAs. *Sci. Rep.* **2017**, *7*, 9993. [CrossRef]
50. Woese, C.R. Bacterial evolution. *Microbiol. Rev.* **1987**, *51*, 221–2711. [CrossRef]
51. Kulkarni, M.; Gorthi, S.; Banerjee, G.; Chattopadhyay, P. Production, characterization and optimization of actinomycin D from Streptomyces hydrogenans ID310, a(n antagonistic bacterium against phytopathogens. *Biocatal. Agric. Biotechnol.* **2017**, *10*, 69–74. [CrossRef]
52. Zhang, L.L.; Wan, C.X.; Luo, X.X.; Lv, L.L.; Wang, X.P.; Xia, Z.F. Manufacture of Actinomycin D with Streptomyces mutabilis for Controlling Plant Pathogenic Bacteria, Cow Mastitis, and Female Colpitis. China Patent CN104450580A, 25 March 2015.
53. Chen, C.; Song, F.; Wang, Q.; Abdel-Mageed, W.M.; Guo, H.; Fu, C.; Hou, W.; Dai, H.; Liu, X.; Yang, N.; et al. A marine-derived Streptomyces sp. MS449 produces high yield of actinomycin X2 and actinomycin D with potent anti-tuberculosis activity. *Appl. Microbiol. Biotechnol.* **2012**, *95*, 919–927. [CrossRef]
54. Chen, Y.F.; Zhou, D.B.; Qi, D.F.; Gao, Z.F.; Xie, J.H.; Luo, Y.P. Growth promotion and disease suppression ability of a *streptomyces* sp. CB-75 from banana rhizosphere soil. *Front. Microbiol.* **2018**, *8*, 2704. [CrossRef]
55. Crawford, D.L.; Lynch, J.M.; Whipps, J.M.; Ousley, M.A. Isolation and characterization of actionomycete antagonists of a fungal root pathogen. *Appl. Environ. Microb.* **1993**, *59*, 3889–3905. [CrossRef]
56. Gao, L.; Qiu, Z.; You, J.; Tan, H.; Zhou, S. Isolation and characterization of endophytic *Streptomyces* strains from surface-sterilized tomato (*Lycopersicon esculentum*) roots. *Lett. Appl. Microbiol.* **2004**, *39*, 425–430.
57. Gopalakrishnan, S.; Pande, S.; Sharma, M.; Humayun, P.; Kiran, B.K.; Sandeep, D.; Vidya, M.S.; Deepthi, K.; Rupela, O. Evaluation of actinomycete isolates obtained from herbal vermicompost for biological control of *Fusarium* wilt of chickpea. *Crop Protect.* **2011**, *30*, 1070–1078. [CrossRef]
58. Bull, C.T. Relationship between root colonization and suppression of *Gaeumannomyces graminis* var tritici by *Pseudomonas fluorescens* strain 2–79. *Phytopathology* **1991**, *81*, 954–959. [CrossRef]
59. Raaijmakers, J.M.; Leeman, M.; van Oorschot, M.M.P.; van der Sluis, I.; Schippers, B.; Bakker, P.A.H.M. Dose-response relationships in biological-control of Fusarium-wilt of radish by *Pseudomonas* spp. *Phytopathology* **1995**, *85*, 1075–1081. [CrossRef]
60. Wang, J.J.; Zhao, Y.; Ruan, Y.Z. Effects of bio-organic fertilizers produced by four *Bacillus amyloliquefaciens* strains on banana *Fusarium* wilt Disease. *Compost Sci. Util.* **2015**, *23*, 185–198. [CrossRef]
61. Hong, K.; Gao, A.H.; Xie, Q.Y.; Gao, H.; Zhuang, L.; Lin, H.P.; Yu, H.P.; Li, J.; Yao, X.S.; Goodfellow, M.; et al. Actino-mycetes for marine drug discovery isolated from man-grove soils and plants in China. *Mar. Drugs* **2009**, *7*, 24–44. [CrossRef]
62. Sasaki, T.; Igarashi, Y.; Ogawa, M.; Furumai, T. Identification of 6-prenylindole as an antifungal metabolite of *Streptomyces* sp. TP-A0595 and synthesis and bioactivity of 6-substituted indoles. *J. Antibiot.* **2002**, *55*, 1009–1012. [CrossRef] [PubMed]
63. Nguyen, X.H.; Naing, K.W.; Lee, Y.S.; Kim, Y.H.; Moon, J.H.; Kim, K.Y. Antagonism of antifungal metabolites from *Streptomyces* griseus H7602 against *Phytophthora capsici*. *J. Basic Microbiol.* **2015**, *55*, 45–53. [CrossRef] [PubMed]
64. Kakinuma, K.; Hanson, C.A.; Rinehart, K.L., Jr. Spectinabilin, a new nitro-containing metabolite isolated from *Streptomyces spectabilis*. *Tetrahedron* **1976**, *32*, 217–222. [CrossRef]
65. Ivanova, V. New macrolactone of the desertomycin family from *Streptomyces spectabilis*. *Prep. Biochem. Biotechnol.* **1997**, *27*, 19–38. [CrossRef]
66. Kim, K.R.; Kim, T.J.; Suh, J.W. The gene cluster for spectinomycin biosynthesis and the aminoglycoside-resistance function of spcm in *Streptomyces spectabilis*. *Curr. Microbiol.* **2008**, *57*, 371–374. [CrossRef] [PubMed]
67. Selvakumar, J.; Chandrasekaran, S.; Vaithilingam, M. Bio prospecting of marine-derived *Streptomyces spectabilis* VITJS10 and exploring its cytotoxicity against human liver cancer celllines. *Pharmacogn. Mag.* **2015**, *11*, 469–473.
68. Waksman, S.A.; Woodruff, H.B. Bacteriostatic and bactericidal substances produced by a soil actinomyces. *Proc. Soc. Exp. Biol. Med.* **1940**, *45*, 609–614. [CrossRef]

69. Vu, T.T.; Kim, H.; Tran, V.K.; Vu, H.D.; Hoang, T.X.; Han, J.W.; Choi, Y.H.; Jang, K.S.; Choi, G.J.; Kim, J.J. Antibacterial activity of tannins isolated from *Sapium baccatum* extract and use for control of tomato bacterial wilt. *PLoS ONE* **2017**, *12*, e0181499. [CrossRef]
70. Zheng, S.J.; Zhu, R.; Tang, B.; Chen, L.Z.; Bai, H.J.; Zhang, J.W. Synthesis and biological evaluations of a series of calycanthaceous analogues as antifungal agents. *Nat. Prod. Res.* **2019**, 1–9. [CrossRef] [PubMed]

 *microorganisms*

Article

New Antimicrobial Phenyl Alkenoic Acids Isolated from an Oil Palm Rhizosphere-Associated Actinomycete, *Streptomyces palmae* CMU-AB204[T]

Kanaporn Sujarit [1,2], **Mihoko Mori** [2,3,*], **Kazuyuki Dobashi** [2], **Kazuro Shiomi** [2,3], **Wasu Pathom-aree** [1,4] **and Saisamorn Lumyong** [1,4,5,*]

[1] Research Center of Microbial Diversity and Sustainable Utilization, Faculty of Science, Chiang Mai University, Chiang Mai 50200, Thailand; k.sujarit@gmail.com (K.S.); wasu215793@gmail.com (W.P.-a.)

[2] Kitasato Institute for Life Sciences, Kitasato University, 5-9-1 Shirokane, Minato-ku, Tokyo 108-8641, Japan; dobashi.kazu@gmail.com (K.D.); shiomi@lisci.kitasato-u.ac.jp (K.S.)

[3] Graduate School of Infection Control Sciences, Kitasato University, 5-9-1 Shirokane, Minato-ku, Tokyo 108-8641, Japan

[4] Department of Biology, Faculty of Science, Chiang Mai University, Chiang Mai 50200, Thailand

[5] Academy of Science, The Royal Society of Thailand, Bangkok 10300, Thailand

* Correspondence: morigon5454@gmail.com (M.M.); scboi009@gmail.com (S.L.); Tel.: +81-35-791-6131 (M.M.); +66-53-941-947 (ext. 144) (S.L.)

Received: 15 January 2020; Accepted: 28 February 2020; Published: 1 March 2020

Abstract: Basal stem rot (BSR), or *Ganoderma* rot disease, is the most serious disease associated with the oil palm plant of Southeast Asian countries. A basidiomycetous fungus, *Ganoderma boninense*, is the causative microbe of this disease. To control BSR in oil palm plantations, biological control agents are gaining attention as a major alternative to chemical fungicides. In the course of searching for effective actinomycetes as potential biological control agents for BSR, *Streptomyces palmae* CMU-AB204[T] was isolated from oil palm rhizosphere soil collected on the campus of Chiang Mai University. The culture broth of this strain showed significant antimicrobial activities against several bacteria and phytopathogenic fungi including *G. boninense*. Antifungal and antibacterial compounds were isolated by antimicrobial activity-guided purification using chromatographic methods. Their structures were elucidated by spectroscopic techniques, including Nuclear Magnetic Resonance (NMR), Mass Spectrometry (MS), Ultraviolet (UV), and Infrared (IR) analyses. The current study isolated new phenyl alkenoic acids **1–6** and three known compounds, anguinomycin A (**7**), leptomycin A (**8**), and actinopyrone A (**9**) as antimicrobial agents. Compounds **1** and **2** displayed broad antifungal activity, though they did not show antibacterial activity. Compounds **3** and **4** revealed a strong antibacterial activity against both Gram-positive and Gram-negative bacteria including the phytopathogenic strain *Xanthomonas campestris* pv. *oryzae*. Compounds **7–9** displayed antifungal activity against *Ganoderma*. Thus, the antifungal compounds obtained in this study may play a role in protecting oil palm plants from *Ganoderma* infection with the strain *S. palmae* CMU-AB204[T].

Keywords: actinomycetes; antimicrobial; phenyl alkenoic acid; rhizosphere; *Streptomyces palmae*

---

1. Introduction

Oil palm (*Elaeis guineensis* Jacq.) is an important economic crop in many tropical areas. In particular, Indonesia, Malaysia, and Thailand are the leading palm oil producing countries of this region. The oil palm plant typically has a productive life of 20 or more years, and oil can be harvested several times each year. Consequently, it holds an advantage over all other oil-producing crops [1]. However, the plant is often damaged by fungal infections, and these can cause a decrease of crop yields and result in the death of oil palm trees.

Fungal pathogens mainly infect the stems and leaves of oil palm trees during all stages of growth, from seedlings to the mature stage, and consequently can affect both the quality and quantity of palm oil. Basal stem rot (BSR), or *Ganoderma* rot disease, is the most severe disease of oil palm trees in Southeast Asian countries, especially Malaysia and Indonesia [2]. In addition to these countries, BSR has also destroyed oil palm plantations in Africa, Colombia [3], Papua New Guinea [4], and Thailand [5]. The causative fungus *Ganoderma boninense* is a basidiomycetous fungus and belongs to the order *Polyporales* and the family of *Ganodermataceae*. Fruiting bodies of *Ganoderma* typically form on the exterior of the oil palm trunk and then release and spread the spores to the soil. The usual method of controlling BSR in oil palm plantations is the use of chemical fungicides. Many fungicides, such as azoxystrobin, benomyl, carbendazim, carboxin, cycloheximide, cyproconazole, drazoxolone, hexaconazole, methfuroxam, nystatin, penconazole, thiram, triadimefon, triadimenol, tridemorph, and quintozene, could inhibit the growth of *Ganoderma* [6–10]. However, the fungicides cannot actually cure infected palm trees; they can only delay the spreading of the disease [9]. Furthermore, the applications of these chemical treatments have some worrying effects on human health and ecosystems. Examples of this would be toxicity to organisms and the suppression of beneficial microbes [9,11–13]. Nowadays, raising concerns about the high cost of chemicals, and the environmental problems they are associated with, have encouraged researchers to seek alternative strategies for BSR suppression.

The use of biological control agents represents a major alternative approach in the management of oil palm diseases. Fungal species, such as *Trichoderma harzianum*, *Trichoderma viride*, and *Gliocladium viride*, have been studied for their anti-*Ganoderma* activity, and their effectiveness against *Ganoderma* in a glasshouse and in a field trial [2,14–16]. Certain *Trichoderma* species are known as mycoparasites and have been utilized to control fungal pathogens. One of the biocontrol mechanisms of *Trichoderma* spp. is the release of glucanase and chitinase enzymes that are involved in the cell-wall degradation of *G. boninense*, and these can be elicitors in inducing a plant defense response [17,18]. Several strains of bacteria, especially *Pseudomonas aeruginosa*, *Pseudomonas syringae*, and *Burkholderia cepacia*, have also been studied for their potential to be applied as biological control agents [19–22]. Their potential abilities to inhibit the spread of *G. boninense* and to reduce the incidence of the disease have been documented [19–22]. Although the control mechanisms of these bacteria have not yet been clarified, they may control *Ganoderma* by producing antifungal secondary metabolites. In addition, several actinomycetes were screened for their antagonistic activity against *G. boninense*. Actinomycetes, especially the genus *Streptomyces*, are well known for their ability to produce a wide variety of bioactive metabolites [23–26]. Many *Streptomyces* species, such as *Streptomyces hygroscopicus*, *Streptomyces ahygroscopicus*, *Streptomyces abikoensis*, and *Streptomyces angustmyceticus*, were found to be promising biocontrol agents for BSR disease [27,28]. *Streptomyces violaceorubidus* released not only secondary metabolites towards *G. boninense* but also released cell-wall degrading enzymes involved in the control of this pathogen [29,30].

Actinomycetes associated with the oil palm rhizosphere may have an important role in protecting plants from *Ganoderma* infection by releasing antibiotics and enzymes. Thus, we isolated actinomycetes from the rhizosphere of healthy oil palm plants and screened the antifungal activities of their culture broth against *G. boninense*. One actinomycete strain, CMU-AB204^T, showed significant antimicrobial activities against, not only *G. boninense* but also phytopathogenic fungi and several bacteria. We had previously identified this strain and proposed that it could serve as a novel species, namely *Streptomyces palmae* CMU-AB204^T [31]. This actinomycete was selected to investigate antimicrobial secondary metabolites. This report describes the results of the isolation, structural elucidation, and antimicrobial activities of six new compounds, AB204-A–F (**1–6**), and three known compounds, anguinomycin A (**7**), leptomycin A (**8**), and actinopyrone A (**9**), that were produced by *S. palmae* CMU-AB204^T.

2. Materials and Methods

2.1. Microbial Material

Streptomyces palmae CMU-AB204^T was previously isolated from the rhizosphere of an oil palm tree collected from the oil palm plantation at Chiang Mai University, Chiang Mai Province, Thailand, in October 2012. This strain has been characterized using a polyphasic approach and was previously proposed as *S. palmae* (type strain CMU-AB204^T = JCM 31289^T = TBRC 1999^T) [31].

2.2. Culture Conditions

S. palmae CMU-AB204^T was grown in the International Streptomyces Project medium 2 (ISP2) agar [32] at 28 °C. For seed culture, 100 mL of ISP2 medium, consisting of 0.4% yeast extract (Becton, Dickinson and Company, Sparks, MD, USA), 1.0% malt extract (Becton, Dickinson and Company, Sparks, MD, USA), and 0.4% glucose, was prepared in an Erlenmeyer flask and the pH was adjusted to 7.0 before sterilization. The slant culture of *S. palmae* was scraped by an inoculating loop and inoculated into ISP2 medium. The inoculated flask was incubated at 30 °C for three days on a rotary shaker at 150 rpm. Two mL portions of this seed culture were transferred into 500 mL Erlenmeyer flasks containing 150 mL of ISP2 medium, which was followed by fermentation using a rotary shaker at 150 rpm, 30 °C for seven days.

2.3. Compound Extraction and Isolation Procedure

The mycelia were separated from fermentation broth (40.0 L) by filtration. The culture filtrate and mycelium were separately extracted twice with an equal volume of EtOAc. The organic layer was evaporated using a rotary evaporator. Extracts from culture filtrate and mycelium were combined and concentrated to dryness in vacuo to obtain a crude extract as a brown oil. The active secondary metabolites were isolated by biological activity-guided purification. The crude extract (4.9 g) was separated using an open column with silica gel (silica gel 60, 0.063–0.200 mm, Merck, Darmstadt, Germany, 150 g of silica gel, Ø40 mm × 240 mm) and eluted with a stepwise gradient of CHCl$_3$/MeOH: 100:0, 99:1, 98:2, 95:5, 90:10, 80:20, 50:50 and 0:100 (v/v), with 1.0 L each. Each eluent was collected in two 500 mL Erlenmeyer flasks (S1–S16) and concentrated in vacuo. The components of each fraction were analyzed using thin-layer chromatography (TLC, silica gel F254, Merck, Darmstadt, Germany) plates with a thickness of 0.25 mm, developed with the CHCl$_3$/MeOH solvent system. Compounds were detected by UV light and phosphomolybdic acid reagent and followed by heating. The active fractions S3 (580.4 mg) and S4 (608.8 mg) eluted with 99:1 (v/v) of CHCl$_3$/MeOH were dissolved in a small amount of MeOH and then separately subjected to Sephadex LH-20 column chromatography (GE Healthcare Bio-Sciences, USA, Ø20 mm × 650 mm) with MeOH as the eluent. The eluate was automatically fractionated into 100 fractions (L1–L100) by a fraction collector (CHF100AA, Advantec, Tokyo, Japan). The active materials were detected from fractions L52–L64. From fractions S3 and S4, 59.6 mg of yellow semi-solid substance was obtained as an active material. Analytical and preparative HPLC of these fractions were carried out on a JASCO HPLC system (JASCO, Tokyo, Japan); pump, PU-2080 Plus; solvent mixer, LG-2808-04; UV detector, MD-1510. The HPLC columns included an analytical column (Pegasil ODS SP100, Ø4.6 mm × 250 mm; Senshu Scientific, Tokyo, Japan) and a preparative column (Pegasil ODS SP100, Ø20 mm × 250 mm; Senshu Scientific). This dried material (59.6 mg) was subjected to preparative HPLC developed with a gradient system of CH$_3$CN aqueous solution containing 0.1% trifluoroacetic acid (60–90% CH$_3$CN for 20 min, 90% CH$_3$CN for 20 min) at flow rate of 7.0 mL/min, and detection was achieved at 254 nm. The eluates at retention times of 16, 21, 32, 33, and 34 min were collected and concentrated in vacuo to dryness in order to afford AB204-A (**1**), AB204-B (**2**), AB204-E (**5**), AB204-F (**6**), and a mixture of AB204-C (**3**) and D (**4**), respectively. Compound **9** was obtained from side fractions (L36–L49) of LH-20 column chromatography of S3. The combined fractions (L36–L49 of S3) were purified by preparative TLC (silica gel, Merck, Darmstadt, Germany) with a developing solvent of CHCl$_3$/MeOH (20:1) to obtain **9**. Compounds **7** and **8** were isolated from the active fraction that was eluted with 98:2 (v/v) of $CHCl_3$/MeOH. The fraction was subjected to silica gel column chromatography with the $CHCl_3$/MeOH solvent system, and active compounds were obtained from the 95:5 (v/v) fraction. This fraction was purified by preparative HPLC with a linear gradient system of 60–90% CH_3CN–H_2O containing 0.1% trifluoroacetic acid for 30 min at a flow rate of 7 mL/min and at room temperature. Detection was achieved at 254 nm. Compounds **7** and **8** were eluted at 24 min and 27 min, respectively.

2.4. Analyses of the Chemical Structure and Physicochemical Properties

The purified compounds were prepared at a concentration of 1 mg/mL in MeOH for the measurement of optical rotation, UV spectra, and IR spectra. An optical rotation $[\alpha]_D$ of the compound suspension was measured using a P-2200 polarimeter (JASCO, Tokyo, Japan). UV spectra of each compound were recorded with a U-2810 spectrophotometer (Hitachi High-Tech Science Co., Tokyo, Japan), and IR spectra (ATR) were measured using a FT–IR 4600 (JASCO, Tokyo, Japan). The isolated compounds were dissolved in chloroform-d ($CDCl_3$) or methanol-d_4 (CD_3OD) for NMR analyses. NMR spectra of each compound were obtained on a JNM ECP500 NMR spectrometer (JEOL, Tokyo, Japan) with 500 MHz for ^{1}H NMR and 125 MHz for ^{13}C NMR. Chemical shifts (ppm) of $CDCl_3$ (δ_H 7.26, δ_C 77.0) and CD_3OD (δ_H 3.30, δ_C 49.0) were used as references. The accurate mass and molecular formulas of the isolated compounds were established by liquid chromatography–mass spectrometry (LC–MS) analyses. Spectra of electron ionization mass spectrometry (EI–MS) were analyzed using a JMS-AX505 HA spectrometer (JEOL, Tokyo, Japan), while the spectra of electrospray ionization mass spectrometry (ESI–MS) were obtained by a JMS-T100LP spectrometer (JEOL, Tokyo, Japan) equipped with an Agilent1100 HPLC system (Agilent, CA, USA).

2.5. Measurement of Antimicrobial Activity

In the purification process of antimicrobial compounds, every fraction obtained from each fractionation step was tested with representative microbes, *Xanthomonas campestris* pv. *oryzae* KB88, *Kocuria rhizophila* ATCC 9341, *Mucor racemosus* IFO 4581, and *G. boninense* BCC 21330, by paper disk diffusion assay. The antimicrobial activity of the purified compounds was analyzed using the paper disk diffusion method (Ø8 mm disk, Advantec, Co., Ltd., Tokyo, Japan) against fourteen microorganisms. Cell suspensions of *Bacillus subtilis* ATCC 6633 (5×10^5 cfu/mL), *Escherichia coli* NIHJ (5×10^5 cfu/mL), *K. rhizophila* ATCC 9341 (2×10^5 cfu/mL), *Mycobacterium smegmatis* ATCC 607 (5×10^5 cfu/mL), *Staphylococcus aureus* ATCC 6538p (5×10^5 cfu/mL), *Klebsiella pneumonia* ATCC 10031 (5×10^5 cfu/mL), *Proteus vulgaris* NBRC 3167 (5×10^5 cfu/mL), *Pseudomonas aeruginosa* IFO 3080 (1×10^6 cfu/mL), and *X. campestris* pv. *oryzae* KB88 (1×10^6 cfu/mL) were individually mixed into the medium containing 0.5% peptone, 0.5% meat extract, and 0.8% agar, while *Aspergillus niger* ATCC 6275 (1×10^6 spores/mL), *Candida albicans* ATCC 64548 (2×10^5 cfu/mL), *G. boninense* BCC 21330 (2×10^5 cfu/mL), *Mu. racemosus* IFO 4581 (2×10^5 spores/mL), and *Saccharomyces cerevisiae* ATCC 9763 (1×10^6 cfu/mL) were individually mixed into the medium containing 1.0% glucose, 0.5% yeast extract, and 0.8% agar, and poured into Petri dishes. After that, paper disks containing the purified compounds at 50 µg/disk were put onto agar plates of each microorganism with three replicates. All bacterial plates, except *M. smegmatis* ATCC 607 and *X. campestris* pv. *oryzae* KB 88, were incubated at 37 °C for 24 h. *M. smegmatis* ATCC 607 was incubated at the same temperature for 48–72 h., whereas *X. campestris*, yeasts, and fungi were incubated at 27 °C for 24–48 h. The diameter of the inhibition zone was measured in mm units.

3. Results

3.1. Biological Activity-Guided Purification of Active Components from Culture Broth of S. palmae CMU-AB204^T and Structure Determination of Active Components

S. palmae CMU-AB204^T was cultured in 40 L of ISP2 medium at 28 °C for seven days, and the broth and mycelia were extracted with EtOAc. The active components in culture broth extract of strain CMU-AB204^T were isolated by biological activities-guided purification using paper disk assay. The extract was purified by silica gel column chromatography, Sephadex LH-20 column chromatography, preparative TLC, and preparative HPLC. The eluates were concentrated in vacuo to yield nine compounds, AB204-A (**1**, 10.2 mg), AB204-B (**2**, 8.2 mg), AB204-E (**5**, 14.2 mg), AB204-F (**6**, 9.9 mg), a mixture (6.5 mg) of AB204-C (**3**) and AB204-D (**4**), **7** (7.0 mg), **8** (1.7 mg), and **9** (17.0 mg) as depicted in Scheme 1.

AB204-A (**1**) was obtained as a pale yellow amorphous solid. It was found to be readily soluble in acetonitrile, MeOH, CHCl$_3$, and was observed to be less soluble in water. As the HREIMS analysis showed *m/z* 190.1000 [M]$^+$, the molecular formula of **1** was elucidated as C$_{12}$H$_{14}$O$_2$ (calculated value of 190.0994, Figure S1). The intense band at 1706 cm^{-1} of the IR spectrum in MeOH solution was assigned as C=O stretching frequency of dimeric carboxylic acid moiety (Figure S2). Based on ^{1}H NMR analysis, **1** revealed four aromatic protons stacked at 7.14–7.18 ppm and a pair of olefinic protons stacked at 5.70 ppm and 6.51 ppm (Figure S3). Coupling constants (*J* = 11.5 Hz) of the two olefinic protons showed Z-configuration of the olefin moiety. Compound **1** had four additional methylene protons at 2.41–2.51 ppm and one methyl singlet signal at 2.24 ppm (Table 1). The ^{13}C NMR spectrum showed 12 carbon signals: one carbonyl carbon at 177.2 ppm that indicated a carboxylic acid, eight aromatic or olefinic carbons, two methylene carbons at 23.5 and 33.7 ppm, and one methyl carbon at 19.8 ppm (Table 1, Figure S4).

Table 1. NMR spectroscopic data of AB204-A (**1**) and B (**2**) (δ_H, 500 MHz; δ_C, 125 MHz).

Carbon No.	1 (in CDCl$_3$)			2 (in CDCl$_3$)		
	δ_C, Type	δ_H, Mult (*J* in Hz)	HMBC	δ_C, type	δ_H, Mult (*J* in Hz)	HMBC
1	177.2, C			177.6, C		
2	33.7, CH$_2$	2.41–2.44, m	C1, C3, C4	33.4, CH$_2$	2.30, t (7.5)	C1, C3, C4
3	23.5, CH$_2$	2.46–2.51, m	C1, C2, C5	24.2, CH$_2$	1.63, tt (7.5, 7.5)	C1, C2, C5
4	129.7 or 129.8, CH	5.70, dt (11.5, 7.0)	C1′	29.1, CH$_2$	1.45, tt (7.5, 7.5)	C2, C5, C6
5	129.7 or 129.8, CH	6.51, d (11.5)	C3, C2′, C6′	27.8, CH$_2$	2.17, dtd (7.5, 7.5, 1.5)	C3, C4, C6, C7
6				132.0, CH	5.69, dt (11.5, 7.5)	C1′
7				128.5, CH	6.45, br.d (11.5)	C5, C6′
1′	136.2, C			136.7, C		
2′	136.2, C			136.2, C		
3′	129.8, CH	7.14–7.18 *, m		129.8, CH	7.13–7.18 *, m	-
4′	125.4 or 127.1, CH	7.14–7.18 *, m		125.3 or 126.8, CH	7.13–7.18 *, m	C2′
5′	125.4 or 127.1, CH	7.14–7.18 *, m	C3′	125.3 or 126.8, CH	7.13–7.18 *, m	C3′
6′	128.8, CH	7.14–7.18 *, m		128.9, CH	7.13–7.18 *, m	C2′
2′-Me	19.8, CH$_3$	2.24, s	C1′, C2′, C3′	19.9, CH$_3$	2.25, s	C1′, C2′, C3′

* overlapped.

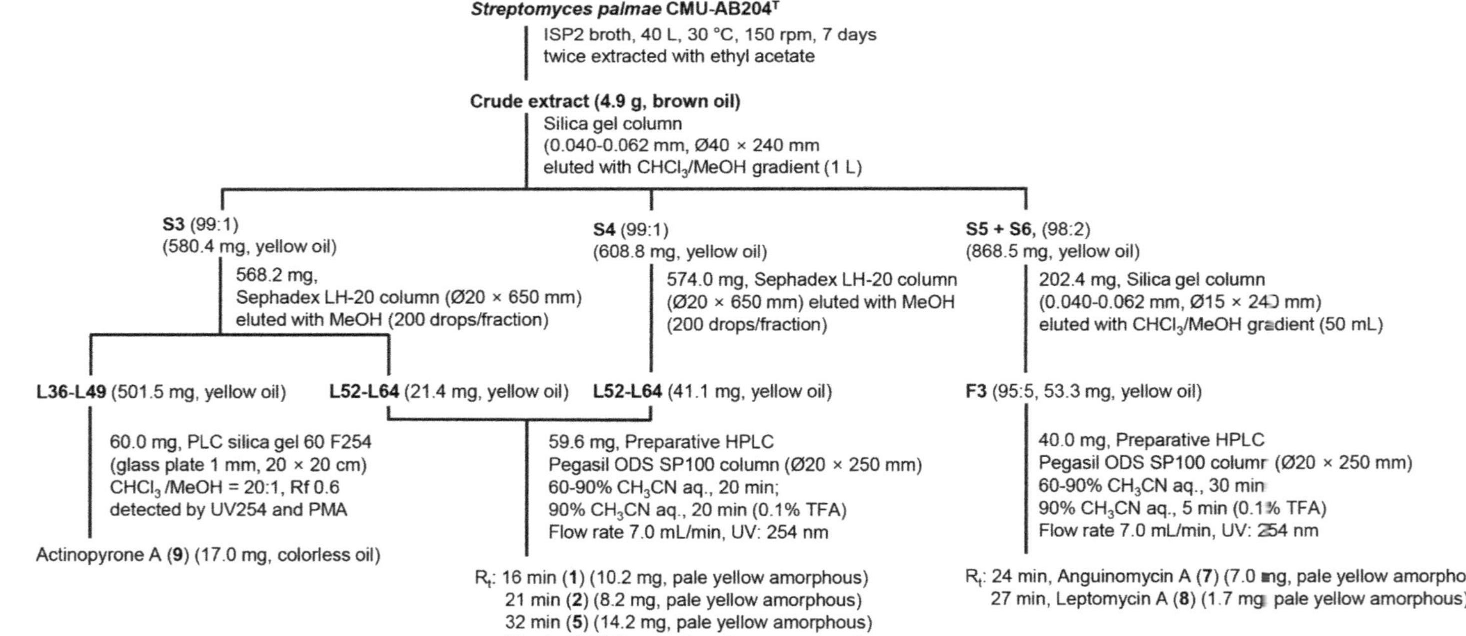

Scheme 1. Purification procedures for compounds 1–9.

MS, HMBC, and HMQC analyses suggested **1** contained one disubstituted aromatic ring, one methyl, and pentenoic acid moieties (Figures S5 and S6). HMBC correlations were observed from two methylene protons (2.41–2.44 ppm and 2.46–2.51 ppm) to a carboxylic carbon at 177.2 ppm, and two olefinic carbons of C-4 and C-5 (129.7 and 129.8 ppm), as are given in Table 1. A correlation between the Z-olefinic proton at 6.51 ppm (H-5) and one methylene carbon (C-3) at 23.5 ppm was also observed; thus **1** was believed to possess 4,5-Z-pentenoic acid moiety in the structure. An HMBC correlation between one methyl proton at 2.24 ppm and three aromatic carbons of C-1′, C-2′, and C-3′ (136.2, 136.2, and 129.8 ppm, respectively), and between Z-olefinic protons and aromatic carbons, H-4 (5.70 ppm) and C-1′ (136.2 ppm), and H-5 (6.51 ppm) and C-6′ (128.8 ppm), indicated **1** was an *ortho*-methyl phenyl alkenoic acid compound, (Z)-5-(2-methylphenyl)-4-pentenoic acid (Figure 1). Differential NOE of **1** was observed between a methyl proton and both an aromatic 3′-proton and an olefinic proton of H-5 as well as between the two olefinic protons (Figure S7). The geometry of two substitutes of the aromatic ring was confirmed by NOE correlations, as is shown in Figure 2. From some *Streptomyces* strains, E-isomer of **1**, (E)-5-(2-methylphenyl)-4-pentenoic acid was identified [33–35]; however, there was no report on the Z-isomer (**1**) obtained from natural sources. Therefore, it was concluded that **1** was a novel natural product.

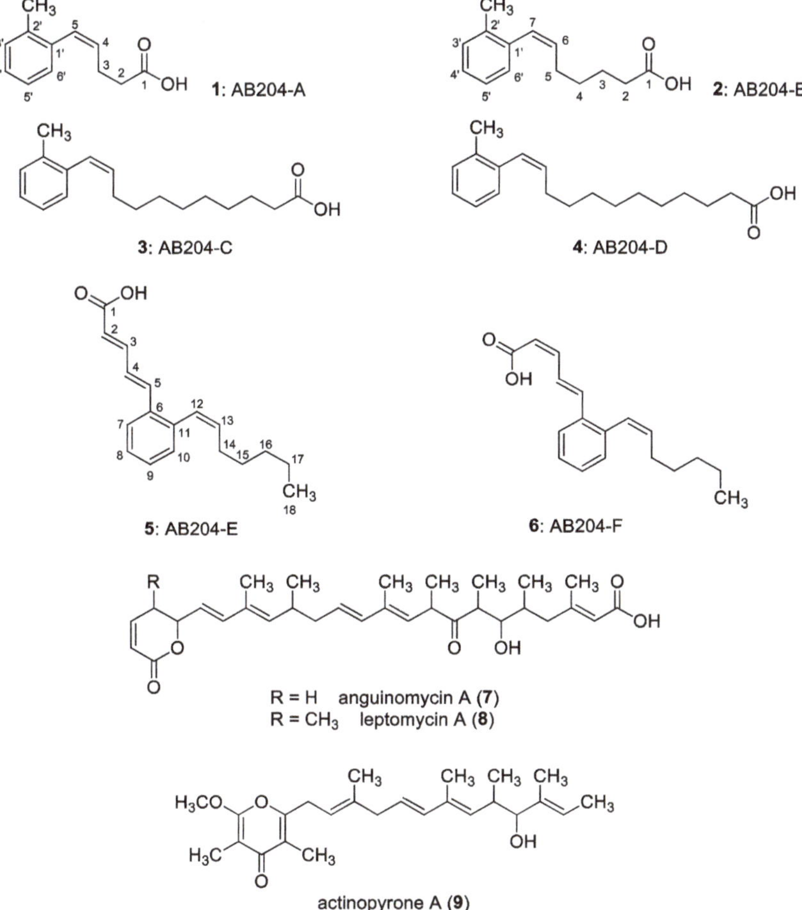

Figure 1. Antimicrobial compounds isolated from the broth extract of *Streptomyces palmae* CMU-AB204[T].

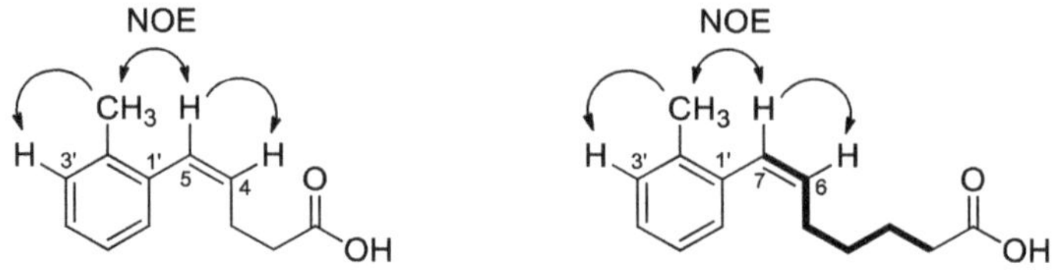

Figure 2. Observed COSY correlation (bold lines, in 2) and NOE correlation in AB204-A (1) and B (2).

AB204-B (**2**) was isolated as a pale yellow amorphous solid. It was readily soluble in the same solvent as **1** and less soluble in water. The molecular formula of **2** was established as $C_{14}H_{18}O_2$ (calculated value of 218.1329) based on NMR data and the HREIMS ion peak of m/z 218.1301 [M]$^+$ (Figure S8), indicating compound **2** had one more C_2H_4 unit when compared to the molecular formula of **1**. The IR spectrum of **2** revealed the presence of C=O stretching frequency of dimeric carboxylic acid moiety at 1706 cm^{-1}, which was similar to the spectrum of **1** (Figure S9). The ^{1}H NMR spectrum of **2** revealed four aromatic protons at 7.14–7.16 ppm, a pair of olefinic protons at 5.69 and 6.45 ppm, eight methylene protons at 1.45, 1.63, 2.17, and 2.30 ppm and one methyl singlet signal at 2.25 ppm (Table 1, Figure S10). The coupling constant of two olefinic protons (11.5 Hz) indicated a Z-configuration. The ^{13}C NMR spectrum of **2** showed 14 carbon signals: one carbonyl carbon at 177.6 ppm, eight aromatic or olefinic carbons, four methylene carbons at 24.2, 27.8, 29.1, and 33.4 ppm, and one methyl carbon at 19.9 ppm (Table 1, Figure S11). These data support the conclusion that the compound had a closely related structure to **1**. Each methylene signal was assigned by COSY, as is shown in Figure 2. Eight methylene protons constructed a C_4 alkyl chain, and COSY correlation confirmed the connection between this C_4 alkyl chain and Z-olefin (Figure S12). This connection was supported by HMBC and HMQC spectra (Figures S13 and S14). HMBC correlations were observed from one olefinic proton H-6 (5.69 ppm) to an aromatic carbon at C-1' (136.7 ppm) and from singlet methyl proton at 2.25 ppm to aromatic carbons at C-1', C-2', and C-3'. Therefore, one methyl moiety and one alkene chain were substituted for an aromatic ring in the *ortho* position. HMBC correlation from two methylene protons of the alkene chain at 1.63 and 2.30 ppm to the carbonyl carbon at 177.6 ppm, and a molecular formula of **2**, suggested that this compound had a carboxylic acid at the end of the alkene chain. The same NOE correlation was observed in **1** and **2** (Figure 2 and Figure S15). Thus, the structure of **2** was assigned as (Z)-7-(2-methylphenyl)-6-heptenoic acid, as is shown in Figure 1.

The structures of AB204-C (**3**) and AB204-D (**4**) were elucidated as a mixture of both compounds because of the difficulty associated with further purification. MS spectra showed the molecular formulas of **3** and **4** were $C_{18}H_{26}O_2$ and $C_{19}H_{28}O_2$, respectively (Figure S16). ^{1}H NMR data suggested compounds **3** and **4** were analogs of **1** and **2**, thus **3** and **4** might be (Z)-11-(2-methylphenyl)-10-undecenoic acid and (Z)-12-(2-methylphenyl)-11-dodecenoic acid, respectively (Figure 1 and Figure S17).

AB204-E (**5**) and AB204-F (**6**) were obtained as a pale yellow amorphous solid. The accurate mass and molecular formula of compounds **5** and **6** were analyzed by both HRESIMS and HREIMS. Molecular ion peaks were exhibited at m/z 271.1705 [M + H]$^+$ and 270.1616 [M]$^+$ for **5**, and m/z 271.1689 [M + H]$^+$ and 270.1633 [M]$^+$ for **6** (Figures S18–S21). These data suggest that both compounds had the same molecular formula as $C_{18}H_{22}O_2$ (calcd. 270.1620 for $C_{18}H_{22}O_2$ and calcd. 271.1698 for $C_{18}H_{23}O_2$). The C=O stretching frequency band at 1685 cm^{-1} in the IR spectrum of **5** and 1684 cm^{-1} in the spectrum of **6** was assigned as a carboxylic acid moiety (Figures S22 and S23). NMR spectra of both compounds demonstrated a structural similarity. The assignment of ^{1}H and ^{13}C NMR spectra of **5** and **6** are given in Table 2.

Table 2. NMR spectroscopic data of AB204-E (**5**) and F (**6**) (δ_H, 500 MHz; δ_C, 125 MHz).

Carbon No.	5 (in CD$_3$OD)			6 (in CDCl$_3$)		
	δ_C, Type	δ_H, Mult (*J* in Hz)	HMBC	δ_C, Type	δ_H, Mult (*J* in Hz)	HMBC
1	170.5, C			169.5, C		
2	122.3, CH	5.99, d (15.5)	C1, C4	115.8, CH	5.74, d (11.5)	C1, C4
3	146.8, CH	7.41, dd (15.5, 11.0)	C1, C5	147.3, CH	6.85, t (11.5)	C1, C5
4	128.2, CH	6.95, dd (15.5, 11.0)	C2, C5, C6	125.4, CH	8.04, dd (15.5, 11.5)	C6
5	140.0, CH	7.13, d (15.5)	C3, C7, C11	140.6, CH	7.08, d (15.5)	C3, C7, C11
6	135.6, C			134.3, C		
7	126.7, CH	7.67, m	C9, C11	126.2, CH	7.72, m	C9, C11
8	128.4, CH	7.26 *, m		127.1, CH	7.26 *, m	C10
9	129.5, CH	7.26 *, m		128.6, CH	7.26 *, m	C11
10	131.0, CH	7.15, m	C8, C12	129.9, CH	7.17, m	C6, C8, C12
11	138.7, C			137.5, C		
12	128.3, CH	6.54, br.d (11.5)	C6, C10, C14	127.0, CH	6.51, br.d (11.5)	C10, C14
13	135.7, CH	5.83, dt (11.5, 7.5)	C11	134.9, CH	5.81, dt (11.5, 7.5)	C11
14	29.4, CH$_2$	2.02, dtd (7.5, 7.5, 1.5)	C12, C13, C15, C16	28.4, CH$_2$	2.04, dtd (7.5, 7.5, 1.0)	C12, C13, C15, C16
15	30.2, CH$_2$	1.38, m	C13, C14, C16, C17	29.2, CH$_2$	1.37, tt (7.5, 7.5)	C14, C16, C17
16	32.6, CH$_2$	1.22, m	C17	31.4, CH$_2$	1.23 *, m	C17
17	23.5, CH$_2$	1.21, m	C16	22.5, CH$_2$	1.23 *, m	C16
18	13.3, CH$_3$	0.83, t (7.0)	C16, C17	14.0, CH$_3$	0.84, t (7.0)	C16, C17

* overlapped.

^{1}H NMR spectrum of **5** showed ten aromatic/olefinic protons, eight methylene protons, and one methyl triplet proton at 0.83 ppm (Figure S24). In ^{13}C NMR spectrum of **5**, one calboxylic carbon at 170.5 ppm, twelve olefinic/aromatic carbons, four methylene carbons, and one methyl carbon were measured (Figure S25). ^{1}H NMR spectrum of **5** suggested the existence of two pairs of *E*-olefin assigned by large coupling constants (15.5 Hz for each) and one pair of *Z*-olefin whose coupling constant was 11.5 Hz. Three partial structures were assigned by COSY correlation; one 1,2-substituted aromatic ring, one 1,2-*Z*-heptene group, and one diene group (Figure S26). HMBC correlations between diene protons of 5.99 and 7.41 ppm and a carbonyl carbon at 170.5 ppm, and between diene protons of 6.95 and 7.13 ppm and three aromatic carbons at positions C-6, C-7, and C-11 (135.6, 126.7, and 138.7 ppm, respectively) indicated that one end of the diene was connected to a carboxylic carbon and the other end of the diene was attached to an aromatic ring at position 6 (Figures S27 and S28). HMBC correlation between *Z*-olefinic protons (5.83 and 6.54 ppm) of 1,2-*Z*-heptene and aromatic ring carbons at C-10 and C-11 (131.0 and 138.7 ppm, respectively) suggested 1,2-*Z*-heptene moiety was connected to the aromatic ring at C-11. These data support the structure of **5** as (2*E*,4*E*)-5-(2-(1*Z*)-heptenylphenyl)-2,4-pentadienoic acid (Figure 1).

^{1}H NMR and ^{13}C NMR spectra of **6** suggested that the structure was almost the same as **5** (Figures S29 and S30). However, the ^{1}H NMR spectrum clarified that only one pair of *E*-olefin existed, while the other two pairs of olefin were identified as *Z*-configuration by analysis of coupling constants. These data indicated that **6** was a stereoisomer of **5**. COSY correlations revealed that two partial structures of **6**, one 1,2-substituted aromatic ring and one 1,2-*Z*-heptene moiety, were identical to those of **5**; however, a diene structure was constituted of both *E* and *Z*-olefins (Figure S31). The connection of 1,2-*Z*-heptene moiety to the aromatic ring at C-11 was confirmed by the HMBC spectrum (Figure 3, Figures S32 and S33). HMBC correlations between *Z*-olefinic protons of the diene moiety and a carboxylic carbon at 169.5 ppm and between one *E*-olefinic proton (7.08 ppm) of the diene moiety and aromatic ring carbons at C-7 and C-11 (126.2 and 137.5 ppm, respectively) established the structure of **6** as (2*Z*,4*E*)-5-(2-(1*Z*)-heptenylphenyl)-2,4-pentadienoic acid, as is depicted in Figure 1.

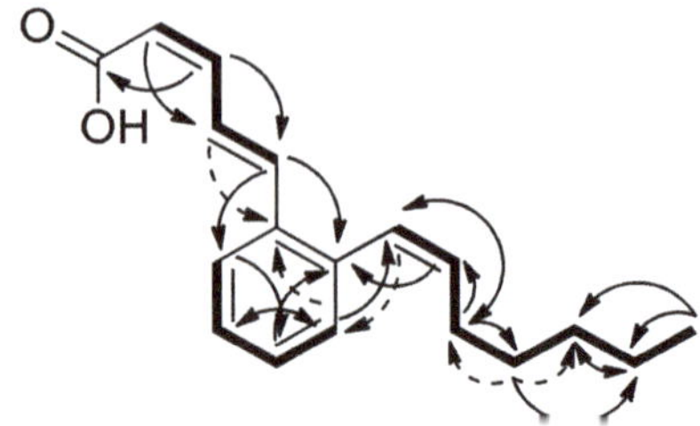

Figure 3. Observed COSY (bold lines) and HMBC correlations (arrows) in AB204-F (**6**). The same correlations were observed in AB204-E (**5**).

To clarify the geometry of two substituted chains of **5** and **6**, the differential NOE experiment was conducted. NOE correlations between H-5 and H-12 were observed in both compounds (Figures S34 and S35), which supported the geometry of two side chains in **5** and **6**, as is shown in Figure 4. This is the first report of **5** and **6** obtained from natural sources. Qureshi et al. [36] found structurally related compounds, MF-EA-705a and b, along with actinopyrone A from a broth extract of *Streptomyces* MF-EA-705. The most related structures were found as cinnamoyl moieties of rare peptide compounds, pepticinnamin E, WS9326A, and RP-1776 (skyllamycin A) [37–39].

Figure 4. NOE correlations in AB204-E (**5**) and AB204-F (**6**).

Compounds **7** and **8** were pale-yellow amorphous solids that were determined by HRESI-MS analyses to have molecular formulas of $C_{31}H_{44}O_6$ and $C_{32}H_{46}O_6$, respectively (Figures S36 and S37). The planar structures of **7** and **8** (Figure 1) were confirmed by NMR spectra (Figures S38 and S39) as known compounds, anguinomycin A (**7**) [40] and leptomycin A (**8**) [41], which were known to be relative structures. Compound **9** was a colorless oil, and the molecular formula of **9** was determined to be $C_{25}H_{36}O_4$ on the basis of HRESI-MS data and the signals of ^{1}H NMR spectrum (Figures S40 and S41). These data supported the structure of **9** to be actinopyrone A (**9**) as is shown in Figure 1 [42].

Physicochemical Properties of **1**, **2**, **5**, and **6**

Compound **1** (AB204-A): pale yellow amorphous solid; $[\alpha]_D$ −3.4 (*c* 0.1, MeOH); UV (MeOH) λ_{max} (log ε) 204.5 (4.09), 235.5 (3.63) nm; IR (ATR) ν_{max} 3014, 2923, 1706, 1654, 1409 cm^{-1}; ^{1}H and ^{13}C NMR (chloroform-*d*) see Table 1; HREIMS *m/z* 190.1000 [M]$^+$ (calcd for $C_{12}H_{14}O_2$, 190.0994).

Compound **2** (AB204-B): pale yellow amorphous solid; $[\alpha]_D$ −0.3 (*c* 0.1, MeOH); UV (MeOH) λ_{max} (log ε) 204.5 (3.98), 235.5 (3.53) nm; IR (ATR) ν_{max} 3011, 2925, 1706, 1653, 1457 cm^{-1}; ^{1}H and ^{13}C NMR (chloroform-*d*) see Table 1; HREIMS *m/z* 218.1301 [M]$^+$ (calcd for $C_{14}H_{18}O_2$, 218.1329).

Compound **5** (AB204-E): pale yellow amorphous solid; $[\alpha]_D$ −13.4 (*c* 0.1, MeOH); UV (MeOH) λ_{max} (log ε) 204 (4.04), 251 (3.90), 309.5 (4.26) nm; IR (ATR) ν_{max} 2956, 2924, 2857, 1684, 1617, 1277, 1000 cm^{-1}; ^{1}H and ^{13}C NMR (methanol-*d₄*) see Table 2; HREIMS *m/z* 270.1616 [M]$^+$ (calcd for $C_{18}H_{22}O_2$, 270.1620).

Compound **6** (AB204-F): pale yellow amorphous solid; $[\alpha]_D$ −14.1 (*c* 0.1, MeOH); UV (MeOH) λ_{max} (log ε) 204 (3.76), 251 (3.51), 309.5 (3.86) nm; IR (ATR) ν_{max} 2955, 2924, 2855, 1684, 1615, 1244, 958 cm^{-1}; ^{1}H and ^{13}C NMR (chloroform-*d*) see Table 2; HREIMS *m/z* 270.1633 [M]$^+$ (calcd for $C_{18}H_{22}O_2$, 270.1620).

3.2. Antimicrobial Activities of Isolated Compounds

Antimicrobial activities of the purified compounds, except for the mixture of AB204-C (**3**) and AB204-D (**4**), were tested against four Gram-positive bacteria, five Gram-negative bacteria, two yeasts, and three fungi using paper disk diffusion assay with an equal amount of each compound at 50 μg/disk. The results are shown in Table 3. All compounds did not show activity against *E. coli* NIHJ, *K. pneumonia* ATCC 10031, *P. vulgaris* NBRC 3167, *Ps. aeruginosa* IFO 3080, and *Sa. cerevisiae* ATCC 9763. AB204-A (**1**) and B (**2**) displayed a weak activity towards *C. albicans* ATCC 64548, *A. niger* ATCC 6275, and *G. boninense* BCC 21330, with a clear zone range from 10.4 to 13.2 mm. However, they did not affect the Gram-positive and Gram-negative bacteria. AB204-E (**5**) and AB204-F (**6**) displayed no antifungal and antiyeast activities but showed good antibacterial activity against Gram-positive bacteria and weak activity against the Gram-negative bacterium *X. campestris* pv. *oryzae* KB88, which is a phytopathogenic strain. AB204-E (**5**) strongly inhibited *K. rhizophila* ATCC 9341, *B. subtilis* ATCC 6633, *M. smegmatis* ATCC 607, and *S. aureus* ATCC 6538p with inhibition zones of 41.3, 35.3, 32.7, and 26.0 mm, respectively, while AB204-F (**6**) showed a slightly lower activity against the same pathogens as is presented in Table 3. Anguinomycin A (**7**) revealed potent inhibitory activity against the Gram-positive bacterium *K. rhizophila* ATCC 9341 (19.1 mm), and two fungi, *Mu. racemosus* IFO 4581 (16.9 mm), and *G. boninense* BCC 21330 (19.6 mm), while leptomycin A (**8**) showed stronger activity against these pathogens at 30.6, 49.0, and 21.2 mm, respectively. Actinopyrone A (**9**) exhibited potent antifungal activity against *C. albicans* ATCC 64548 and three fungal strains with the inhibition zone in a range of 11.9 to 23.9 mm (Table 3). These results suggest that the antibacterial and antifungal activities of the *S. palmae* CMU-AB204$^\text{T}$ may have been displayed as a consequence of the contribution of all these antimicrobial secondary metabolites.

Table 3. Antimicrobial activities of the pure compounds against fourteen microorganisms using an equal amount of each compound at 50 μg/disk. Inhibition zone (mm) (Mean ± SD; n = 3) including the diameter of the paper disk (8 mm) was measured after 24 and 48 h of incubation. **1**, AB204-A; **2**, AB204-B; **5**, AB204-E; **6**, AB204-F; **7**, anguinomycin A; **8**, leptomycin A; **9**, actinopyrone A.

Microorganism	Inhibition Zone (mm) of Seven Pure Compounds						
	1	2	5	6	7	8	9
Gram-positive bacteria							
Bacillus subtilis ATCC 6633	-	-	35.3 ± 1.4	12.3 ± 1.8	-	-	-
Kocuria rhizophila ATCC 9341	-	-	41.3 ± 2.0	17.5 ± 1.5	19.1 ± 1.9	30.6 ± 2.3	-
Mycobacterium smegmatis ATCC 607	-	-	32.7 ± 1.2	14.0 ± 2.2	-	-	-
Staphylococcus aureus ATCC 6538p	-	-	26.0 ± 1.6	13.2 ± 1.9	-	-	-
Gram-negative bacteria							
Escherichia coli NIHJ	-	-	-	-	-	-	-
Klebsiella pneumonia ATCC 10031	-	-	-	-	-	-	-
Proteus vulgaris NBRC 3167	-	-	-	-	-	-	-
Pseudomonas aeruginosa IFO 3080	-	-	-	-	-	-	-
Xanthomonas campestris pv. *oryzae* KB 88	-	-	10.6 ± 1.3	11.0 ± 2.2		-	-
Yeasts							
Candida albicans ATCC 64548	13.1 ± 1.6	10.4 ± 0.9	-	-	-	-	20.8 ± 1.5
Saccharomyces cerevisiae ATCC 9763	-	-	-	-	-	-	-
Fungi							
Mucor racemosus IFO 4581	-	-	-	-	16.9 ± 1.7	49.0 ± 2.3	23.1 ± 1.6
Aspergillus niger ATCC 6275	11.5 ± 1.1	11.1 ± 1.2	-	-	-	-	23.9 ± 1.8
Ganoderma boninense BCC 21330	11.0 ± 1.4	13.2 ± 2.0	-	-	19.6 ± 1.6	22.1 ± 1.7	11.9 ± 1.0

4. Discussion

Several mechanisms have been proposed to control *G. boninense* causing BSR disease in oil palm trees. However, none of them have successfully been treated or been shown to suppress the disease [9]. The search for antifungal alternatives is representative of a potential solution that has drawn significant interest. In this study, new compounds were identified during the isolation of anti-*Ganoderma* substances from *S. palmae* CMU-AB204$^\text{T}$. The assessment of antimicrobial activity of four new phenyl alkenoic acids showed that AB204-A (**1**) and B (**2**) mildly inhibited the growth of fungi, *G. boninense*

BCC 21330, *Mu. racemosus* IFO 4581, and *A. niger* ATCC 6275, while AB204-E (**5**) and F (**6**) displayed a positive degree of activity against Gram-positive bacteria, *B. subtilis* ATCC 6633, *K. rhizophila* ATCC 9341, *M. smegmatis* ATCC 607, and *S. aureus* ATCC 6538p. New antifungal compounds, AB204-A (**1**) and B (**2**), possessed similar structures to phenylethyl alcohol (PEA), an antifungal aromatic compound that was obtained from *Trichoderma virens* 7b, which had significant potential as a biological control agent for BSR [43]. These compounds may exhibit a mechanism in inhibiting fungi similar to PEA which inhibits protein, DNA, RNA, and aminoacyl tRNA syntheses of fungi [44,45]. However, a mode of action of novel compounds in controlling fungi should be confirmed in the future.

AB204-B (**2**) contained more C_2H_4 units than AB204-A (**1**), but it displayed antimicrobial activity against the same pathogenic strains with similar inhibition zone sizes. This result indicated that the presence of a longer chain of the carboxylic group in **2** had not been involved in the antimicrobial activity. Biological activities of the *E*-isomer of AB204-A (**1**), (*E*)-5-(2-methylphenyl)-4-pentenoic acid was previously reported to be an inactive compound against bacteria and fungi but the tested concentration and the strain of tested microorganisms have not been indicated [33]. However, the difference of an antimicrobial activity between (*E*)-5-(2-methylphenyl)-4-pentenoic acid and new compounds, AB204-A (**1**) and AB204-B (**2**), revealed that the existence of *Z*-olefin in **1** and **2** had been involved in their antifungal activity. The structures of a mixture of AB204-C (**3**) and AB204-D (**4**) were predicted based on HREI-MS and ^{1}H NMR spectra. In the future, the mixture should be reseparated using other techniques, and additional data is needed to confirm their structures and antimicrobial activities. AB204-E (**5**) and F (**6**) have an 1,2-*Z*-heptene moiety connected to the aromatic ring. These metabolites have not shown antifungal activity but exhibited strong antibacterial activity when associated with this moiety. Moreover, the existence of one pair of *Z*-olefin in the chain of the carboxylic group of AB204-E (**5**), instead of the *E* and *Z*-olefins of AB204-F (**6**), increased the antibacterial activity of this compound.

Previously, Thong et al. [35] found two closely related compounds of **1**–**4**, and they were isolated from a *Streptomyces* that had been spontaneously acquired rifampicin resistance. These compounds contained *E*-olefins and have both a methylbenzene unit and a 2-amino-3-hydroxycyclopent-2-enone (C_5N) moiety. The phenyl alkenoic acid-associated metabolites discovered by Thong et al. did not display antibacterial activity against *E. coli*, *M. luteus*, *S. aureus*, and *B. subtilis* in testing with a microplate assay at 100 µM or approximately 28.5 and 33.5 mg/mL, thus revealing similar results to AB204-A (**1**) and B (**2**). Notably, the presence of a carboxylic acid moiety in the novel compounds, and a C_5N moiety in the known compounds, may not be involved in the antimicrobial activity. Based on draft genome sequences of the rifampicin-resistant mutant (TW-R50–13), the methylbenzene moiety may be biosynthesized by the expression of polyketide synthase (PKS) genes that are located at a different locus from the biosynthetic genes for the C_5N moiety [35]. The genes encoding for PKS have been disclosed to complex biosynthetic mechanisms, which were involved in the production of many metabolites in microorganisms [46]. Genome sequences would provide the data of potential gene clusters to understand the metabolic pathways of *S. palmae* CMU-AB204^T. Thus, the genome sequences of this strain should be further studied to determine the presence of both silent and cryptic secondary metabolite biosynthetic gene clusters that are able to synthesize the corresponding novel natural products.

In addition to **1** and **2**, other antifungal compounds, anguinomycin A (**7**), leptomycin A (**8**), and actinopyrone A (**9**) obtained from the same broth of *S. palmae* CMU-AB204^T, also displayed anti-*Ganoderma* activity. The ability of *S. palmae* to produce a variety of antifungal compounds was proven. This strain might produce each antifungal secondary metabolite depending on the prevailing environmental conditions, such as nutritional source, incubation period, pH value, and temperature [47,48]. Hence, the optimization of culture conditions should be studied in order to obtain high yields of the antifungal metabolites. The protecting effect of *S. palmae* CMU-AB204^T against BSR has also been confirmed in a glasshouse experiment [49]. The results obtained from this study strongly suggest that the antimicrobial secondary metabolites were involved in the mechanism exhibiting anti-BSR effects by this *Streptomyces* strain.

Although the new compounds obtained in this study showed moderate activity towards *G. boninense*, they inhibited clinical bacterial pathogens and other phytopathogenic fungi, suggesting a possible utility of the four new antimicrobial substances in both agricultural and medical treatments. However, cytotoxicity to mammalian cell of both new compounds and three known compounds should be tested before being applied to these compounds. The recovery of novel actinomycetes species, especially the genus *Streptomyces*, has the potential to be a rich source of both new and known natural products [50,51]. Notably, *S. palmae* CMU-AB204^T was found to produce various bioactive metabolites.

Supplementary Materials: The following are available online at http://www.mdpi.com/2076-2607/8/3/350/s1.

Author Contributions: The project approach was designed by K.S., W.P.-a., S.L.; *Streptomyces* strain was selected by K.S.; activity-guided purification was conducted by K.S., M.M., K.D.; structural determination was conducted by K.S., M.M., K.D.; antimicrobial activity was measured by K.S.; the research was supervised by K.S., W.P.-a., S.L.; funding was acquired by M.M. and S.L. All authors have read and agreed to the published version of the manuscript.

Funding: This research work was partially supported by Chiang Mai University, the Thailand Research Fund (TRF) within the program entitled Research and Researchers for Industries (RRi) PHD56I0048 with the cooperation of Trang Palm Oil Co., Ltd., Thailand, and the Institute for Fermentation, Osaka (IFO), Japan.

Acknowledgments: We thank Kenichiro Nagai, School of Pharmacy, Kitasato University, for help in obtaining mass data, and Toshiyuki Tokiwa, Kitasato Institute for Life Sciences, Kitasato University, for providing the microbes used to measure antimicrobial activity.

Conflicts of Interest: The authors declare no conflicts of interest.

References

1. Rees, R.W.; Flood, J.; Hasan, Y.; Cooper, R.M. Effects of inoculum potential, shading and soil temperature on root infection of oil palm seedlings by the basal stem rot pathogen *Ganoderma boninense*. *Plant. Pathol.* **2007**, *56*, 862–870. [CrossRef]

2. Susanto, A.; Sudharto, P.S.; Purba, R.Y. Enhancing biological control of basal stem rot disease (*Ganoderma boninense*) in oil palm plantations. *Mycopathologia* **2005**, *159*, 153–157. [CrossRef] [PubMed]

3. Nieto, L. Incidence of oil palm stem rots in Colombia. *Palmas* **1995**, *16*, 227–232.

4. Turner, P.D. *Oil Palm Diseases and Disorders*; Oxford University Press: Oxford, UK, 1981; pp. 88–110.

5. Likhitekaraj, S.; Tummakate, A. Basal stem rot of oil palm in Thailand caused by *Ganoderma*. In *Ganoderma Diseases of Perennial Crops*; Flood, J., Bridge, P.D., Holderness, M., Eds.; CABI Publishing: Wallingford, UK, 2000; pp. 66–70.

6. Jollands, P. Laboratory investigations on fungicides and biological agents to control three diseases of rubber and oil palm and their potential applications. *Trop. Pest. Manag.* **1983**, *29*, 33–38. [CrossRef]

7. Idris, A.S.; Ismail, S.; Ariffin, D.; Ahmed, D. Control of *Ganoderma* infected palm-development of pressure injection and field application. *MPOB Info. Ser.* **2002**, *148*, 2.

8. Idris, A.S.; Arifurrahman, R. Determination of 50% effective concentration (EC$_{50}$) of fungicides against pathogenic *Ganoderma*. *MPOB Info. Ser.* **2008**, *449*, 2.

9. Naher, L.; Siddiquee, S.; Yusuf, U.K.; Mondal, M.M.A. Issues of *Ganoderma* spp. and basal stem rot disease management in oil palm. *Am. J. Agri. Sci.* **2015**, *2*, 103–107.

10. Shariffah-Muzaimah, S.; Idris, A.; Madihah, A.; Dzolkhifli, O.; Kamaruzzaman, S.; Cheong, P. Isolation of actinomycetes from rhizosphere of oil palm (*Elaeis guineensis* Jacq.) for antagonism against *Ganoderma boninense*. *J. Oil Palm Res.* **2015**, *27*, 19–29.

11. Wightwick, A.; Walters, R.; Allinson, G.; Reichman, S.; Menzies, N. Environmental risks of fungicides used in horticultural production systems. In *Fungicides*; Carisse, O., Ed.; IntechOpen: London, UK, 2010; pp. 273–304.

12. Ji, X.Y.; Zhong, Z.J.; Xue, S.T.; Meng, S.; He, W.Y.; Gao, R.M.; Li, Y.H.; Li, Z.R. Synthesis and antiviral activities of synthetic Glutarimide derivatives. *Chem. Pharm. Bull.* **2010**, *58*, 1436–1441. [CrossRef]

13. Gupta, P.K. Toxicity of fungicides. In *Veterinary Toxicology, Basic and Clinical Principles*, 3rd ed.; Gupta, R.C., Ed.; Academic Press: London, UK, 2018; pp. 569–580.

14. Nur Ain Izzati, M.Z.; Abdullah, F. Disease suppression in *Ganoderma*-infected oil palm seedlings treated with *Trichoderma harzianum*. *Plant. Protect. Sci.* **2008**, *44*, 101–107. [CrossRef]

15. Sundram, S.; Abdullah, F.; Ahmad, Z.A.M.; Yusuf, U.K. Efficacy of single and mixed treatments of *Trichoderma harzianum* as biocontrol agents of *Ganoderma* basal stem rot in oil palm. *J. Oil Palm Res.* **2008**, *20*, 470–483.

16. Naher, L.; Tan, S.G.; Yusuf, U.K.; Ho, C.-L.; Abdullah, F. Biocontrol agent *Trichoderma harzianum* strain FA 1132 as an enhancer of oil palm growth. *Pertanika J. Trop. Agric. Sci.* **2012**, *35*, 173–182.

17. Harman, G.E.; Howell, C.R.; Viterbo, A.; Chet, I.; Lorito, M. *Trichoderma* species-opportunistic, avirulent plant symbionts. *Nat. Rev. Microbiol.* **2004**, *2*, 43–56. [CrossRef]

18. Naher, L.; Ho, C.-L.; Tan, S.G.; Yusuf, U.K.; Abdullah, F. Cloning of transcripts encoding chitinases from *Elaeis guineensis* Jacq. and their expression profiles in response to fungal infections. *Physiol. Mol. Plant. Pathol.* **2011**, *76*, 96–103. [CrossRef]

19. Sapak, Z.; Meon, S.; Ahmad, Z.A.M. Effect of endophytic bacteria on growth and suppression of *Ganoderma* infection in oil palm. *Int. J. Agric. Biol.* **2008**, *10*, 127–132.

20. Bivi, M.R.; Farhana, M.; Khairulmazmi, A.; Idris, A. Control of *Ganoderma boninense*: A causal agent of basal stem rot disease in oil palm with endophyte bacteria in vitro. *Int. J. Agric. Biol.* **2010**, *12*, 833–839.

21. Sundram, S.; Meon, S.; Seman, I.A.; Othman, R. Symbiotic interaction of endophytic bacteria with arbuscular mycorrhizal fungi and its antagonistic effect on *Ganoderma boninense*. *J. Microbiol.* **2011**, *49*, 551–557. [CrossRef]

22. Nurrashyeda, R.; Maizatul, S.; Idris, A.; Madihah, A.; Nasyaruddin, M. The potential of endophytic bacteria as a biological control agent for *Ganoderma* disease in oil palm. *Sains Malaysiana* **2016**, *45*, 401–409.

23. Katz, L.; Baltz, R.H. Natural product discovery: past, present, and future. *J. Ind. Microbiol. Biotechnol.* **2016**, *43*, 155–176. [CrossRef]

24. Procópio, R.E.; Silva, I.R.; Martins, M.K.; Azevedo, J.L.; Araújo, J.M. Antibiotics produced by *Streptomyces*. *Braz. J. Infect. Dis.* **2012**, *16*, 466–471. [CrossRef]

25. Ōmura, S.; Ikeda, H.; Ishikawa, J.; Hanamoto, A.; Takahashi, C.; Shinose, M.; Takahashi, Y.; Horikawa, H.; Nakazawa, H.; Osonoe, T.; et al. Genome sequence of an industrial microorganism *Streptomyces avermitilis*: Deducing the ability of producing secondary metabolites. *PNAS* **2001**, *98*, 12215–12220.

26. Watve, M.G.; Tickoo, R.; Jog, M.M.; Bhole, B.D. How many antibiotics are produced by the genus *Streptomyces*? *Arch. Microbiol.* **2001**, *176*, 386–390. [CrossRef]

27. Pithakkit, S.; Petcharat, V.; Chuenchit, S.; Pornsuriya, C.; Sunpapao, A. Isolation of antagonistic actinomycetes species from rhizosphere as effective biocontrol against oil palm fungal diseases. *Walailak J. Sci. Tech.* **2014**, *12*, 481–490.

28. Shariffah-Muzaimah, S.; Idris, A.; Madihah, A.; Dzolkhifli, O.; Kamaruzzaman, S.; Maizatul-Suriza, M. Characterization of *Streptomyces* spp. isolated from the rhizosphere of oil palm and evaluation of their ability to suppress basal stem rot disease in oil palm seedlings when applied as powder formulations in a glasshouse trial. *World J. Microbiol. Biotech.* **2018**, *34*, 15. [CrossRef]

29. Ting, A.S.Y.; Hermanto, A.; Peh, K.L. Indigenous actinomycetes from empty fruit bunch compost of oil palm: evaluation on enzymatic and antagonistic properties. *Biocat. Agric. Biotech.* **2014**, *3*, 310–315. [CrossRef]

30. Pal, K.K.; Gardener, B.M. Biological control of plant pathogens. *Plant. Health Instructor* **2006**, *2*, 1117–1142. [CrossRef]

31. Sujarit, K.; Kudo, T.; Ohkuma, M.; Pathom-Aree, W.; Lumyong, S. *Streptomyces palmae* sp. nov., isolated from oil palm (*Elaeis guineensis*) rhizosphere soil. *Int. J. Syst. Evol. Microbiol.* **2016**, *66*, 3983–3988. [CrossRef]

32. Shirling, E.B.; Gottlieb, D. Methods for characterization of *Streptomyces* species. *Int. J. Syst. Bacteriol.* **1966**, *16*, 313–340. [CrossRef]

33. Mukku, V.J.R.V.; Maskey, R.P.; Monecke, P.; Grün-Wollny, I.; Laatsch, H. 5-(2-Methylphenyl)-4-pentenoic acid from a terrestrial Streptomycete. *Z. Naturforsch.* **2002**, *57b*, 335–337. [CrossRef]

34. Shaaban, K.A.; Helmke, E.; Kelter, G.; Fiebig, H.H.; Laatsch, H. Glucopiericidin C: A cytotoxic piericidin glucoside antibiotics produced by a marine *Streptomyces* isolate. *J. Antibiot.* **2011**, *64*, 205–209. [CrossRef]

35. Thong, W.L.; Shin-ya, K.; Nishiyama, M.; Kuzuyama, T. Methylbenzene-containing polyketides from a *Streptomyces* that spontaneously acquired rifampicin resistance: Structural elucidation and biosynthesis. *J. Nat. Prod.* **2016**, *79*, 857–864. [CrossRef] [PubMed]

36. Qureshi, A.; Mauger, J.B.; Cano, R.J.; Galazzo, J.L.; Lee, M.D. MF-EA-705a & MF-EA-705b, new metabolites from microbial fermentation of a *Streptomyces* sp. *J. Antibiot.* **2001**, *54*, 1100–1103. [PubMed]

37. Shiomi, K.; Yang, H.; Inokoshi, J.; Van der Pyl, D.; Nakagawa, A.; Takeshima, H.; Ōmura, S. Pepticinnamins, new farnesyl-protein transferase inhibitors produced by an actinomycete. II. Structural elucidation of pepticinnamin E. *J. Antibiot.* **1993**, *46*, 229–234. [CrossRef] [PubMed]

38. Shigematsu, N.; Hayashi, K.; Kayakiri, N.; Takase, S.; Hashimoto, M.; Tanaka, H. Structure of WS9326A, a novel tachykinin antagonist from a *Streptomyces*. *J. Org. Chem.* **1993**, *58*, 170–175. [CrossRef]
39. Toki, S.; Agatsuma, T.; Ochiai, K.; Saitoh, Y.; Ando, K.; Nakanishi, S.; Lokker, N.; Giese, N.A.; Matsuda, Y. RP-1776, a novel cyclic peptide produced by *Streptomyces* sp., inhibits the binding of PDGF to the extracellular domain of its receptor. *J. Antibiot.* **2001**, *54*, 405–414. [CrossRef]
40. Hayakawa, Y.; Adachi, K.; Komeshima, N. New antitumor antibiotics, anguinomycins A and B. *J. Antibiot.* **1987**, *40*, 1349–1352. [CrossRef]
41. Hamamoto, T.; Seto, H.; Beppu, T. Leptomycins A and B, new antifungal antibiotics. II. Structure elucidation. *J. Antibiot.* **1983**, *36*, 646–650. [CrossRef]
42. Yano, K.; Yokoi, K.; Sato, J.; Oono, J.; Kouda, T.; Ogawa, Y.; Nakashima, T. Actinopyrones A, B, and C, new physiologically active substances. II. Physicochemical properties and chemical structures. *J. Antibiot.* **1986**, *39*, 38–43. [CrossRef]
43. Angel, L.P.L.; Yusof, M.T.; Ismail, I.S.; Ping, B.T.Y.; Mohamed Azni, I.N.A.; Kamarudin, N.H.; Sundram, S. An in vitro study of the antifungal activity of *Trichoderma virens* 7b and a profile of its non-polar antifungal components related against *Ganoderma boninense*. *J. Microbiol.* **2016**, *54*, 732–744. [CrossRef]
44. Lester, G. Inhibition of growth, synthesis, and permeability in *Neurospora crassa* by phenethyl alcohol. *J. Bacteriol.* **1965**, *90*, 29–37. [CrossRef]
45. Liu, P.; Cheng, Y.; Yang, M.; Liu, Y.; Chen, K.; Long, C.A.; Deng, X. Mechanisms of action for 2-phenylethanol isolated from *Kloeckera apiculata* in control of *Penicillium* molds of citrus fruits. *BMC Microbiol.* **2014**, *14*, 242. [CrossRef]
46. Staunton, J.; Weissman, K.J. Polyketide biosynthesis: a millennium review. *Nat. Prod. Rep.* **2001**, *18*, 380–416. [CrossRef]
47. Banga, J.; Praveen, V.; Singh, V.; Tripathi, C.K.M.; Bihari, V. Studies on medium optimization for the production of antifungal and antibacterial antibiotics from a bioactive soil actinomycete. *Med. Chem. Res.* **2008**, *17*, 425–436. [CrossRef]
48. Ruiz, B.; Chávez, A.; Forero, A.; García-Huante, Y.; Romero, A.; Sánchez, M.; Rocha, D.; Sánchez, B.; Rodríguez-Sanoja, R.; Sánchez, S.; et al. Production of microbial secondary metabolites: Regulation by the carbon source. *Crit. Rev. Microbiol.* **2010**, *36*, 146–167. [CrossRef]
49. Sujarit, K. Selection and Characterization of Actinomycetes for Growth Promotion of Oil Palm and Biological Control of Oil Palm Diseases. Ph.D. Thesis, Chiang Mai University, Chiang Mai, Thailand, 2018.
50. Bérdy, J. Thoughts and facts about antibiotics: where we are now and where we are heading. *J. Antibiot.* **2012**, *65*, 385–395. [CrossRef]
51. Lucas, X.; Senger, C.; Erxleben, A.; Grüning, B.A.; Döring, K.; Mosch, J.; Flemming, S.; Günther, S. StreptomeDB: A resource for natural compounds isolated from *Streptomyces* species. *Nucleic Acids Res.* **2013**, *41*, D1130–D1136. [CrossRef]

 *microorganisms*

Article

Identification and Heterologous Expression of the Albucidin Gene Cluster from the Marine Strain *Streptomyces Albus* Subsp. *Chlorinus* NRRL B-24108

Maksym Myronovskyi [1], Birgit Rosenkränzer [1], Marc Stierhof [1], Lutz Petzke [2], Tobias Seiser [2] and Andriy Luzhetskyy [1,3,*]

[1]   Pharmazeutische Biotechnologie, Universität des Saarlandes, 66123 Saarbrücken,
     Germany; maksym.myronovskyi@uni-saarland.de (M.M.); b.rosenkraenzer@mx.uni-saarland.de (B.R.);
     s8mcstie@stud.uni-saarland.de (M.S.)
[2]   BASF SE, 67056 Ludwigshafen, Germany; lutz.petzke@basf.com (L.P.); tobias.seiser@basf.com (T.S.)
[3]   Helmholtz-Institut für Pharmazeutische Forschung Saarland, 66123 Saarbrücken, Germany
*   Correspondence: a.luzhetskyy@mx.uni-saarland.de; Tel.: +49-681-302-70200

Received: 14 January 2020; Accepted: 7 February 2020; Published: 10 February 2020

Abstract: Herbicides with new modes of action and safer toxicological and environmental profiles are needed to manage the evolution of weeds that are resistant to commercial herbicides. The unparalleled structural diversity of natural products makes these compounds a promising source for new herbicides. In 2009, a novel nucleoside phytotoxin, albucidin, with broad activity against grass and broadleaf weeds was isolated from a strain of *Streptomyces albus* subsp. *chlorinus* NRRL B-24108. Here, we report the identification and heterologous expression of the previously uncharacterized albucidin gene cluster. Through a series of gene inactivation experiments, a minimal set of albucidin biosynthetic genes was determined. Based on gene annotation and sequence homology, a model for albucidin biosynthesis was suggested. The presented results enable the construction of producer strains for a sustainable supply of albucidin for biological activity studies.

Keywords: albucidin; herbicide; nucleoside; biosynthetic gene cluster; heterologous expression; *Streptomyces albus* Del14

---

1. Introduction

Pesticides play an important role in modern agriculture. Among all chemicals being produced, pesticides are in second place after fertilizers in their extent of use. A total of 2.4 billion kilograms of pesticides were applied worldwide in 2007 [1]. Nevertheless, lack of weed control is still the most topical issue. Among all pests, weeds have the largest negative effect on crop productivity [2,3]. In light of the rapidly increasing evolution of herbicide resistance, the need for new herbicides with new modes of action (MOAs) and safer ecological profiles is growing [4,5].

From all new pesticide active ingredients registered by the Environmental Protection Agency from 1997 to 2010, almost 70% have origins in natural products. Interestingly, only 8% of conventional herbicides are natural product-derived [3,6]. The wide structural diversity of natural products and their small amount of overlap with synthetic compounds imply their potential as lead structures for the development of new pesticides [7–9]. This is further confirmed by the phytotoxin literature, which suggests that natural products have many more MOAs than the commercial herbicides currently possess [3].

A novel bleaching herbicide, albucidin, from the strain *Streptomyces albus* subsp. *chlorinus* NRRL B-24108 was discovered in 2009 [10]. In this paper, we present the identification, heterologous expression, and engineering of the albucidin gene cluster. We also propose the biosynthetic route that leads to the production of albucidin. The identified minimal set of biosynthetic genes allows for the straightforward construction of overproducing strains for a high yield albucidin supply for biological activity studies.

2. Materials and Methods

2.1. General Experimental Procedures

All strains, plasmids and BACs used in this work are listed in Tables S1 and S2. *Escherichia coli* strains were cultured in LB medium [11]. *Streptomyces* strains were grown on soya flour mannitol agar (MS agar) [12] and in liquid tryptic soy broth (TSB; Sigma-Aldrich, St. Louis, MO, USA). For albucidin production, liquid SG medium [13] was used. The antibiotics kanamycin, apramycin, hygromycin, ampicillin and nalidixic acid were supplemented when required.

2.2. Isolation and Manipulation of DNA

The previously constructed BAC library of *Streptomyces albus* subsp. *chlorinus* NRRL B-24108 was used [14]. DNA manipulation, *E. coli* transformation and *E. coli/Streptomyces* intergeneric conjugation were performed according to standard protocols [11,12,15]. BAC DNA was purified with the BACMAX™ DNA purification kit (Lucigen, Middleton, WI, USA). Restriction endonucleases were used according to the manufacturer's recommendations (New England Biolabs, Ipswich, MA, USA). All the strains and plasmids are listed in the Tables S1 and S2, respectively.

2.3. Metabolite Extraction and Analysis

For metabolite extraction, *Streptomyces* strains were grown in 15 mL of TSB in a 100 mL baffled flask for 1 day, and 1 mL of seed culture was used to inoculate 100 mL of SG production medium in a 500 mL baffled flask. Cultures were grown for 7 days at 28 °C and 180 rpm in an Infors multitron shaker. Albucidin was extracted from the culture supernatant with an equal amount of butanol, evaporated, and dissolved in methanol. Albucidin production was analysed on a Bruker Amazon Speed mass spectrometer coupled to UPLC Thermo Dionex Ultimate 3000 RS. Analytes were separated either on a Waters ACQUITY BEH C18 column (1.7 μm, 2.1 mm × 30 mm) or on a Waters ACQUITY BEH C18 column (1.7 μm, 2.1 mm × 100 mm). Water + 0.1% formic acid and methanol + 0.1% formic acid were used as the mobile phases. For the determination of high-resolution mass, analytes were analysed with a Thermo LTQ Orbitrap XL coupled to UPLC Thermo Dionex Ultimate 3000 RS. Analytes were separated on a Waters ACQUITY BEH C18 column (1.7 μm, 2.1 mm × 100 mm) with water + 0.1% formic acid and methanol + 0.1% formic acid as the mobile phase.

2.4. Chemical Mutagenesis

One millilitre of spore suspension of *Streptomyces albus* subsp. *chlorinus* NRRL B-24108 was inoculated into 100 mL of SG medium in a 500 mL baffled flask and cultivated overnight at 28 °C and 180 rpm. The pH of the culture was adjusted to 8.5 with 1 M NaOH. Ten millilitres of culture was transferred into three 50 mL falcon tubes. Then, 64 mg of wet NTG was dissolved in 16 mL of water. Next, 6.666 mL, 4.285 mL and 1.765 mL of NTG stock solution were added to the tubes containing culture to reach final NTG concentrations of 800 μg/mL, 600 μg/mL and 300 μg/mL, respectively. The samples were incubated at 28 °C for 30 min in the overhead shaker. The mycelium was precipitated by centrifugation, and the supernatant was discarded. The mycelium was washed twice with 5% sodium thiosulfate solution. The treated samples were plated on MS agar plates and cultivated for 14 days at 28 °C. The spores were washed with water and plated in dilutions on MS agar. The plates with spore dilutions were incubated for 10 days at 28 °C. Single colonies were picked on 30 mm plates with SG agar. The plates were incubated for 14 days at 28 °C. Agar blocks were cut out from the plates and transferred into 2 mL tubes. Albucidin was extracted from the agar blocks with 500 μL of butanol for 48 h. The extracts were analysed using HPLC-MS.

2.5. Albucidin Isolation and ^{1}H-NMR Spectroscopy

Streptomyces albus 1K1 was grown in 10 L of SG medium, and albucidin was extracted with butanol. The dry extract was dissolved in 50 mL of water containing 5% acetonitrile and 0.1% formic acid. The extract was loaded onto 3 C18 SPE columns (Discovery DSC-18 SPE 52607-U) equilibrated with 5% acetonitrile in water with 0.1% formic acid. The flowthrough was collected. The columns were washed twice with 12 mL of 5% acetonitrile containing 0.1% formic acid. The flowthrough and the wash fractions were combined and evaporated in a rotary evaporator. The presence of albucidin was detected by HPLC-MS.

The dry material after the SPE purification step was dissolved in methanol and used for size-exclusion chromatography. Separation was performed on a glass column (30 mm × 1000 mm) packed with Sephadex LH-20 and methanol as the mobile phase. Fractions containing albucidin were identified by HPLC-MS. Albucidin-containing fractions were combined and evaporated. The dry extract was dissolved in 5 mL of 5% methanol in water containing 10 mM potassium phosphate buffer pH 6.4.

HPLC separation was performed on a preparative HPLC Thermo Dionex Ultimate 3000 equipped with a Macherey Nagel Nucleodur HTec C18 column (5 μm, 21 mm × 150 mm). A 10 mM potassium phosphate buffer (pH 6.4) was used as solvent A, and 50% methanol in 10 mM phosphate buffer (pH 6,4) was used as solvent B. The following gradient at a flowrate of 15 mL/min was used for separation: 0 min–13% B, 20 min—25% B, 21 min–25% B, 24 min–100% B, 25 min–100% B, 28 min–13% B, 29 min–13% B. Albucidin eluted at 18 min. The albucidin-containing fractions were pooled and evaporated.

For the final purification step, the dry material after HPLC purification was dissolved in 5 mL of water and loaded onto a Sephadex LH-20 column (30 mm × 450 mm) previously equilibrated with water. Water was used as the mobile phase. The fractions containing albucidin were identified by HPLC-MS, pooled and evaporated.

The ^{1}H-NMR spectra were recorded on a Bruker Avance 500 spectrometer (Bruker, BioSpin GmbH, Rheinstetten, Germany) at 300 K equipped with a 5 mm BBO probe using deuterated trifluoroacetic acid (Deutero, Kastellaun, Germany) as the solvent containing tetramethylsilane (TMS) as a reference. Albucidin was measured in deuterated water (Deutero, Kastellaun, Germany). The chemical shifts are reported in parts per million (ppm) relative to TMS. All spectra were recorded with the standard ^{1}H pulse program using 128 scans. The structure of albucidin was confirmed by comparison of the recorded ^{1}H NMR data (Figure S1) with published data [10].

2.6. Construction of the 1K1 BAC Derivatives

The derivatives of 1K1 BAC with gene deletions were constructed using the RedET approach. For this, the antibiotic resistance marker was amplified by PCR with primers harbouring overhang regions complementary to the boundaries of the DNA to be deleted. The amplified fragment was used for recombineering of the BAC. The recombinant BACs were analysed by restriction mapping and sequencing. The primers used for recombineering purposes are listed in Table S3.

For the construction of BAC 1K1_LS, the ampicillin marker from pUC19 was amplified with the primers LS-F/LS-R. For the construction of the BAC 1K1_RS, the hygromycin marker from pACS-hyg [16] was amplified with the primers RS-F/RS-R. For the construction of the BACs 1K1_KO14, 1K1_KO15 and 1K1_KO16, the ampicillin cassette was amplified with the pairs of primers KO14-F/KO14-R, KO15-F/KO15-R and KO16-F/KO16-R, respectively. For the construction of the BACs 1K1_KO7, 1K1_KO8, 1K1_KO9, 1K1_KO10, 1K1_KO11, 1K1_KO12 and 1K1_KO13, the ampicillin cassette was amplified with the pairs of primers KO7-F/KO7-R, KO8-F/KO8-R, KO9-F/KO9-R, KO10-F/KO10-R, KO11-F/KO11-R, KO12-F/KO12-R and KO13-F/KO13-R, respectively.

BAC 1K1_alb_act was constructed in two steps. First, 1K1_RS2 BAC was constructed from 1K1 using an ampicillin marker amplified with primers RS2-F/RS2-R. Then BAC 1K1_alb_act was constructed by recombineering the BAC 1K1_RS2 using a hygromycin marker from pACS-hyg amplified with the primers ACT-F/ACT-R.

2.7. Genome Mining and Bioinformatics Analysis

The *S. albus* subsp. *chlorinus* genome was screened for secondary metabolite biosynthetic gene clusters using the antiSMASH online tool [17] and the software Geneious [18]. The genomic sequence of the albucidin producer *S. albus* subsp. *chlorinus* NRRL B-24108 was deposited in GenBank under accession number VJOK00000000 [14].

3. Results and Discussion

3.1. Identification of the Albucidin Biosynthetic Gene Cluster

The aim of this study was to identify the biosynthetic genes leading to the production of the nucleoside phytotoxin albucidin. For this purpose, the genome sequence of the producer strain of *Streptomyces albus* subsp. *chlorinus* NRRL B-24108 was analysed by genome-mining software [17]. This analysis led to the identification of several putative nucleoside clusters. To prove the involvement of these candidate clusters in albucidin production, they were heterologously expressed in a genetically engineered cluster-free strain *Streptomyces albus* Del14 [19] and in *Streptomyces lividans* TK24 [20]. No albucidin production was detected in the extracts of the obtained strains, indicating that either the expressed clusters were not involved in the biosynthesis of albucidin or they were not expressed in the heterologous host environment. The inactivation of the candidate clusters in the natural albucidin producer was not feasible because the strain is refractory to genetic manipulation. Considering the difficulties in identifying the albucidin gene cluster using conventional methods, an alternative approach using chemical mutagenesis was chosen.

For chemical mutagenesis of the albucidin-producing strain *S. albus* subsp. *chlorinus* NRRL B-24108, 1-methyl-3-nitro-1-nitrosoguanidine (NTG) was used. The strain in the exponential growth stage was treated with various NTG concentrations (800 µg/mL, 600 µg/mL and 300 µg/mL) for 30 min. After mutagenesis, the cells were washed with 5% thiosulfate solution and plated in dilutions on MS-agar medium for segregation of mutations. The spores of the obtained mutant populations were washed and plated on MS-agar plates in dilutions to obtain single colonies. Altogether, 4000 individual mutants were analysed for albucidin production. The mutants were cultivated on individual plates with the production medium SG agar. The metabolites were extracted with butanol, and albucidin production was assayed by HPLC-MS. Eight mutants that lost the ability to produce albucidin were identified in the course of this screening: 6-238, 6-260, 6-389, 6-444, 6-612, 6-892, 8-610 and 8-639. The genomic DNA of the obtained zero mutants was sequenced using Illumina technology. The point mutations in the genomes of the mutants were detected by mapping the sequencing reads to the reference genome of the wild type albucidin producer. Up to 100 transition mutations were identified in the genomes of the mutant strains. By comparing the mutation patterns of the separate mutants, a short genomic region was identified that was affected by point mutations in all analysed zero mutants, implying its potential involvement in albucidin production (Figure S2). The identified region contains two genes, *SACHL2_05525* and *SACHL2_05524*, which encode putative radical SAM proteins and were named *albA* and *albB* (Table 1, Figure 1b). The genes constitute a putative operon with the third gene *SACHL2_05523*, which was named *albC*. The *albC* gene encodes a putative ribonucleoside-triphosphate reductase and was not affected by point mutations in the analysed zero mutants of *S. albus* subsp. *chlorinus* NRRL B-24108. The identified genes *albA* and *albB* were not a part of the nucleoside gene clusters previously identified by genome mining and analysed in this study. Interestingly, these genes were located within the DNA fragment annotated by genome mining software as a putative NRPS gene cluster.

Table 1. Genes encoded within the chromosomal fragment cloned in BAC 1K1.

Gene	Locus Tag [1]	Putative Function
1	SACHL2_05539	Hypothetical protein
2	SACHL2_05538	ABC transporter
3	SACHL2_05537	Transcriptional regulatory protein LiaR
4	SACHL2_05536	Hypothetical protein
5	SACHL2_05535	Hypothetical protein
6	SACHL2_05534	beta-lactamase/D-alanine carboxypeptidase
7	SACHL2_05533	Chondramide synthase
8	SACHL2_05532	Hypothetical protein
9	SACHL2_05531	Hypothetical protein
10	SACHL2_05530	Thymidylate kinase
11	SACHL2_05529	Pyruvate, phosphate dikinase
12	SACHL2_05528	Hypothetical protein
13	SACHL2_05527	Hypothetical protein
14	SACHL2_05526	Hypothetical protein
15; *albA*	SACHL2_05525	Biotin synthase, radical SAM protein
16; *albB*	SACHL2_05524	Radical SAM protein
17; *albC*	SACHL2_05523	Ribonucleoside-triphosphate reductase
18	SACHL2_05522	Tyrocidine synthase 3
19	SACHL2_05521	Plipastatin synthase, subunit A
20	SACHL2_05520	Acyl carrier protein
21	SACHL2_05519	Demethylmenaquinone methyltransferase
22	SACHL2_05518	Linear gramicidin synthase, subunit D
23	SACHL2_05517	Hypothetical protein
24	SACHL2_05516	Acyl carrier protein
25	SACHL2_05515	Fatty-acid–CoA ligase
26	SACHL2_05514	Ribonucleotide-diphosphate reductase
27	SACHL2_05513	Hypothetical protein
28	SACHL2_05512	Tyrocidine synthase 3
29	SACHL2_05511	Hypothetical protein

[1] The locus tags refer to the genome sequence of *S. albus* subsp. *chlorinus* NRRL B-24108 available under GenBank accession number VJOK00000000.

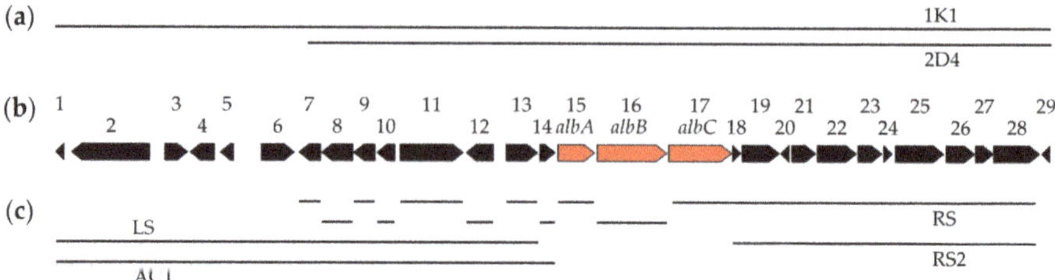

Figure 1. Chromosomal fragment of *S. albus* subsp. *chlorinus* NRRL B-24108 with the albucidin biosynthetic genes. (**a**) Schematic representations of DNA fragments cloned in BACs 1K1 and 2D4; (**b**) The genes encoded within the fragment cloned in BAC 1K1. The *albA–C* operon is marked in red; and (**c**) Overview of the performed deletions within BAC 1K1.

To determine whether the identified genes *albA* and *albB* encode albucidin biosynthetic enzymes, a BAC 1K1 containing the abovementioned genes (Figure 1a) was selected from the genomic library of *S. albus* subsp. *chlorinus* NRRL B-24108. BAC 1K1 was transferred into the heterologous host strains *S. albus* Del14 and *S. lividans* TK24 by conjugation, and the production profile of the obtained strains *S. albus* 1K1 and *S. lividans* 1K1 was analysed by HPLC-MS. The production of the compound with a high-resolution mass corresponding to albucidin could be detected in the extracts of *S. albus* 1K1 (Figure S3). No production could be detected in *S. lividans* 1K1.

Due to the lack of an albucidin standard, we set out to purify the compound identified in the extracts of *S. albus* 1K1 for structure elucidation studies by NMR spectroscopy. The *S. albus* 1K1 strain was cultivated in 10 L of SG medium for 7 days. The culture supernatant was extracted with equal amount of butanol, and the obtained extract was concentrated under vacuum. Four milligrams of the compound was purified using size exclusion and reverse phase chromatography and used for subsequent NMR studies. Analysis of the recorded NMR spectra of the purified compound unequivocally demonstrated its identity as albucidin (Figure S1 and Figure 2a).

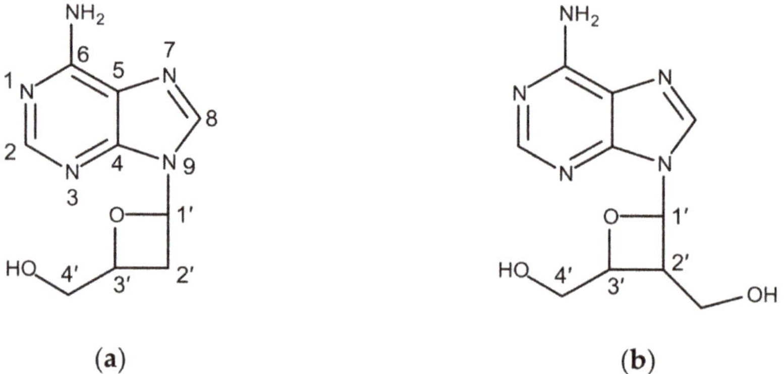

(a)

(b)

Figure 2. The structures of (**a**) albucidin and (**b**) oxetanocin A.

The production of albucidin by *S. albus* 1K1 gives evidence that the genes *albA* and *albB* identified by chemical mutagenesis encode albucidin biosynthetic genes. The lack of albucidin production by *S. lividans* 1K1 can be explained by differences in regulatory networks of the *S. albus* Del14 and *S. lividans* TK24 strains.

3.2. Identification of the Minimal Set of Albucidin Biosynthetic Genes

BAC 1K1, which leads to the production of albucidin under expression in the heterologous host *S. albus* Del14, contains a 32 kb chromosomal fragment from the natural albucidin producer *S. albus* subsp. *chlorinus* NRRL B-24108. Twenty-nine open reading frames were annotated in this 32 kb region (Table 1, Figure 1b). Of these genes, only two, *albA* and *albB*, were affected by point mutations in albucidin zero mutants identified in the course of the chemical mutagenesis studies. These two genes constitute a putative operon with the gene *albC*, implying that either only the genes *albA* and *albB* are necessary for albucidin production or that all three gene within the operon are required. To experimentally determine the minimal set of albucidin biosynthetic genes, a series of gene deletions was performed within the cloned region of 1K1 BAC.

The genes *albA–C* are located in the middle part of the chromosomal fragment cloned in 1K1 BAC. The *alb* operon is preceded by the genes *SACHL2_05539–SACHL2_05526*, followed by the genes *SACHL2_05522–SACHL2_05511* (Table 1). For the sake of simplicity, the 29 genes *SACHL2_05539–SACHL2_05511* cloned in the BAC 1K1 will be designated in the text according to their sequence number (1 to 29). (Figure 1b) To determine which genes within the 1K1 cloned fragment are essential for albucidin production, five deletions (LS, KO14, KO15, KO16 and RS) were performed in the 1K1 BAC yielding the BACs 1K1_LS, 1K1_KO14, 1K1_KO15, 1K1_KO16 and 1K1_RS (Figure 1c). In BAC 1K1_LS, the left shoulder encompassing genes 1–13 was substituted by an ampicillin resistance marker (Figure 1c). The genes 14, *albA* (gene 15) and *albB* (gene 16) were substituted with the ampicillin resistance gene in BACs 1K1_KO14, 1K1_KO15 and 1K1_KO16, respectively (Figure 1c). In the BAC 1K1_RS, the right shoulder encompassing genes *albC* (gene 17)-28 was substituted by the hygromycin resistance marker (Figure 1c). The constructed BACs were transferred separately into the *S. albus* Del14 strain by conjugation, and the albucidin production of the resulting strains was assayed by HPLC-MS.

The deletion of gene 14 did not affect albucidin production in *S. albus* 1K1_KO14 (Figure S4B). As expected from the results of chemical mutagenesis, inactivation of the genes *albA* (gene 15) and *albB* (gene 16) completely abolished albucidin production in the strains *S. albus* 1K1_KO15 and *S. albus* 1K1_KO16 (Figure S4C,D). This unambiguously demonstrates the essential role of the genes *albA* and *albB* in albucidin biosynthesis.

Deletion of the genes *albC* (gene 17)-28 did not affect albucidin production by the *S. albus* RS strain (Figure S4F). It was expected from the gene annotation and results of chemical mutagenesis that the genes 18–28 do not participate in albucidin biosynthesis. However, the dispensability of the gene *albC* (gene 17) is surprising since it belongs to the same operon as the essential genes *albA* (gene 15) and *albB* (gene 16). The deletion of the *albC* gene (gene 17) might be cross-complemented by an unidentified gene in the genome of the host strain *S. albus* Del14.

Albucidin production was heavily abolished in the strain *S. albus* 1K1_LS (Figure S4E), implying that at least one of the genes 1–13 that were deleted in the BAC 1K1_LS might be essential for albucidin biosynthesis. No genes encoding regulatory proteins or structural enzymes that might participate in nucleoside biosynthesis were identified in close proximity to the *albA–C* operon. To identify the genes within the deleted LS region that influence albucidin production, a BAC 2D4 was isolated from the genomic library of *S. albus* subsp. *chlorinus* NRRL B-24108. The chromosomal fragment cloned in BAC 2D4 overlaps with the fragment cloned in BAC 1K1 and covers the *albA–C* operon (Figure 1a). In contrast to 1K1, 2D4 BAC lacks genes 1–6, which are present in the deleted LS region. BAC 2D4 was transferred into *S. albus* Del14. Albucidin production could be detected in the extracts of the obtained strain *S. albus* 2D4 by HPLC-MS (Figure S5). This indicates that the genes 1–6 within the LS region are not involved in albucidin production and that one of the genes among 7–13 is responsible for the abolishment of albucidin production in *S. albus* 1K1_LS.

To identify which of the genes 7–13 is involved in albucidin biosynthesis, each of them was individually substituted by an ampicillin resistance marker in 1K1 BAC yielding 1K1_KO7, 1K1_KO8, 1K1_KO9, 1K1_KO10, 1K1_KO11, 1K1_KO12 and 1K1_KO13 (Figure 1b,c). The constructed BACs were transferred into the *S. albus* Del14 strain, and the albucidin production was analysed. The albucidin production levels of all obtained strains, except *S. albus* 1K1_KO12, were in the range of *S. albus* 1K1 harbouring the unmodified BAC (Figure S6). Albucidin production was abolished in *S. albus* 1K1_KO12 (Figure S6G), indicating that gene 12 is responsible for the detrimental effect of the LS deletion on albucidin biosynthesis. No enzymatic activity could be assigned to the peptide product of gene 12 using blast analysis. The product also did not show homology to any known regulatory protein. Considering this, it was proposed that only the genes *albA* and *albB* encode structural enzymes essential for albucidin production in the heterologous host *S. albus* Del14 and that the product of the gene 12 elicits a regulatory effect on transcription of the *albA–C* operon through a mechanism that is not understood. To prove this, a BAC 1K1_alb_act was constructed containing only *albA–C* genes under the control of a strong promoter. The genes downstream of the *albA–C* operon (genes 18–28) were substituted with the ampicillin resistance gene and the genes upstream of the operon (genes 1–14) were substituted with the hygromycin resistance gene (Figure 1c). The hygromycin resistance gene used was under the control of the strong synthetic promoter TS81 and did not contain a terminator at its 3'-end [21]. The insertion of the hygromycin marker in front of the *albA–C* genes was performed in the orientation, which enabled their read-through from the TS81 promoter and their transcriptional activation. The constructed BAC 1K1_alb_act was transferred into the heterologous host strain *S. albus* Del14. The production of albucidin was detected in the extracts of the obtained strain *S. albus* 1K1_alb_act by HPLC-MS (Figure S7). Three times increase of albucidin production was observed in the strain *S. albus* 1K1_alb_act compared to *S. albus* 1K1 containing non-modified albucidin cluster (Figure S7). Taking into account that the total recovered albucidin yield from the *S. albus* 1K1 strain was approximately 0.4 mg/L, the calculated albucidin production by *S. albus* 1K1_alb_act corresponded to 1.2 mg/L. The albucidin production rate of 2 mg/L was reported for the original producer *Streptomyces albus* subsp. *chlorinus* NRRL B-24108 [10].

Albucidin production by the strain *S. albus* 1K1_alb_act clearly demonstrates that the genes *albA* and *albB* constitute the minimal set of the genes required for albucidin biosynthesis in heterologous host *S. albus* Del14. The role of the gene *albC* in albucidin biosynthesis is not completely understood. Because *albC* constitutes a single operon with *albA* and *albB* and its product shows homology to nucleotide biosynthetic enzymes, it cannot be completely excluded that the *albC* is involved in albucidin production in the natural producer. However, the deletion of *albC* has no effect on albucidin production in heterologous host.

The identification of the minimal set of albucidin biosynthetic genes allows its expression in various heterologous chassis strains as well as rational construction of albucidin overproducers. The engineering of the albucidin biosynthetic genes can be performed in *E. coli* and the obtained constructs can be heterologously expressed in *Streptomyces* hosts. In contrast to the genetically intractable original albucidin producer *Streptomyces albus* subsp. *chlorinus* NRRL B-24108, commonly used heterologous strains possess a well-established toolkit for their genetic manipulation. This opens the possibility to engineer their metabolic network to increase the intracellular levels of biosynthetic precursors and therefore to increase the production yields. The heterologous strains are often characterized by the simplified metabolic background which provides better detection limits for heterologously expressed compounds than the original producers, higher product yields and simplified downstream processing. Construction of the albucidin overproducers based on heterologous expression hosts is not necessarily limited to a rational approach. The chassis strains expressing heterologous cluster may be also subjected to classical mutagenesis and screening for overproducing clones.

3.3. Proposed Biosynthetic Pathway of Albucidin

Structurally, albucidin is closely related to oxetanocin A (Figure 2b), which has been isolated from the culture of *Bacillus megaterium* NK84-0218 [22]. Both compounds are the only known naturally occurring nucleosides featuring four membered oxetane rings in their structure. From a structural view, albucidin is 2'-dehydroxymethyl oxetanocin A. Two genes, *oxsA* and *oxsB*, encoding a putative HD domain phosphohydrolase and a cobalamin-dependent S-adenosylmethionine radical enzyme have been reported to be responsible for oxetanocin biosynthesis [23]. dAMP, dADP and dATP were identified as direct oxetanocin precursors [24]. The product of *oxsB* catalyses the contraction of the deoxyribose ring, while the product of *oxsA* is responsible for the removal of one or multiple phosphates from a phosphorylated 2′-deoxyadenosine derivative [24,25]. Through the simultaneous actions of OxsA and OxsB, the phosphorylated 2′-deoxyadenosine is converted to the oxetanocin A precursor, its aldehyde form, which must be reduced to complete biosynthesis [24]. This reaction is not encoded by the genes within the oxetanocin A cluster and is likely to be carried by an unidentified enzyme of *B. megaterium* NK84-0218.

Gene inactivation studies have given evidence that two genes, *albA* and *albB*, are required for the production of albucidin. Both genes encode putative SAM radical proteins. At the protein level, the *albA* gene shows homology to biotin synthases and the *albB* gene shows homology to the product of the oxetanocin biosynthetic gene *oxsB*. Despite the high structural similarity of albucidin and oxetanocin A, the homologue of the second oxetanocin biosynthetic gene *oxsA* cannot be found within the albucidin cluster or in the genome of albucidin producer *S. albus* subsp. *chlorinus* NRRL B-24108. The homology of the *albB* gene to *oxsB* implies that the product of *albB* might also be responsible for the ring contraction reaction in albucidin biosynthesis. However, the structural differences between albucidin and oxetanocin and the absence of an *oxsA* homologue imply substantial differences in biosynthetic routes leading to the biosynthesis of the nucleosides. Due to the lack of an *oxsA* homologue that is responsible for the dephosphorylation of adenine deoxyribonucleotides during oxetanocin biosynthesis, we propose that deoxyadenosine is used instead of dAMP, dADP or dATP as a precursor for albucidin production. The product of *albB* is likely responsible for the contraction of the deoxyribose ring of deoxyadenosine (Figure 3) in a similar manner as its homologue OxsB catalyses oxetane ring formation in oxetanocin A biosynthesis [24]. As a result of this reaction, the aldehyde form of oxetanocine A is formed. The conversion of the latter into albucidin is likely to be catalysed by the product of *albA*, which removes the aldehyde group from the 2′-position (Figure 3).

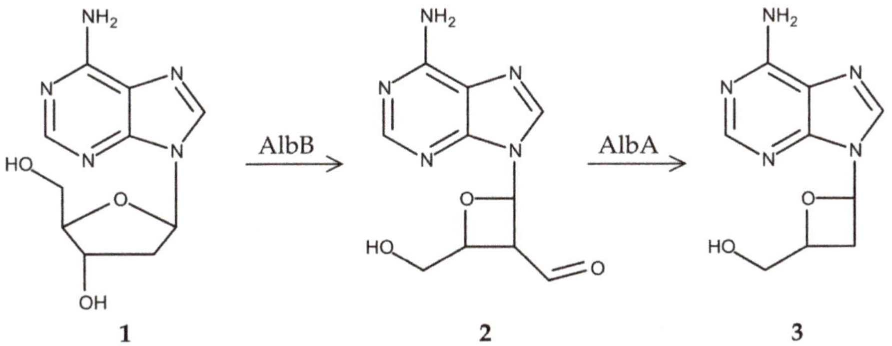

Figure 3. The proposed scheme of albucidin biosynthesis. 2′-Deoxyadenosine (**1**) is converted into the aldehyde form of oxetanocin A (**2**) by the product of *albB*. The latter is then converted into albucidin (**3**) by the product of *albA*.

In this paper, we report the identification, cloning, and heterologous expression of the albucidin biosynthetic gene cluster from *Streptomyces albus* subsp. *chlorinus* NRRL B-24108. Albucidin is a nucleoside phytotoxin featuring a rare oxetane ring in its structure. This metabolite shows herbicidal activity against a broad spectrum of grass and broadleaf weeds. In treated plants, albucidin induces metabolic perturbation, chlorosis, and bleaching [10]. The exact MOA of the compound remains unknown. The identification of the albucidin cluster presented in this paper enables biosynthetic studies of albucidin, optimization of its production as well as albucidin supply for the determination of its MOA.

4. Patents

(WO2018224939) Gene cluster for the biosynthesis of albucidin.

Supplementary Materials: The following are available online at http://www.mdpi.com/2076-2607/8/2/237/s1, Figure S1: ^{1}H NMR spectrum of albucidin (500 MHz), Figure S2: The genes *albA* and *albB* with the mapped point mutations identified in the course of NTG-mutagenesis, Figure S3: High resolution HPLC-MS analysis of albucidin production by *Streptomyces albus* 1K1, Figure S4: HPLC-MS analysis of albucidin production by *Streptomyces albus* strain harboring 1K1 BAC and its derivatives with gene deletions, Figure S5: HPLC-MS analysis of albucidin production by *Streptomyces albus* 1K1 BAC (**a**) and *Streptomyces albus* 2D4 BAC (**b**), Figure S6: HPLC-MS analysis of albucidin production by *Streptomyces albus* strain harboring 1K1 BAC and its derivatives with gene deletions, Figure S7: HPLC-MS analysis of albucidin production by *Streptomyces albus* strain harboring full length 1K1 BAC (**a**) and minimized 1K1_alb_act BAC, containing only transcriptionally activated *albA–C* operon (**b**), Table S1: Bacterial strains used in the study, Table S2: Plasmids and BACs used in the study, Table S3: Primers used in this study. References [26,27] are cite d in the supplementary materials.

Author Contributions: L.P., T.S., M.M., and A.L. designed the experiments; M.M. and B.R. performed the experiments; M.S. performed structure elucidation studies; M.M., B.R., and A.L. analysed the data and wrote the manuscript, and all authors reviewed the manuscript. All authors have read and agreed to the published version of the manuscript.

Funding: This work was partially supported by the BMBF grant "MyBio" 031B0344B.

Conflicts of Interest: The authors declare no conflict of interest.

References

1. Malakof, D.; Stokstad, E. Infographic: Pesticide Planet. *Science* **2013**, *341*, 730–731.
2. Pimentel, D.; Zuniga, R.; Morrison, D. Update on the environmental and economic costs associated with alien-invasive species in the United States. *Ecol. Econ.* **2005**, *52*, 273–288. [CrossRef]
3. Dayan, F.E.; Duke, S.O. Natural compounds as next-generation herbicides. *Plant Physiol.* **2014**, *166*, 1090–1105. [CrossRef]
4. Duke, S.O. Why have no new herbicide modes of action appeared in recent years? *Pest Manag. Sci.* **2012**, *68*, 505–512. [CrossRef]
5. Heap, I. Global perspective of herbicide-resistant weeds. *Pest Manag. Sci.* **2014**, *70*, 1306–1315. [CrossRef]
6. Cantrell, C.L.; Dayan, F.E.; Duke, S.O. Natural products as sources for new pesticides. *J. Nat. Prod.* **2012**, *75*, 1231–1242. [CrossRef]
7. Duke, S.O.; Stidham, M.A.; Dayan, F.E. A novel genomic approach to herbicide and herbicide mode of action discovery. *Pest Manag. Sci.* **2019**, *75*, 314–317. [CrossRef]
8. Harvey, A.L. Natural products as a screening resource. *Curr. Opin. Chem. Biol.* **2007**, *11*, 480–484. [CrossRef]
9. Koch, M.A.; Schuffenhauer, A.; Scheck, M.; Wetzel, S.; Casaulta, M.; Odermatt, A.; Ertl, P.; Waldmann, H. Charting biologically relevant chemical space: A structural classification of natural products (SCONP). *Proc. Natl. Acad. Sci. USA* **2005**, *102*, 17272–17277. [CrossRef]
10. Hahn, D.R.; Graupner, P.R.; Chapin, E.; Gray, J.; Heim, D.; Gilbert, J.R.; Gerwick, B.C. Albucidin: A novel bleaching herbicide from *Streptomyces albus* subsp. *chlorinus* NRRL B-24108. *J. Antibiot.* **2009**, *62*, 191–194. [CrossRef]
11. Green, M.R.; Sambrook, J. *Molecular Cloning: A Laboratory Manual*, 4th ed.; Cold Spring Harbor Laboratory Press: New York, NY, USA, 2012.
12. Kieser, T.; Bibb, M.J.; Buttner, M.J.; Chater, K.F.; Hopwood, D.A. *Practical Streptomyces Genetics*; John Innes Foundation: Norwich, UK, 2000.
13. Rebets, Y.; Ostash, B.; Luzhetskyy, A.; Hoffmeister, D.; Brana, A.; Mendez, C.; Salas, J.A.; Bechthold, A.; Fedorenko, V. Production of landomycins in *Streptomyces globisporus* 1912 and *S. cyanogenus* S136 is regulated by genes encoding putative transcriptional activators. *FEMS Microbiol. Lett.* **2003**, *222*, 149–153. [CrossRef]
14. Rodríguez Estévez, M.; Myronovskyi, M.; Gummerlich, N.; Nadmid, S.; Luzhetskyy, A. Heterologous Expression of the Nybomycin Gene Cluster from the Marine Strain *Streptomyces albus* subsp. *chlorinus* NRRL B-24108. *Mar. Drugs* **2018**, *16*, 435.
15. Mazodier, P.; Petter, R.; Thompson, C. Intergeneric conjugation between Escherichia coli and *Streptomyces* species. *J. Bacteriol.* **1989**, *171*, 3583–3585. [CrossRef] [PubMed]

16. Myronovskyi, M.; Brötz, E.; Rosenkränzer, B.; Manderscheid, N.; Tokovenko, B.; Rebets, Y.; Luzhetskyy, A. Generation of new compounds through unbalanced transcription of landomycin A cluster. *Appl. Microbiol. Biotechnol.* **2016**, *100*, 9175–9186. [CrossRef] [PubMed]

17. Blin, K.; Medema, M.H.; Kottmann, R.; Lee, S.Y.; Weber, T. The antiSMASH database, a comprehensive database of microbial secondary metabolite biosynthetic gene clusters. *Nucleic Acids Res.* **2017**, *45*, D555–D559. [CrossRef]

18. Kearse, M.; Moir, R.; Wilson, A.; Stones-Havas, S.; Cheung, M.; Sturrock, S.; Buxton, S.; Cooper, A.; Markowitz, S.; Duran, C.; et al. Geneious Basic: An integrated and extendable desktop software platform for the organization and analysis of sequence data. *Bioinformatics* **2012**, *28*, 1647–1649. [CrossRef] [PubMed]

19. Myronovskyi, M.; Rosenkränzer, B.; Nadmid, S.; Pujic, P.; Normand, P.; Luzhetskyy, A. Generation of a cluster-free *Streptomyces albus* chassis strains for improved heterologous expression of secondary metabolite clusters. *Metab. Eng.* **2018**, *49*, 316–324. [CrossRef]

20. Rückert, C.; Albersmeier, A.; Busche, T.; Jaenicke, S.; Winkler, A.; Friðjónsson, Ó.H.; Hreggviðsson, G.Ó.; Lambert, C.; Badcock, D.; Bernaerts, K.; et al. Complete genome sequence of *Streptomyces lividans* TK24. *J. Biotechnol.* **2015**, *199*, 21–22. [CrossRef]

21. Siegl, T.; Tokovenko, B.; Myronovskyi, M.; Luzhetskyy, A. Design, construction and characterisation of a synthetic promoter library for fine-tuned gene expression in actinomycetes. *Metab. Eng.* **2013**, *19*, 98–106. [CrossRef]

22. Shimada, N.; Hasegawa, S.; Harada, T.; Tomisawa, T.; Fujii, A.; Takita, T. Oxetanocin, a novel nucleoside from bacteria. *J. Antibiot.* **1986**, *39*, 1623–1625. [CrossRef]

23. Morita, M.; Tomita, K.; Ishizawa, M.; Takagi, K.; Kawamura, F.; Takahashi, H.; Morino, T. Cloning of oxetanocin A biosynthetic and resistance genes that reside on a plasmid of *Bacillus megaterium* strain NK84-0128. *Biosci. Biotechnol. Biochem.* **1999**, *63*, 563–566. [CrossRef] [PubMed]

24. Bridwell-Rabb, J.; Zhong, A.; Sun, H.G.; Drennan, C.L.; Liu, H.-W. A B12-dependent radical SAM enzyme involved in oxetanocin A biosynthesis. *Nature* **2017**, *544*, 322–326. [CrossRef] [PubMed]

25. Bridwell-Rabb, J.; Kang, G.; Zhong, A.; Liu, H.-W.; Drennan, C.L. An HD domain phosphohydrolase active site tailored for oxetanocin-A biosynthesis. *Proc. Natl. Acad. Sci. USA* **2016**, *113*, 13750–13755. [CrossRef] [PubMed]

26. Grant, S.G.; Jessee, J.; Bloom, F.R.; Hanahan, D. Differential plasmid rescue from transgenic mouse DNAs into *Escherichia coli* methylation-restriction mutants. *Proc. Natl. Acad. Sci. USA* **1990**, *87*, 4645–4649. [CrossRef]

27. Flett, F.; Mersinias, V.; Smith, C.P. High efficiency intergeneric conjugal transfer of plasmid DNA from *Escherichia coli* to methyl DNA-restricting streptomycetes. *FEMS Microbiol. Lett.* **1997**, *155*, 223–229. [CrossRef]

Article

Diversity and Bioactive Potential of Actinobacteria from Unexplored Regions of Western Ghats, India

Saket Siddharth [1], Ravishankar Rai Vittal [1,*], Joachim Wink [2] and Michael Steinert [3]

[1] Department of Studies in Microbiology, University of Mysore, Manasagangotri, Mysore 570 006, India; saketsiddharth@gmail.com

[2] Microbial Strain Collection, Helmholtz Centre for Infection Research GmbH (HZI), Inhoffenstrasse 7, 38124 Braunschweig, Germany; joachim.wink@helmholtz-hzi.de

[3] Institute of Microbiology, Technische Universität Braunschweig, 38106 Braunschweig, Germany; m.steinert@tu-bs.de

* Correspondence: raivittal@gmail.com

Received: 3 February 2020; Accepted: 6 February 2020; Published: 7 February 2020

Abstract: The search for novel bioactive metabolites continues to be of much importance around the world for pharmaceutical, agricultural, and industrial applications. Actinobacteria constitute one of the extremely interesting groups of microorganisms widely used as important biological contributors for a wide range of novel secondary metabolites. This study focused on the assessment of antimicrobial and antioxidant activity of crude extracts of actinobacterial strains. Western Ghats of India represents unique regions of biologically diverse areas called "hot spots". A total of 32 isolates were obtained from soil samples of different forest locations of Bisle Ghat and Virajpet situated in Western Ghats of Karnataka, India. The isolates were identified as species of *Streptomyces*, *Nocardiopsis*, and *Nocardioides* by cultural, morphological, and molecular studies. Based on preliminary screening, seven isolates were chosen for metabolites extraction and to determine antimicrobial activity qualitatively (disc diffusion method) and quantitatively (micro dilution method) and scavenging activity against DPPH (2,2-diphenyl-1-picrylhydrazyl) and ABTS (2,2′-Azino-bis(3-ethylbenzothiazoline-6-sulfonic acid) radicals. Crude extracts of all seven isolates exhibited fairly strong antibacterial activity towards MRSA strains (MRSA ATCC 33591, MRSA ATCC NR-46071, and MRSA ATCC 46171) with MIC varying from 15.62 to 125 µg/mL, whereas showed less inhibition potential towards Gram-negative bacteria *Salmonella typhi* (ATCC 25241) and *Escherichia coli* (ATCC 11775) with MIC of 125–500 µg/mL. The isolates namely S1A, SS5, SCA35, and SCA 11 inhibited *Fusarium moniliforme* (MTCC 6576) to a maximum extent with MIC ranging from 62.5 to 250 µg/mL. Crude extract of SCA 11 and SCA 13 exhibited potent scavenging activities against DPPH and ABTS radicals. The results from this study suggest that actinobacterial strains of Western Ghats are an excellent source of natural antimicrobial and antioxidant compounds. Further research investigations on purification, recovery, and structural characterization of the active compounds are to be carried out.

Keywords: Western Ghats; diversity; actinobacteria; antimicrobial; MRSA; antioxidants

1. Introduction

The quest for novel biologically active secondary metabolites from microorganisms continues to rise due to emergence of drug resistance in pathogens causing life threatening diseases around the globe [1]. Particularly, methicillin resistant *Staphylococcus aureus* (MRSA) and methicillin resistant *Staphylococcus epidermidis* (MRSE) strains are not only exposed to hospital-acquired infections but also to community-acquired infections [2]. The mortality and morbidity associated with these infections are largely affecting economic conditions of patients and hospitals [3]. Therefore, there is an urgent need for developing novel and effective antimicrobial agents to overcome or delay acquired resistance to existing drugs. Reactive oxygen species (ROS) play an important role as signaling molecules involved in mitogenesis. However, high generation of ROS during aerobic metabolism creates oxidative stress within the intracellular milieu causing oxidative damage to cells [4]. The oxidative stress caused is often associated with many human diseases including cancer, diabetes [5], cardiovascular [6], and neurodegenerative diseases [7]. In order to withstand the oxidative stress caused, cells or organisms make use of antioxidants, that are able to block or delay the damage caused by several possible mechanisms such as halting chain reactions, preventing the formation of free radicals, neutralizing the singlet oxygen molecule, promoting anti-oxidant enzymes, and inhibiting pro-oxidative enzymes [8]. Formation of free radicals can be prevented by antioxidant systems present within the cells. However, these defense mechanisms are insufficient to prevent the damages that arise, therefore exogenous antioxidants through dietary intake and supplements are required [9]. Natural antioxidants are found abundantly in metabolites produced by microorganisms. These products have consistently been considered as mainstay for drugs with various interesting biological activities [10]. They are considered to be an excellent scaffold for the formulation and development of antibiotics, antioxidants, immunomodulators, enzyme inhibitors, anticancer agents, plant growth hormones, and insect control agents [11]. With many improved techniques under combinatorial chemistry for high throughput findings of novel compounds, natural products from microbial sources have been screened extensively and gained much attention owing to their massive chemical and biological diversity [12]. Under various screening strategies, the rate of discovery of natural products has increased many folds, of which around 22,250 bioactive compounds are of microbial origin [13]. Among microorganisms, actinomycetes have contributed nearly 45% of all the reported metabolites [14].

A major group of natural products from microbial origin have been identified from organisms that inhabit the soil. Since soil itself is a mixture of minerals and organic matter, the filamentous bacteria are predominantly more present in the gaps between the soil particles than their unicellular counterparts [15]. Actinobacteria are ubiquitous in soils. They are responsible for biodegradation and biodeterioration processes in nature. Their flexile and proven abilities have prompted biologists to screen these organisms from unexplored niche habitats in order to obtain novel molecules [16].

Western Ghats of India is considered as one of the global biodiversity hotspots covering an area of 180,000 km^2 and harbors numerous species of plants, animals, and microbes [17]. The unique biodiversity of Western Ghats is conserved and protected by wildlife sanctuaries, national parks, and biosphere reserves situated in states where hill ranges run through, like Karnataka, Gujarat, Tamil Nadu, Maharashtra, and Kerala [18]. The forest regions in Western Ghats are largely underexplored, though in recent times few studies were carried out for bioprospection. Ganesan et al. [19] reported larvicidal, ovicidal, and repellent activities of *Streptomyces enissocaesilis* (S12–17) isolated from Western Ghats of Tamil Nadu, India. Actinobacterial strains isolated from Western Ghats soil of Tamil Nadu were reported to produce antimicrobial compounds against range of pathogens [20]. In the present study, the forest range in Western Ghats of Karnataka was studied for microbial population and taxonomical identification of potential actinobacteria. An attempt was also made to characterize microbial diversity for the potential to produce antioxidants and antimicrobial compounds.

2. Materials and Methods

2.1. Site, Sampling, Pre-Treatment, and Selective Isolation

Soil samples were collected from different forest locations of Bisle Ghat and Virajpet in Western Ghats regions of Karnataka, India. The samples were collected from a depth of 15–25 cm in dry sterile insulated containers and stored aseptically at 4 °C until subjected to plating. Samples were air dried in a hot air oven (Equitron, India) at 50 °C for 72 h. Pre-treated samples were ground aseptically with mortar and pestle and serially diluted up to 10^{-6} in 10-fold dilution. The aliquots of each dilution (100 µL) were spread evenly on starch casein agar (SCA, Himedia, India) plates in triplicates supplemented with cycloheximide (30 µg/mL) and nalidixic acid (25 µg/mL). The plates were incubated at 28 ± 2 °C for 14 days. Emerging colonies with different morphological characters were selected and the purified strains were maintained on International *Streptomyces* Project (ISP-2, Himedia, India) agar slants and stored at 4 °C as stock for further use.

2.2. Morphological Characterization of Isolates

Morphological characteristics of isolates were assessed by scanning electron microscopy (SEM). Bacterial colonies were inoculated in ISP-2 medium and incubated at 28 ± 2 °C for 7 days. Cells were centrifuged (Eppendorf, USA) at 8000× g for 10 min and pellet was resuspended in 2%–5% gluteraldehyde (Sigma, Burlington, VT, USA) prepared in 0.1M phosphate buffer, pH 7.2. After incubating samples for 30 min, supernatant was discarded and pellet was resuspended in 1% osmium tetraoxide (Sigma, Burlington, VT, USA), incubated for 1 h and centrifuged at 5000 × g. To the pellet, sterile water was added and centrifuged twice for 10 min at 5000 × g. For dehydration, the pellet was resuspended in 35% ethanol for 10 min, 50% ethanol for 10 min, 75% ethanol for 10 min, 95% ethanol for 10 min, and a final wash with 100% ethanol for 10 min. For SEM analysis, sterilized aluminum stubs and cover slips were inserted into the SCA plates at an angle of about 45 °C. The plates with stubs and coverslips were incubated at 37 °C for 24 h to check any contamination. After 24 h, isolates were introduced along the line where the surface of the stub met the agar medium and incubated at 28 ± 2 °C for 7 days. The stubs were then carefully removed and coated under vacuum, with a film of gold for 25–30 min and viewed on the scanning electron microscope (Zeiss Evo 40 EP, Germany).

2.3. Molecular Identification and Phylogenetic Analysis

The total genomic DNA of bacteria was extracted by phenol-chloroform method, quality checked by agarose gel electrophoresis and quantified using NanoDrop1000 (Thermo-Scientific, USA). The PCR amplification of 16S rRNA gene was carried out with universal primers: 27F (5′-AGA GTT TGA TCC TGG CTC AG-3′) and 1492R (5′- ACG GCT ACC TTG TTA CGA CTT-3′) using the following conditions: initial denaturation temperature was set at 95 °C for 5 min, followed by 35 cycles at same temperature for 1 min, primer annealing at 54 °C for 1 min, and primer extension at 72 °C for 2 min. The reaction mixture was kept at 72 °C for 10 min subsequently and then cooled to 4 °C. The PCR products were checked in 1.5% agarose gel and visualized in a UV transilluminator (Tarsons, India) and the gel imaging was done using a Gel documentation system (Bio-Rad, USA). The amplified PCR products were sequenced using same set of primers (27F′ and 1492R′) on Applied Biosystems 3130 Genetic Analyzer (Applied Biosystems, USA). The genetic relationship between the strains was determined by neighbor-joining tree algorithm method. The phylogenetic tree was constructed with a bootstrapped database containing 1000 replicates in MEGA 7.0 software (Mega, Raynham, MA, USA). The nearly complete 16S rRNA consensus sequences were deposited in the GenBank database.

2.4. Isolates Cultivation and Metabolites Extraction

Pure isolates were subcultured in Tryptone Yeast Extract broth as seed medium (ISP-1 medium, Himedia, India) at 28 °C for 2 weeks prior to fermentation process. The production medium, ISP-2 was autoclaved at 121 °C and 1.5 atm for 15 min. Fermentation was carried out in 750 mL of (in 7 nos. conical flasks −1000 mL) ISP-2 medium, shaking at 140 rev min^{-1} for 14 days at 28 °C, inoculated with 250 µL of seed medium. After incubation, the culture medium was split into mycelium and filtrate by centrifugation at 12,000× g for 15 min. The cell free supernatant from each flask was subjected to extraction thrice with equal volume of ethyl acetate (Qualigens Fine Chemicals Pvt. Ltd., San Diego, USA) and the organic phase was concentrated by rotary vacuum evaporator (Hahn-Shin, Bucheon, South Korea) at 50 °C. The crude concentrate was dried in a desiccator and suspended in methanol prior to bioactivity screening assays.

2.5. Antimicrobial Susceptibility Test

2.5.1. Disc Diffusion Assay

Antimicrobial susceptibility assay was carried out by disc diffusion method against methicillin-resistance *Staphylococcus aureus* (MRSA ATCC 33591, MRSA ATCC NR-46071 and MRSA ATCC 46171), Gram-negative bacteria (*Salmonella typhi* (ATCC 25241) and *Escherichia coli* (ATCC 11775)) and fungus (*Fusarium moniliforme* (MTCC 6576)). Gentamicin and Nystatin discs were used as positive control. The sterile discs (6mm, Himedia) were impregnated with 30 µL of crude extract dissolved in 0.5% DMSO. Discs impregnated with 0.5% DMSO (Qualigens Fine Chemicals Pvt. Ltd., San Diego, USA) were used as solvent control. The plates were left for 30 min at 4 °C to allow the diffusion of extracts before they were incubated for 24–48 h at 37 °C. The clear zones of inhibition observed around discs suggested antagonistic activity against test organisms and diameter of inhibition zones were measured subsequently. The test was performed in triplicate.

2.5.2. Minimum Inhibitory Concentration (MIC)

The minimum inhibitory concentration assay was determined by micro broth dilution method as previously reported by Siddharth and Rai [21]. The serially diluted fraction of extracts with sterile Mueller Hinton broth (Himedia, India) was added to pre coated microbial cultures in 96-well micro titer plates to give a final concentration of 1–3.8 µg/mL. The titer plate was incubated for 24 h at 37 °C. The lowest concentration of extract which completely inhibited the bacterial growth (no turbidity) was considered as MIC. Each test was done in triplicate.

2.6. Antioxidant Assays

2.6.1. 2,2-diphenyl-1-picrylhydrazyl Radical Scavenging activity (DPPH)

DPPH radical scavenging activity of crude extracts was examined based on a previously described method by Siddharth and Rai [22]. Crude extracts at varying concentrations (7.81–1000 µg/mL) were reacted with freshly prepared DPPH in methanol (60 mM, Sigma, USA). Reaction mixture was incubated at room temperature for 30 min in the dark prior to the measurement of absorbance at 520 nm. The radical scavenging activity was expressed as IC_{50} (µg/mL). The percentage scavenging of DPPH radicals was computed by the following equation:

$$\text{DPPH scavenging effect (\%)} = [(A_o - A_1)/A_o] \times 100 \qquad (1)$$

where, A_o = Absorbance of control, A_1 = Absorbance of crude extracts. Trolox (Sigma) was used as a reference compound whereas methanol was used as blank.

2.6.2. 2,2′-Azino-bis(3-ethylbenzothiazoline-6-sulfonic acid) Radical Scavenging Activity (ABTS)

ABTS radical scavenging activity was performed according to the method developed by Ser et al. [23] with slight modifications. Crude extracts at concentrations (7.81–1000 µg/mL) were mixed with ABTS (Sigma, USA) cation complex and incubated in dark at room temperature for 30 min. The absorbance was measured at 415 nm. The radical scavenging activity was expressed as IC_{50} (µg/mL). The percentage inhibition of $ABTS^{*+}$ radicals were computed by using following equation:

$$ABTS^{*+}\text{ scavenging effect (\%)} = [(A_o - A_1)/A_o] \times 100 \qquad (2)$$

where, A_o = Absorbance of control, A_1 = Absorbance of crude extracts. Trolox (Sigma) was used as a positive reference.

3. Results and Discussion

Soil is among the most productive habitat colonized by a large number of organisms. The rhizosphere soil in the vicinity of plant roots provides essential nutrients in the form of exudates which favors the growth of microbial communities [24]. Soil microbes are a major source of a number of natural products including clinical important antibiotics, immunomodulators, enzyme inhibitors, antioxidants, anti-tumor and anticancer agents. Actinobacteria are abundant in soil, species of *Streptomyces* in particular represent the dominance over other microbes present in soil and play a vital role in recycling of materials and production of important metabolites [25]. Rare actinobacteria genera such as *Nocardia*, *Nocardiopsis*, and *Nocardioides* are also encountered in soils, though their presence is subjected to conditions of soil such as salinity and alkalinity [26]. The rapid emergence of drug resistant pathogens urges the exploration of new niche habitats for the isolation of new microbial species which can contribute to the uncovering of novel, safe, effective, and broad spectrum bioactive compounds [27].

3.1. Isolation, Morphological, and Molecular Characterization of Actinobacterial Isolates

In this study, from forest soil of Bisle Ghat and Virajpet of Western Ghats region of Karnataka, we targeted the isolation of different genera of actinobacteria in search of new natural products (Table 1). A total of 32 actinobacterial isolates were recovered on Starch casein agar and Actinomycetes isolation agar. Of them, seven isolates grown on starch casein agar medium showing marked antimicrobial activity against test organisms in primary screening were characterized on the basis of cultural, morphological, and molecular characteristics. The colonies of isolates revealed diverse morphological appearances with varied spore color, aerial and substrate mycelium, and colony morphology (Table 2 and Figure 1). Scanning electron microscope examination showed chains of smooth and spiny spores in oval, round, and spiral ornamentation (Figure 2). The molecular identification of isolates by amplification of 16S-rRNA gene was done by using universal primers 27F′ and 1492R′ (Figure S1). In 16S-rRNA sequencing, alignment of the nucleotide sequences of strain S1A, SS4, SS5, SS6, and SCA35 exhibited a similarity of 98.32%, 99.71%, 99.70%, 99.72%, and 99.35% with closely related *Streptomyces* species, respectively. The strain SCA11 was considered to represent a species of the genus *Nocardiopsis*, since it was closely related to *Nocardiopsis* species with 98.99% sequence similarity. The nucleotide sequence of strain SCA13 showed 98.28% similarity with the closely related *Nocardioides* sp. (Table 3). The phylogenetic relatedness of strains with their closely related species obtained by neighbor-joining method is shown in Figures 3–5.

Table 1. Actinobacterial isolates and sampling areas in Western Ghats regions of Karnataka, India.

Name	Sample Code	Sampling Area	Latitude (N)	Longitude (E)	Elevation (m)
Streptomyces Sp.	S1A	Bisle Ghat, Hassan District	12°71′88.04″	75°68′70.02″	802
Streptomyces Sp.	SS4	Bisle Ghat, Hassan District	12°72′00.89″	75°68′41.42″	752
Streptomyces Sp.	SS5	Bisle Ghat, Hassan District	12°71′24.30″	75°68′04.74″	710
Streptomyces Sp.	SS6	Virajpet, Madikeri District	12°19′75.83″	75°79′52.93″	885
Streptomyces Sp.	SCA35	Virajpet, Madikeri District	12°18′83.38″	75°83′06.79″	864
Nocardiopsis Sp.	SCA11	Virajpet, Madikeri District	12°21′21.64″	75°80′24.84″	830
Nocardioides Sp.	SCA13	Virajpet, Madikeri District	12°18′47.27″	75°76′24.17″	798

Table 2. Morphological characteristics of isolated actinobacterial strains.

Isolate	Medium	Diffusible Pigment	Colony Morphology	Aerial Mycelium	Substrate Mycelium
Streptomyces Sp. S1A	SCA	None	Powdery	White	Cream
Streptomyces Sp. SS4	SCA	None	Cottony	Dark Grey	Grey
Streptomyces Sp. SS5	SCA	None	Cottony	Grey	Cream
Streptomyces Sp. SS6	SCA	Pink	Rough	Pale red	Pink
Streptomyces Sp. SCA35	SCA	None	Powdery	White	Cream
Nocardiopsis Sp. SCA11	SCA	None	Powdery	Cream	Light brown
Nocardioides Sp. SCA13	SCA	None	Raised	White	White

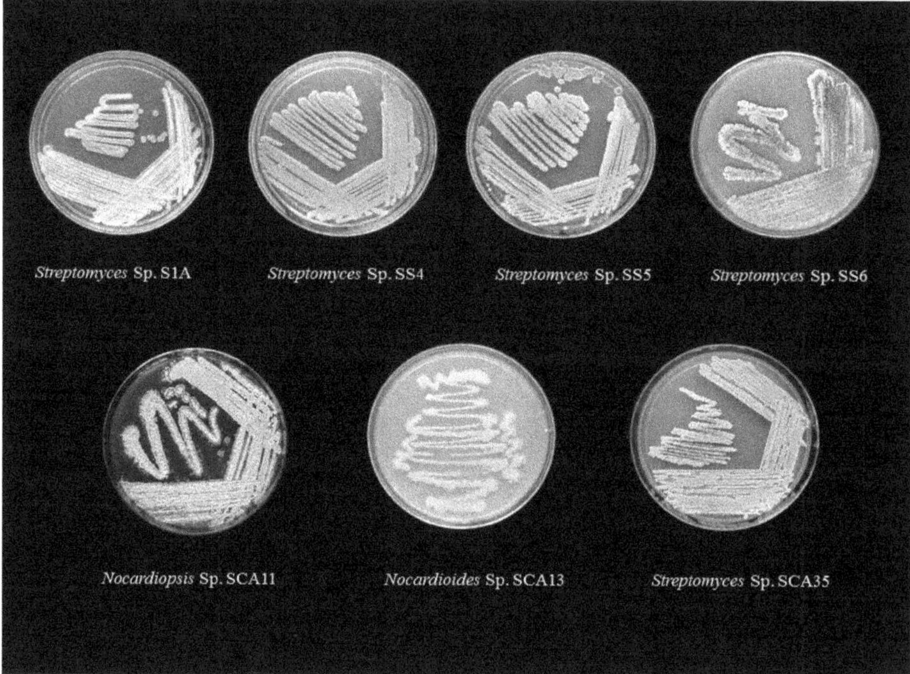

Figure 1. Morphological characterization of actinobacterial isolates on starch casein agar plates.

Table 3. Molecular identification (based on 16S rRNA amplification) of actinobacterial strains isolated from Western Ghats.

Source	Organism	GenBank Accession No	% Similarity
Soil	*Streptomyces* Sp. S1A	KU921223	98.32%
Soil	*Streptomyces* Sp. SS4	MF668120	99.71%
Soil	*Streptomyces* Sp. SS5	MF925722	99.70%
Soil	*Streptomyces* Sp. SS6	MF925723	99.72%
Soil	*Streptomyces* Sp. SCA35	MN176654	99.35%
Soil	*Nocardiopsis* Sp. SCA11	MG934272	98.99%
Soil	*Nocardioides* Sp. SCA13	MG934273	98.28%

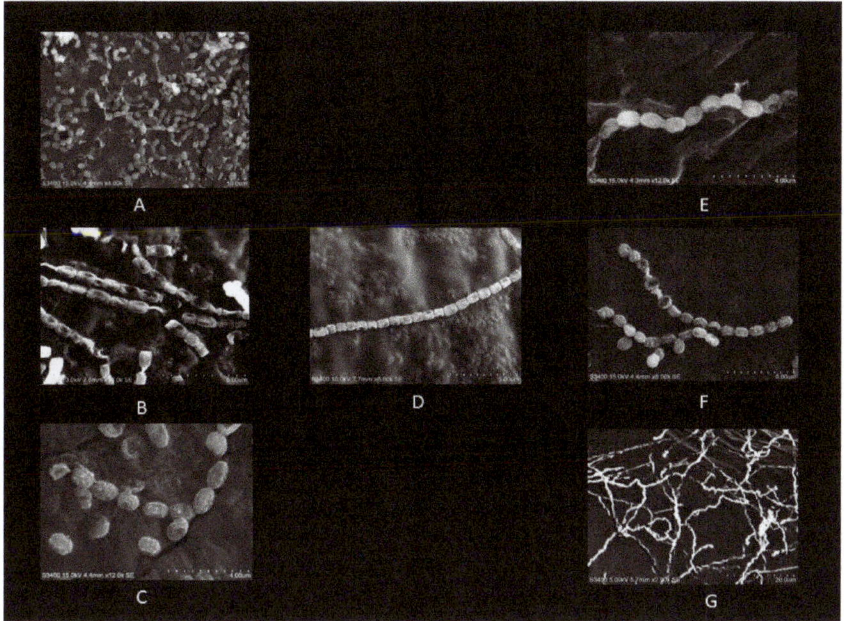

Figure 2. Scanning electron micrograph of strains (**A**) S1A, (**B**) SCA11, (**C**) SS4, (**D**) SS5, (**E**) SS6, (**F**) SCA35, (**G**) SCA13; scale bar represents 5 μm.

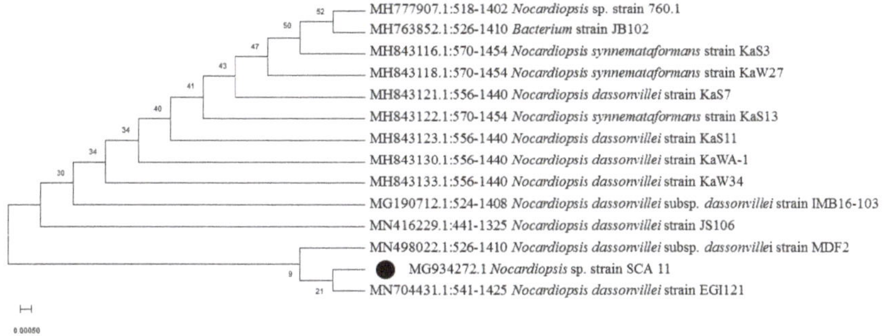

Figure 3. Phylogenetic analysis of isolate *Nocardiopsis* sp. strain SCA11. Neighbor-joining phylogenetic tree showing evolutionary relationship of selected isolate based on 16S r-RNA sequence alignments. Bootstrap values at the nodes indicate collated values based on 1000 resampled datasets. Bar indicates 0.0005 substitutions per nucleotide position.

3.2. Antimicrobial and Antioxidant Potential of Isolates

In this study, all seven isolates showed antibacterial activity against at least one test bacterium. All isolates inhibited MRSA strains significantly (MRSA ATCC 33591, MRSA ATCC NR-46071, and MRSA ATCC 46171), whereas showed less inhibition potential against Gram-negative bacteria *Salmonella typhi* (ATCC 25241) and *Escherichia coli* (ATCC 11775). The isolates namely S1A, SS5, SCA35, and SCA 11 inhibited *Fusarium moniliforme* (MTCC 6576) to a maximum extent (Figure 6). The minimum inhibition concentration (MIC) ranges from 15.62 to 125 μg/mL for MRSA strains, 125–500 μg/mL for Gram-negative bacteria, and 62.5–250 μg/mL for fungi (Table 4). Numerous studies have reported antimicrobial activity of actinobacterial species. Sengupta et al. [28] reported potential antimicrobial activity of three isolates against *Pseudomonas aueroginosa*, *Enterobacter aueroginosa*, *Salmonella typhi*, *Salmonella typhimurium*, *Escherichia coli*, *Bacillus subtilis*, and *Vibrio cholera*. Satheeja and Jebakumar [29] reported isolation of *Streptomcyes* species from a mangrove ecosystem for antibacterial activity against clinical isolates of MRSA, methicillin-susceptible *Staphylococcus aureus* (MSSA), and *Salmonella typhi*. Vu et al. [30] reported antimicrobial activity of *Streptomyces cavourensis* YBQ59 against methicillin-resistant *Staphylococcus aureus* ATCC 33591 and methicillin-resistant *Staphylococcus epidermidis* ATCC 358984. Dashti et al. [31] reported the co-cultivation of *Actinokineospora* sp.EG49 and *Nocardiopsis* sp.RV163 and metabolites produced were tested for antimicrobial activity against the range of pathogens. The bioactive metabolite from *Streptomyces cyaneofuscatus* M-169 showed significant inhibition of Gram-positive bacteria with MIC value of 0.03 µg/mL [32]. It has been shown that actinobacterial isolates from Western Ghats exhibit antimicrobial activity. The *Streptomyces* species from Agumbe [33], *Streptomyces* sp. RAMPP-065 from Kudremukh [34], and *Streptomyces* sp. GOS1 isolated from Western Ghats of Agumbe, Karnataka [35] exhibited remarkable antimicrobial activity. In earlier studies, *Streptomyces* species isolated from Kodachadri were found to possess antifungal activity [33]. Each extract was evaluated for scavenging activity against DPPH and ABTS radicals for antioxidant activity. Crude extract of strain SCA11 showed potent scavenging activity against DPPH radicals with IC_{50} (µg/mL) 30.91 ± 0.25, whereas SCA13 showed remarkable scavenging activity against ABTS radicals with IC_{50} (µg/mL) 37.91 ± 0.17. Trolox as a standard showed significant activity with IC_{50} (µg/mL) 11.07 ± 0.06 and 9.87 ± 0.01 against DPPH and ABTS radicals, respectively (Table 5). Similar studies were carried out for the detection of compounds with antioxidant activity from *Streptomyces* spp., dihydroherbimycin A [36], 5-(2,4-dimethylbenzyl)pyrrolidin-2-one [37]. Ser et al. [38] reported antioxidant activity of pyrrolo[1,2-a]pyrazine-1,4-dione, hexahydro extracted from *Streptomyces mangrovisoli*, a novel *Streptomyces* species isolated from a mangrove forest in Malaysia. Narendhran et al. [39] successfully reported antioxidant potential of phenol, 2,4-bis(1,1-dimethylethyl) in *Streptomyces cavouresis* KUV39 isolated from vermicompost samples. Tian et al. [40] reported antioxidant, antifungal, and antibacterial activity of *p*-Terphenyls isolated from halophilic actinobacteria *Nocardiopsis gilva* YIM 90087. Current findings indicated the bioactive potential of actinobacterial isolates from Western Ghats region of Karnataka. The isolates were found to possess significant antimicrobial activity against Gram-positive MRSA bacteria, Gram-negative bacteria, and fungal pathogens. They also exhibited potent scavenging activity against DPPH and ABTS radicals suggesting their antioxidant potential. It is anticipated that findings of the study will be useful in the discovery of novel species of actinobacteria for a potential source of bioactive compounds from underexplored environments.

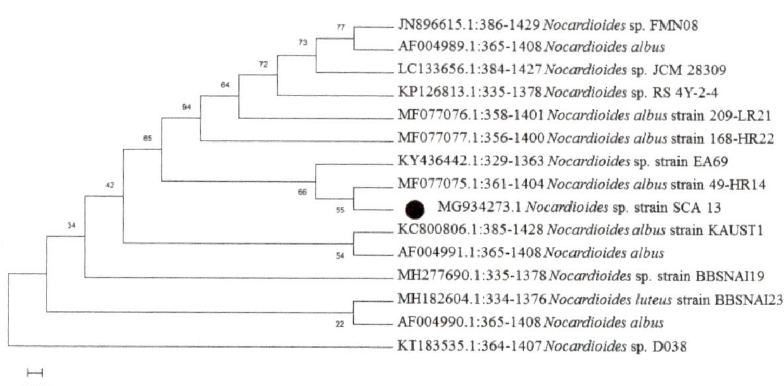

Figure 4. Phylogenetic analysis of isolate *Nocardioides* sp. strain SCA13. Neighbor-joining phylogenetic tree showing evolutionary relationship of selected isolate based on 16S r-RNA sequence alignments. Bootstrap values at the nodes indicate collated values based on 1000 resampled datasets. Bar indicates 0.001 substitutions per nucleotide position.

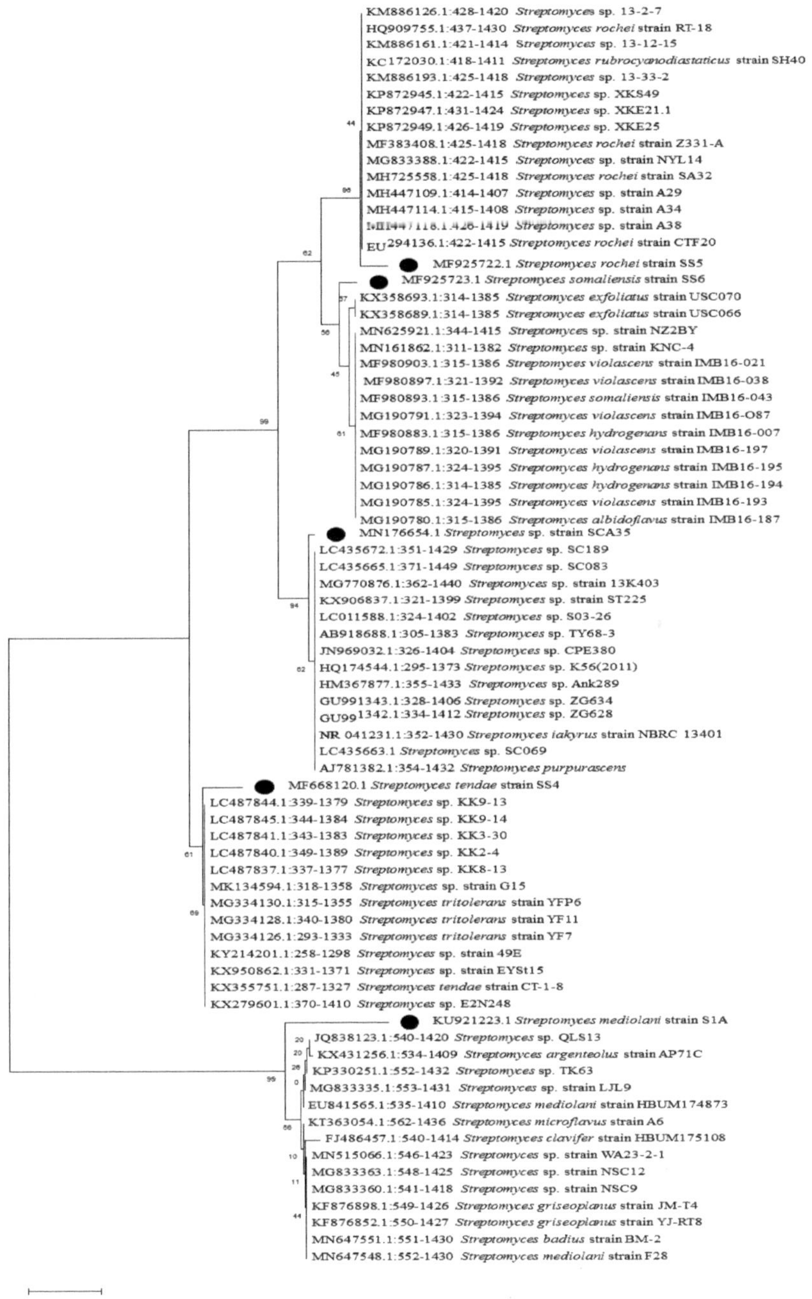

Figure 5. Neighbor-joining phylogenetic tree showing evolutionary relationship between isolates S1A, SS4, SS5, SS6, and SCA35 based on 16S r-RNA sequence alignments. Bootstrap values at the nodes indicate collated values based on 1000 resampled datasets. Bar indicates 0.005 substitutions per nucleotide position.

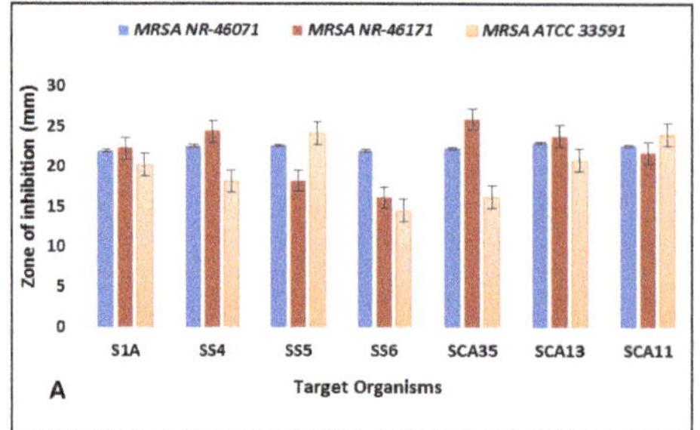
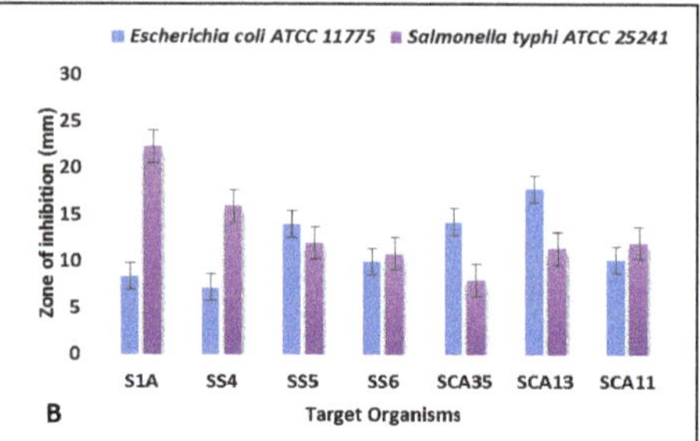
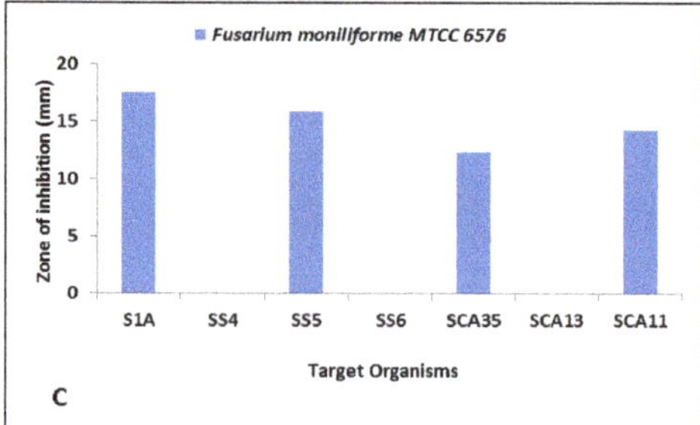

Figure 6. Antimicrobial activity of actinobacterial isolates against (**A**) MRSA ATCC NR 46071, MRSA ATCC NR 46171, and MRSA ATCC 33591 (**B**) *Escherichia coli* ATCC 11775 and *Salmonella typhi* ATCC 25241, (**C**) *Fusarium moniliforme* MTCC 6576.

Table 4. Minimum inhibitory concentration (MIC in µg/mL) of actinobacterial isolates against pathogenic test organisms.

Organisms	Minimum Inhibition Concentration (µg/mL)					
	Test Organisms					
	MRSA ATCC NR-46071	MRSA ATCC NR-46171	MRSA ATCC 33591	*Salmonella typhi* ATCC 25241	*Escherichia coli* ATCC 11775	*Fusarium moniliforme* MTCC 6576
Streptomyces Sp. S1A	15.62	15.62	62.5	125	>500	62.5
Streptomyces Sp. SS4	62.5	31.25	125	250	>500	-
Streptomyces Sp. SS5	31.25	125	15.62	250	250	125
Streptomyces Sp. SS6	31.25	62.5	62.5	250	125	-
Streptomyces Sp. SCA35	62.5	15.62	125	>500	125	250
Nocardiopsis Sp. SCA11	31.25	15.62	31.25	250	>500	125
Nocardioides Sp. SCA13	62.5	62.5	125	250	125	-

Table 5. Comparison of IC$_{50}$ (µg/mL) of crude extracts and trolox for DPPH and ABTS radical scavenging activity.

Isolates	IC$_{50}$ (µg/mL) Crude Extracts	
	DPPH	**ABTS**
Streptomyces Sp. S1A	189.40 ± 0.12	156.81 ± 0.06
Streptomyces Sp. SS4	98.29 ± 0.32	123.48 ± 0.13
Streptomyces Sp. SS5	86.45 ± 0.04	114.87 ± 0.29
Streptomyces Sp. SS6	114.15 ± 0.03	164.04 ± 0.07
Streptomyces Sp. SCA35	65.86 ± 0.49	49.11 ± 0.73
Nocardiopsis Sp. SCA11	30.91 ± 0.25	48.24 ± 0.30
Nocardioides Sp. SCA13	42.30 ± 0.10	37.91 ± 0.17
Trolox (Standard)	11.07 ± 0.06	9.87 ± 0.01

4. Conclusions

The present study was successful in determining the diversity and bioactive potential of actinobacterial isolates from Western Ghats region of Karnataka. Relatively underexplored forest regions of Western Ghats of Karnataka are found to be promising resources for the discovery of natural bioactive metabolites. The isolates showed significant antimicrobial activity against pathogenic Gram-positive MRSA bacteria, Gram-negative bacteria, and fungi. They were also found to possess antioxidant potential. Our studies encourage the exploration of diverse ecosystems for the isolation of new species for novel and biologically active compounds. Further studies are under progress to purify and characterize the crude extracts that may result in the economic production of bioactive compounds for pharmaceutical applications.

Supplementary Materials: Supplementary materials can be found at http://www.mdpi.com/2076-2607/8/2/225/s1.

Author Contributions: Conceptualization, S.S., R.R.V., W.J. and M.S.; Methodology, S.S. and R.R.V.; Formal analysis, S.S. and R.R.V.; Investigation, S.S., Original Draft Preparation, S.S. and R.R.V.; Writing—Reviewing and Editing, S.S., R.R.V., W.J. and M.S.; Visualization, S.S. and R.R.V.; Supervision, R.R.V. All authors have read and agreed to the published version of the manuscript.

Funding: This research was funded by India Council of Medical Research, New Delhi (ICMR-SRF/AMR/FELLOWSHIPS/5/2019-ECD-II, dated: 27.06.2019).

Acknowledgments: The authors would like to acknowledge the facilities provided by University of Mysore. The authors gratefully acknowledge the financial support grant to Mr. Saket Siddharth, ICMR-SRF from India Council of Medical Research, New Delhi (ICMR-SRF/AMR/FELLOWSHIPS/5/2019-ECD-II, dated: 27.06.2019).

Conflicts of Interest: The authors declare that they have no conflict of interest. The funders had no role in the design of the study; in the collection, analyses, or interpretation of data; in the writing of the manuscript, or in the decision to publish the results.

References

1. Liu, H.; Xiao, L.; Wei, J.; Schmitz, J.C.; Liu, M.; Wang, C.; Cheng, L.; Wu, N.; Chen, L.; Zhang, Y.; et al. Identification of *Streptomyces* sp. nov. WH26 producing cytotoxic compounds isolated from marine solar saltern in China. *World J. Microbiol. Biotechnol.* **2013**, *29*, 1271–1278. [CrossRef]

2. David, M.Z.; Daum, R.S. Community-associated methicillin-resistant *Staphylococcus aureus*: Epidemiology and clinical consequences of an emerging epidemic. *Clin. Microbiol. Rev.* **2010**, *23*, 616–687. [CrossRef] [PubMed]

3. Chalasani, A.G.; Dhanarajan, G.; Nema, S.; Sen, R.; Roy, U. An antimicrobial metabolite from *Bacillus* sp.: Significant activity against pathogenic bacteria including multidrug-resistant clinical strains. *Front. Microbiol.* **2015**, *6*, 1335. [CrossRef] [PubMed]

4. Torres, M.A.; Jones, J.D.; Dangl, J.L. Reactive oxygen species signalling in response to pathogens. *Plant. Physiol.* **2006**, *141*, 373–378. [CrossRef] [PubMed]

5. Giacco, F.; Brownlee, M. Oxidative stress and diabetic complications. *Circ. Res.* **2010**, *107*, 1058–1070. [CrossRef] [PubMed]

6. Fearon, I.M.; Faux, S.P. Oxidative stress and cardiovascular disease: Novel tools give (free) radical insight. *J. Mol. Cell. Cardiol.* **2009**, *47*, 372–381. [CrossRef] [PubMed]

7. Barnham, K.J.; Masters, C.L.; Bush, A.I. Neurodegenerative diseases and oxidative stress. *Nat. Rev. Drug Discov.* **2004**, *3*, 205–214. [CrossRef]

8. Zhang, H.; Tsao, R. Dietary polyphenols, oxidative stress and antioxidant and anti-inflammatory effects. *Curr. Opin. Food Sci.* **2016**, *8*, 33–42. [CrossRef]

9. Carocho, M.; Ferreira, I.C. A review on antioxidants, prooxidants and related controversy: Natural and synthetic compounds, screening and analysis methodologies and future perspectives. *Food Chem. Toxicol.* **2013**, *51*, 15–25. [CrossRef]

10. Thompson, C.C.; Kruger, R.H.; Thompson, F.L. Unlocking marine biotechnology in the developing world. *Trends Biotechnol.* **2017**, *35*, 1119–1121. [CrossRef]

11. Berdy, J. Thoughts and facts about antibiotics: Where we are now and where we are heading. *J. Antibiot.* **2012**, *65*, 385–395. [CrossRef] [PubMed]

12. Qin, S.; Xing, K.; Jiang, J.H.; Xu, L.H.; Li, W.J. Biodiversity, bioactive natural products and biotechnological potential of plant-associated endophytic actinobacteria. *Appl. Microbiol. Biotechnol.* **2011**, *89*, 457–473. [CrossRef] [PubMed]

13. Manivasagan, P.; Venkatesan, J.; Sivakumar, K.; Kim, S.K. Pharmaceutically active secondary metabolites of marine actinobacteria. *Microbiol. Res.* **2013**, *169*, 262–278. [CrossRef] [PubMed]

14. Nagpure, A.; Choudhary, B.; Kumar, S.; Gupta, R.K. Isolation and characterization of chitinolytic *Streptomyces* sp. MT7 and its antagonism towards wood-rotting fungi. *Ann. Microbiol.* **2014**, *64*, 531–541. [CrossRef]

15. Che, Q.; Zhu, T.; Qi, X.; Mandi, A.; Mo, X.; Li, J. Hybrid isoprenoids from a reed rhizosphere soil derived actinomycete *Streptomyces* sp. CHQ-64. *Org. Lett.* **2012**, *14*, 3438–3441. [CrossRef]

16. Ibeyaima, A.; Rana, J.; Dwivedi, A.K.; Saini, N.; Gupta, S.; Sarethy, I.P. *Pseudonocardiaceae* sp. TD-015 from the Thar Desert, India: Antimicrobial activity and identification of antimicrobial compounds. *Curr. Bioact. Compd.* **2017**, *14*, 112–118. [CrossRef]

17. Gautham, S.A.; Shobha, K.S.; Onkarappa, R.; Kekuda, P.T.R. Isolation, characterization and antimicrobial potential of *Streptomyces* species from Western Ghats of Karnataka, India. *Res. J. Pharm. Technol.* **2012**, *5*, 233–238.

18. Nampoothiri, M.K.; Ramkumar, B.; Pandey, A. Western Ghats of India: Rich source of microbial diversity. *J. Sci. Ind. Res.* **2013**, *72*, 617–623.

19. Ganesan, P.; Anand, S.; Sivanandhan, S.; David, R.H.A.; Paulraj, M.G.; Al-Dhabi, N.A.; Ignacimuthu, S. Larvicidal, ovicidal and repellent activities of *Streptomyces enissocaesilis* (S12-17) isolated from Western Ghats of Tamil Nadu, India. *J. Entomol.* **2018**, *6*, 1828–1835.

20. Ganesan, P.; Reegan, A.D.; David, R.H.A.; Gandhi, M.R.; Paulraj, M.G.; Al-Dhabi, N.A.; Ignacimuthu, S. Antimicrobial activity of some actinomycetes from Western Ghats of Tamil Nadu, India. *Alexandria J.* **2017**, *53*, 101–110. [CrossRef]

21. Siddharth, S.; Vittal, R.R. Evaluation of antimicrobial, enzyme inhibitory, antioxidant and cytotoxic activities of partially purified volatile metabolites of marine *Streptomyces* sp. S2A. *Microorganisms* **2018**, *6*, 72. [CrossRef]

22. Siddharth, S.; Vittal, R.R. Isolation and characterization of bioactive compounds with antibacterial, antioxidant and enzyme inhibitory activities from marine-derived rare actinobacteria, *Nocardiopsis* sp. SCA21. *Microb. Pathog.* **2019**, 103775. [CrossRef] [PubMed]

23. Ser, H.L.; Tan, L.T.; Palanisamy, U.D.; Abd Malek, S.N.; Yin, W.F.; Chan, K.G. *Streptomyces antioxidans* sp. nov., a novel mangrove soil actinobacterium with antioxidative and neuroprotective potentials. *Front. Microbiol.* **2016**, *7*, 899. [CrossRef] [PubMed]

24. Ceylan, O.; Okmen, G.; Ugur, A. Isolation of soil *Streptomyces* as source of antibiotics active against antibiotic-resistant bacteria. *EurAsian J. Biosci.* **2008**, *2*, 73–82.

25. Junaid, S.; Dileep, N.; Rakesh, K.N.; Kekuda, P.T.R. Antimicrobial and antioxidant efficacy of *Streptomyces* species SRDP-TK-07 isolated from a soil of Talakaveri, Karnataka, India. *Pharmanest* **2013**, *4*, 736–750.

26. Bennur, T.; Kumar, A.R.; Zinjarde, S.; Javdekar, V. *Nocardiopsis* species as potential sources of diverse and novel extracellular enzymes. *Appl Microbiol Biotechnol.* **2014**, *98*, 9173–9185. [CrossRef]

27. Fiedler, H.P.; Bruntner, C.; Bull, A.T. Marine actinomycetes as a source of novel secondary metabolites. *Ant. V. Leeuw.* **2004**, *87*, 37–42. [CrossRef]

28. Sengupta, S.; Pramanik, A.; Ghosh, A.; Bhattacharyya, M. Antimicrobial activities of actinomycetes isolated from unexplored regions of Sundarbans mangrove ecosystem. *BMC Microbiol.* **2015**, *15*, 170. [CrossRef]

29. Satheeja, S.; Jebakumar, S.R.D. Phylogenetic analysis and antimicrobial activities of *Streptomyces* isolates from mangrove sediment. *J. Basic Microbiol.* **2011**, *51*, 71–79. [CrossRef]

30. Vu, H.T.; Nguyen, D.T.; Nguyen, H.Q. Antimicrobial and cytotoxic properties of bioactive metabolites produced by *Streptomyces cavourensis* YBQ59 isolated from *Cinnamomum cassia Prels* in Yen Bai province of Vietnam. *Curr. Microbiol.* **2018**, *75*, 1247–1255. [CrossRef]

31. Dashti, Y.; Grkovic, T.; Abdelmohsen, U.R. Production of induced secondary metabolites by a co-culture of sponge-associated actinomycetes, *Actinokineospora* sp. EG49 and *Nocardiopsis* sp. RV163. *Mar. Drugs* **2014**, *12*, 3046–3059. [CrossRef] [PubMed]

32. Rodríguez, V.; Martín, J.; Sarmiento-Vizcaíno, A. Anthracimycin B, a potent Antibiotic against gram-positive bacteria isolated from cultures of the deep-sea actinomycetes *Streptomyces cyaneofuscatus* M-169. *Mar. Drugs* **2018**, *16*, 406. [CrossRef] [PubMed]

33. Shobha, K.S.; Onkarappa, R. In vitro susceptibility of C. albicans and C. neoformens to potential metabolites from *Streptomycetes*. *Indian J. Microbiol.* **2011**, *51*, 445–449. [CrossRef] [PubMed]

34. Manasa, M.; Poornima, G.; Abhipsa, V.; Rekha, C.; Prashith, K.T.R.; Onkarappa, R.; Mukunda, S. Antimicrobial and antioxidant potential of Streptomyces sp. RAMPP-065 isolated from Kudremukh soil, Karnataka, India. *Sci. Technol. Arts Res. J.* **2012**, *1*, 39–44. [CrossRef]

35. Gautham, S.A.; Onkarappa, R. Pharmacological activities of metabolite from *Streptomyces fradiae* strain GOS1. *Int. J. Chem.* **2013**, *11*, 583–590.

36. Chang, H.B.; Kim, J.-H. Antioxidant properties of dihydroherbimycin A from a newly isolated *Streptomyces* sp. *Biotechnol. Lett.* **2007**, *29*, 599–603. [CrossRef]

37. Saurav, K.; Kannabiran, K. Cytotoxicity and antioxidant activity of 5-(2, 4-dimethylbenzyl) pyrrolidin-2-one extracted from marine *Streptomyces* VITSVK5 spp. *Saudi J. Biol. Sci.* **2012**, *19*, 81–86. [CrossRef]

38. Ser, H.L.; Palanisamy, U.D.; Yin, W.F.; Malek, S.N.A.; Chan, K.G.; Goh, B.H. Presence of antioxidative agent, Pyrrolo[1,2-a] pyrazine-1,4-dione, hexahydro-in newly isolated *Streptomyces mangrovisoli* sp. nov. *Front. Microbiol.* **2015**, *6*, 854. [CrossRef]

39. Narendhran, S.; Rajiv, P.; Vanathi, P.; Sivaraj, R. Spectroscopic analysis of bioactive compounds from Streptomyces cavouresis kuv39: Evaluation of antioxidant and cytotoxicity activity. *Int. J. Pharm. Pharm. Sci.* **2014**, *6*, 319–322.

40. Tian, S.Z.; Pu, X.; Luo, G.; Zhao, L.X.; Xu, L.H.; Li, W.J.; Luo, Y. Isolation and characterization of new p-Terphenyls with antifungal, antibacterial, and antioxidant activities from halophilic actinomycete *Nocardiopsis gilva* YIM 90087. *J. Agric. Food Chem.* **2013**, *61*, 3006–3012. [CrossRef]

 *microorganisms*

Article

Taxonomic Characterization, and Secondary Metabolite Analysis of *Streptomyces triticiradicis* sp. nov.: A Novel Actinomycete with Antifungal Activity

Zhiyin Yu [1,2,†], Chuanyu Han [1,†], Bing Yu [1], Junwei Zhao [1], Yijun Yan [2], Shengxiong Huang [2], Chongxi Liu [1,2,*] and Wensheng Xiang [1,3,*]

[1] Key Laboratory of Agricultural Microbiology of Heilongjiang Province, Northeast Agricultural University, Harbin 150030, China; zhiyinyu@yeah.net (Z.Y.); chuanyuhan@yeah.net (C.H.); yubing95yb@163.com (B.Y.); guyan2080@126.com (J.Z.)
[2] State Key Laboratory of Phytochemistry and Plant Resources in West China, Kunming Institute of Botany, Chinese Academy of Sciences, Kunming 650201, China; yanyijun@mail.kib.ac.cn (Y.Y.); sxhuang@mail.kib.ac.cn (S.H.)
[3] State Key Laboratory for Biology of Plant Diseases and Insect Pests, Institute of Plant Protection, Chinese Academy of Agricultural Sciences, Beijing 100193, China
* Correspondence: xizi-ok@163.com (C.L.); xiangwensheng@neau.edu.cn (W.X.)
† These authors have contributed equally to this work.

Received: 14 November 2019; Accepted: 2 January 2020; Published: 5 January 2020

Abstract: The rhizosphere, an important battleground between beneficial microbes and pathogens, is usually considered to be a good source for isolation of antagonistic microorganisms. In this study, a novel actinobacteria with broad-spectrum antifungal activity, designated strain NEAU-H2[T], was isolated from the rhizosphere soil of wheat (*Triticum aestivum* L.). 16S rRNA gene sequence similarity studies showed that strain NEAU-H2[T] belonged to the genus *Streptomyces*, with high sequence similarities to *Streptomyces rhizosphaerihabitans* NBRC 109807[T] (98.8%), *Streptomyces populi* A249[T] (98.6%), and *Streptomyces siamensis* NBRC 108799[T] (98.6%). Phylogenetic analysis based on 16S rRNA, *atpD*, *gyrB*, *recA*, *rpoB*, and *trpB* gene sequences showed that the strain formed a stable clade with *S. populi* A249[T]. Morphological and chemotaxonomic characteristics of the strain coincided with members of the genus *Streptomyces*. A combination of DNA–DNA hybridization results and phenotypic properties indicated that the strain could be distinguished from the abovementioned strains. Thus, strain NEAU-H2[T] belongs to a novel species in the genus *Streptomyces*, for which the name *Streptomyces triticiradicis* sp. nov. is proposed. In addition, the metabolites isolated from cultures of strain NEAU-H2[T] were characterized by nuclear magnetic resonance (NMR) and mass spectrometry (MS) analyses. One new compound and three known congeners were isolated. Further, genome analysis revealed that the strain harbored diverse biosynthetic potential, and one cluster showing 63% similarity to natamycin biosynthetic gene cluster may contribute to the antifungal activity. The type strain is NEAU-H2[T] (= CCTCC AA 2018031[T] = DSM 109825[T]).

Keywords: *Streptomyces triticiradicis* sp. nov.; antifungal activity; rhizosphere soil; new compound; genome analysis

---

1. Introduction

It is well known that plant pathogenic fungi can cause a tremendous loss of global agricultural production [1]. Despite synthetic fungicides being effective and playing an indispensable role against pathogenic fungi, the available antifungal agents are far from satisfactory as a result of several drawbacks, such as severe drug resistance, drug-related toxicity, and many other problems [2]. Therefore, novel antifungal agents and antagonistic microorganisms are needed to effectively control the fungal diseases of agricultural crops. Natural products and their derivatives, in particular secondary metabolites derived from *Streptomyces*, have always been valuable sources for lead discovery in medicinal and agricultural chemistry because their novel scaffolds can provide new modes of action [3,4]. Members of the genus *Streptomyces* species are gram-positive, filamentous, and sporulating actinobacteria containing a number of biosynthetic gene clusters, indicating their potential ability to produce large numbers of secondary metabolites with diverse biological activities [5,6], and they represent the source of 75% of clinically useful antibiotics presently available [7]. Many *Streptomyces* species have been successfully developed as commercial biofungicides based on *Streptomyces griseoviridis* [8]. Thus, *Streptomyces* are still an attractive and indispensable resource for drug discovery.

The rhizosphere is an environment where pathogenic and beneficial microorganisms constitute a major influential force on plant growth and health, which differs from the bulk soil [9]. Plants not only provide nutrients for microbial growth but also change the microbial diversity and increase the numbers of bioactive microorganisms in the rhizosphere. The rhizosphere provides an excellent place for pursuing actinobacteria producing novel antibiotics [10,11]. During our search for antagonistic actinobacteria from the rhizosphere soil of wheat (*Triticum aestivum* L.), the strain NEAU-H2^T was isolated, which showed broad inhibitory activities against phytopathogenic fungi. Based on the polyphasic taxonomy analysis, this strain was classified as representative of a novel species in the genus *Streptomyces*. In addition, the secondary metabolites of this strain were investigated by spectroscopic and genomic analyses.

2. Materials and Methods

2.1. Isolation of Actinobacterial Strain

Strain NEAU-H2^T was isolated from the rhizosphere soil of wheat (*Triticum aestivum* L.) collected from Zhumadian, Henan Province, Central China (32°98′ N, 114°02′ E). The root sample was air-dried for 24 h at room temperature, and then the surface soil was shaken off gently. After, the sample was shaken at 250 rpm in 100 mL of sterile water with glass beads for 30 min at 20 °C and then filtered with a single layer of gauze to obtain the rhizosphere soil suspension. The suspension was serially diluted and spread on cellulose-proline agar (CPA) [12] supplemented with cycloheximide (50 mg·L^{-1}) and nalidixic acid (20 mg·L^{-1}), and cultured at 28 °C for 3 weeks. Strain NEAU-H2^T was isolated and purified on the International *Streptomyces* Project (ISP) medium 3 [13], and maintained as glycerol suspensions (20%, *v/v*) at −80 °C.

2.2. Morphological and Biochemical Characteristics

Gram staining was performed by the Hucker method [14]. Morphological characteristics were observed by light microscopy (Nikon ECLIPSE E200, Nikon Corporation, Tokyo, Japan) and scanning electron microscopy (Hitachi SU8010, Hitachi Co., Tokyo, Japan) using cultures grown on ISP 3 agar at 28 °C for 2 weeks. Samples for scanning electron microscopy were prepared as described by Jin et al. [15]. Cultural characteristics were determined on the ISP media 1–7 [13], Bennett's agar [16], Czapek's agar [17], and Nutrient agar [18] after 2 weeks at 28 °C. The color of substrate mycelium, aerial mycelium, and diffusible pigment on the different tested media were determined using color chips from the ISCC-NBS color charts [19]. Temperature tolerance for growth was evaluated at 4, 10, 15, 20, 25, 28, 35, 37, 40, and 45 °C on ISP 3 agar after incubation for 2 weeks. The pH range for growth (pH 3.0–12.0, at intervals of 1.0 pH unit) was tested in GY broth [20] using the buffer system described by Zhao et al. [21] and NaCl tolerance (0%–15% (*w/v*) in 1% intervals) for growth was determined after 2 weeks growth in GY broth at 28 °C with shaking at 250 rpm. Hydrolysis of Tweens (20, 40, and 80) and production of urease were tested according to the method of Smibert and Krieg [14]. The utilization of sole carbon and nitrogen sources were determined following the methods of Gordon et al. [22]. The decomposition of cellulose, hydrolysis of starch, coagulation of milk, aesculin, reduction of nitrate, liquefaction of gelatin, and production of H$_2$S were examined as described previously [23].

2.3. Chemotaxonomic Analysis

The freeze-dried cells used for chemotaxonomic analysis were obtained from cultures grown in GY medium on a rotary shaker for seven days at 28 °C. Cells were acquired and washed twice with sterile distilled water and freeze-dried. The isomer of diaminopimelic acid (DAP) in the cell wall hydrolysates was derivatized and analyzed by HPLC (Agilent TC-C_{18} Column, 250 × 4.6 mm, i.d. 5 μm) with a mobile phase consisting of acetonitrile/phosphate buffer (0.05 mol·L^{-1}, pH 7.2, 15:85, *v/v*), and a flow rate of 0.5 mL·min^{-1} at a column temperature of 28 °C [24]. An Agilent G1321A fluorescence detector was used to detect the peak with a 365 nm excitation and 455 nm longpass emission filters. The whole-cell sugars were analyzed according to Lechevalier [25]. The polar lipids were extracted and examined by two-dimensional TLC (thin-layer chromatography, Qingdao Marine Chemical Inc., Qingdao, China) and identified according to the method of Minnikin et al. [26]. Menaquinones were extracted and purified from freeze-dried biomass following the methods of Collins [27]. The extracts were analyzed by HPLC-UV (Agilent Extend-C_{18} Column, 150 × 4.6 mm, i.d. 5 μm, 1.0 mL·min^{-1} acetonitrile: iso-propyl alcohol = 60:40) at 270 nm [28]. Fatty acid methyl esters were performed by GC-MS according to the method of Xiang et al. [29] and identified with the NIST 14 database.

2.4. Phylogenetic Analysis

Strain NEAU-H2[T] was grown on ISP 3 agar plates for one week at 28 °C. Then, it was inoculated into 250-mL baffle Erlenmeyer flasks containing 50 mL of GY broth and cultivated for two days at 28 °C with shaking at 250 rpm. After that, the total DNA was extracted according to the lysozyme-sodium dodecyl sulfate-phenol/chloroform method [30]. The primers and procedure for PCR amplification were carried out as described by Yi et al. [31]. The PCR product was purified and cloned into the vector pMD19-T (Takara, Shiga, Japan) and sequenced using an Applied Biosystems DNA sequencer (model 3730XL, Applied Biosystems Inc., Foster City, CA, USA). Almost full-length 16S rRNA gene sequence (1519 bp) was multiply aligned in MEGA (Molecular Evolutionary Genetics Analysis) using the Clustal W algorithm and trimmed manually if necessary. Phylogenetic trees were constructed with neighbor-joining [32] and maximum likelihood [33] algorithms using MEGA software version 7.0 (Kumar S, Philly, PA, USA) [34]. The stability of the topology of the phylogenetic tree was assessed using the bootstrap method with 1000 repetitions [35]. A distance matrix was calculated using Kimura's two-parameter model [36]. All positions containing gaps and missing data were eliminated from the dataset (complete deletion option). The calculation of 16S rRNA gene sequence similarities between strains was carried out on the basis of pairwise alignment using the EzBioCloud server (https://www.ezbiocloud.net/) [37]. Phylogenetic relationships of strain NEAU-H2[T] were also confirmed using sequences for five individual housekeeping genes (*atpD*, *gyrB*, *recA*, *rpoB*, and *trpB*). These sequences of housekeeping genes of strain NEAU-H2[T] were obtained from the Whole Genome sequences. The sequences of each locus were aligned using the software package MEGA version 7.0 and trimmed manually at the same position before being used for further analysis. Phylogenetic analysis was performed as described above.

2.5. DNA–DNA Relatedness Tests

The total DNA was extracted according to the method in the Section 2.4. The harvested DNA was detected by agarose gel electrophoresis and quantified by a Qubit 2.0 Fluorometer (Thermo Scientific, Ashville, NC, USA). The Illumina Novaseq PE150 (Illumina, San Diego, CA, USA) platform was used to perform whole-genome sequencing. A-tailed, ligated to paired-end adaptors, and PCR-amplified samples with a 350-bp insert were used for the library construction at the Beijing Novogene Bioinformatics Technology Co., Ltd. Illumina PCR adapter reads and low-quality reads from the paired end were filtered with a quality control step by our own compling pipeline. All good-quality paired reads were assembled by the SOAP (Short Oligonucleotide Alignment Program) denovo [38,39] (https://github.com/aquaskyline) into a number of contigs. After that, the filter reads were handled by the next step of the gap closing.

Because of a lack of the whole genome sequence of strains *Streptomyces rhizosphaerihabitans* NBRC 109807^T and *Streptomyces siamensis* NBRC 108799^T, a DNA–DNA relatedness test was carried out as described by De Ley et al. [40] under consideration of modifications [41] with a model Cary 100 Bio UV/VIS-spectrophotometer (Hitachi U-3900, Hitachi Co., Tokyo, Japan) and a temperature controller. The DNA hybridization samples were diluted to OD_{260} around 1.0 using 0.1 × SSC (saline sodium citrate buffer), then sheared using a JY92-II ultrasonic cell disruptor (ultrasonic time 3s, interval time 4 s, 90 times). The DNA renaturation rates were measured in 2 × SSC at 70 °C. This experiment was repeated three times to calculate the average value. The DNA–DNA relatedness value was determined between the genomes of strain NEAU-H2^T and *Streptomyces populi* A249^T (PJOS01000000) using the genome-to-genome distance calculator (GGDC 2.0) at http://ggdc.dsmz.de [42]. Genome mining analysis was performed with antiSMASH (version 4.0, Blin K, Oxford, UK) [43].

2.6. In Vitro Antifungal Activity Test

Antifungal screening was performed against 10 different phytopathogenic fungi: *Sclerotinia sclerotiorum, Exserohilum turcicum, Colletotrichum orbiculare, Corynespora cassiicola, Rhizoctonia solani, Fusarium graminearum, Fusarium oxysporum, Sphacelotheca reiliana, Curvularia lunata,* and *Helminthosporium maydis.* Ten phytopathogenic fungi were preserved in the Key Laboratory of Agricultural Microbiology within the Heilongjiang province, China. Antifungal activity of strain NEAU-H2^T was assessed using the dual culture plate assay [44]. The strain was point-inoculated at the margin of potato dextrose agar (PDA) [45] plates and cultivated for three days at 28 °C, after which a fresh mycelial PDA agar plug of the fungus was transferred into the opposite margin of the corresponding plate. Inhibition of hyphal growth of the fungus was recorded after incubated for seven days at 28 °C. The percentage inhibition rates were calculated using the formula: Inhibition rate (%) = Wi/W × 100%, where Wi is the width of inhibition and W is the width between the pathogen and actinobacteria. The assay was repeated three times and the average was calculated.

2.7. Isolation and Characterization of Secondary Metabolites

Strain NEAU-H2^T was grown on ISP 3 agar plates for five days at 28 °C. Then, it was inoculated into 250-mL baffle Erlenmeyer flasks containing 50 mL of tryptone soy broth (TSB) and cultivated for one day at 28 °C with shaking at 250 rpm. After that, aliquots (15 mL) of the culture were transferred into 1-L baffled Erlenmeyer flasks filled with 250 mL of the production medium (tryptone 0.1%, glucose 3%, beef extract 0.5%, 0.25% $CaCO_3$, 0.5% NaCl, 0.1% minor elements concentrate ($FeSO_4 \cdot 7H_2O$ 1.0 g, $CuSO_4 \cdot 5H_2O$ 0.45 g, $ZnSO_4 \cdot 7H_2O$ 1.0 g, $MnSO_4 \cdot 4H_2O$ 0.1 g, K_2MoO_4 0.1 g, distilled water 1 L), pH 7.2–7.4), and cultured at 30 °C for six days with shaking at 250 rpm.

The fermentation broth (25 L) was centrifuged (4000 rev/min, 20 min), and the supernatant was extracted with ethylacetate three times. The ethylacetate extract was evaporated under reduced pressure at temperatures within 40 °C to yield an oily crude extract (5.0 g). The mycelia were extracted with methanol (1 L) and then concentrated in vacuo to remove the methanol to yield the aqueous concentrate. The mycelia concentrate was extracted with ethylacetate (1 L × 3) to afford 1.0 g of crude extract after removing the ethylacetate. Both extracts displayed most of the similar secondary metabolites based on HPLC analyses. Thus, they were combined for further purification. The samples were applied to reverse-phase HPLC analysis eluted with a flow rate of 1 mL·min^{-1} over a 28 min gradient with water and methanol (T = 0 min, 10% methanol; T = 20.0 min, 100% methanol; T = 24.0 min, 100% methanol; T = 24.1 min, 10% methanol; T = 28.0 min, 10% methanol) and at 25 °C.

The crude extract (6.1 g) was subjected to silica gel CC using a successive elution of petroleum ether/ethylacetate (1:0, 10:1, 5:1, 1:1 and 0:1, *v/v*) to yield A–F fractions. Fraction D (petroleum ether/ethylacetate = 1:1, *v/v*) was subjected to semipreparative HPLC (YMC- Hydrosphere C_{18} column, 250 mm × 10 mm i.d., 5 μm, 0–20.0 min $CH_3OH:H_2O$ = 35:55, *v/v*, 3 mL/min) to afford **1** (t_R = 25.4 min,

2.2 mg) and **2** (t_R = 30.4 min, 2.6 mg). Compound **4** (t_R = 17.2 min, 5.5 mg) was obtained from the fraction C (petroleum ether/ethylacetate = 5:1, *v/v*) by semipreparative HPLC (YMC-Triart C_{18} column, 250 mm × 10 mm i.d., 5 μm, $CH_3OH:H_2O$ = 25:75, *v/v*, 3 mL/min). Fraction E (petroleum ether/ ethylacetate = 0:1, *v/v*) was further purified by semipreparative HPLC (YMC-Hydrosphere C_{18} column, 250 mm × 10 mm i.d., 5 μm, 0–15.0 min, $CH_3OH:H_2O$ = 25:75; 15.1–30 min, $CH_3OH:H_2O$ = 35:65, *v/v*, 3 mL/min) to give compound **3** (t_R = 21.4 min, 2.0 mg).

NMR spectra were recorded in methanol-d4 or DMSO-d6 using a Bruker AVANCE III-600 or a Bruker AVANCE III-400 spectrometer (Bruker Corp., Karlsruhe, Germany) and TMS was used as the internal standard. HR-ESI-MS data were obtained using an Agilent G6230 Q-TOF mass instrument (Agilent Technologies Inc. Santa Clara, CA, USA). Thin-layer chromatography (TLC) was performed using precoated silica gel GF254 plates (Qingdao Marine Chemical Inc., Qingdao, China), and spots were visualized by UV light (254 nm) and colored by iodine, or by spraying heated silica gel plates with 10% H_2SO_4 in ethanol. Semipreparative HPLC was conducted on a HITACHI Chromaster system (Hitachi-DAD, Tokyo, Japan).

2.8. In Vitro Antifungal Activity Test of Compounds

The fungi were retrieved from the storage tube and cultured for seven days at 28 °C on PDA. All fungi were further cultured for one week to get new mycelium for the antifungal assays in PDA at 28 °C. The medium was mixed with the pathogenic fungi suspension at about 45 °C, ensuring the abundance of the strains was about 10^8 cfu/mL. Next, the mixture was poured on 9-cm Petri dishes. The tested compounds were dissolved in DMSO at a concentration of 2 mg/mL. Each filter paper (5 mm in diameter) was impregnated with 10 μL of the tested compounds. The inoculated Petri dishes were cultured for 3 to 4 d at 28 °C. DMSO served as a blank control. The assay was measured three times. The inhibition diameter was measured by the cross bracketing method [46].

3. Results and Discussion

3.1. Polyphasic Taxonomic Characterization of NEAU-H2^T

Morphological observation of the two-week culture of strain NEAU-H2^T grown on ISP 3 medium revealed that it had characteristics typical of the genus *Streptomyces* [47]. Aerial and substrate mycelium were well developed without fragmentation. Spiral spore chains with spiny surfaced spores (0.8–1.0 × 1.0–1.3 μm) were borne on the aerial mycelium (Figure 1). Strain NEAU-H2^T exhibited good growth on ISP 1–4, ISP 7, and Nutrition agar media; moderate growth on Bennett's and Czapek's agar media; and poor growth on ISP 5 and ISP 6 media. The colony colors varied from white to moderate yellow. A dark grayish olive pigment was produced on ISP 6 medium. The detailed cultural characteristics of strain NEAU-H2^T are shown in Table S1. Strain NEAU-H2^T was found to grow at a temperature range of 4 to 40 °C (optimum temperature 28 °C), pH 5 to 10 (optimum pH 7), and NaCl tolerance of 0% to 9% (optimum NaCl of 0% to 1%). The physiological and biochemical properties of strain NEAU-H2^T are given in Table 1 and the species description.

Table 1. Differential characteristics of strain NEAU-H2[T] and its most closely related *Streptomyces* species. Strains: 1, NEAU-H2[T]; 2, *S. rhizosphaerihabitans* NBRC 109807[T]; 3, *S. populi* A249[T]; 4, *S. siamensis* NBRC 108799[T]. All data are from this study except where marked. +, positive; –, negative. [‡] Data from Lee et al. [48]; [†] Data from Wang et al. [49]; [§] Data from Sripreechasak et al. [50].

Characteristic	1	2	3	4
Spore chain	Spiral	Straight [‡]	Straight [†]	Spiral [§]
Spore surface	Spiny	Hairy [‡]	Rough [†]	Smooth [§]
Growth temperature range (°C)	4–40	10–40	10–37	10–40
Growth pH range	5–10	5–11	6–12	5–11
NaCl tolerance range (*w/v*, %)	0–9.0	0–10.0	0–4.0	0–10.0
Cellulose decomposition	+	+	–	–
Gelatin liquefaction	–	+	+	–
Catalase production	–	+	+	–
H_2S production	+	–	+	+
Milk coagulation	+	+	+	–
Nitrate reduction	–	–	+	–
Starch hydrolysis	+	–	–	–
Tween 20 hydrolysis	+	+	+	–
Tween 80 hydrolysis	+	+	–	+
Nitrogen source utilization				
L-serine	–	+	+	+
L-threonine	+	+	–	+
L-tyrosine	–	+	+	+
Carbon source utilization				
L-arabinose	–	+	+	+
Dulcitol	–	–	–	+
meso-inositol	–	+	+	+
D-mannitol	–	+	–	+
L-rhamnose	+	–	+	–
D-ribose	–	+	–	+
D-sorbitol	–	+	–	+
D-xylose	–	+	+	–
Phospholipids *	DPG, PE, PL, PI, PIM	AL, DPG, GL, PE, PG, PI, PIM, 2PLs [‡]	AL, APL, DPG, 2Ls, PE, PL, PIM [†]	DPG, PE, PG, PI, PL [§]
Menaquinones	MK-9(H$_6$), MK-9(H$_8$), MK-9(H$_4$)	MK-9(H$_6$), MK-9(H$_8$) [‡]	MK-9(H$_6$), MK-9(H$_8$), MK-9(H$_2$), MK-9(H$_4$) [†]	MK-9(H$_6$), MK-9(H$_4$), MK-9(H$_8$) [§]
Whole cell-wall sugars	Glucose	Glucose, ribose [‡]	Xylose, galactose [†]	ND

* APL, aminophopholipid; DPG, diphosphatidylglycerol; PE, phosphatidylethanolamine; PI, phosphatidylinositol; PIM, phosphatidylinositol mannoside; AL, unidentified aminolipid; GL, unknown glycolipid; L, unidentified lipid; PL, unidentified phospholipid; MK, menaquinone; ND, no detection.

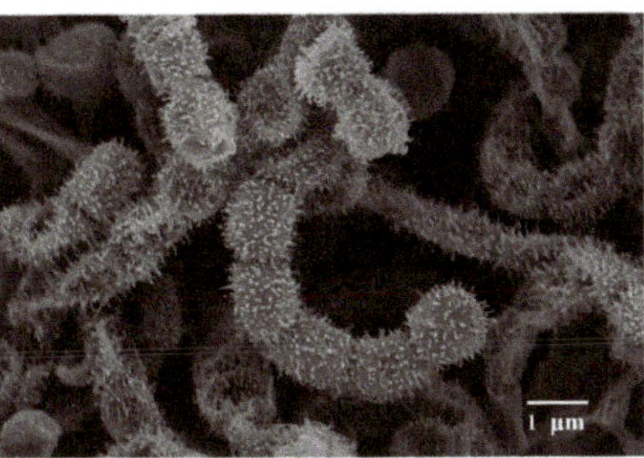

Figure 1. Scanning electron micrograph of strain NEAU-H2^T grown on International *Streptomyces* Project (ISP) medium 3 (ISP 3) for 2 weeks at 28 °C; Bar 1 µm.

Chemotaxonomic analyses revealed that strain NEAU-H2^T also exhibited the typical characteristics of the genus *Streptomyces* [47]. It contained LL-diaminopimelic acid as cell wall diamino acid, indicating that the strain is of cell wall chemotype I [51]. The whole-cell sugar was found to contain glucose. The phospholipid profile consisted of diphosphatidylglycerol (DPG), phosphatidylethanolamine (PE), phosphatidylinositol (PI), phosphatidylinositolmannosides (PIM), and an unidentified phospholipid (PL), corresponding to phospholipid type II [52] (Figure S1). The major cellular fatty acids (>10%) were iso-$C_{16:0}$ (21.6%), anteiso-$C_{15:0}$ (19.4%), iso-$C_{15:0}$ (16.9%), and anteiso-$C_{17:0}$ (13.0 %), which is fatty acid type IIc [53]; minor amounts of $C_{16:1}\omega7c$ (8.5 %), $C_{16:0}$ (7.5 %), iso-$C_{14:0}$ (7.2%), $C_{17:0}$ cyclo (2.1%), $C_{17:1}\omega8c$ (2.1%), $C_{18:0}$ (1.2%), and $C_{15:0}$ (0.5%) were also present. The menaquinones detected were MK-9(H_8) (57.5%), MK-9(H_6) (32.3%), and MK-9(H_4) (10.2%), which have been reported for most species of the genus *Streptomyces* [47].

EzBioCloud analysis suggests that strain NEAU-H2^T belongs to the genus *Streptomyces*. The novel strain shared the highest 16S rRNA gene sequence similarities with *S. rhizosphaerihabitans* NBRC 109807^T (98.8%), *S. populi* A249^T (98.6%), and *S. siamensis* NBRC 108799^T (98.6%). In the neighbor-joining phylogenetic tree based on 16S rRNA gene sequences, strain NEAU-H2^T formed a separate clade with *S. populi* A249^T (Figure 2), a relationship also recovered by the maximum likelihood algorithm (Figure S2). To further clarify the affiliation of strain NEAU-H2^T to closely related strains, phylogenetic trees were constructed from the concatenated sequence alignment of the five housekeeping genes based on the neighbor-joining and maximum likelihood algorithms (Figure 3 and Figure S3), which showed the same topology as the 16S rRNA gene tree. Furthermore, the concatenated sequences of *atp*D-*gyr*B-*rec*A-*rpo*B-*trp*B were used to calculate pairwise distances well above 0.007 (Table S2) for the related species, which was considered to be the threshold for species determination [54]. Based on the 16S rRNA gene sequence similarities and phylogenetic trees, *S. rhizosphaerihabitans* NBRC 109807^T, *S. populi* A249^T, and *S. siamensis* NBRC 108799^T were selected as the closely related strains for subsequent comparative analysis.

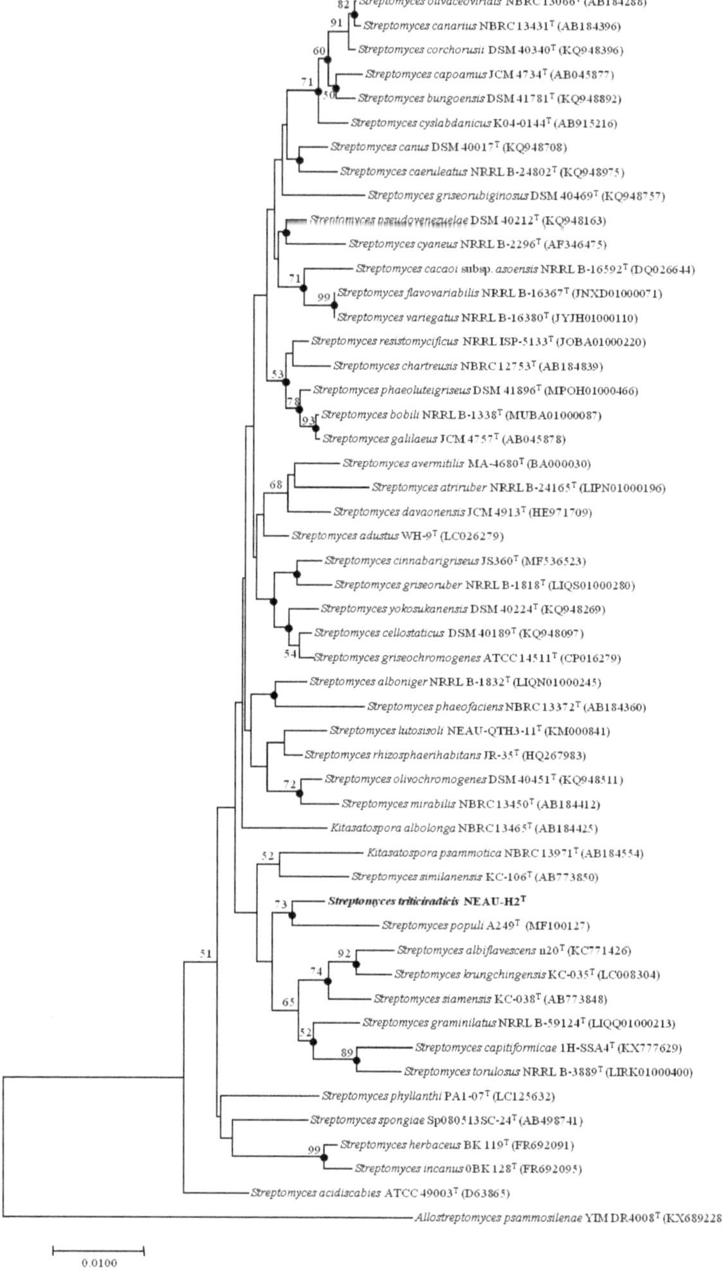

0.0100

Figure 2. Neighbor-joining tree based on 16S rRNA gene sequences (1418 bp) showing the relationship of strain NEAU-H2^T (in bold) with related taxa, which are the top 50 type strains of *Streptomyces* species of gene sequence similarities based on analysis using EzTaxon-e. Filled circles indicate branches that were also recovered using the maximum likelihood methods. Only bootstrap values above 50% (percentages of 1000 replications) are indicated. *Allostreptomyces psammosilenae* YIM DR4008^T (KX689228) was used as an outgroup. Bar, 0.01 nucleotide substitutions per site.

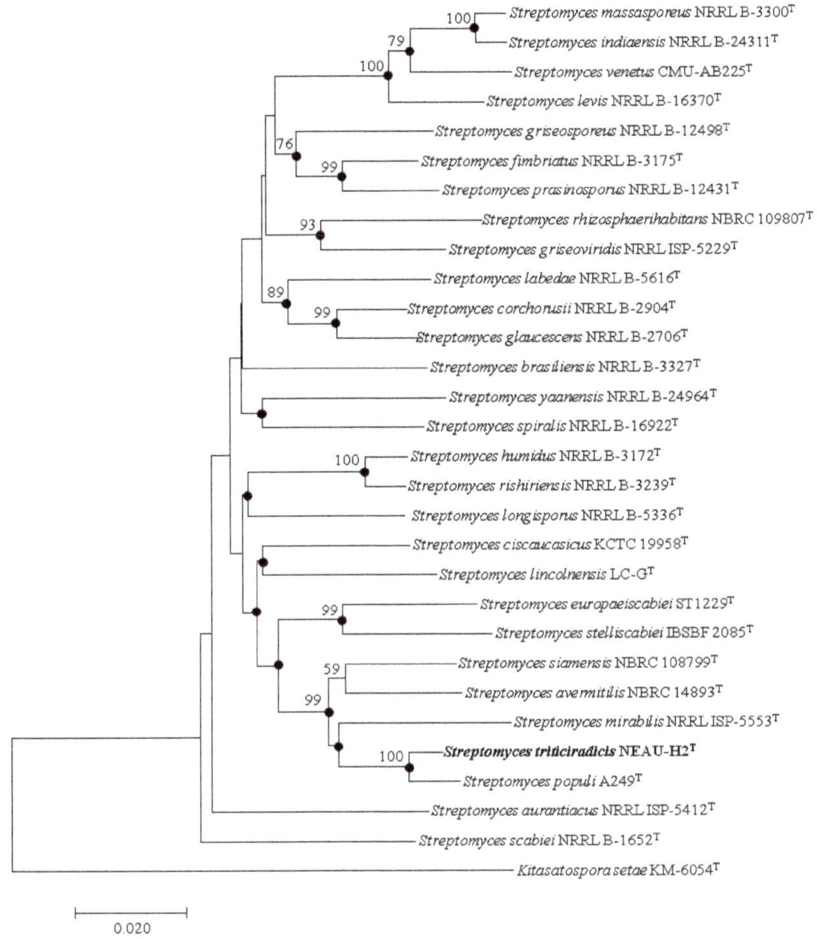

Figure 3. Neighbor-joining tree based on multilocus sequence analysis (MLSA)analysis of the concatenated partial sequences (2060 bp) from five housekeeping genes (*atpD*, *gyrB*, *recA*, *rpoB*, and *trpB*) of strain NEAU-H2^T (in bold) with related taxa. Filled circles indicate branches that were also recovered using the maximum likelihood methods. Only bootstrap values above 50% (percentages of 1000 replications) are indicated. *Kitasatospora setae* KM-6054^T was used as an outgroup. Bar, 0.02 nucleotide substitutions per site.

The assembled genome sequence of strain NEAU-H2^T was found to be 9,921,301 bp long and composed of 135 contigs with an N50 of 167,996 bp, a DNA G+C content of 71.5 mol%, and a coverage of 152.0×. It was deposited into GenBank under the accession number WBKG00000000. Detailed genomic information is presented in the Table S3. DNA–DNA hybridization was employed to further clarify the relatedness between strain NEAU-H2^T and *S. rhizosphaerihabitans* NBRC 109807^T and *S. siamensis* NBRC 108799^T. The DNA–DNA relatedness values were 33.3 ± 2.5% and 44.5 ± 3.5%, respectively. Digital DNA–DNA hybridization was employed to clarify the relatedness between strain NEAU-H2^T and *S. populi* A249^T. The level of DNA–DNA relatedness between them was 56.5 to 62.1%. According to the description proposed by Wayne et al. [55], the relatedness values are below the threshold value of 70% for assigning bacterial strains to the same genomic species.

Besides the genotypic evidence above, some obvious differences can also be found between strain NEAU-H2^T with its closely related strains regarding several phenotypic and chemotaxonomic characteristics. Strain NEAU-H2^T could be easily distinguished from its most closely related species by cultural characteristics, such as colony colors and diffusible pigment production (Table S1 and Figure S4). Morphological characteristics, including spore chain and surface ornamentation, could also distinguish the isolate from its closely related strains (Table 1). In addition, the isolate was able to grow at 4 °C, in contrast to its closely related strains, which could not. The novel strain could not utilize L-serine, L-tyrosine, L-arabinose, and meso-inositol while the closely related species could. Strain NEAU-H2^T was found to contain both PI and PIM in its phospholipid profile, which could distinguish it from *S. siamensis* NBRC 108799^T and *S. populi* A249^T. The presence of MK-9(H$_4$) could differentiate the isolate from *S. rhizosphaerihabitans* NBRC 109807^T. Most notably, the whole-cell sugar of strain NEAU-H2^T was evidently different from that of *S. rhizosphaerihabitans* NBRC 109807^T and *S. populi* A249^T, with the only presence of glucose. The detailed characteristics of strain NEAU-H2^T in comparison with its closely related strains are listed in Table 1.

Therefore, it is evident from the genotypic, phenotypic, and chemotaxonomic data that strain NEAU-H2^T represents a novel species of the genus *Streptomyces*, for which the name *Streptomyces triticiradicis* sp. nov. is proposed.

3.2. Description of Streptomyces triticiradicis sp. nov.

Streptomyces triticiradicis (tri.ti.ci.ra'di.cis. L. neut. n. *triticum* wheat; L. fem. n. *radix* a root; N.L. gen. n. *triticiradicis* of a wheat root).

This is an aerobic gram-staining-positive actinomycete that forms well-developed, branched substrate hyphae and aerial mycelium that differentiate into spiral spore chains consisting of spiny surfaced spores. It has good growth on ISP 1–4, ISP 7 and Nutrient agar media, moderate growth on Bennett's and Czapek's agar media, and poor growth on ISP 5 and ISP 6 media. A dark grayish olive pigment is produced on ISP 6 medium. Growth is observed at temperatures between 4 and 40 °C, with an optimum temperature of 28 °C. Growth occurs in the pH range from 5.0 to 10.0 (optimum pH 7.0) with 0% to 9.0% (*w/v*) NaCl tolerance (optimum 0%–1%). It is positive for coagulation of milk; decomposition of cellulose; hydrolysis of aesculin, starch, and Tweens (20, 40, and 80); and production of H$_2$S and urease; but negative for liquefaction of gelatin, production of catalase, and reduction of nitrate. L-alanine, L-arginine, L-asparagine, L-aspartic acid, creatine, L-glutamic acid, L-glutamine glycine, L-proline, and L-threonine are utilized as sole nitrogen sources but not L-serine or L-tyrosine. D-Fructose, D-galactose, D-glucose, lactose, D-maltose, D-mannose, D-raffinose, L-rhamnose, and D-sucrose are utilized as sole carbon sources but not L-arabinose, dulcitol, meso-inositol, D-mannitol, D-ribose, D-sorbitol, or D-xylose. The cell wall contains LL-diaminopimelic acid as diagnostic diamino acid and the whole cell hydrolysate contains glucose. The major menaquinones are MK-9(H$_8$), MK-9(H$_6$), and MK-9(H$_4$). The polar lipids profile contains DPG, PE, PI, PIM, and PL. Major fatty acids (>10%) are iso-C$_{16:0}$, anteiso-C$_{15:0}$, iso-C$_{15:0}$, and anteiso-C$_{17:0}$. The DNA G + C content of the type strain is 71.5 mol%.

The type strain is NEAU-H2^T (= CCTCC AA 2018031^T = DSM 109825^T), isolated from the rhizosphere soil of wheat (*Triticum aestivum* L.) collected from Zhumadian, Henan Province, Central China. The GenBank/EMBL/DDBJ accession number for the 16S rRNA gene sequence of strain NEAU-H2^T is MN512450. This Whole Genome Shotgun project has been deposited at DDBJ/ENA/GenBank under the accession WBKG00000000. The version described in this paper is version WBKG01000000.1.

3.3. Antifungal Activity Evaluation

Strain NEAU-H2^T showed a wide range of inhibitory effects on the mycelial growth of the 10 tested phytopathogenic fungi (Figure 4A). It displayed significant inhibitory effects against four phytopathogenic fungi, including *C. orbiculare*, *C. cassiicola*, *S. sclerotiorum*, and *E. turcicum*, with the inhibition rate ranging from 48.4% to 69.0% (Figure 4B).

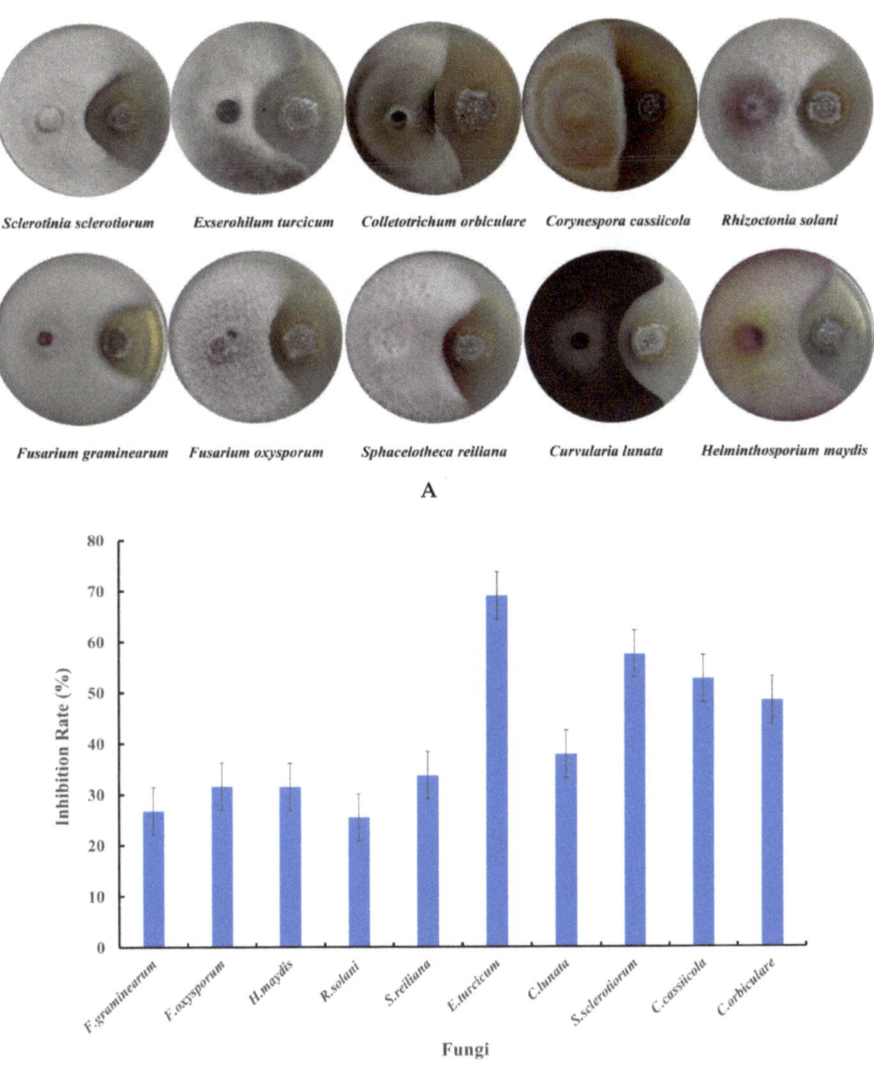

Figure 4. Antifungal activity of strain NEAU-H2[T] against the tested fungi. (**A**) Dual culture plate assay against tested fungi; (**B**) Inhibition rate against the tested fungi.

3.4. Identified of Secondary Metabolites from Strain NEAU-H2[T]

Only major components were identified from the liquid fermentation extract. Compound **1** was obtained as white amorphous powder, and its molecular formula, $C_{12}H_{13}NO_3$, was determined by high resolution electrospray ionization mass spectrometry (HRESIMS) data (m/z 242.0792 [M + Na]$^+$, calculatedd for 242.0788), corresponding to 7 degrees of unsaturation (Figure S5). The ^{1}H NMR showed the presence of five aromatic protons with signals at δ_H 8.29 (s, 1H), 8.27 (d, *J* = 7.3 Hz, 1H), 7.46 (d, *J* = 7.4 Hz, 1H), 7.24 (td, *J* = 7.2, 1.2 Hz, 1H), and 7.21 (td, *J* = 7.2, 1.2 Hz, 1H), which indicated a three-substituted indole moiety (Table 2, Figure S5). The ^{13}C NMR and HSQC spectra revealed 12 carbons, which were classified into one methyl (δ_C 17.9), five sp^2 methines (δ_C 135.8, 124.5, 123.4, 122.9, 112.9), three sp^2 quaternary carbons (δ_C 138.2, 127.3, 116.3), and two oxygenated tertiary carbons (δ_C 79.7 and 71.1) and a carbonyl carbon (δ_C 197.2) (Figure S5).

Table 2. ^{1}H (600 MHz) and ^{13}C (150 MHz) NMR Data of **1** in CD$_3$OD.

No.	δ_C	δ_H (mult, J in Hz)	^{1}H-^{1}H COSY	HMBC (H→C)
2	135.8	6.29 (s, 1H)		C-3, 3a, 7a, 8
3	116.3			
3a	127.3			
4	122.9	8.27 (d, J = 7.3 Hz, 1H)	H-5	C-6
5	123.4	7.21 (td, J = 7.2, 1.2 Hz, 1H)	H-5, H-6	C-4, 6, 3a, 7
6	124.5	7.24 (td, J = 7.2, 1.2 Hz, 1H)	H-5, H-7	C-4, 7a
7	112.9	7.46 (d, J = 7.4 Hz, 1H)	H-6	C-5, 3a
7a	138.2			
8	197.2			
9	79.7	4.74 (d, J = 4.8 Hz, 1H)	H-10	C-11, 10, 8
10	71.1	4.10 (m, 1H)	H-11, H-9	C-11, 8
11	17.9	1.16 (d, J = 6.4 Hz, 3H)	H-10	C-10, 9

* $\delta_{C \text{ or } H}$: chemical shift; J: coupling constant; COSY: correlated spectroscopy; HMBC: ^{1}H detected heteronuclear multiple bond correlation.

The ^{1}H-^{1}H COSY and HSQC spectra of **1** showed two spin-coupling systems, H-9/H-10/H-11 and H-4/H-5/H-6/H-7 (Figure 5B). The HMBC cross-peaks from H-5 to C-3a, from H-6 to C-7a, and from H-2 to C-3/3a/C-7a further revealed the presence of an indole moiety. Cross-peaks from H-9 to C-8 and from H-10 to C-8 were observed in the HMBC spectrum, which suggested a 2,3-dihydroxybutanone connected with indole moiety at C-3 (Figure 5B). Therefore, the planar structure **1** was elucidated as depicted in Figure 5A.

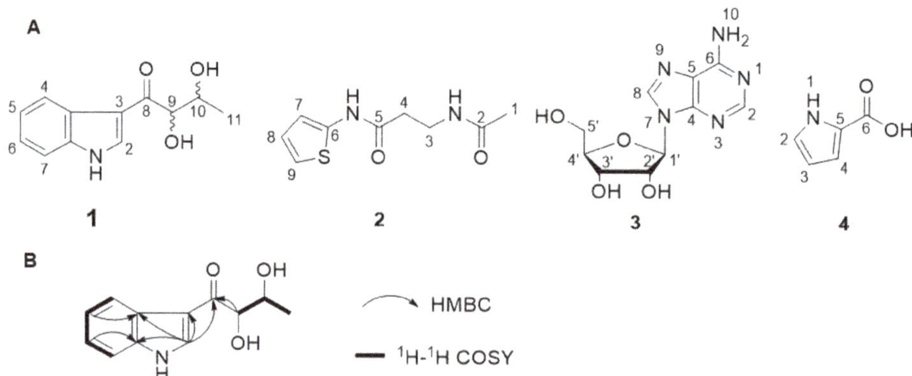

Figure 5. (**A**)The structure of compounds **1–4**; (**B**) 2D NMR correlations of **1**.

Compound **2** was isolated as a colorless powder, HRESIMS m/z 235.0532 [M + Na]$^+$ (calculated for C$_9$H$_{12}$N$_2$O$_2$S, 235.0512); ^{1}H NMR data (600 MHz) δ_H 7.03 (1H, dd, J = 2.4, 1.4 Hz, H-7), 6.96 (1H, dd, J = 3.9, 1.3 Hz, H-9), 6.20 (1H, dd, J = 3.8, 2.5 Hz, H-8), 3.13 (2H, t, J = 6.7 Hz, H-4), 3.38 (2H, t, J = 6.7 Hz, H-3), 1.92 (3H, s, H-1); ^{13}C NMR data (150 MHz, CD$_3$OD) δ_C 181.9 (C-5), 173.5 (C-2), 131.2 (C-6), 125.7 (C-7), 116.4 (C-9), 111.1 (C-8), 40.7 (C-3), 28.2 (C-4), and 22.5 (C-1) (Figure S6). Compound **2** was proven to be 3-Acetylamino-N-2-thienyl-propanamide by direct comparison of these data with those from the literature [56].

Compound **3**: ^{1}H NMR data (600 MHz, DMSO-d_6) δ_H 8.37 (1H, s, H-8), 8.21 (1H, s, H-2), 5.90 (1H, d, J = 5.9 Hz, H-1'), 4.55 (2H, t, J = 5.4 Hz, H-2'), 4.14 (1H, m, H-3'), 3.95 (1H, q, J = 3.3 Hz, H-4'), 3.67 (1H, dd, J = 12.1, 3.5 Hz, H-5'), 3.55 (1H, dd, J = 12.1, 3.5 Hz, H-5'); ^{13}C NMR data (150 MHz, DMSO-d_6) δ_C 154.3 (C-6), 151.7 (C-2), 149.9 (C-4), 138.6 (C-8), 119.8 (C-5), 87.8 (C-1'), 85.4 (C-4'), 73.6 (C-2'), 70.5 (C-3'), and 61.6 (C-5') (Figure S7). Compound **3** was proven to be ß-adenosine by direct comparison of these data with those from the literature [57].

Compound **4**: ^{1}H NMR data (400 MHz, CD$_3$OD) δ_H 6.94 (1H, s, H-4), 6.85 (1H, d, J = 2.7 Hz, H-2), 6.18 (1H, m, H-3); ^{13}C NMR (100 MHz, CD$_3$OD) δ_C 164.6 (C-6), 124.4 (C-2), 124.1 (C-5), 116.6 (C-4), and 110.6 (C-3) (Figure S8). Compound **4** was proven to be 2-minaline by direct comparison of these data with those from the literature [58].

3.5. Mining the Biosynthetic Potential of the Strain

All the compounds were evaluated for their antifungal activity, which showed no significant inhibitory activity. In order to further discover the biosynthetic potential of the strain, we performed draft genome sequencing analysis. AntiSMASH analysis led to the identification of 38 putative gene clusters in the genome of strain NEAU-H2^T. Eleven clusters were identified belonging to a family of polyketide synthases (PKSs), including four type I PKSs, one type II PKS, and three type III PKSs. Likewise, further genome sequence analysis revealed eight additional gene clusters comprising modular enzyme coding genes, such as non-ribosomal peptide synthetase (NRPS, four clusters) and hybrid PKS-NRPS genes (four clusters). Other gene clusters included seven terpene gene clusters, three bacteriocin gene clusters, two siderophore gene clusters, one lanthipeptide gene cluster, one lassopeptide gene cluster, one melanin gene cluster, one ectoine gene cluster, and three butyrolactone gene clusters.

The important feature of NRPs is their ability to use nonproteinogenic amino acids as building blocks. By using such building blocks, NRPSs are able to produce peptides with diverse structures and bioactivities. As such, many NRPSs have been developed into pharmaceuticals, such as vancomycin, daptomycin, and β-lactam [59].

However, only a few metabolites were isolated and identified in culture broth under laboratory conditions from strain NEAU-H2^T. One answer is that most biosynthetic gene clusters of secondary metabolites are cryptic in culture broth under conventional laboratory culture conditions [60]. In addition, an active ingredient has not been isolated, possibly due to the low production of these metabolites under our culture conditions.

One of the secondary metabolite biosynthetic gene clusters of strain NEAU-H2^T shows a 63% similarity to the biosynthetic gene cluster of natamycin, which is a 26-membered polyene macrolide antifungal agent produced by *Streptomyces chattanoogensis* L10, and the macrolide core was synthesized by five PKSs (ScnS0, ScnS1, ScnS2, ScnS3, and ScnS4) in turn [61]. Natamycin is currently widely used as an antifungal agent in human therapy and the food industry [62]. However, considering the poor quality of the genome sequence, with a large number of contigs, this may not be related to the antifungal active components identified with antibiotics and secondary metabolite analysis shell–antiSMASH. In the following research, we will focus on the study of secondary metabolites using activity tracking, amplification fermentation, and other approaches involving modification of the nutrient conditions in the medium and the genetic recombination of biosynthetic gene clusters.

4. Conclusions

A novel strain, NEAU-H2^T, with antifungal activity was isolated from the rhizosphere soil of wheat (*Triticum aestivum* L.). Four compounds, including one new compound, along with three known congeners (3-Acetylamino-N-2-thienyl-propanamide, β-adenosine, 2-minaline), were isolated. Morphological and chemotaxonomic features together with phylogenetic analysis and genomes suggested that strain NEAU-H2^T belonged to the genus *Streptomyces*. Cultural and biochemical characteristics combined with DNA–DNA relatedness values clearly revealed that strain NEAU-H2^T

was differentiated from its closely related strains. Based on the polyphasic taxonomic analysis, it is suggested that strain NEAU-H2^T represents a novel species of the genus *Streptomyces*, for which the name *Streptomyces triticiradicis* sp. nov. is proposed. The type strain is NEAU-H2^T (=CCTCC AA 2018031^T = DSM 109825^T).

Supplementary Materials: The following are available online at http://www.mdpi.com/2076-2607/8/1/77/s1.

Author Contributions: Z.Y., C.H., B.Y. and J.Z. performed the experiments. S.H. and Y.Y. analyzed the data. Z.Y. and C.H. wrote the paper. C.L. and W.X. designed the experiments and reviewed the manuscript. All authors have read and agreed to the published version of the manuscript.

Funding: This work was supported by the Academic Backbone Project of Northeast Agricultural University (Grant No. 17XG17) and Open Project of Key Laboratory of Agricultural Microbiology of Heilongjiang Provincial (NW 2017003).

Acknowledgments: We are grateful to Aharon Oren for helpful advice on the species epithet.

Conflicts of Interest: The authors declare that they have no conflicts of interest.

References

1. Volova, T.; Prudnikova, S.; Boyandin, A.; Zhila, N.; Kiselev, E.; Shumilova, A.; Baranovskiy, S.; Demidenko, A.; Shishatskaya, A.; Thomas, S. Constructing slow-release fungicide formulations based on poly (3-hydroxybutyrate) and natural materials as a degradable matrix. *J. Agric. Food Chem.* **2019**, *67*, 9220–9231. [CrossRef] [PubMed]

2. Bai, Y.B.; Gao, Y.Q.; Nie, X.D.; Tuong, T.M.; Li, D.; Gao, J.M. Antifungal activity of griseofulvin derivatives against phytopathogenic fungi in vitro and in vivo and three-dimensional quantitative structure-activity relationship analysis. *J. Agric. Food Chem.* **2019**, *67*, 6125–6132. [CrossRef] [PubMed]

3. Onaka, H. Novel antibiotic screening methods to awaken silent or cryptic secondary metabolic pathways in actinomycetes. *J. Antibiot.* **2017**, *70*, 865–870. [CrossRef] [PubMed]

4. Liu, B.; Li, R.; Li, Y.A.; Li, S.Y.; Yu, J.; Zhao, B.F.; Liao, A.C.; Wang, Y.; Wang, Z.W.; Lu, A.D.; et al. Discovery of pimprinine alkaloids as novel agents against a plant virus. *J. Agric. Food Chem.* **2019**, *67*, 1795–1806. [CrossRef] [PubMed]

5. Chen, Y.; Zhou, D.; Qi, D.; Gao, Z.; Xie, J.; Luo, Y. Growth promotion and disease suppression ability of a *Streptomyces* sp. CB-75 from Banana rhizosphere soil. *Front. Microbiol.* **2018**, *8*, 2704. [CrossRef] [PubMed]

6. Kemung, H.M.; Tan, L.T.; Khan, T.M.; Chan, K.G.; Pusparajah, P.; Goh, B.H.; Lee, L.H. *Streptomyces* as a prominent resource of future anti-MRSA drugs. *Front. Microbiol.* **2018**, *9*, 2221. [CrossRef] [PubMed]

7. Janardhan, A.; Kumar, A.P.; Viswanath, B.; Saigopal, D.V.; Narasimha, G. Production of bioactive compounds by Actinomycetes and their antioxidant properties. *Biotechnol. Res. Int.* **2014**, *2014*, 217030. [CrossRef]

8. Minuto, A.; Spadaro, D.; Garibaldi, A.; Gullino, M.L. Control of soilborne pathogens of tomato using a commercial formulation of *Streptomyces griseoviridis* and *solarization*. *Crop Protect.* **2006**, *25*, 468–475. [CrossRef]

9. Raaijmakers, J.M.; Paulitz, T.C.; Steinberg, C.; Alabouvette, C.; Moënne-Loccoz, Y. The rhizosphere: A playground and battlefield for soilborne pathogens and beneficial microorganisms. *Plant. Soil.* **2009**, *321*, 341–361. [CrossRef]

10. Che, Q.; Zhu, T.J.; Keyzers, R.A.; Liu, X.F.; Li, J.; Gu, Q.Q.; Li, D.H. Polycyclic hybrid isoprenoids from a reed rhizosphere soil derived *Streptomyces* sp. CHQ-64. *J. Nat. Prod.* **2013**, *76*, 759–763. [CrossRef]

11. Zhao, K.; Penttinen, P.; Chen, Q.; Guan, T.W.; Lindström, K.; Ao, X.L.; Zhang, L.L.; Zhang, X.P. The rhizospheres of traditional medicinal plants in Panxi, China, host a diverse selection of actinobacteria with antimicrobial properties. *Appl. Microbiol. Biotechnol.* **2012**, *94*, 1321–1335. [CrossRef]

12. Zhao, J.W.; Shi, L.L.; Li, W.C.; Wang, J.B.; Wang, H.; Tian, Y.Y.; Xiang, W.S.; Wang, X.J. *Streptomyces tritici* sp. nov. a novel actinomycete isolated from rhizosphere soil of wheat (*Triticum aestivum* L.). *Int. J. Syst. Evol. Microbiol.* **2018**, *68*, 492–497. [CrossRef]

13. Shirling, E.B.; Gottlieb, D. Methods for characterization of *Streptomyces* species. *Int. J. Syst. Bacteriol.* **1966**, *16*, 313–340. [CrossRef]

14. Smibert, R.M.; Krieg, N.R. Phenotypic Characterization. In *Methods for General and Molecular Bacteriology*; Gerhardt, P., Murray, R.G.E., Wood, W.A., Krieg, N.R., Eds.; American Society for Microbiology: Washington, DC, USA, 1994; pp. 607–654.

15. Jin, L.Y.; Zhao, Y.; Song, W.; Duan, L.P.; Jiang, S.W.; Wang, X.J.; Zhao, J.W.; Xiang, W.S. *Streptomyces inhibens* sp. nov., a novel actinomycete isolated from rhizosphere soil of wheat (*Triticum aestivum* L.). *Int. J. Syst. Evol. Microbiol.* **2019**, *69*, 688–695. [CrossRef]

16. Jones, K.L. Fresh isolates of actinomycetes in which the presence of sporogenous aerial mycelia is a fluctuating characteristic. *J. Bacteriol.* **1949**, *57*, 141–145. [CrossRef]

17. Waksman, S.A. *The Actinomycetes. A Summary of Current Knowledge*; The Ronald Press Co.: New York, NY, USA, 1967; p. 286.

18. Waksman, S.A. *The Actinomycetes, Volume 2, Classification, Identification and Descriptions of Genera and Species*; Williams and Wilkins Company: Philadelphia, PA, USA, 1961; p. 363.

19. Kelly, K.L. Color-Name Charts Illustrated with Centroid Colors. In *Inter-Society Color Council-National Bureau of Standards*; U.S. National Bureau of Standards: Washington, DC, USA, 1965.

20. Jia, F.Y.; Liu, C.X.; Wang, X.J.; Zhao, J.W.; Liu, Q.F.; Zhang, J.; Gao, R.X.; Xiang, W.S. *Wangella harbinensis* gen. nov., sp. nov., a new member of the family *Micromonosporaceae*. *Antonie Leeuwenhoek* **2013**, *103*, 399–408. [CrossRef]

21. Zhao, J.W.; Han, L.Y.; Yu, M.Y.; Cao, P.; Li, D.; Guo, X.W.; Liu, Y.Q.; Wang, X.J.; Xiang, W.S. Characterization of *Streptomyces sporangiiformans* sp. nov., a novel soil actinomycete with antibacterial activity against Ralstonia solanacearum. *Microorganisms* **2019**, *7*, 360–376.

22. Gordon, R.E.; Barnett, D.A.; Handerhan, J.E.; Pang, C. *Nocardia coeliaca, Nocardia autotrophica*, and the nocardin strain. *Int. J. Syst. Bacteriol.* **1974**, *24*, 54–63. [CrossRef]

23. Yokota, A.; Tamura, T.; Hasegawa, T.; Huang, L.H. *Catenuloplanes japonicas* gen. nov., sp. nov., nom. rev., a new genus of the order actinomycetales. *Int. J. Syst. Bacteriol.* **1993**, *43*, 805–812. [CrossRef]

24. McKerrow, J.; Vagg, S.; McKinney, T.; Seviour, E.M.; Maszenan, A.M.; Brooks, P.; Sevious, R.J. A simple HPLC method for analysing diaminopimelic acid diastereomers in cell walls of Gram-positive bacteria. *Lett. Appl. Microbiol.* **2000**, *30*, 178–182. [CrossRef]

25. Lechevalier, M.P.; Lechevalier, H.A. The Chemotaxonomy of Actinomycetes. In *Actinomycete Taxonomy*; Dietz, A., Thayer, D.W., Eds.; Special Publication for Society of Industrial Microbiology: Arlington, TX, USA, 1980; pp. 227–291.

26. Minnikin, D.E.; O'Donnell, A.G.; Goodfellow, M.; Alderson, G.; Athalye, M.; Schaal, A.; Parlett, J.H. An integrated procedure for the extraction of bacterial isoprenoid quinones and polar lipids. *J. Microbiol. Methods.* **1984**, *2*, 233–241. [CrossRef]

27. Collins, M.D. Isoprenoid Quinone Analyses in Bacterial Classification and Identification. In *Chemical Methods in Bacterial Systematics*; Goodfellow, M., Minnikin, D.E., Eds.; Academic Press: Cambridge, MA, USA, 1985; pp. 267–284.

28. Wu, C.; Lu, X.; Qin, M.; Wang, Y.; Ruan, J. Analysis of menaquinone compound in microbial cells by HPLC. *Microbiology* **1989**, *16*, 76–178.

29. Xiang, W.S.; Liu, C.X.; Wang, X.J.; Du, J.; Xi, L.J.; Huang, Y. *Actinoalloteichus nanshanensis* sp. nov., isolated from the rhizosphere of a fig tree (*Ficus religiosa*). *Int. J. Syst. Evol. Microbiol.* **2011**, *61*, 1165–1169. [CrossRef]

30. Zhou, J.Z.; Bruns, M.A.; Tiedje, J.M. DNA recovery from soils of diverse composition. *Appl. Environ. Microbiol.* **1996**, *62*, 316–322.

31. Yi, R.K.; Tan, F.; Liao, W.; Wang, Q.; Mu, J.F.; Zhou, X.R.; Yang, Z.N.; Zhao, X. Isolation and identification of *Lactobacillus plantarum* HFY05 from natural fermented Yak Yogurt and Its effect on alcoholic liver injury in mice. *Microorganisms* **2019**, *7*, 530. [CrossRef]

32. Saitou, N.; Nei, M. The neighbor-joining method: A new method for reconstructing phylogenetic trees. *Mol. Biol. Evol.* **1987**, *4*, 406–425.

33. Felsenstein, J. Evolutionary trees from DNA sequences: A maximum likelihood approach. *J. Mol. Evol.* **1981**, *17*, 368–376. [CrossRef]

34. Kumar, S.; Stecher, G.; Tamura, K. Mega7: Molecular evolutionary genetics analysis version 7.0 for bigger datasets. *Mol. Biol. Evol.* **2016**, *33*, 1870–1874. [CrossRef]

35. Felsenstein, J. Confidence limits on phylogenies: An approach using the bootstrap. *Evolution* **1985**, *39*, 83–791. [CrossRef]

36. Kimura, M. A simple method for estimating evolutionary rates of base substitutions through comparative studies of nucleotide sequences. *J. Mol. Evol.* **1980**, *16*, 111–120. [CrossRef]
37. Yoon, S.H.; Ha, S.M.; Kwon, S.; Lim, J.; Kim, Y.; Seo, H.; Chun, J. Introducing EzBioCloud: A taxonomically united database of 16S rRNA and whole genome assemblies. *Int. J. Syst. Evol. Microbiol.* **2017**, *67*, 1613–1617. [CrossRef] [PubMed]
38. Li, R.Q.; Zhu, H.M.; Ruan, J.; Qian, W.B.; Fang, X.D.; Shi, Z.B.; Li, Y.R.; Li, S.T.; Shan, G.; Kristiansen, K.; et al. De novo assembly of human genomes with massively parallel short read sequencing. *Genome Res.* **2010**, *20*, 265–272. [CrossRef] [PubMed]
39. Li, R.; Li, Y.; Kristiansen, K.; Wang, J. SOAP. Short oligonucleotide alignment program. *Bioinformatics* **2008**, *24*, 713–714. [CrossRef]
40. Ley, J.D.; Cattoir, H.; Reynaerts, A. The quantitative measurement of DNA hybridization from renaturation rates. *Eur. J. Biochem.* **1970**, *12*, 133–142. [CrossRef]
41. Huss, V.A.R.; Festl, H.; Schleifer, K.H. Studies on the spectrometric determination of DNA hybridisation from renaturation rates. *Syst. Appl. Microbiol.* **1983**, *4*, 184–192. [CrossRef]
42. Meier-Kolthoff, J.P.; Auch, A.F.; Klenk, H.P.; Goker, M. Genome sequence-based species delimitation with confidence intervals and improved distance functions. *BMC Bioinform.* **2013**, *14*, 60. [CrossRef]
43. Blin, K.; Wolf, T.; Chevrette, M.G.; Lu, X.; Schwalen, C.J.; Kautsar, S.A.; Duran, H.G.S.; Santos, E.L.C.D.L.; Kim, H.U.; Nave, M.; et al. Antismash 4.0-improvements in chemistry prediction and gene cluster boundary identification. *Nucleic Acids Res.* **2017**, *45*, W36–W41. [CrossRef]
44. Liu, C.X.; Zhuang, X.X.; Yu, Z.Y.; Wang, Z.Y.; Wang, Y.J.; Guo, X.W.; Xiang, W.S.; Huang, S.X. Community structures and antifungal activity of root-associated endophytic actinobacteria of healthy and diseased soybean. *Microorganisms* **2019**, *7*, 243. [CrossRef]
45. Zhang, J.; Wang, X.J.; Yan, Y.J.; Jiang, L.; Wang, J.D.; Li, B.J.; Xiang, W.S. Isolation and identification of 5-hydroxyl-5-methyl-2-hexenoic acid from *Actinoplanes* sp. HBDN08 with antifungal activity. *Bioresour. Technol.* **2010**, *101*, 8383–8388. [CrossRef]
46. Yu, Z.Y.; Wang, L.; Yang, J.; Zhang, F.; Sun, Y.; Yu, M.M.; Yan, Y.J.; Ma, Y.T.; Huang, S.X. A new antifungal macrolide from *Streptomyces* sp. KIB-H869 and structure revision of halichomycin. *Tetrahedron Lett.* **2016**, *57*, 1375–1378. [CrossRef]
47. Kämpfer, P.; Genus, I. *Streptomyces* Waksman and Henrici 1943, 339 [AL]. In *Bergey's Manual of Systematic Bacteriology*, 2nd ed.; Springer: New York, NY, USA, 2012; pp. 1679–1680.
48. Lee, H.J.; Whang, K.S. Streptomyces rhizosphaerihabitans sp. nov. and Streptomyces adustus sp. nov. isolated from bamboo forest soil. *Int. J. Syst. Microbiol.* **2016**, *66*, 3573–3578.
49. Wang, Z.K.; Jiang, B.J.; Li, X.G.; Gan, L.Z.; Long, X.F.; Zhang, Y.Q.; Tian, Y.Q. Streptomyces populi sp. nov. a novel endophytic actinobacterium isolated from stem of Populus adenopoda Maxim. *Int. J. Syst. Microbiol.* **2018**, *68*, 2568–2573.
50. Sripreechasak, P.; Matsumoto, A.; Suwanborirux, K.; Inahashi, Y.; Shiomi, K.; Tanasupawat, S.; Takahashi, Y. Streptomyces siamensis sp. nov., and Streptomyces similanensis sp. nov., isolated from Thai soils. *J. Antibiot.* **2013**, *66*, 633–640.
51. Kim, S.B.; Lonsdale, J.; Seong, C.N.; Goodfellow, M. *Streptacidiphilus* gen. nov., acidophilic actinomycetes with wall chemotype I and emendation of the family *Streptomycetaceae* (Waksman and Henrici (1943) AL) emend. Rainey et al. 1997. *Antonie. Leeuwenhoek.* **2003**, *83*, 107–116. [CrossRef]
52. Kroppenstedt, R.M. Fatty Acid and Menaquinone Analysis of Actinomycetes and Related Organisms. In *Chemical Methods in Bacterial Systematics*; Goodfellow, M., Minnikin, D.E., Eds.; Academic Press: London, UK, 1985; pp. 173–199.
53. Lechevalier, M.P.; Lechevalier, H. Chemical composition as a criterion in the classification of aerobic actinomycetes. *Int. J. Syst. Bacteriol.* **1970**, *20*, 435–443. [CrossRef]
54. Rong, X.Y.; Huang, Y. Taxonomic evaluation of the *Streptomyces hygroscopicus* clade using multilocus sequence analysis and DNA-DNA hybridization, validating the MLSA scheme for systematics of the whole genus. *Syst. Appl. Microbiol.* **2012**, *35*, 7–18. [CrossRef]
55. Krichevsky, M.I.; Moore, L.H.; Moore, W.E.C.; Murray, R.G.E.; Stackebrandt, E.; Starr, M.P.; Trper, H.G. International Committee on Systematic Bacteriology. Report of the ad hoc committee on reconciliation of approaches to bacterial systematics. *Int. J. Syst. Bacteriol.* **1987**, *37*, 463–464.

56. Ye, X.W.; Chai, W.Y.; Lian, X.Y.; Zhang, Z.Z. Novel propanamide analogue and antiproliferative diketopiperazines from mangrove *Streptomyces* sp. Q 24. *Nat. Prod. Res.* **2017**, *31*, 1390–1396. [CrossRef]

57. Domondon, D.L.; He, W.; De Kimpe, N.; Höfte, M.; Poppe, J. β-Adenosine, a bioactive compound in grass chaff stimulating mushroom production. *Phytochemistry* **2004**, *65*, 181–187. [CrossRef]

58. Cheng, Y.X.; Zhou, J.; Teng, R.W.; Tan, N.H. Nitrogen-containing compounds from *Brachystemma calycinum*. *Acta Bot. Yunnanica* **2001**, *23*, 527–530.

59. Katsuyama, Y. Mining novel biosynthetic machineries of secondary metabolites from actinobacteria. *Biosci. Biotechnol. Biochem.* **2019**, *83*, 1606–1615. [CrossRef] [PubMed]

60. Rutledge, P.J.; Challis, G.L. Discovery of microbial natural products by activation of silent biosynthetic gene clusters. *Nat. Rev. Microbiol.* **2015**, *13*, 509–523. [CrossRef] [PubMed]

61. Liu, S.P.; Yuan, P.H.; Wang, Y.Y.; Liu, X.F.; Zhou, Z.X.; Bu, Q.T.; Yu, P.; Jiang, H.; Li, Y.Q. Generation of the natamycin analogs by gene engineering of natamycin biosynthetic genes in *Streptomyces chattanoogensis* L10. *Microbiol. Res.* **2015**, *173*, 25–33. [CrossRef] [PubMed]

62. Du, Y.L.; Li, S.Z.; Zhou, Z.; Chen, S.F.; Fan, W.M.; Li, Y.Q. The pleitropic regulator AdpAch is required for natamycin biosynthesis and morphological differentiation in *Streptomyces chattanoogensis*. *Microbiology* **2011**, *157*, 1300–1311. [CrossRef] [PubMed]

 *microorganisms*

Article

Genome Mining Coupled with OSMAC-Based Cultivation Reveal Differential Production of Surugamide A by the Marine Sponge Isolate *Streptomyces* sp. SM17 When Compared to Its Terrestrial Relative *S. albidoflavus* J1074

Eduardo L. Almeida [1], Navdeep Kaur [2], Laurence K. Jennings [2], Andrés Felipe Carrillo Rincón [1], Stephen A. Jackson [1,3], Olivier P. Thomas [2] and Alan D.W. Dobson [1,3,*]

[1] School of Microbiology, University College Cork, T12 YN60 Cork, Ireland; e.leaodealmeida@umail.ucc.ie (E.L.A.); andres-felipe.carrillo@alumnos.unican.es (A.F.C.R.); stevejackson71@hotmail.com (S.A.J.)

[2] Marine Biodiscovery, School of Chemistry and Ryan Institute, National University of Ireland Galway (NUI Galway), University Road, H91 TK33 Galway, Ireland; navdeep.kaur@nuigalway.ie (N.K.); laurence.jennings@nuigalway.ie (L.K.J.); olivier.thomas@nuigalway.ie (O.P.T.)

[3] Environmental Research Institute, University College Cork, T23 XE10 Cork, Ireland

* Correspondence: a.dobson@ucc.ie

Received: 12 August 2019; Accepted: 24 September 2019; Published: 26 September 2019

Abstract: Much recent interest has arisen in investigating *Streptomyces* isolates derived from the marine environment in the search for new bioactive compounds, particularly those found in association with marine invertebrates, such as sponges. Among these new compounds recently identified from marine *Streptomyces* isolates are the octapeptidic surugamides, which have been shown to possess anticancer and antifungal activities. By employing genome mining followed by an one strain many compounds (OSMAC)-based approach, we have identified the previously unreported capability of a marine sponge-derived isolate, namely *Streptomyces* sp. SM17, to produce surugamide A. Phylogenomics analyses provided novel insights on the distribution and conservation of the surugamides biosynthetic gene cluster (*sur* BGC) and suggested a closer relatedness between marine-derived *sur* BGCs than their terrestrially derived counterparts. Subsequent analysis showed differential production of surugamide A when comparing the closely related marine and terrestrial isolates, namely *Streptomyces* sp. SM17 and *Streptomyces albidoflavus* J1074. SM17 produced higher levels of surugamide A than *S. albidoflavus* J1074 under all conditions tested, and in particular producing >13-fold higher levels when grown in YD and 3-fold higher levels in SYP-NaCl medium. In addition, surugamide A production was repressed in TSB and YD medium, suggesting that carbon catabolite repression (CCR) may influence the production of surugamides in these strains.

Keywords: genome mining; OSMAC; phylogenomics; secondary metabolites; surugamides; surugamide A; marine sponge-associated bacteria; *Streptomyces*; *albidoflavus* phylogroup

---

1. Introduction

Members of the *Streptomyces* genus are widely known to be prolific producers of natural products. Many of these compounds have found widespread use in the pharmaceutical industry as antibiotics, immunosuppressant, antifungal, anticancer, and anti-parasitic drugs [1]. However, there continues to be an urgent need to discover new bioactive compounds, and especially antibiotics, primarily due to the emergence of antibiotic resistance in clinically important bacterial pathogens [2,3]. In particular, the increase in multi-resistant ESKAPE pathogens (*Enterococcus faecium, Staphylococcus aureus, Klebsiella pneumoniae, Acinetobacter baumannii, Pseudomonas aeruginosa,* and *Enterobacter* species) has focused research efforts to develop new antibiotics to treat these priority antibiotic-resistant bacteria [4].

Up until relatively recently, marine ecosystems had largely been neglected as a potential source for the discovery of novel bioactive compounds, in comparison to terrestrial environments, primarily due to issues of accessibility [5]. Marine sponges are known to host a variety of different bacteria and fungi, which produce a diverse range of natural products, including compounds with antiviral, antifungal, antiprotozoal, antibacterial, and anticancer activities [5,6]. Marine sponge-associated *Streptomyces* spp. are a particularly important source of bioactive compounds, with examples including *Streptomyces* sp. HB202, isolated from the sponge *Halichondria panicea,* which produces mayamycin, a compound with activity against *Staphylococcus aureus* [7]; and streptophenazines G and K, with activity against *Bacillus subtilis* [8]; together with *Streptomyces* sp. MAPS15, which was isolated from *Spongia officinalis,* which produces 2-pyrrolidine, with activity against *Klebsiella pneumoniae* [9]. Additionally, our group has reported the production of antimycins from *Streptomyces* sp. SM8 isolated from the sponge *Haliclona simulans,* with antifungal and antibacterial activities [10,11]. In further work, we genetically characterised 13 *Streptomyces* spp. that were isolated from both shallow and deep-sea sponges, which displayed antimicrobial activities against a number of clinically relevant bacterial and yeast species [12,13]. Amongst these strains, the *Streptomyces* sp. SM17 demonstrated an ability to inhibit the growth of *E. coli* NCIMB 12210, methicillin-resistant *S. aureus* (MRSA), and *Candida* spp., when employing deferred antagonism assays [12,13].

Among other clinically relevant natural products derived from marine *Streptomyces* isolates are the recently identified surugamides family of molecules. The cyclic octapeptide surugamide A and its derivatives were originally identified in the marine-derived *Streptomyces* sp. JAMM992 [14], and have been shown to belong to a particularly interesting family of compounds due not only to their relevant bioactivity, but also due to their unusual metabolic pathway involving D-amino acids [14–16]. Since their discovery, concerted efforts have been employed in order to chemically characterise these compounds and determine the genetic mechanisms involved in their production [14,15,17–20]. The surugamides and their derivatives have been shown to possess a number of bioactivities, with the surugamides A–E and the surugamides G–J being shown to possess anticancer activity by inhibiting bovine cathepsin B, a cysteine protease reported to be involved in the invasion of metastatic tumour cells [14,16]; while another derivative, namely acyl-surugamide A, has been shown to possess anti-fungal activity [16]. It has been determined that the non-ribosomal peptide synthase-encoding *surABCD* genes are the main biosynthetic genes involved in the biosynthesis of surugamides and their derivatives [19], with these genes being involved in the production of at least 20 different compounds [16]. Surugamides A–E have been reported to be produced by the *surA* and *surD* genes, while the linear decapeptide surugamide F has been shown to be produced by the *surB* and *surC* genes, involving a unique pattern of intercalation of the biosynthetic genes [19]. Further metabolic pathways studies have reported that the expression of the *surABCD* gene cluster is strongly regulated by the *surR* transcriptional repressor [16], while the cyclisation of the cyclic surugamides has been shown to involve a penicillin binding protein (PBP)-like thioesterase encoded by the *surE* gene [17,18,21].

Although apparently widespread in marine-derived *Streptomyces* isolates [18,19], the production of surugamides has also been reported in the *S. albidoflavus* strain J1074 [16,22], a derivative of the soil isolate *S. albus* G [23,24]. The *S. albidoflavus* strain J1074 is a well-characterised *Streptomyces* isolate, which is frequently used as a model for the genus and has commonly been successfully employed in the heterologous expression of biosynthetic gene clusters (BGCs) [25–29]. This strain was originally classified as an *S. albus* isolate, however, due to more recent taxonomy studies, it has been reclassified as a *S. albidoflavus* species isolate [30,31]. Interestingly, surugamides and their derivatives have been shown to only be produced by *S. albidoflavus* J1074 under specific conditions, such as when employing chemical stress elicitors [16], and more recently when cultivating the strain in a soytone-based liquid-based medium SG2 [22].

In a previous study [32], we reported that the *S. albidoflavus* J1074 and *Streptomyces* sp. SM17 possessed morphological and genetic similarities. Differences were observed, however, when both strains were exposed to high salt concentrations using culture media, such as TSB or ISP2, in which the marine sponge-derived strain SM17 grew and differentiated more rapidly in comparison with the soil strain *S. albidoflavus* J1074, which appeared to have trouble growing and differentiating when salts were present in the growth medium [32]. Genome mining based on the prediction of secondary metabolites BGCs also showed many similarities between the two strains [32]. Among these predicted BGCs, both the *S. albidoflavus* J1074 and *Streptomyces* sp. SM17 isolates appeared to possess the *sur* BGC, encoding for the production of surugamides A/D. Due to the fact that marine-derived *Streptomyces* isolates have been shown to produce good levels of surugamides when grown under standard conditions [18,19], and that production of surugamides and derivatives can be induced in the presence of chemical stress elicitors in *S. albidoflavus* J1074 [16], it appears likely that marine-derived *Streptomyces* isolates and their *sur* BGCs could share genetic similarities that might help to optimise production of the compound. To investigate this possibility we 1) employed genome mining approaches together with phylogenomics in order to better characterise the SM17 strain, and to investigate the distribution and differences/similarities between marine- (or aquatic saline-) and terrestrial-derived *sur* BGCs and *sur* BGC-harbouring microorganisms; and 2) experimentally compared the metabolic profiles of surugamide A production between a marine (SM17) and a terrestrial (J1074) *Streptomyces* isolate. With respect to the latter, we employed an "one strain many compounds" (OSMAC)-based approach, which has been shown to be a useful strategy in eliciting production of natural products from silent gene clusters by employing different culture conditions [33,34]; together with analytical chemistry methods such as liquid chromatography–mass spectrometry to monitor production of surugamide A in both *S. albidoflavus* J1074 and *Streptomyces* sp. SM17.

2. Materials and Methods

2.1. Bacterial Strains and Nucleotide Sequences

The *Streptomyces* sp. SM17 strain was isolated from the marine sponge *Haliclona simulans*, from the Kilkieran Bay, Galway, Ireland, as previously described [13]. The *Streptomyces albidoflavus* J1074 strain was provided by Dr Andriy Luzhetskyy (Helmholtz Institute for Pharmaceutical Research Saarland, Saarbrücken, Germany). Their complete genome sequences are available from the GenBank database [35] under the accession numbers NZ_CP029338 and NC_020990, for *Streptomyces* sp. SM17 and *S. albidoflavus* J1074, respectively. The surugamides biosynthetic gene cluster (*sur* BGC) sequence used as a reference for this study was the one previously described in *Streptomyces albidoflavus* LHW3101 (GenBank accession number: MH070261) [18]. Other genomes used in this study's analyses were obtained from the GenBank RefSeq database [35].

2.2. Phylogenetic Analyses

The NCBI BLASTN tool [36,37] was used to determine the closest 30 *Streptomyces* strains with complete genome available in the GenBank RefSeq database [35] to the *Streptomyces* sp. SM17. Then, phylogeny analysis was performed with the concatenated sequences of the 16S rRNA, and the housekeeping genes *atpD*, *gyrB*, *recA*, *rpoB*, and *trpB*. The sequences were aligned using the MAFFT program [38], and the phylogeny analysis was performed using the MrBayes program [39]. In MrBayes, the general time reversible (GTR) model of nucleotide substitution was used [40], with gamma-distributed rates across sites with a proportion of invariable sites, with 1 million generations sampled every 100 generations. Final consensus phylogenetic tree generated by MrBayes was processed using MEGA X [41], with a posterior probability cut-off of 95%.

Phylogeny analysis of the surugamides biosynthetic gene cluster (*sur* BGC) was performed by using the *S. albidoflavus* LHW3101 *sur* BGC nucleotide sequence as reference [18] and searching for similar sequences on the GenBank RefSeq database using the NCBI BLASTN tool [35–37], only taking into account complete genomes. The genome regions with similarity to the *S. albidoflavus* LHW3101 *sur* BGC undergone phylogeny analysis using the same aforementioned tools and parameters.

2.3. Prediction of Secondary Metabolites Biosynthetic Gene Clusters

In order to assess the similarities and differences between the *Streptomyces* isolates belonging to the *albidoflavus* phylogroup, in regard to their potential to produce secondary metabolites, BGCs were predicted in their genomes, using the antiSMASH (version 5 available at https://docs.antismash. secondarymetabolites.org/) program [42]. The predicted BGCs were then processed using the BiG-SCAPE program (version 20190604, available at https://git.wageningenur.nl/medema-group/BiG-SCAPE) [43], with the MiBIG database (version 1.4 available at https://mibig.secondarymetabolites.org/) as reference [44], and similarity clustering of gene cluster families (GCFs) was performed. The similarity network was processed using Cytoscape (version 3.7.1, available at https://cytoscape.org/) [45].

2.4. Gene Synteny Analysis

The genome regions previously determined to share similarities with the *S. albidoflavus* LHW3101 *sur* BGC were manually annotated, for the known main biosynthetic genes (*surABCD*), the penicillin binding protein (PBP)-like peptide cyclase and hydrolase *surE* gene, and the gene with regulatory function *surR* [15–19,21]. This was performed using the UniPro UGENE toolkit (version 1.32.0, available at http://ugene.net/) [46], the GenBank database, and the NCBI BLASTN tool (available at https://blast.ncbi.nlm.nih.gov/Blast.cgi, accessed on June 2019) [35–37]. The gene synteny and reading frame analysis was performed using the UniPro UGENE toolkit [46] and the Artemis genome browser (version 18.0.0, available at https://www.sanger.ac.uk/science/tools/artemis) [47].

2.5. Diagrams and Figures

All the Venn diagrams presented in this study were generated using the Venn package in R [48,49], and RStudio [50]. All the images presented in this study were edited using the Inkscape program (available from https://inkscape.org/).

2.6. Strains Culture, Maintenance, and Secondary Metabolites Production

The same culture media and protocols were employed for both isolates *Streptomyces* sp. SM17 and *Streptomyces albidoflavus* J1074. Glycerol stocks were prepared from spores collected from soya-mannitol (SM) medium after 8 days of cultivation at 28 °C and preserved at −20 °C. To verify the secondary metabolites production profile, spores were cultivated for 7 days on SM agar medium at 28 °C, then pre-inoculated in 5 mL TSB medium, and cultivated at 28 °C and 220 rpm for 2 days. Then, 10% (*v/v*) of the pre-inoculum was transferred to 30 mL of the following media: TSB; SYP-NaCl (1% starch, 0.4% yeast extract, 0.2% peptone, and 0.1% NaCl); YD (0.4% yeast extract, 1% malt extract, and 4% dextrin pH 7.0); P1 (2% glucose, 1% soluble starch, 0.1% meat extract, 0.4% yeast extract, 2.5% soy flour, 0.2% NaCl, and 0.005% K_2HPO_4 pH 7.3); P2 (1% glucose, 0.6% glycerol, 0.1% yeast extract, 0.2% malt extract, 0.6% $MgCl_2.6H_2O$, 0.03% $CaCO_3$, and 10% sea water); P3 (2.5% soy flour, 0.75% starch, 2.25% glucose, 0.35% yeast extract, 0.05% $ZnSO_4 \times 7H_2O$, and 0.6% $CaCO_3$ pH 6.0); CH-F2 (2% soy flour, 0.5% yeast extract, 0.2% $CaCO_3$, 0.05% citric acid, 5% glucose, and pH 7.0); SY (2.5% soluble starch, 1.5% soy flour, 0.2% yeast extract, and 0.4% $CaCO_3$ pH 7.0); Sporulation medium (2% soluble starch and 0.4 yeast extract); and Oatmeal medium (2% oatmeal). These were cultivated at 28 °C and 220 rpm for 4 days in TSB; and for 8 days in SYP-NaCl, YD, SY, P1, P2, P3, CH-F2, Sporulation, and Oatmeal media. Once the bioprocess was completed, the broth was frozen at −20 °C for further chemical analysis.

2.7. Metabolic Profiling, Compound Isolation, and Chemical Structure Analysis

The *Streptomyces* broth of TSB, SYP-NaCl, and YD medium cultures (180 mL) was exhaustively extracted using a solvent mixture of 1:1 MeOH:DCM yielding a crude extract (3.89 g). This crude extract was first separated using SPE on C_{18} bonded silica gel (Polygoprep C18 (Fisher Scientific, Dublin, Ireland) 12%C, 60 Å, 40–63 µm), eluting with varying solvent mixtures to produce five fractions: H_2O (743.62 mg), 1:1 H_2O:MeOH (368.6 mg), MeOH (15.4 mg), 1:1 MeOH:DCM (10.9 mg), DCM (8.2 mg). The final three fractions (MeOH, 1:1 MeOH:DCM, DCM, 34.5 mg) were then combined and subject to analytical reverse phase HPLC on a Waters Symmetry (VWR, Dublin, Ireland) C18 5 µm, 4.6 × 250 mm column. The column was eluted with 10% MeCN (0.1% TFA)/90% H2O (0.1% TFA) for 5 min, then a linear gradient to 100% MeCN (0.1% TFA) over 21 min was performed. The column was further eluted with 100% MeCN (0.1% TFA) for 6 min. After the HPLC was complete, a linear gradient back to 10% MeCN (0.1% TFA)/90% H_2O (0.1% TFA) over 1 min and then further elution of 10% MeCN (0.1% FA)/90% H_2O (0.1% FA) for 4 min was performed. This yielded pure surugamide A (0.8 mg). Surugamide A was characterised using MS and NMR data to confirm the structure for use as an analytical standard.

Surugamide A was quantified in the broth using LC-MS analysis on an Agilent UHR-qTOF 6540 (Agilent Technologies, Cork, Ireland) mass spectrometer. The column used for separation was Waters equity UPLC BEH (Apex Scientific, Kildare, Ireland) C18 1.7 µm 2.1 × 75 mm. The column was eluted with 10% MeCN (0.1% FA)/90% H_2O (0.1% FA) for 2 min, then a linear gradient to 100% MeCN (0.1% FA) over 6 min was performed. The column was further eluted with 100% MeCN (0.1% FA) for 4 min. After the UPLC was complete, a linear gradient back to 10% MeCN (0.1% FA)/90% H_2O (0.1% FA) over 1 min and then further elution of 10% MeCN (0.1% FA)/90% H_2O (0.1% FA) for 3 min was performed before the next run. The MS detection method was positive ion. A calibration curve was produce using the LC-MS method above and injecting the pure surugamide A at seven concentrations (100, 25, 10, 2, 1, 0.2, 0.1 mg/L). Thirty millilitres of each *Streptomyces* strain in broth were extracted using a solvent mixture of 1:1 MeOH:DCM three times to yield a crude extract. These extracts were resuspended in MeOH and filtered through PTFE 0.2 µm filters (Sigma Aldrich, Arklow, Ireland) before being subject to the above LC-MS method.

The surugamide A calibration standards 1–7 and the six extracts were analysed using the Agilent MassHunter Quantification software package. This allowed the quantification of surugamide A in the extracts based on the intensity of peaks in the chromatogram with matching retention time and exact mass.

3. Results and Discussion

3.1. Multi-locus Sequence Analysis and Taxonomy Assignment of the Streptomyces sp. SM17 Isolate

In order to taxonomically characterise the *Streptomyces* sp. SM17 isolate based on genetic evidence, multi-locus sequence analysis (MLSA) [51] employing the 16S rRNA sequence, in addition to five housekeeping genes, namely *atpD* (ATP synthase subunit beta), *gyrB* (DNA gyrase subunit B), *recA* (recombinase RecA), *rpoB* (DNA-directed RNA polymerase subunit beta), and *trpB* (tryptophan synthase beta chain) was performed, in a similar manner to a previous report [32]. A similarity search was performed in the GenBank database [35], using the NCBI BLASTN tool [36,37], based on the 16S rRNA nucleotide sequence of the SM17 isolate. The top 30 most similar *Streptomyces* species for which complete genome sequences were available in GenBank were selected for further phylogenetic analysis.

The concatenated nucleotide sequences [51,52] of the 16S rRNA and the aforementioned five housekeeping genes, were first aligned using the MAFFT program [38], and the phylogeny analysis was performed using the MrBayes program [39]. The general time reversible (GTR) model of nucleotide substitution with gamma-distributed rates across sites with a proportion of invariable sites was applied [40], with 1 million generations sampled every 100 generations. The final phylogenetic tree was then processed using MEGA X [41], with a posterior probability cut-off of 95% (Figure 1).

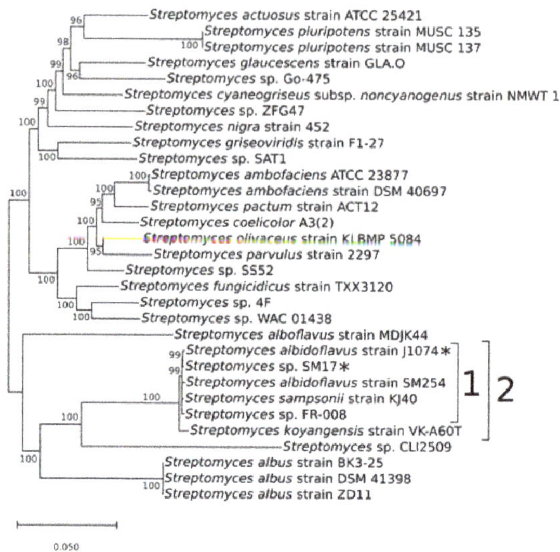

Figure 1. Phylogenetic tree of the concatenated sequences of the 16S rRNA and the housekeeping genes *atpD*, *gyrB*, *recA*, *rpoB*, and *trpB*, from the *Streptomyces* sp. SM17 together with 30 *Streptomyces* isolates for which complete genome sequences were available in the GenBank database. Analysis was performed using MrBayes, with a posterior probability cut-off of 95%. 1) *albidoflavus* phylogroup. 2) Clade including the neighbour isolate *Streptomyces koyangensis* strain VK-A60T. The strains SM17 and J1074 are indicated with asterisks.

The resulting phylogenetic tree clearly indicates the presence of a clade that includes the isolates *Streptomyces albidoflavus* strain J1074; *Streptomyces* sp. SM17; *Streptomyces albidoflavus* strain SM254; *Streptomyces sampsonii* strain KJ40; *Streptomyces* sp. FR-008; and *Streptomyces koyangensis* strain VK-A60T (clade 2 in Figure 1). In addition, this larger clade contains a sub-clade (clade 1 in Figure 1) that includes *Streptomyces* isolates similar to the strain *Streptomyces albidoflavus* J1074. The J1074 strain is a well-studied *Streptomyces* isolate widely used as a model for the genus and for various biotechnological applications, including the heterologous expression of secondary metabolites biosynthetic gene clusters (BGCs) [25–29]. This isolate was originally classified as "*Streptomyces albus* J1074", but due to recent taxonomy data, it has been reclassified as *Streptomyces albidoflavus* J1074 [30,31]. Hence, in this study, this strain will be referred to as *Streptomyces albidoflavus* J1074, and this clade will from now on be referred to as the *albidoflavus* phylogroup (Figure 1).

Interestingly, members of the *albidoflavus* phylogroup were all isolated from quite different environments. The *Streptomyces albidoflavus* strain J1074 stems from the soil isolate *Streptomyces albus* G [23,24]. The *Streptomyces sampsonii* strain KJ40 was isolated from rhizosphere soil in a poplar plantation [53]. The *Streptomyces* sp. strain FR-008 is a random protoplast fusion derivative of two *Streptomyces hygroscopicus* isolates [54]. On the other hand, two of these strains were isolated from aquatic saline environments, with *Streptomyces* sp. SM17 being isolated from the marine sponge *Haliclona simulans* [13]; while the *Streptomyces albidoflavus* strain SM254 strain was isolated from copper-rich subsurface fluids within an iron mine, following growth on artificial sea water (ASW) [55]. The fact that these isolates, although derived from quite distinct environmental niches, simultaneously share significant genetic similarities is interesting, and raises questions about their potential evolutionary relatedness.

3.2. Analysis of Groups of Orthologous Genes in the Albidoflavus Phylogroup

In an attempt to provide further genetic evidence with respect to the similarities shared among the members of the *albidoflavus* phylogroup (Figure 1), a pan-genome analysis was performed to determine the number of core genes, accessory genes, and unique genes present in this group of isolates. The Roary program was employed for this objective [56], which allowed the identification of groups of orthologous and paralogous genes (which from now on will be referred to simply as "genes") present in the set of *albidoflavus* genomes, with a protein identity cut-off of 95%, which is the identity value recommended by the Roary program manual when analysing organisms belonging to the same species.

A total of 7565 genes were identified in the *albidoflavus* pan-genome, and among these a total of 5177 were determined to be shared among all the *albidoflavus* isolates (i.e., the core genome) (Figure 2). This represents a remarkably high proportion of genes that appear to be highly conserved between all the isolates, representing approximately 68.4% of the pan-genome. Additionally, when considering the genomes individually (Table S1), the core genome accounts for approximately 84.5% of the FR-008 genome; 88.5% of J1074; 85.5% of KJ40; 86.7% of SM17; and 83.7% of the SM254 genome. On the other hand, the accessory genome (i.e., genes present in at least two isolates) was determined to consist of 1055 genes (or ~13.9% of the pan-genome); while the unique genome (i.e., genes present in only one isolate) was determined to consist of 1333 genes (or ~17.6% of the pan-genome). This strikingly high conservation of genes present in their genomes together with the previous multi-locus phylogeny analysis are very strong indicators that these microorganisms may belong to the same species.

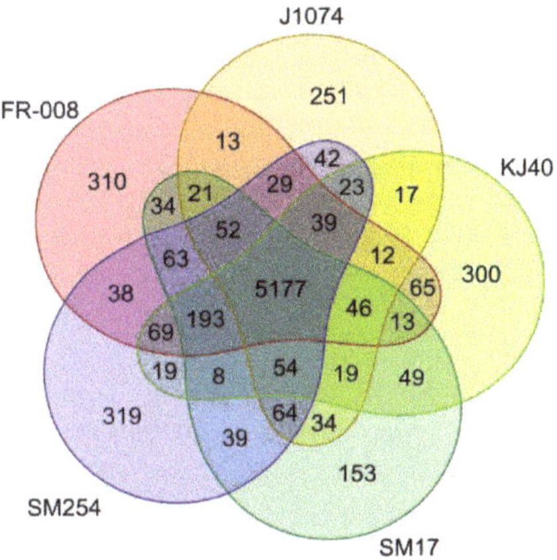

Figure 2. Venn diagram representing the presence/absence of groups of orthologous genes in the organisms belonging to the *albidoflavus* phylogroup.

An additional pan-genome analysis similar to the aforementioned analysis was also performed including the *Streptomyces koyangensis* strain VK-A60T in the dataset (Figure S1), which was an isolate shown to be a closely related neighbour to the *albidoflavus* phylogroup (Figure 1, clade 2). When compared to the previous analysis, the pan-genome analysis including the VK-A60T isolate showed significant changes in the values representing the core genome, which changed from 5177 genes (Figure 2) to 3912 genes (Figure S1), with an additional 1273 genes also shared among all of the *albidoflavus* isolates (Figure S1). The results also showed a much larger number of genes uniquely present in the VK-A60T genome than in the other genomes, with 2059 unique genes identified from a total of 6245 CDSs present in the VK-A60T genome in total, or approximately a third of its total number of genes (Figure S1). This proportion of unique genes present in the VK-A60T genome is considerably higher than the proportions of unique genes observed in the other *albidoflavus* phylotype genomes (Figure 2), which accounted for approximately only 2.5% of the total number of genes in

SM17; 4.2% in J1074; 4.9% in KJ40; 5% in FR-008; and 5.1% in SM254. Taken together, these results further demonstrate the similarities between the isolates belonging to the *albidoflavus* phylogroup, while the VK-A60T isolate is clearly more distantly related.

Thus, from previous studies [30,31] and in light of the phylogeny analysis and further genomic evidence presented in this study, it is likely that all the isolates belonging to the *albidoflavus* phylogroup are in fact members of the same species. It is reasonable to infer that, for example, the isolates in the *albidoflavus* phylogroup that possess no species assignment thus far (i.e., strains SM17 and FR-008) are indeed members of the *albidoflavus* species. Also, it is possible that the *Streptomyces sampsonii* KJ40 has been misassigned, and possibly requires reclassification as an *albidoflavus* isolate.

Misassignment and reclassification of *Streptomyces* species is a common issue, and an increase in the quantity and the quality of available data from these organisms (e.g., better-quality genomes available in the databases) will provide better support for taxonomy claims, or correction of these when new information becomes available [31,57–59].

3.3. Prediction of Secondary Metabolites Biosynthetic Gene Clusters in the Albidoflavus Phylogroup

Isolates belonging to the *albidoflavus* phylogroup have been reported to produce bioactive compounds of pharmacological relevance, such as antibiotics. As mentioned previously, the *Streptomyces albidoflavus* strain J1074 is the best described member of the *albidoflavus* phylogroup to date. As such, several of secondary metabolites produced by this isolate have been identified, including acyl-surugamides and surugamides with antifungal and anticancer activities, respectively [16]; together with paulomycin derivatives with antibacterial activity [60]. The *Streptomyces* sp. FR-008 isolate has been shown to produce the antimicrobial compound FR-008/candicidin [61,62]; while the *Streptomyces sampsonii* KJ40 isolate has been shown to produce a chitinase that possesses anti-fungal activity against plant pathogens [53]. On the other hand, although no bioactive compounds have been characterised from *Streptomyces albidoflavus* SM254, this isolate has been shown to possess anti-fungal activity, specifically against the fungal bat pathogen *Pseudogymnoascus destructans*, which is responsible for the White-nose Syndrome [55,63]. The *Streptomyces* sp. SM17 isolate has also previously been shown to possess antibacterial and antifungal activities against clinically relevant pathogens, including methicillin-resistant *Staphylococcus aureus* (MRSA) [13]. However, no natural products derived from this strain have been identified and isolated until now.

In order to further *in silico* assess the potential of these *albidoflavus* phylogroup isolates to produce secondary metabolites, and also to determine how potentially similar or diverse they are within this phylogroup, prediction of secondary metabolites biosynthetic gene clusters (BGCs) was performed using the antiSMASH (version 5) program [42]. The antiSMASH prediction was processed using the BiG-SCAPE program [43], in order to cluster the BGCs into gene cluster families (GCFs), based on sequence and Pfam [64] protein families similarity, and also by comparing them to the BGCs available from the minimum information about a biosynthetic gene cluster (MiBIG) repository [44] (Figure 3). When compared to known BGCs from the MiBIG database, a significant number of BGCs predicted to be present in the *albidoflavus* phylogroup genomes could potentially encode for the production of novel compounds, including those belonging to the non-ribosomal peptide synthetase (NRPS) and bacteriocin families of compounds (Figure 3). The presence/absence of homologous BGCs in the *albidoflavus* isolates' genomes was determined using BiG-SCAPE and is represented in Figure 4. Interestingly, the vast majority of the BGCs predicted in the *albidoflavus* phylogroup are shared among all of its members (15 BGCs); while another large portion (8 BGCs) are present in at least two isolates (Figure 4). Among the five members of the *albidoflavus* phylogroup, only the J1074 strain and the SM17 strain appeared to possess unique BGCs when compared to the other strains. Three unique BGCs were predicted to be present in the J1074 genome: a predicted type I polyketide synthase (T1PKS)/NRPS without significant similarity to the BGCs from the MiBIG database; a predicted bacteriocin, which also did not show any significant similarity to the BGCs from the MiBIG database; and a BGC predicted to encode for the production of the antibiotic paulomycin, with similarity to the paulomycin-encoding

BGCs from *Streptomyces paulus* and *Streptomyces* sp. YN86 [65], which has also been experimentally shown to be produced by the J1074 strain [60]. One BGC predicted to encode a type III polyketide synthase (T3PKS)—with no significant similarity to the BGCs from the MiBIG database—was also identified as being unique to the SM17 genome.

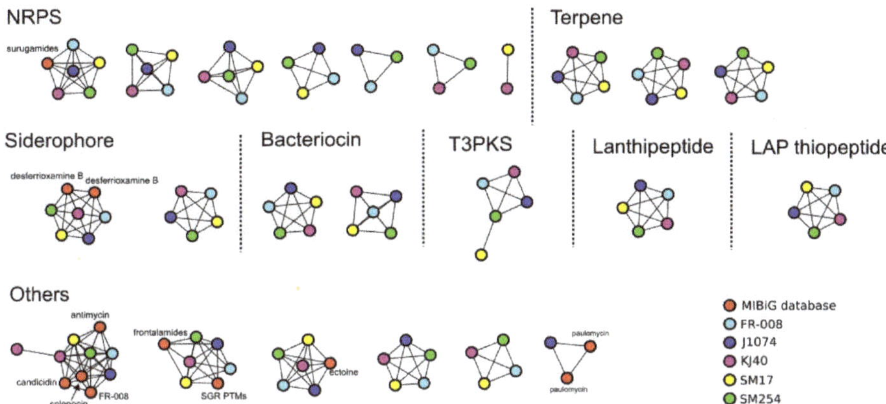

Figure 3. Biosynthetic gene clusters (BGCs) similarity clustering using BiG-SCAPE. Singletons, i.e., BGCs without significant similarity with the BGCs from the minimum information about a biosynthetic gene cluster (MiBIG) database or with the BGCs predicted in other genomes, are not represented.

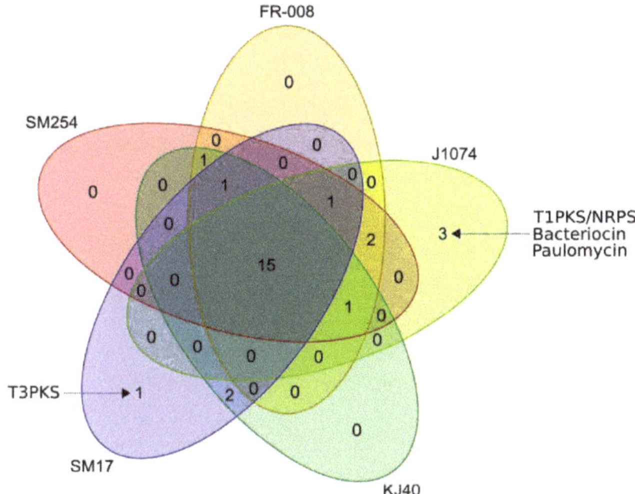

Figure 4. Venn diagram representing BGCs presence/absence in the genomes of the members of the *albidoflavus* phylogroup, determined using antiSMASH and BiG-SCAPE.

Importantly, BGCs with similarity to the surugamide A/D BGC from "*Streptomyces albus* J1074" (now classified as *S. albidoflavus*) from the MiBIG database [16] were identified in all the other genomes of the members of the *albidoflavus* phylogroup. This raises the possibility that this BGC may be commonly present in *albidoflavus* species isolates. However, as only a few complete genomes of isolates belonging to this phylogroup are currently available, further data will be required to support this hypothesis. Nevertheless, these results further highlight the genetic similarities of the isolates belonging to the *albidoflavus* phylogroup, even with respect to their potential to produce secondary metabolites.

3.4. Phylogeny and Gene Synteny Analysis of Sur BGC Homologs

In parallel to the previous phylogenomics analysis performed with the *albidoflavus* phylogroup isolates, sequence similarity and phylogenetic analyses were performed, using the previously described and experimentally characterised *Streptomyces albidoflavus* LHW3101 surugamides biosynthetic gene cluster (*sur* BGC, GenBank accession number: MH070261) as a reference [18]. The aim was to assess how widespread in nature the *sur* BGC might be, and the degree of genetic variation, if any; that might be present in *sur* BGCs belonging to different microorganisms.

Nucleotides sequence similarity to the *sur* BGC was performed in the GenBank database [35], using the NCBI BLASTN tool [36,37]. It is important to note that, since the quality of the data is crucial for sequence similarity, homology, and phylogeny inquiries, only complete genome sequences were employed in this analysis. For this reason, for example, the marine *Streptomyces* isolate in which surugamides and derivatives were originally identified, namely *Streptomyces* sp. JAMM992 [14], was not included, since its complete genome is not available in the GenBank database.

The sequence similarity analysis identified five microorganisms that possessed homologs to the *sur* BGC and had their complete genome sequences available in the GenBank database: *Streptomyces* sp. SM17; *Streptomyces albidoflavus* SM254; *Streptomyces* sp. FR-008; *Streptomyces albidoflavus* J1074; and *Streptomyces sampsonii* KJ40. Notably, these results overlapped with the isolates belonging to the previously discussed *albidoflavus* phylogroup (Figure 1), further highlighting the possibility that the *sur* BGC may be commonly present in and potentially exclusive to the *albidoflavus* species.

Phylogenetic analysis was performed in the genomic regions determined to be homologs to the *Streptomyces albidoflavus* LHW3101 *sur* BGC, using the MrBayes program [39] (Figure 5). Although a larger number of sequences should ideally be employed in this type of analysis, these results suggest the possibility of a clade with aquatic saline environment-derived *sur* BGCs (Figure 5). Thus, these aquatic saline environment-derived *sur* BGCs are likely to share more genetic similarities amongst each other, rather than with those derived from terrestrial environments. Since this analysis took into consideration the whole genome regions that contained the *sur* BGCs of each isolate, it is likely that the similarities and differences present in these regions involve not only coding sequences (CDSs) for biosynthetic genes and/or transcriptional regulators, but also could include promoter regions and other intergenic sequences.

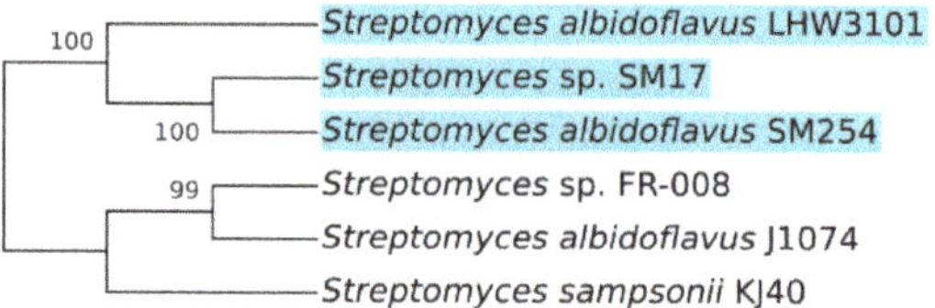

Figure 5. Consensus phylogenetic tree of the *sur* BGC region of the *S. albidoflavus* LHW3101 reference *sur* BGC sequence, plus five *Streptomyces* isolates determined to have *sur* BGC homologs, generated using MrBayes and Mega X, with a 95% posterior probability cut-off. Aquatic saline environment-derived isolates are highlighted in cyan.

With this in mind, the genomic regions previously determined to share homology with the *sur* BGC from *S. albidoflavus* LHW3101 were further analysed, with respect to the genes present in the surrounding region, the organisation of the BGCs, together with the overall gene synteny (Figure 6). Translated CDSs predicted in the region were manually annotated using the NCBI BLASTP tool [36,37], together with GenBank [35] and the CDD [66] databases. These included the main biosynthetic genes, namely *surABCD*, the transcriptional regulator *surR*, and the thioesterase *surE*—all of which had previously been reported to have roles in the biosynthesis of surugamides and their derivatives [15–19] (Figure 6).

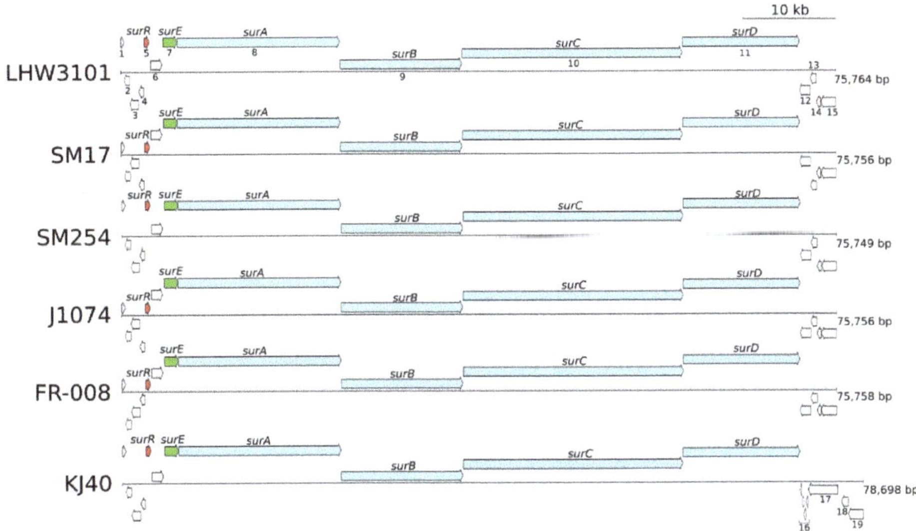

Figure 6. Gene synteny of the *sur* BGC region, including the reference *sur* BGC nucleotide sequence (LHW3101) and each of the *albidoflavus* phylogroup genomes. Arrows at different positions represent genes transcribed in different reading frames.

Interestingly, this result indicated that the gene synteny of the biosynthetic genes as well as the flanking genes is highly conserved, with the exception to the 3′ flanking region of the BGC from *S. sampsonii* KJ40. Notably, even the reading frames of the *surE* gene and the *surABCD* genes are conserved amongst all the genomes. As indicated by the numbers in Figure 6, the 5′ region in all the genomic regions consisted of: 1) A MbtH-like protein, which have been reported to be involved in the synthesis of non-ribosomal peptides, antibiotics, and siderophores, in *Streptomyces* species [67,68]; 2) a putative ABC transporter, which is a family of proteins with varied biological functions, including conferring resistance to drugs and other toxic compounds [69,70]; 3) a BcrA family ABC transporter, which is a family commonly involved in peptide antibiotics resistance [71,72]; 4) a hypothetical protein; followed by 5) the transcriptional repressor SurR, which has been experimentally demonstrated to repress the production of surugamides [16]; 6) a hypothetical membrane protein; 7) the thioesterase SurE, which is homologous to the penicillin binding protein, reported to be responsible for the cyclisation of surugamides molecules [21]; and finally 8–11) the main surugamides biosynthetic genes *surABCD*, all of which encode non-ribosomal peptide synthetase (NRPS) proteins [19]. The 3′ flanking region consisted of: 12) A predicted multi-drug resistance (MDR) transporter belonging to the major facilitator superfamily (MFS) of membrane transport proteins [73,74]; 13) a predicted TetR/AcrR transcriptional regulator, which is a family of regulators reported to be involved in antibiotic resistance [75]; 14) a hypothetical protein; and 15) another predicted MDR transporter belonging to the MFS superfamily. In contrast, the 3′ flanking region of the KJ40 strain *sur* BGC, consisted of: 16) A group of four hypothetical proteins, which may represent pseudogene versions of the first MDR transporter identified in the other isolates (gene number 12 in Figure 6); 17) a predicted rearrangement hotspot (RHS) repeat protein, which is a family of proteins reported to be involved in mediating intercellular competition in bacteria [76]; 18) a hypothetical protein; and 19) a MDR transporter belonging to the MFS superfamily, which, interestingly, is a homolog of protein number 15, which is present in all the other isolates.

The conserved gene synteny observed in the *sur* BGC genomic region, particularly those positioned upstream of the main biosynthetic *surABCD* genes, together with the observation that even the reading frames of the *surE* and the *surABCD* genes are conserved among all the genomes analysed, coupled with the previous phylogenetic and pan-genome analyses, suggest the following. Firstly, it is very likely that these strains share a common ancestry and that the *sur* BGC genes had a common origin. Secondly, there is a strong evolutionary pressure ensuring the maintenance of not only gene synteny, but also of the reading frames of the main biosynthetic genes involved in the production of surugamides. The latter raises the question of which other genes in this region may be involved in the production of these compounds, or potentially conferring mechanisms of self-resistance to surugamides in the isolates, particularly since many of the genes have predicted functions that are compatible with the transport of small molecules and with multi-drug resistance. These observations are particularly interesting considering that these strains are derived from quite varied environments and geographic locations.

3.5. Growth, Morphology, Phenotype, and Metabolism Assessment of Streptomyces sp. SM17 in Complex Media

In order to assess the metabolic potential of the SM17 strain [77], particularly with respect to the production of surugamide A, the isolate was cultivated in a number of different growth media, within an OSMAC-based approach [33,34]. While the SM17 strain was able to grow in SYP-NaCl, YD, SY, P1, P2, P3, and CH-F2 liquid media, the strain was unable to grow in Oatmeal and Sporulation media. The latter indicated an inability to metabolise oat and starch when nutrients other than yeast extract are not present. Morphologically, the SM17 strain formed cell aggregates or pellets in TSB, YD, and SYP-NaCl, while this differentiation was not observed in the other media. Preliminary chemical analyses of these samples, employing liquid chromatography–mass spectrometry (UPLC-DAD-HRMS), indicated that secondary metabolism in SM17 was not very active when the strain was cultivated in SY, P1, P2, P3, and CH-F2 media. In contrast, significant production of surugamides was evidenced in the extracts from TSB, SYP-NaCl, and YD media, with characteristic ions at *m/z* 934.6106 (surugamide A) and 920.5949 (surugamide B) [M + Na]$^+$, which correlated with the formation of cell pellets and the production of natural products, as previously described in other *Streptomyces* strains [77,78].

3.6. Differential Production of Surugamide A by Streptomyces sp. SM17 and S. albidoflavus J1074

To confirm the production of surugamide A by the SM17 isolate, extracts from the TSB, SYP-NaCl, and YD media were combined and purified using high-performance liquid chromatography (HPLC). The structures of the major compounds of the extract were subsequently analysed using nuclear magnetic resonance (NMR) spectroscopy, which allowed for the identification of the chemical structure of the surugamide A molecule as major metabolite by comparison with reference NMR data (Figure 7) [14].

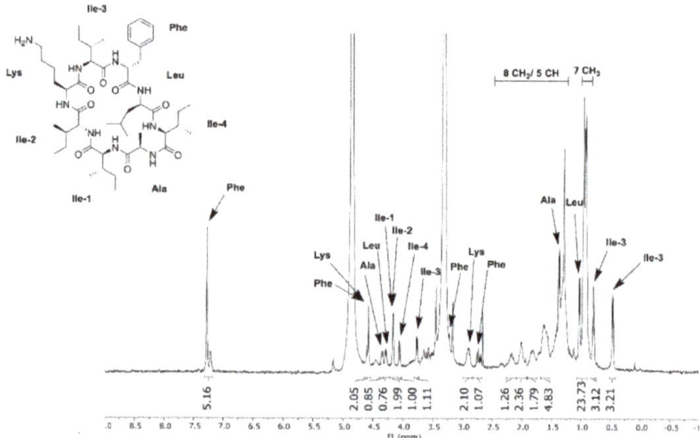

Figure 7. Structure of surugamide A isolated from SM17 grown in TSB, SYP-NaCl, and YD medium with annotated ^{1}H NMR spectrum obtained in CD$_3$OD at 500 MHz.

The isolates *Streptomyces* sp. SM17 and *S. albidoflavus* J1074 were subsequently cultivated in the aforementioned media in which the SM17 strain had been shown to be metabolically active, namely the TSB, SYP-NaCl, and YD media. This was performed in order to assess whether there were any significant differences in the production of surugamide A when different growth media are employed for the production of this compound, and to compare the levels of surugamide A produced by the SM17 and the J1074 isolates. The MeOH/DCM (1:1) extracts from the aforementioned cultures of SM17 and J1074 were subjected to liquid chromatography–mass spectrometry (UPLC-HRMS) to quantify the levels of surugamide A being produced under each condition (Table 1), using a surugamide A standard calibration curve (Figure S2).

Table 1. Surugamide A production by SM17 and J1074 measured using different media.

Strain	Media	Percent (*w/w*) of Extract	Concentration of Surugamide A (mg/L) Corrected in 5 mg/mL of Extract
SM17	TSB	2.44%	122.01
SM17	SYP-NaCl	10.60%	530.15
SM17	YD	1.13%	56.27
J1074	TSB	0.27%	13.32
J1074	SYP-NaCl	3.55%	176.82
J1074	YD	0.09%	4.26

The LC-MS quantification analysis (Table 1) indicated that both strains were capable of producing surugamide A in all the conditions tested. However, the SM17 strain appeared to produce considerably higher yields of the compound when compared to J1074, in all the conditions analysed. In addition, the *S. albidoflavus* J1074 isolate appeared to produce quite low levels of surugamide A when grown in TSB and YD media, accounting for less than 1% (*w/w*) of the extracts from these media. Interestingly, higher yields of surugamide A were produced in the SYP-NaCl medium in both strains, when compared with the levels of surugamide A produced by these strains when grown in TSB and the YD media (Table 1). In the SM17 culture in SYP-NaCl, surugamide A accounted for 10.60% (*w/w*) of the extract, compared to 2.44% and 1.13% from TSB and YD, respectively; while in J1074 it accounted for 3.55% (*w/w*) of the extract from the SYP-NaCl culture, compared to 0.27% and 0.09% from TSB and YD, respectively (Table 1). These results provide further insights into factors that are potentially involved in regulation the biosynthesis of surugamide A, in the *albidoflavus* phylogroup and in *Streptomyces* sp. SM17 in particular.

Firstly, it appears likely that surugamide A biosynthesis may be regulated, at least in part, by carbon catabolite repression (CCR). Carbon catabolite repression is a well-described regulatory mechanism in bacteria that controls carbon metabolism [79–82], and which has also been reported to regulate the biosynthesis of secondary metabolites in a number of different bacterial species, including in *Streptomyces* isolates [83–86]. While the TSB and the YD media contain glucose and dextrins as carbon sources, respectively; the complex polysaccharide starch is the carbon source in the SYP-NaCl medium. Therefore, it is reasonable to infer that glucose and dextrin may repress the production of surugamide A in *Streptomyces* sp. SM17 and in *Streptomyces albidoflavus* J1074, while starch does not. Further evidence for this can be found when considering the different production media previously employed in the production of surugamides by different *Streptomyces* isolates. For example, in the original research that led to the discovery of surugamides in *Streptomyces* sp. JAMM992 [14], the PC-1 medium (1% starch, 1% polypeptone, 1% meat extract, 1% molasses, pH 7.2) was employed for production of these compounds. Similar to the SYP-NaCl medium employed in our study, the PC-1 medium also contains starch as the carbon source, together with another complex carbon source, namely molasses. Likewise, for the production of surugamides in *S. albidoflavus* strain LHW3101 [18], the TSBY medium (3% tryptone soy broth, 10.3% sucrose, 0.5% yeast extract) was employed, which utilises sucrose as its main carbon source. In contrast, when elicitors were employed to induce the production of surugamides and their derivatives in the J1074 strain [16], by activating the *sur* BGC, which appeared to be silent in this isolate, the R4 medium (0.5% glucose, 0.1% yeast extract, among other non-carbon related components) was employed, which utilises glucose as its main carbon supply, and, as shown in this study, it potentially represses the production of surugamide A. Thus, from these previous reports and from our observations, it appears likely that CCR plays an important role in regulating the biosynthesis of surugamides.

Secondly, it is important to note the presence of salts in the form of NaCl in the SYP-NaCl medium. As previously mentioned, genetic and phylogenetic analyses of the *sur* BGC indicated similarities between those BGCs belonging to aquatic saline-derived *Streptomyces* isolates (Figure 5), together with the likelihood that these *sur* BGCs might have had a common origin. Thus, it is plausible that this origin may have been marine, and hence the presence of salts in the growth medium may also have an influence on the biosynthesis of surugamide A. Different concentrations of salts in the form of NaCl in the culture medium have also previously been shown to impact on the chemical profile of metabolites produced in the marine-obligate bacteria *Salinispora arenicola* [87].

Nevertheless, it is interesting to observe that, despite the repression/induction of the biosynthesis of surugamide A observed when different media were employed, the SM17 isolate clearly produces considerably higher amounts of surugamide A when compared to *S. albidoflavus* J1074—reaching yields up to >13-fold higher in the YD medium, and around 3-fold higher when grown in the SYP-NaCl medium (Table 1).

4. Conclusions

Marine-derived bacteria, particularly those isolated in association with marine invertebrates, such as sponges, have been shown to be reservoirs of bioactive molecules, including those with antibacterial, antifungal, and anticancer activities. Among these newly identified bioactive compounds, the surugamides and their derivatives are of particular interest due to their clinically relevant bioactivities, i.e., anticancer and antifungal, and their original metabolic pathway.

Based on genome mining, this study identified the previously unreported capability of the marine sponge-derived isolate *Streptomyces* sp. SM17 to produce surugamide A and also sheds new light on factors such as the carbon catabolite repression (CCR) that may be involved in regulating production of this molecule. Phylogenomics analysis indicated that the *sur* BGC is commonly present in members of the proposed *albidoflavus* phylogroup, and that the *sur* BGCs present in different isolates derived from varied environmental niches may possess a common ancestry. Although high quality genomic data from this proposed *albidoflavus* phylogroup are still lacking, results presented here suggest that the *sur* BGCs derived from *Streptomyces* isolated from aquatic saline environment are more similar to each other, when compared to those isolated from terrestrial environments.

Chemical analysis was performed in order to assess differential production of surugamide A when comparing a marine *Streptomyces* isolate with a terrestrial *Streptomyces* isolate, namely SM17 and J1074 strains, respectively, following an OSMAC-based approach employing different culture media. This analysis showed that not only the marine-derived isolate SM17 was capable of producing more surugamide A when compared to J1074 under all the conditions tested, but also that the biosynthesis of surugamide A is likely to be influenced by the CCR, and potentially by the presence of salts in the growth medium. These results also highlight the importance of employing an OSMAC-based approach even when analysing the production of known compounds, since there is a clear difference in the yields of surugamide A obtained when employing different culture media. Thus, it is possible to gain further insights into the production of bacterial types of compounds by 1) discovering strains that possess a higher capability to produce these compounds; 2) establishing optimal conditions for the biosynthesis of their production; and 3) providing a better understanding of the genetic and regulatory mechanisms potentially underpinning the production of these compounds.

Supplementary Materials: The following are available online at http://www.mdpi.com/2076-2607/7/10/394/s1, Figure S1: Groups of orthologous genes in the genomes of the *albidoflavus* phylogroup, Figure S2: Calibration curve for surugamide A, Table S1: Genome statistics of the *albidoflavus* phylogroup.

Author Contributions: Conceived and designed the experiments: E.L.A., A.F.C.R, O.P.T, A.D.W.D. Performed the experiments: E.L.A., A.F.C.R., N.K., L.K.J. Analysed the data: E.L.A., A.F.C.R., N.K., L.K.J., O.P.T., S.A.J., A.D.W.D. Wrote the paper: E.L.A., O.P.T., A.D.W.D.

Funding: E.L.A. acknowledges the funding support of the Brazilian National Council for Scientific and Technological Development (CNPq). This project was supported by a research grant from the Marine Institute through the NMBLI project (Grant-Aid Agreement PBA/MB/16/01) and by the Marine Biotechnology ERA/NET, NEPTUNA project (contract No. PBA/MB/15/02) and by Science Foundation Ireland (SSPC-2, 12/RC/2275).

Conflicts of Interest: The authors declare no conflict of interest.

References

1. Hwang, K.-S.; Kim, H.U.; Charusanti, P.; Palsson, B.Ø.; Lee, S.Y. Systems biology and biotechnology of *Streptomyces* species for the production of secondary metabolites. *Biotechnol. Adv.* **2014**, *32*, 255–268. [CrossRef] [PubMed]

2. Thabit, A.K.; Crandon, J.L.; Nicolau, D.P. Antimicrobial resistance: Impact on clinical and economic outcomes and the need for new antimicrobials. *Expert Opin. Pharmacother.* **2015**, *16*, 159–177. [CrossRef] [PubMed]

3. Tommasi, R.; Brown, D.G.; Walkup, G.K.; Manchester, J.I.; Miller, A.A. ESKAPEing the labyrinth of antibacterial discovery. *Nat. Rev. Drug Discov.* **2015**, *14*, 529–542. [CrossRef] [PubMed]

4. Demers, D.; Knestrick, M.; Fleeman, R.; Tawfik, R.; Azhari, A.; Souza, A.; Vesely, B.; Netherton, M.; Gupta, R.; Colon, B.; et al. Exploitation of Mangrove Endophytic Fungi for Infectious Disease Drug Discovery. *Mar. Drugs* **2018**, *16*, 376. [CrossRef] [PubMed]

5. Indraningrat, A.; Smidt, H.; Sipkema, D. Bioprospecting Sponge-Associated Microbes for Antimicrobial Compounds. *Mar. Drugs* **2016**, *14*, 87. [CrossRef] [PubMed]

6. Calcabrini, C.; Catanzaro, E.; Bishayee, A.; Turrini, E.; Fimognari, C. Marine Sponge Natural Products with Anticancer Potential: An Updated Review. *Mar. Drugs* **2017**, *15*, 310. [CrossRef]

7. Schneemann, I.; Kajahn, I.; Ohlendorf, B.; Zinecker, H.; Erhard, A.; Nagel, K.; Wiese, J.; Imhoff, J.F. Mayamycin, a Cytotoxic Polyketide from a *Streptomyces* Strain Isolated from the Marine Sponge *Halichondria panicea*. *J. Nat. Prod.* **2010**, *73*, 1309–1312. [CrossRef] [PubMed]

8. Kunz, A.; Labes, A.; Wiese, J.; Bruhn, T.; Bringmann, G.; Imhoff, J. Nature's Lab for Derivatization: New and Revised Structures of a Variety of Streptophenazines Produced by a Sponge-Derived *Streptomyces* Strain. *Mar. Drugs* **2014**, *12*, 1699–1714. [CrossRef]

9. Sathiyanarayanan, G.; Gandhimathi, R.; Sabarathnam, B.; Seghal Kiran, G.; Selvin, J. Optimization and production of pyrrolidone antimicrobial agent from marine sponge-associated *Streptomyces* sp. MAPS15. *Bioprocess Biosyst. Eng.* **2014**, *37*, 561–573. [CrossRef]

10. Almeida, E.L.; Margassery, L.M.; Kennedy, J.; Dobson, A.D.W. Draft Genome Sequence of the Antimycin-Producing Bacterium *Streptomyces* sp. Strain SM8, Isolated from the Marine Sponge *Haliclona simulans*. *Genome Announc.* **2018**, *6*, e01535-17. [CrossRef]

11. Viegelmann, C.; Margassery, L.; Kennedy, J.; Zhang, T.; O'Brien, C.; O'Gara, F.; Morrissey, J.; Dobson, A.; Edrada-Ebel, R. Metabolomic Profiling and Genomic Study of a Marine Sponge-Associated *Streptomyces* sp. *Mar. Drugs* **2014**, *12*, 3323–3351. [CrossRef] [PubMed]

12. Jackson, S.; Crossman, L.; Almeida, E.; Margassery, L.; Kennedy, J.; Dobson, A. Diverse and Abundant Secondary Metabolism Biosynthetic Gene Clusters in the Genomes of Marine Sponge Derived *Streptomyces* spp. Isolates. *Mar. Drugs* **2018**, *16*, 67. [CrossRef] [PubMed]

13. Kennedy, J.; Baker, P.; Piper, C.; Cotter, P.D.; Walsh, M.; Mooij, M.J.; Bourke, M.B.; Rea, M.C.; O'Connor, P.M.; Ross, R.P.; et al. Isolation and Analysis of Bacteria with Antimicrobial Activities from the Marine Sponge *Haliclona simulans* Collected from Irish Waters. *Mar. Biotechnol.* **2009**, *11*, 384–396. [CrossRef] [PubMed]

14. Takada, K.; Ninomiya, A.; Naruse, M.; Sun, Y.; Miyazaki, M.; Nogi, Y.; Okada, S.; Matsunaga, S. Surugamides A–E, Cyclic Octapeptides with Four d-Amino Acid Residues, from a Marine *Streptomyces* sp.: LC–MS-Aided Inspection of Partial Hydrolysates for the Distinction of d- and l-Amino Acid Residues in the Sequence. *J. Org. Chem.* **2013**, *78*, 6746–6750. [CrossRef] [PubMed]

15. Matsuda, K.; Kuranaga, T.; Sano, A.; Ninomiya, A.; Takada, K.; Wakimoto, T. The Revised Structure of the Cyclic Octapeptide Surugamide A. *Chem. Pharm. Bull.* **2019**, *67*, 476–480. [CrossRef] [PubMed]

16. Xu, F.; Nazari, B.; Moon, K.; Bushin, L.B.; Seyedsayamdost, M.R. Discovery of a Cryptic Antifungal Compound from *Streptomyces albus* J1074 Using High-Throughput Elicitor Screens. *J. Am. Chem. Soc.* **2017**, *139*, 9203–9212. [CrossRef] [PubMed]

17. Thankachan, D.; Fazal, A.; Francis, D.; Song, L.; Webb, M.E.; Seipke, R.F. A trans-Acting Cyclase Offloading Strategy for Nonribosomal Peptide Synthetases. *ACS Chem. Biol.* **2019**, *14*, 845–849. [CrossRef] [PubMed]

18. Zhou, Y.; Lin, X.; Xu, C.; Shen, Y.; Wang, S.-P.; Liao, H.; Li, L.; Deng, H.; Lin, H.-W. Investigation of Penicillin Binding Protein (PBP)-like Peptide Cyclase and Hydrolase in Surugamide Non-ribosomal Peptide Biosynthesis. *Cell Chem. Biol.* **2019**, *26*, 737–744.e4. [CrossRef] [PubMed]

19. Ninomiya, A.; Katsuyama, Y.; Kuranaga, T.; Miyazaki, M.; Nogi, Y.; Okada, S.; Wakimoto, T.; Ohnishi, Y.; Matsunaga, S.; Takada, K. Biosynthetic Gene Cluster for Surugamide A Encompasses an Unrelated Decapeptide, Surugamide F. *ChemBioChem* **2016**, *17*, 1709–1712. [CrossRef]

20. Kuranaga, T.; Matsuda, K.; Sano, A.; Kobayashi, M.; Ninomiya, A.; Takada, K.; Matsunaga, S.; Wakimoto, T. Total Synthesis of the Nonribosomal Peptide Surugamide B and Identification of a New Offloading Cyclase Family. *Angew. Chemie Int. Ed.* **2018**, *57*, 9447–9451. [CrossRef] [PubMed]

21. Matsuda, K.; Kobayashi, M.; Kuranaga, T.; Takada, K.; Ikeda, H.; Matsunaga, S.; Wakimoto, T. SurE is a trans-acting thioesterase cyclizing two distinct non-ribosomal peptides. *Org. Biomol. Chem.* **2019**, *17*, 1058–1061. [CrossRef] [PubMed]

22. Koshla, O.T.; Rokytskyy, I.V.; Ostash, I.S.; Busche, T.; Kalinowski, J.; Mösker, E.; Süssmuth, R.D.; Fedorenko, V.O.; Ostash, B.O. Secondary Metabolome and Transcriptome of *Streptomyces albus* J1074 in Liquid Medium SG2. *Cytol. Genet.* **2019**, *53*, 1–7. [CrossRef]

23. Chater, K.F.; Wilde, L.C. Restriction of a bacteriophage of *Streptomyces albus* G involving endonuclease *Sal*I. *J. Bacteriol.* **1976**, *128*, 644–650. [PubMed]

24. Chater, K.F.; Wilde, L.C. *Streptomyces albus* G Mutants Defective in the *Sal*GI Restriction-Modification System. *Microbiology* **1980**, *116*, 323–334. [CrossRef] [PubMed]

25. Huang, C.; Yang, C.; Zhang, W.; Zhu, Y.; Ma, L.; Fang, Z.; Zhang, C. Albumycin, a new isoindolequinone from *Streptomyces albus* J1074 harboring the fluostatin biosynthetic gene cluster. *J. Antibiot. (Tokyo)* **2019**, *72*, 311–315. [CrossRef] [PubMed]

26. Zaburannyi, N.; Rabyk, M.; Ostash, B.; Fedorenko, V.; Luzhetskyy, A. Insights into naturally minimised *Streptomyces albus* J1074 genome. *BMC Genomics* **2014**, *15*, 97. [CrossRef] [PubMed]

27. Myronovskyi, M.; Rosenkränzer, B.; Nadmid, S.; Pujic, P.; Normand, P.; Luzhetskyy, A. Generation of a cluster-free *Streptomyces albus* chassis strains for improved heterologous expression of secondary metabolite clusters. *Metab. Eng.* **2018**, *49*, 316–324. [CrossRef]

28. Bilyk, O.; Sekurova, O.N.; Zotchev, S.B.; Luzhetskyy, A. Cloning and Heterologous Expression of the Grecocycline Biosynthetic Gene Cluster. *PLoS ONE* **2016**, *11*, e0158682. [CrossRef]

29. Jiang, G.; Zhang, Y.; Powell, M.M.; Zhang, P.; Zuo, R.; Zhang, Y.; Kallifidas, D.; Tieu, A.M.; Luesch, H.; Loria, R.; et al. High-Yield Production of Herbicidal Thaxtomins and Thaxtomin Analogs in a Nonpathogenic *Streptomyces* Strain. *Appl. Environ. Microbiol.* **2018**, *84*, e00164-18. [CrossRef]

30. Labeda, D.P.; Doroghazi, J.R.; Ju, K.-S.; Metcalf, W.W. Taxonomic evaluation of *Streptomyces albus* and related species using multilocus sequence analysis and proposals to emend the description of *Streptomyces albus* and describe *Streptomyces pathocidini* sp. nov. *Int. J. Syst. Evol. Microbiol.* **2014**, *64*, 894–900. [CrossRef]

31. Labeda, D.P.; Dunlap, C.A.; Rong, X.; Huang, Y.; Doroghazi, J.R.; Ju, K.-S.; Metcalf, W.W. Phylogenetic relationships in the family Streptomycetaceae using multi-locus sequence analysis. *Antonie Van Leeuwenhoek* **2017**, *110*, 563–583. [CrossRef] [PubMed]

32. Almeida, E.L.; Carillo Rincón, A.F.; Jackson, S.A.; Dobson, A.D. Comparative genomics of marine sponge-derived *Streptomyces* spp. isolates SM17 and SM18 with their closest terrestrial relatives provides novel insights into environmental niche adaptations and secondary metabolite biosynthesis potential. *Front. Microbiol.* **2019**, *10*, 1713. [CrossRef] [PubMed]

33. Romano, S.; Jackson, S.; Patry, S.; Dobson, A. Extending the "One Strain Many Compounds" (OSMAC) Principle to Marine Microorganisms. *Mar. Drugs* **2018**, *16*, 244. [CrossRef] [PubMed]

34. Pan, R.; Bai, X.; Chen, J.; Zhang, H.; Wang, H. Exploring Structural Diversity of Microbe Secondary Metabolites Using OSMAC Strategy: A Literature Review. *Front. Microbiol.* **2019**, *10*, 294. [CrossRef] [PubMed]

35. Benson, D.A.; Cavanaugh, M.; Clark, K.; Karsch-Mizrachi, I.; Ostell, J.; Pruitt, K.D.; Sayers, E.W. GenBank. *Nucleic Acids Res.* **2018**, *46*, D41–D47. [CrossRef] [PubMed]
36. Johnson, M.; Zaretskaya, I.; Raytselis, Y.; Merezhuk, Y.; McGinnis, S.; Madden, T.L. NCBI BLAST: A better web interface. *Nucleic Acids Res.* **2008**, *36*, W5–W9. [CrossRef] [PubMed]
37. Camacho, C.; Coulouris, G.; Avagyan, V.; Ma, N.; Papadopoulos, J.; Bealer, K.; Madden, T.L. BLAST+: Architecture and applications. *BMC Bioinform.* **2009**, *10*, 421. [CrossRef] [PubMed]
38. Katoh, K.; Standley, D.M. MAFFT Multiple Sequence Alignment Software Version 7: Improvements in Performance and Usability. *Mol. Biol. Evol.* **2013**, *30*, 772–780. [CrossRef] [PubMed]
39. Ronquist, F.; Teslenko, M.; van der Mark, P.; Ayres, D.L.; Darling, A.; Höhna, S.; Larget, B.; Liu, L.; Suchard, M.A.; Huelsenbeck, J.P. MrBayes 3.2: Efficient Bayesian Phylogenetic Inference and Model Choice Across a Large Model Space. *Syst. Biol.* **2012**, *61*, 539–542. [CrossRef] [PubMed]
40. Waddell, P.J.; Steel, M. General Time-Reversible Distances with Unequal Rates across Sites: Mixing Γ and Inverse Gaussian Distributions with Invariant Sites. *Mol. Phylogenet. Evol.* **1997**, *8*, 398–414. [CrossRef] [PubMed]
41. Kumar, S.; Stecher, G.; Li, M.; Knyaz, C.; Tamura, K. MEGA X: Molecular Evolutionary Genetics Analysis across Computing Platforms. *Mol. Biol. Evol.* **2018**, *35*, 1547–1549. [CrossRef] [PubMed]
42. Blin, K.; Shaw, S.; Steinke, K.; Villebro, R.; Ziemert, N.; Lee, S.Y.; Medema, M.H.; Weber, T. antiSMASH 5.0: Updates to the secondary metabolite genome mining pipeline. *Nucleic Acids Res.* **2019**, *47*, 81–87. [CrossRef] [PubMed]
43. Navarro-Muñoz, J.C.; Selem-Mojica, N.; Mullowney, M.W.; Kautsar, S.; Tryon, J.H.; Parkinson, E.I.; Santos, E.L.C.D.L.; Yeong, M.; Cruz-Morales, P.; Abubucker, S.; et al. A computational framework for systematic exploration of biosynthetic diversity from large-scale genomic data. *bioRxiv* **2018**, *10*, 445270.
44. Medema, M.H.; Kottmann, R.; Yilmaz, P.; Cummings, M.; Biggins, J.B.; Blin, K.; de Bruijn, I.; Chooi, Y.H.; Claesen, J.; Coates, R.C.; et al. Minimum Information about a Biosynthetic Gene cluster. *Nat. Chem. Biol.* **2015**, *11*, 625–631. [CrossRef]
45. Shannon, P.; Markiel, A.; Ozier, O.; Baliga, N.S.; Wang, J.T.; Ramage, D.; Amin, N.; Schwikowski, B.; Ideker, T. Cytoscape: A software environment for integrated models of biomolecular interaction networks. *Genome Res.* **2003**, *13*, 2498–2504. [CrossRef] [PubMed]
46. Okonechnikov, K.; Golosova, O.; Fursov, M. Unipro UGENE: A unified bioinformatics toolkit. *Bioinformatics* **2012**, *28*, 1166–1167. [CrossRef] [PubMed]
47. Rutherford, K.; Parkhill, J.; Crook, J.; Horsnell, T.; Rice, P.; Rajandream, M.-A.; Barrell, B. Artemis: Sequence visualization and annotation. *Bioinformatics* **2000**, *16*, 944–945. [CrossRef] [PubMed]
48. Dusa, A. Venn: Draw Venn Diagrams 2018. Available online: https://cran.r-project.org/web/packages/venn/index.html (accessed on 25 September 2019).
49. *R Core Team R: A Language and Environment for Statistical Computing*, version 3.5.3; Vienna, Austria, 2018. Available online: https://www.R-project.org/ (accessed on 25 September 2019).
50. RStudio Team. *RStudio Team RStudio: Integrated Development Environment for RStudio Inc.*; RStudio Inc.: Boston, MA, USA, 2015; Volume 14.
51. Glaeser, S.P.; Kämpfer, P. Multilocus sequence analysis (MLSA) in prokaryotic taxonomy. *Syst. Appl. Microbiol.* **2015**, *38*, 237–245. [CrossRef]
52. Gadagkar, S.R.; Rosenberg, M.S.; Kumar, S. Inferring species phylogenies from multiple genes: Concatenated sequence tree versus consensus gene tree. *J. Exp. Zool. Part B Mol. Dev. Evol.* **2005**, *304B*, 64–74. [CrossRef]
53. Li, S.; Zhang, B.; Zhu, H.; Zhu, T. Cloning and Expression of the Chitinase Encoded by *ChiKJ406136* from *Streptomyces sampsonii* (Millard & Burr) Waksman KJ40 and Its Antifungal Effect. *Forests* **2018**, *9*, 699.
54. Liu, Q.; Xiao, L.; Zhou, Y.; Deng, K.; Tan, G.; Han, Y.; Liu, X.; Deng, Z.; Liu, T. Development of *Streptomyces* sp. FR-008 as an emerging chassis. *Synth. Syst. Biotechnol.* **2016**, *1*, 207–214. [CrossRef] [PubMed]
55. Badalamenti, J.P.; Erickson, J.D.; Salomon, C.E. Complete Genome Sequence of *Streptomyces albus* SM254, a Potent Antagonist of Bat White-Nose Syndrome Pathogen *Pseudogymnoascus destructans*. *Genome Announc.* **2016**, *4*, e00290-16. [CrossRef] [PubMed]
56. Page, A.J.; Cummins, C.A.; Hunt, M.; Wong, V.K.; Reuter, S.; Holden, M.T.G.; Fookes, M.; Falush, D.; Keane, J.A.; Parkhill, J. Roary: Rapid large-scale prokaryote pan genome analysis. *Bioinformatics* **2015**, *31*, 3691–3693. [CrossRef] [PubMed]

57. Rong, X.; Doroghazi, J.R.; Cheng, K.; Zhang, L.; Buckley, D.H.; Huang, Y. Classification of *Streptomyces* phylogroup *pratensis* (Doroghazi and Buckley, 2010) based on genetic and phenotypic evidence, and proposal of *Streptomyces pratensis* sp. nov. *Syst. Appl. Microbiol.* **2013**, *36*, 401–407. [CrossRef] [PubMed]

58. Ward, A.; Allenby, N. Genome mining for the search and discovery of bioactive compounds: The *Streptomyces* paradigm. *FEMS Microbiol. Lett.* **2018**, *365*, fny240. [CrossRef]

59. Li, Y.; Pinto-Tomás, A.A.; Rong, X.; Cheng, K.; Liu, M.; Huang, Y. Population Genomics Insights into Adaptive Evolution and Ecological Differentiation in Streptomycetes. *Appl. Environ. Microbiol.* **2019**, *85*, e02555-18. [CrossRef]

60. Hoz, J.F.-D.L.; Méndez, C.; Salas, J.A.; Olano, C. Novel Bioactive Paulomycin Derivatives Produced by *Streptomyces albus* J1074. *Molecules* **2017**, *22*, 1758. [CrossRef]

61. Chen, S.; Huang, X.; Zhou, X.; Bai, L.; He, J.; Jeong, K.J.; Lee, S.Y.; Deng, Z. Organizational and Mutational Analysis of a Complete FR-008/Candicidin Gene Cluster Encoding a Structurally Related Polyene Complex. *Chem. Biol.* **2003**, *10*, 1065–1076. [CrossRef]

62. Zhao, Z.; Deng, Z.; Pang, X.; Chen, X.-L.; Zhang, P.; Bai, L.; Li, H. Production of the antibiotic FR-008/candicidin in *Streptomyces* sp. FR-008 is co-regulated by two regulators, FscRI and FscRIV, from different transcription factor families. *Microbiology* **2015**, *161*, 539–552.

63. Hamm, P.S.; Caimi, N.A.; Northup, D.E.; Valdez, E.W.; Buecher, D.C.; Dunlap, C.A.; Labeda, D.P.; Lueschow, S.; Porras-Alfaro, A. Western Bats as a Reservoir of Novel *Streptomyces* Species with Antifungal Activity. *Appl. Environ. Microbiol.* **2017**, *83*, e03057-16. [CrossRef]

64. El-Gebali, S.; Mistry, J.; Bateman, A.; Eddy, S.R.; Luciani, A.; Potter, S.C.; Qureshi, M.; Richardson, L.J.; Salazar, G.A.; Smart, A.; et al. The Pfam protein families database in 2019. *Nucleic Acids Res.* **2019**, *47*, D427–D432. [CrossRef] [PubMed]

65. Li, J.; Xie, Z.; Wang, M.; Ai, G.; Chen, Y. Identification and Analysis of the Paulomycin Biosynthetic Gene Cluster and Titer Improvement of the Paulomycins in *Streptomyces paulus* NRRL 8115. *PLoS ONE* **2015**, *10*, e0120542. [CrossRef] [PubMed]

66. Marchler-Bauer, A.; Derbyshire, M.K.; Gonzales, N.R.; Lu, S.; Chitsaz, F.; Geer, L.Y.; Geer, R.C.; He, J.; Gwadz, M.; Hurwitz, D.I.; et al. CDD: NCBI's conserved domain database. *Nucleic Acids Res.* **2015**, *43*, D222–D226. [CrossRef] [PubMed]

67. Quadri, L.E.; Sello, J.; Keating, T.A.; Weinreb, P.H.; Walsh, C.T. Identification of a *Mycobacterium tuberculosis* gene cluster encoding the biosynthetic enzymes for assembly of the virulence-conferring siderophore mycobactin. *Chem. Biol.* **1998**, *5*, 631–645. [CrossRef]

68. Lautru, S.; Oves-Costales, D.; Pernodet, J.-L.; Challis, G.L. MbtH-like protein-mediated cross-talk between non-ribosomal peptide antibiotic and siderophore biosynthetic pathways in *Streptomyces coelicolor* M145. *Microbiology* **2007**, *153*, 1405–1412. [CrossRef]

69. Glavinas, H.; Krajcsi, P.; Cserepes, J.; Sarkadi, B. The role of ABC transporters in drug resistance, metabolism and toxicity. *Curr. Drug Deliv.* **2004**, *1*, 27–42. [CrossRef]

70. Polgar, O.; Bates, S.E. ABC transporters in the balance: Is there a role in multidrug resistance? *Biochem. Soc. Trans.* **2005**, *33*, 241–245. [CrossRef]

71. Podlesek, Z.; Comino, A.; Herzog-Velikonja, B.; Zgur-Bertok, D.; Komel, R.; Grabnar, M. *Bacillus licheniformis* bacitracin-resistance ABC transporter: Relationship to mammalian multidrug resistance. *Mol. Microbiol.* **1995**, *16*, 969–976. [CrossRef]

72. Ohki, R.; Tateno, K.; Okada, Y.; Okajima, H.; Asai, K.; Sadaie, Y.; Murata, M.; Aiso, T. A Bacitracin-Resistant *Bacillus subtilis* Gene Encodes a Homologue of the Membrane-Spanning Subunit of the *Bacillus licheniformis* ABC Transporter. *J. Bacteriol.* **2003**, *185*, 51–59. [CrossRef]

73. Kumar, S.; He, G.; Kakarla, P.; Shrestha, U.; Ranjana, K.C.; Ranaweera, I.; Willmon, T.M.; Barr, S.R.; Hernandez, A.J.; Varela, M.F. Bacterial Multidrug Efflux Pumps of the Major Facilitator Superfamily as Targets for Modulation. *Infect. Disord. Drug Targets* **2016**, *16*, 28–43. [CrossRef]

74. Yan, N. Structural Biology of the Major Facilitator Superfamily Transporters. *Annu. Rev. Biophys.* **2015**, *44*, 257–283. [CrossRef] [PubMed]

75. Cuthbertson, L.; Nodwell, J.R. The TetR Family of Regulators. *Microbiol. Mol. Biol. Rev.* **2013**, *77*, 440–475. [CrossRef] [PubMed]

76. Koskiniemi, S.; Lamoureux, J.G.; Nikolakakis, K.C.; de Roodenbeke, C.T.K.; Kaplan, M.D.; Low, D.A.; Hayes, C.S. Rhs proteins from diverse bacteria mediate intercellular competition. *Proc. Natl. Acad. Sci. USA* **2013**, *110*, 7032–7037. [CrossRef] [PubMed]

77. Manteca, Á.; Yagüe, P. *Streptomyces* as a Source of Antimicrobials: Novel Approaches to Activate Cryptic Secondary Metabolite Pathways. In *Antimicrobials, Antibiotic Resistance, Antibiofilm Strategies and Activity Methods*; IntechOpen: London, UK, 2019. [CrossRef]

78. Manteca, A.; Alvarez, R.; Salazar, N.; Yague, P.; Sanchez, J. Mycelium Differentiation and Antibiotic Production in Submerged Cultures of *Streptomyces coelicolor*. *Appl. Environ. Microbiol.* **2008**, *74*, 3877–3886. [CrossRef] [PubMed]

79. Stülke, J.; Hillen, W. Carbon catabolite repression in bacteria. *Curr. Opin. Microbiol.* **1999**, *2*, 195–201. [CrossRef]

80. Kremling, A.; Geiselmann, J.; Ropers, D.; de Jong, H. Understanding carbon catabolite repression in *Escherichia coli* using quantitative models. *Trends Microbiol.* **2015**, *23*, 99–109. [CrossRef]

81. Deutscher, J. The mechanisms of carbon catabolite repression in bacteria. *Curr. Opin. Microbiol.* **2008**, *11*, 87–93. [CrossRef] [PubMed]

82. Brückner, R.; Titgemeyer, F. Carbon catabolite repression in bacteria: Choice of the carbon source and autoregulatory limitation of sugar utilization. *FEMS Microbiol. Lett.* **2002**, *209*, 141–148. [CrossRef]

83. Romero-Rodríguez, A.; Ruiz-Villafán, B.; Tierrafría, V.H.; Rodríguez-Sanoja, R.; Sánchez, S. Carbon Catabolite Regulation of Secondary Metabolite Formation and Morphological Differentiation in *Streptomyces coelicolor*. *Appl. Biochem. Biotechnol.* **2016**, *180*, 1152–1166. [CrossRef]

84. Inoue, O.O.; Schmidell Netto, W.; Padilla, G.; Facciotti, M.C.R. Carbon catabolite repression of retamycin production by *Streptomyces olindensis* ICB20. *Braz. J. Microbiol.* **2007**, *38*, 58–61. [CrossRef]

85. Gallo, M.; Katz, E. Regulation of secondary metabolite biosynthesis: Catabolite repression of phenoxazinone synthase and actinomycin formation by glucose. *J. Bacteriol.* **1972**, *109*, 659–667. [PubMed]

86. Magnus, N.; Weise, T.; Piechulla, B. Carbon Catabolite Repression Regulates the Production of the Unique Volatile Sodorifen of *Serratia plymuthica* 4Rx13. *Front. Microbiol.* **2017**, *8*, 2522. [CrossRef] [PubMed]

87. Bose, U.; Hewavitharana, A.; Ng, Y.; Shaw, P.; Fuerst, J.; Hodson, M. LC-MS-Based Metabolomics Study of Marine Bacterial Secondary Metabolite and Antibiotic Production in *Salinispora arenicola*. *Mar. Drugs* **2015**, *13*, 249–266. [CrossRef] [PubMed]

 *microorganisms*

Article

Characterization of *Streptomyces sporangiiformans* sp. nov., a Novel Soil Actinomycete with Antibacterial Activity against *Ralstonia solanacearum*

Junwei Zhao [1,†], Liyuan Han [1,†], Mingying Yu [1], Peng Cao [1], Dongmei Li [1], Xiaowei Guo [1], Yongqiang Liu [1], Xiangjing Wang [1,*] and Wensheng Xiang [1,2,*]

[1]   Key Laboratory of Agricultural Microbiology of Heilongjiang Province, Northeast Agricultural University, No. 59 Mucai Street, Xiangfang District, Harbin 150030, China
[2]   State Key Laboratory for Biology of Plant Diseases and Insect Pests, Institute of Plant Protection, Chinese Academy of Agricultural Sciences, Beijing 100193, China
*   Correspondence: wangneau2013@163.com (X.W.); xiangwensheng@neau.edu.cn (W.X.); Tel.: +86-451-55190413 (X.W. & W.X.); Fax: +86-451-55190413 (X.W. & W.X.)
†   These authors contributed equally to this work.

Received: 27 August 2019; Accepted: 13 September 2019; Published: 17 September 2019

Abstract: *Ralstonia solanacearum* is a major phytopathogenic bacterium that attacks many crops and other plants around the world. In this study, a novel actinomycete, designated strain NEAU-SSA 1[T], which exhibited antibacterial activity against *Ralstonia solanacearum*, was isolated from soil collected from Mount Song and characterized using a polyphasic approach. Morphological and chemotaxonomic characteristics of the strain coincided with those of the genus *Streptomyces*. The 16S rRNA gene sequence analysis showed that the isolate was most closely related to *Streptomyces aureoverticillatus* JCM 4347[T] (97.9%). Phylogenetic analysis based on 16S rRNA gene sequences indicated that the strain formed a cluster with *Streptomyces vastus* JCM4524[T] (97.4%), *S. cinereus* DSM43033[T] (97.2%), *S. xiangluensis* NEAU-LA29[T] (97.1%) and *S. flaveus* JCM3035[T] (97.1%). The cell wall contained *LL*-diaminopimelic acid and the whole-cell hydrolysates were ribose, mannose and galactose. The polar lipids were diphosphatidylglycerol (DPG), phosphatidylethanolamine (PE), hydroxy-phosphatidylethanolamine (OH-PE), phosphatidylinositol (PI), two phosphatidylinositol mannosides (PIMs) and an unidentified phospholipid (PL). The menaquinones were MK-9(H_4), MK-9(H_6), and MK-9(H_8). The major fatty acids were *iso*-$C_{17:0}$, $C_{16:0}$ and $C_{17:1}$ ω9c. The DNA G+C content was 69.9 mol %. However, multilocus sequence analysis (MLSA) based on five other house-keeping genes (*atp*D, *gyr*B, *rec*A, *rpo*B, and *trp*B), DNA–DNA relatedness, and physiological and biochemical data showed that the strain could be distinguished from its closest relatives. Therefore, it is proposed that strain NEAU-SSA 1[T] should be classified as representatives of a novel species of the genus *Streptomyces*, for which the name *Streptomyces sporangiiformans* sp. nov. is proposed. The type strain is NEAU-SSA 1[T] (=CCTCC AA 2017028[T] = DSM 105692[T]).

Keywords: *Streptomyces sporangiiformans* sp. nov.; antibacterial activity; multilocus sequence analysis; *Ralstonia solanacearum*

---

1. Introduction

Ralstonia solanacearum is the causal agent of bacterial wilt, one of the most devastating plant pathogenic bacteria around the world [1], which has an unusually wide host range, infecting over 200 plant species [2], including many important agricultural crops such as potato, tomato, banana and pepper. Even though different approaches have been developed to control bacterial wilt, we still lack an efficient and environmentally friendly control measure for most of the host crops [3].

Therefore, the search and discovery of novel, environmentally friendly, commercially significant, naturally bioactive compounds are in demand to control this disease at present.

The actinobacteria are known to produce biologically active secondary metabolites, including antibiotics, enzymes, enzyme inhibitors, antitumour agents and antibacterial compounds [4–6]. The genus *Streptomyces*, within the family *Streptomycetaceae*, is the largest genus of the phylum *Actinobacteria*, first proposed by Waksman and Henrici (1943) [7] and currently encompasses more than 800 species with valid published names (http://www.bacterio.net/streptomyces.html), which are widely distributed in soils throughout the world. Therefore, members of novel *Streptomyces* species are in demand as sources of novel, environmentally friendly, commercially significant, naturally bioactive compounds [8,9]. During our search for antagonistic actinobacteria from soil in Mount Song, an aerobic actinomycete, strain NEAU-SSA 1^T with inhibitory activity against phytopathogenic bacterium *Ralstonia solanacearum* was isolated and subjected to the polyphasic taxonomy analysis. Results demonstrated that the strain represents a novel species of the genus *Streptomyces*, for which the name *Streptomyces sporangiiformans* sp. nov. is proposed.

2. Materials and Methods

2.1. Isolation of Actinomycete Strain

Strain NEAU-SSA 1^T was isolated from soil collected from Mount Song (34°29′ N, 113°2′ E), Dengfeng, Henan Province, China. The soil sample was air-dried at room temperature for 14 days before isolation for actinomycetes. After drying, the soil sample was ground into powder and then suspended in sterile distilled water, followed by a standard serial dilution technique. The diluted soil suspension was spread on humic acid-vitamin agar (HV) [10] supplemented with cycloheximide (50 mg L^{-1}) and nalidixic acid (20 mg L^{-1}). After 28 days of aerobic incubation at 28 °C, colonies were transferred and purified on the International *Streptomyces* Project (ISP) medium 3 [11], and maintained as glycerol suspensions (20%, *v/v*) at −80 °C for long-term preservation.

2.2. Morphological and Physiological and Biochemical Characteristics of NEAU-SSA 1^T

Gram staining was carried out by using the standard Gram stain, and morphological characteristics were observed using light microscopy (Nikon ECLIPSE E200, Nikon Corporation, Tokyo, Japan) and scanning electron microscopy (Hitachi SU8010, Hitachi Co., Tokyo, Japan) using cultures grown on ISP 3 agar at 28 °C for 6 weeks. Samples for scanning electron microscopy were prepared as described by Jin et al. [12]. Cultural characteristics were determined on the ISP 1 agar [11], ISP media 2–7 [8], Czapek's agar [13], Bennett's agar [14], and Nutrient agar [15] after 14 days at 28 °C. Color determination was done with color chips from the ISCC-NBS (Inter-Society Color Council-National Bureau of Standards) color charts [16]. Growth at different temperatures (10, 15, 20, 25, 28, 32, 35, 40, 45, and 50 °C) was determined on ISP 3 medium after incubation for 14 days. Growth tests for pH range (pH 4.0–12.0, at intervals of 1.0 pH unit) and NaCl tolerance (0, 1, 2, 3, 4, 5, 6, 7, 8, 9, 10, 15, and 20%, *w/v*) were tested in GY (Glucose-Yeast extract) medium [17] at 28 °C for 14 days on a rotary shaker. The buffer systems were: pH 4.0–5.0, 0.1 M citric acid/0.1 M sodium citrate; pH 6.0–8.0, 0.1 M KH$_2$PO$_4$/0.1 M NaOH; pH 9.0–10.0, 0.1 M NaHCO$_3$/0.1 M Na$_2$CO$_3$; and pH 11.0–12.0, 0.2 M KH$_2$PO$_4$/0.1 M NaOH. Hydrolysis of Tweens (20, 40, and 80) and production of urease were tested as described by Smibert and Krieg [18]. The utilization of sole carbon and nitrogen sources, decomposition of cellulose, hydrolysis of starch and aesculin, reduction of nitrate, coagulation and peptonization of milk, liquefaction of gelatin, and production of H$_2$S were examined as described previously [19,20].

2.3. Chemotaxonomic Analysis of NEAU-SSA 1^T

Biomass for chemotaxonomic studies was prepared by growing the organisms in GY medium in shake flasks at 28 °C for 5 days. Cells were harvested using centrifugation, washed with distilled water, and freeze-dried. The isomer of diaminopimelic acid (DPA) in the cell wall hydrolysates was derivatized and analyzed using an HPLC (High Performance Liquid Chromatography) method [21] with an Agilent TC-C$_{18}$ Column (250 × 4.6 mm i.d. 5 µm; Agilent Technologies, Santa Clara, CA, USA) that had a mobile phase consisting of acetonitrile: 0.05 mol L^{-1} phosphate buffer pH 7.2 (15:85, *v/v*) at a flow rate of 0.5 mL min^{-1}. The peak detection used an Agilent G1321A fluorescence detector (Agilent Technologies, Santa Clara, CA, USA) with a 365 nm excitation and 455 nm longpass emission filters. The whole-cell sugars were analyzed according to the procedures developed by Lechevalier and Lechevalier [22]. The polar lipids were examined using two-dimensional TLC (Thin-Layer Chromatography) and identified using the method of Minnikin et al. [23]. Menaquinones were extracted from the freeze-dried biomass and purified according to Collins [24]. Extracts were analyzed using a HPLC-UV method [25] with an Agilent Extend-C$_{18}$ Column (150 × 4.6 mm, i.d. 5 µm; Agilent Technologies, Santa Clara, CA, USA) at 270 nm. The mobile phase was acetonitrile-*iso*-propyl alcohol (60:40, *v/v*). To determine cellular fatty acid compositions, the strain NEAU-SSA 1^{T} was cultivated in GY medium in shake flasks at 28 °C for 4 days. Fatty acid methyl esters were extracted from the biomass as described by Gao et al. [26] and analyzed using GC-MS according to the method of Xiang et al. [27].

2.4. Phylogenetic Analysis of NEAU-SSA 1^{T}

For DNA extraction, strain NEAU-SSA 1^{T} was cultured in GY medium for 3 days to the early stationary phase and harvested using centrifugation. The chromosomal DNA was extracted according to the method of sodium dodecyl sulfate (SDS)-based DNA extraction [28]. PCR amplification of the 16S rRNA gene sequence was carried out using the universal bacterial primers 27F (5′-AGAGTTTGATCCTGGCTCAG-3′) and 1541R (5′-AAGGAGGTGATCCAGCC-3′) under conditions described previously [29,30]. The PCR product was purified and cloned into the vector pMD19-T (Takara) and sequenced using an Applied Biosystems DNA sequencer (model 3730XL, Applied Biosystems Inc., Foster City, California, USA). The almost complete 16S rRNA gene sequence of strain NEAU-SSA 1^{T} (1412bp) was obtained and compared with type strains available at the EzBioCloud server (https://www.ezbiocloud.net/), retrieved using NCBI BLAST (National Center for Biotechnology Information, Basic Local Alignment Search Tool; https://blast.ncbi.nlm.nih.gov/Blast.cgi;) and then submitted to the GenBank database. Phylogenetic trees were constructed based on the 16S rRNA gene sequences of strain NEAU-SSA 1^{T} and related reference species. Sequences were multiply aligned in Molecular Evolutionary Genetics Analysis (MEGA) software version 7.0 using the Clustal W algorithm and trimmed manually where necessary. Phylogenetic trees were constructed with neighbor-joining [31] and maximum likelihood [32] algorithms using MEGA [33]. The stability of the topology of the phylogenetic tree was assessed using the bootstrap method with 1000 repetitions [34]. A distance matrix was generated using Kimura's two-parameter model [35]. All positions containing gaps and missing data were eliminated from the dataset (complete deletion option). 16S rRNA gene sequence similarities between strains were calculated on the basis of pairwise alignment using the EzBioCloud server [36]. To further clarify the affiliation of strain NEAU-SSA 1^{T} to its closely related strains, phylogenetic relationships of the strain NEAU-SSA 1^{T} were also confirmed using sequences of five individual housekeeping genes (*atpD*, *gyrB*, *recA*, *rpoB*, and *trpB*) for core-genome analysis. The sequences of NEAU-SSA 1^{T} and its related strains were obtained from the genomes or GenBank/EMBL/DDBJ (European Molecular Biology Laboratory/DNA Data Bank of Japan). GenBank accession numbers of the sequences used are given in Table 1. The sequences of each locus were aligned using MEGA 7.0 software and trimmed manually at the same position before being used for further analysis. Trimmed sequences of the five housekeeping genes were concatenated head-to-tail in-frame in the order *atpD-gyrB-recA-rpoB-trpB*. Phylogenetic analysis was performed as described above. Genome mining for bioactive secondary metabolites was performed using "antibiotics and secondary metabolite analysis shell" (antiSMASH) version 4.0 [37].

Table 1. GenBank Accession Numbers of the Sequences Used in MLSA.

Strain	Type Strain	Whole Genome	*atp*D	*gyr*B	*rec*A	*rpo*B	*trp*B
Streptomyces sporangiiformans	NEAU-SSA 1[T]	VCHX00000000		–	–	–	–
Streptomyces coelescens	DSM 40421[T]	–	GU383344	AY508508	KT385220	GU383768	KT389192
Streptomyces violaceolatus	DSM 40438[T]	–	GU383347	AY508509	KT385451	GU383771	KT389418
Streptomyces anthocyanicus	NBRC 14892[T]	–	KT384465	KT384814	KT385162	KT388784	KT389134
Streptomyces humiferus	DSM 43030[T]	–	KT384598	KT384947	KT385296	KT388918	KT389267
Streptomyces violaceoruber	NBRC 12826[T]	CP020570	KT384751	KT385099	KT385453	KT389071	KT389420
Streptomyces rubrogriseus	LMG 20318[T]	BEWD00000000	KT384715	KT385065	KT385416	KT389036	KT389384
Streptomyces tendae	ATCC 19812[T]	–	KT384733	KT385082	KT385434	KT389053	KT389402
Streptomyces violaceorubidus	LMG 20319[T]	JODM00000000	–	–	–	–	–
Streptomyces lienomycini	LMG 20091[T]	–	KT384622	KT384971	KT385321	KT388942	KT389291
Streptomyces diastaticus subsp. *ardesiacus*	NRRL B-1773[T]	BEWC00000000	KT384534	KT384883	KT385231	KT388853	KT389203
Streptomyces albaduncus	JCM 4715[T]	–	KT384449	KT384798	KT385146	EJ996741	KT389118
Streptomyces matensis	NBRC 12889[T]	–	KT384637	KT384986	KT385337	KT388957	KT389306
Streptomyces althioticus	NRRL B-3981[T]	–	KT384460	KT384809	KT385157	KT388779	KT389129
Streptomyces davaonensis	JCM 4913[T]	HE971709	–	–	–	–	–
Streptomyces canus	DSM 40017[T]	LMWO00000000	KT384500	KT384849	KT385197	KT388819	KT389169
Streptomyces lincolnensis	NRRL 2936[T]	CP016438	–	–	–	–	–
Streptomyces pseudovenezuelae	DSM 40212[T]	LMWM00000000	KT384695	KT385045	KT385396	KT389016	KT389364
Streptomyces xiangluensis	NEAU-LA29[T]	–	MH291276	MH345670	MH291277	MH291275	MH291278
Streptomyces vastus	NBRC 13094[T]	–	KU323834	KT385093	KU975607	KT389065	KT389414
Streptomyces cinereus	NBRC 12247[T]	–	KT384513	KT384862	KT385210	KJ996667	KT389182
Streptomyces flaveus	NRRL B-16074[T]	JOCU00000000	KT384551	KT384900	KT385249	KT388870	KT389220
Streptomyces chilikensis	RC 1830[T]	LWCC00000000	–	–	–	–	–
Streptomyces coeruleorubidus	ISP 5145[T]	–	KT384528	KT384877	KT385225	KT388847	KT389197
Streptomyces misionensis	DSM 40306[T]	FNTD00000000	KT384647	KT384996	KT385347	KT388967	KT389316
Streptomyces phaeoluteichromatogenes	NRRL 5799[T]	–	KT384680	KT385030	KT385381	KT389001	KT389350
Streptomyces tricolor	NBRC 15461[T]	MUMF00000000	KT384741	KT385089	KT385443	KT389061	KT389410
Streptomyces achromogenes subsp. *achromogenes*	NBRC 12735[T]	JODT00000000	–	–	–	–	–
Streptomyces eurythermus	ATCC 14975[T]	–	KT384544	KT384893	KT385242	KT388863	KT389213
Streptomyces nogalater	JCM 4799[T]	–	KT384664	KT385014	KT385365	KT388984	KT389333
Streptomyces jietaisiensis	FXJ46[T]	FNAX00000000	KT384605	KT384954	KT385304	KT388925	KT389274
Streptomyces griseoaurantiacus	NBRC 15440[T]	AEYX00000000	–	–	–	–	–
Streptomyces lavenduligriseus	NRRL ISP-5487[T]	JOBD00000000	KT384620	AB072859	KT385319	KT388940	KT389289
Streptomyces uncialis	DCA2648[T]	LFBV00000000	–	–	–	–	–

Table 1. *Cont.*

Strain	Type Strain	Whole Genome	*atp*D	*gyr*B	*rec*A	*rpo*B	*trp*B
Streptomyces alboniger	NRRL B-1832[T]	–	KT384455	KT384804	KT385152	KT388774	KT389124
Streptomyces alfalfae	XY25[T]	CP015588	–	–	–	–	–
Streptomyces lasiicapitis	3H-HV17(2)[T]	–	MH651782	KY229066	MH651785	MH651788	MH651791
Streptomyces aureocirculatus	NRRL ISP-5386[T]	JOAP00000000	KT384476	KT384825	KT385173	KT388795	KT389145
Streptomyces aureoverticillatus	NRRL B-3326[T]	–	KT384478	KT384827	KT385175	KT388797	KT389147
Streptomyces alboflavus	NRRL B-2373[T]	CP021748	–	–	–	–	–
Streptomyces rutgersensis	NBRC 12819[T]	–	KT384716	KT385066	KT385417	KT389037	KT389385
Streptomyces intermedius	NBRC 13049[T]	–	KT384602	KT384951	KT385301	KT388922	KT389271
Streptomyces gougerotii	NBRC 3198[T]	–	KT384572	KT384921	KT385270	KT388891	KT389241
Streptomyces diastaticus subsp. *diastaticus*	NBRC 3714[T]	–	KT384535	KT384884	KT385232	KT388854	KT389204
Kitasatospora setae	LM-6054[T]	AP010968	–	–	–	–	–

2.5. Draft Genome Sequencing and Assembly of NEAU-SSA 1^T

For draft genome sequencing and assembly, the genomic DNA of strain NEAU-SSA 1^T was extracted using the method of SDS-based DNA extraction [28]. The harvested DNA was detected using agarose gel electrophoresis and quantified using Qubit® 2.0 Fluorometer (Thermo Scientific). Whole-genome sequencing was performed on the Illumina HiSeq PE150 (Illumina, San Diego, CA, USA) platform. A-tailed, ligated to paired-end adaptors, and PCR amplified samples with a 350 bp insert were used for the library construction at the Beijing Novogene Bioinformatics Technology Co., Ltd. Illumina PCR adapter reads and low quality reads from the paired-end were filtered using a quality control step using our own compling pipeline. All good-quality paired reads were assembled using the SOAP (Short Oligonucleotide Alignment Program) denovo [38,39] (https://github.com/aquaskyline) into a number of scaffolds. Then, the filter reads were handled by the next step of the gap-closing.

2.6. DNA–DNA Relatedness Tests

Because of a lacking number of genome sequences of *Streptomyces aureoverticillatus* JCM4347^T, *Streptomyces vastus* JCM4524^T, *S. cinereus* DSM43033^T, and *S. xiangluensis* NEAU-LA29^T, DNA–DNA relatedness tests between strain NEAU-SSA 1^T and those strains were carried out as described by De Ley et al. [40] under consideration of the modifications described by Huss et al. [41], using a model Cary 100 Bio UV/VIS-spectrophotometer (Hitachi U-3900, Hitachi Co., Tokyo, Japan) equipped with a Peltier-thermostatted 6 × 6 multicell changer and a temperature controller with in situ temperature probe (Varian). The genomic DNAs of strain NEAU-SSA 1^T and its closely related species—*S. aureoverticillatus* JCM4347^T, *S. vastus* JCM4524^T, *S. cinereus* DSM43033^T, and *S. xiangluensis* NEAU-LA29^T—were extracted using the method of SDS-based DNA extraction [28]. The concentration and purity of these DNA samples were determined by measuring the optical density (OD) at 260, 280, and 230 nm. The DNA samples used for hybridization were diluted to OD_{260} around 1.0 using 0.1 × SSC (saline sodium citrate buffer), then sheared using a JY92-II ultrasonic cell disruptor (ultrasonic time 3 s, interval time 4 s, 90 times; Ningbo Scientz Biotechnology Co., Ltd, Ningbo, China). The DNA renaturation rates were determined in 2 × SSC at 70 °C. The experiments were performed with three replications and the DNA–DNA relatedness value was expressed as a mean of the three values. Several genomic metrics are now available to distinguish between orthologous genes of closely related prokaryotes, including the calculation of average nucleotide identity (ANI) and digital DNA–DNA hybridization (dDDH) values [42,43]. In the present study, ANI and dDDH values were determined from the genomes of strain NEAU-SSA 1^T and *S. flaveus* JCM3035^T (JOCU00000000) using the ortho-ANIu algorithm from Ezbiotaxon and the genome-to-genome distance calculator (GGDC 2.0) at http://ggdc.dsmz.de.

2.7. In Vitro Antibacterial Activity Test

The antibacterial activity of strain NEAU-SSA 1^T against two pathogenic bacteria (*Micrococcus luteus* and *Ralstonia solanacearum*) was evaluated using the agar well diffusion method [44] with the cultures growth on ISP 3 medium at 28 °C for four weeks as follows: All the spores and mycelia were collected from one ISP 3 plate (diameter, 9mm) and then extracted using 1 mL methanol with an ultrasonic step (300 W, 30–60 min). Afterwards, 200 µL methanol extract or methanol was added to the agar well, and methanol was used as the control. To further investigate the antibacterial components produced by NEAU-SSA 1^T, the strain was cultured in tryptone-glucose-soluble starch-yeast extract medium (tryptone 0.2%, glucose 1%, soluble starch 0.5%, yeast extract 0.2%, NaCl 0.4%, K_2HPO_4 0.05%, $MgSO_4.7H_2O$ 0.05%, $CaCO_3$ 0.2%, *w/v*, pH 7.0–7.4), and the inhibitory activity was tested. Briefly, strain NEAU-SSA 1^T was inoculated into MB medium and incubated at 28 °C for seven days in a rotary shaker. The supernatant (100 mL for this study) was obtained via centrifugation at 8000 rpm and 4 °C for 10 min and subsequently extracted by using an equal volume of ethyl acetate. Then, the extract was dried in a rotary evaporator at 40 °C and eluted with proper volume methanol (1 mL used in this study). The cell precipitate was extracted with an equal volume of methanol and also condensed as above. After that, the antibacterial activity was evaluated using the agar well diffusion method, and each well contained 200 μL of the methanol extract. To examine the effect of temperature on antibacterial activity, the ten-fold dilution methanol extract was placed in a water bath at 40, 60, 80, and 100 °C for 30 min, and then cooled to room temperature. The antibacterial activity was evaluated using the agar well diffusion method.

3. Result and Discussion

3.1. Polyphasic Taxonomic Characterization of NEAU-SSA 1^T

The morphological characteristics of strain NEAU-SSA 1^T showed that the strain had the typical characteristics of the genus *Streptomyces*. Observation of 6-week cultures of strain NEAU-SSA 1^T grown on ISP 3 medium revealed that it formed well-developed, branched substrate hyphae and aerial mycelia. Sporangia consisted of cylindrical, and rough-surfaced spores (0.6–0.8 μm × 0.9–1.6 μm) were produced on aerial mycelia, but spore chains were not observed (Figure 1). Strain NEAU-SSA 1^T exhibited good growth on ISP 3, ISP 4, ISP 7, and Nutrient agar media; moderate growth on ISP 1, ISP 2, ISP 5, ISP 6, and Czapek's agar media; and poor growth on Bennett's agar medium. The cultural characteristics of strain NEAU-SSA 1^T is shown in Table S1. Strain NEAU-SSA 1^T grew well between pH 6.0 and 11.0, with an optimum pH of 7.0. The range of temperature of the strain was determined to be 15–45 °C, with the optimum growth temperature being 28 °C. The strain grew in the presence of 0–6% NaCl (*w/v*) with an optimal level of 0–1% (*w/v*). Detailed physiological characteristics are presented in the species description (Table 2 and Table S1).

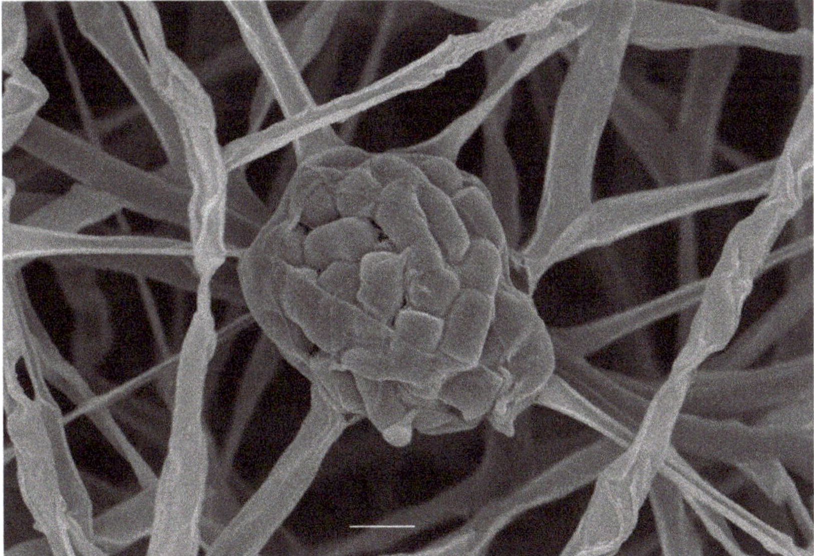

Figure 1. Scanning electron micrograph of strain NEAU-SSA 1^T grown on ISP 3 agar for 6 weeks at 28 °C; Scale bar represents 1 μm.

Chemotaxonomic analyses revealed that strain NEAU-SSA 1^T exhibited characteristics that are typical of representatives of the genus *Streptomyces*. The strain was found to contain *LL*-diaminopimelic acid as diamino acid. The whole-cell hydrolysates of the strain were determined to contain ribose, mannose, and galactose. The menaquinones of strain NEAU-SSA 1^T were MK-9(H$_4$) (29.5%), MK-9(H$_6$) (41.2%), and MK-9(H$_8$) (29.4%). The cellular fatty acid profile of strain NEAU-SSA 1^T was composed of *iso*-C$_{17:0}$ (30.9%), C$_{16:0}$ (26.4%), C$_{17:1}$ω9c (19.9%), C$_{15:0}$ (7.8%), C$_{17:0}$ (4.4%), C$_{14:0}$ (3.3%), *iso*-C$_{16:0}$

(1.7%), *anteiso*-$C_{15:0}$ (1.7%), $C_{18:1}\omega9c$ (1.7%), $C_{16:0}$ 1-OH (1.2%), and *iso*-$C_{18:0}$ (1.1%). The polar lipids of the strain consisted of diphosphatidylglycerol (DPG), phosphatidylethanolamine (PE), hydroxy-phosphatidylethanolamine (OH-PE), phosphatidylinositol (PI), two phosphatidylinositol mannosides (PIMs), and an unidentified phospholipid (PL) (Supplementary Figure S1). All the chemotaxonomic data are consistent with the assignment of strain NEAU-SSA 1[T] to the genus *Streptomyces*.

Table 2. Differential characteristics of strain NEAU-SSA 1[T], *S. aureoverticillatus* JCM 4347[T], *S. vastus* JCM4524[T], *S. cinereus* DSM43033[T], *S. xiangluensis* NEAU-LA29[T], and *S. flaveus* JCM3035[T].

Characteristic	1	2	3	4	5	6[a]
Decomposition of cellulose	+	−	−	−	+	ND
Production of H_2S	−	−	−	−	−	+
Tween 20	−	+	−	−	−	ND
Tween 40	+	+	−	−	+	ND
Tween 80	+	+	−	−	+	ND
Liquefaction of gelatin	+	−	−	−	−	ND
Growth temperature (°C)	15–45	10–45	20–40	20–40	20–40	10–37
pH range for growth	6–11	5–12	6–10	6–9	6–9	ND
NaCl tolerance range (*w/v*, %)	0–6	0–15	0–5	0–5	0–6	0–7
Carbon source utilization						
D-fructose	+	+	−	−	+	+
D-galactose	+	+	−	−	+	+
Lactose	+	+	−	−	+	+
D-maltose	+	+	−	−	+	ND
L-rhamnose	+	+	−	+	+	+
D-ribose	−	+	−	−	−	−
D-sorbitol	−	+	−	+	−	ND
D-mannose	+	+	+	−	+	+
Raffinose	+	+	+	−	+	+
L-arabinose	−	−	−	−	−	+
D-xylose	−	−	−	−	−	+
Myo-inositol	+	+	+	−	+	+
Nitrogen source utilization						
L-glutamine	+	+	−	−	+	ND
Glycine	−	+	+	+	−	ND
L-threonine	+	+	−	−	−	+
L-tyrosine	−	−	−	+	+	ND
L-arginine	+	+	+	+	−	+
L-asparagine	+	+	+	+	−	ND
L-serine	+	+	+	+	−	+
L-proline	+	+	+	+	−	−

Strains: 1—NEAU-SSA 1[T]; 2—*S. aureoverticillatus* JCM 4347[T]; 3—*S. vastus* JCM4524[T]; 4—*S. cinereus* DSM43033[T]; 5—*S. xiangluensis* NEAU-LA29[T]; 6—*S. flaveus* JCM3035[T]. Abbreviation: +, positive; −, negative. All data are from this study except where marked. [a] Data from Michael Goodfellow et al. [45].

Sequence analysis of the 16S rRNA gene showed that strain NEAU-SSA 1[T] were affiliated with the genus *Streptomyces* and most closely related to *S. aureoverticillatus* JCM 4347[T] (97.9%). Phylogenetic analysis based on 16S rRNA gene sequences indicated that the strain formed a cluster with *S. vastus* JCM4524[T] (97.4%), *S. cinereus* DSM43033[T] (97.2%), *S. xiangluensis* NEAU-LA29[T] (97.1%), and *S. flaveus* JCM3035[T] (97.1%) in the neighbor-joining tree (Figure 2), a relationship also recovered by the maximum-likelihood algorithm (Figure S2). Phylogenetic trees based on the neighbor-joining and maximum-likelihood algorithms were constructed from the concatenated sequence alignment of the five housekeeping genes (Figure 3 and Figure S3), and had the same topology as the 16S rRNA gene tree. Moreover, pairwise distances calculated for NEAU-SSA 1[T] and the related species using the concatenated sequences of *atp*D-*gyr*B-*rec*A-*rpo*B-*trp*B were well above 0.007 (Table S2), which is considered to be the threshold for species determination by Rong et al. [46]. DNA–DNA hybridization was employed to further clarify the relatedness between the strain and *S. aureoverticillatus* JCM 4347[T], *S. vastus* JCM4524[T], *S. cinereus* DSM 43033[T], and *S. xiangluensis* NEAU-LA29[T]. Results showed that strain NEAU-SSA 1[T] shared DNA–DNA relatedness of 37.1 ± 3.4% with *S. aureoverticillatus* JCM 4347[T], 35.4 ± 4.3% with *S. vastus* JCM 4524[T], 33.1 ± 4.1% with *S. cinereus* DSM 43033[T], and 29.0 ± 4.9% with *S. xiangluensis* NEAU-LA29[T]. Digital DNA–DNA hybridization was employed to clarify the relatedness between strain NEAU-SSA 1[T] and *S. flaveus* JCM 3035[T]. The level of digital DNA–DNA hybridization between them was 24.9 ± 2.4%. These five values are all below the threshold value of 70% recommended by Wayne et al. [47] for assigning strains to the same genomic species. Similarly, a low ANI value of 80.99% was found between strain NEAU-SSA 1[T] and *S. flaveus* JCM 3035[T], a result well below the threshold used to delineate prokaryote species [48,49].

The assembled genome sequence of strain NEAU-SSA 1[T] was found to be 10,364,704 bp long and composed of 352 contigs with an N50 of 59,982 bp, a DNA G+C content of 69.9 mol % and a coverage of 200x. It was deposited into GenBank under the accession number VCHX00000000. The 16S rRNA gene sequence from the whole genome sequence shared a 100% similarity with that from PCR sequencing, suggesting that the genome sequence was not contaminated. Detailed genomic information is presented in the Table S3.

Comparison of phenotypic characteristics between strain NEAU-SSA 1[T] and its closely related species—*S. aureoverticillatus* JCM 4347[T], *S. vastus* JCM4524[T], *S. cinereus* DSM 43033[T], *S. xiangluensis* NEAU-LA29[T], and *S. flaveus* JCM 3035[T]—was performed to differentiate these strains (Table 2). Differential cultural characteristics included: NaCl tolerance of the strain was up to 5.0%, which is lower than that of *S. aureoverticillatus* JCM 4347[T] (15%) and *S. flaveus* JCM 3035[T] (7%); and the strain could grow at pH 11.0, while *S. vastus* JCM4524[T], *S. cinereus* DSM 43033[T], and *S. xiangluensis* NEAU-LA29[T] could not. Other phenotypic differences included the production of H_2S; decomposition of cellulose; liquefaction of gelatin; growth temperature; hydrolysis of Tweens (20, 40, and 80); and utilization of L-arabinose, D-galactose, D-fructose, D-maltose, lactose, L-rhamnose, D-ribose, D-sorbitol, D-mannose, raffinose, D-xylose, *myo*-inositol, L-glutamine, glycine, L-threonine, L-tyrosine, L-serine, L-proline, L-asparagine, and L-arginine.

On the basis of morphological, physiological, chemotaxonomic, and phylogenetic results, strain NEAU-SSA 1[T] is considered to represent a novel species within the genus *Streptomyces*, for which the name *Streptomyces sporangiiformans* is proposed.

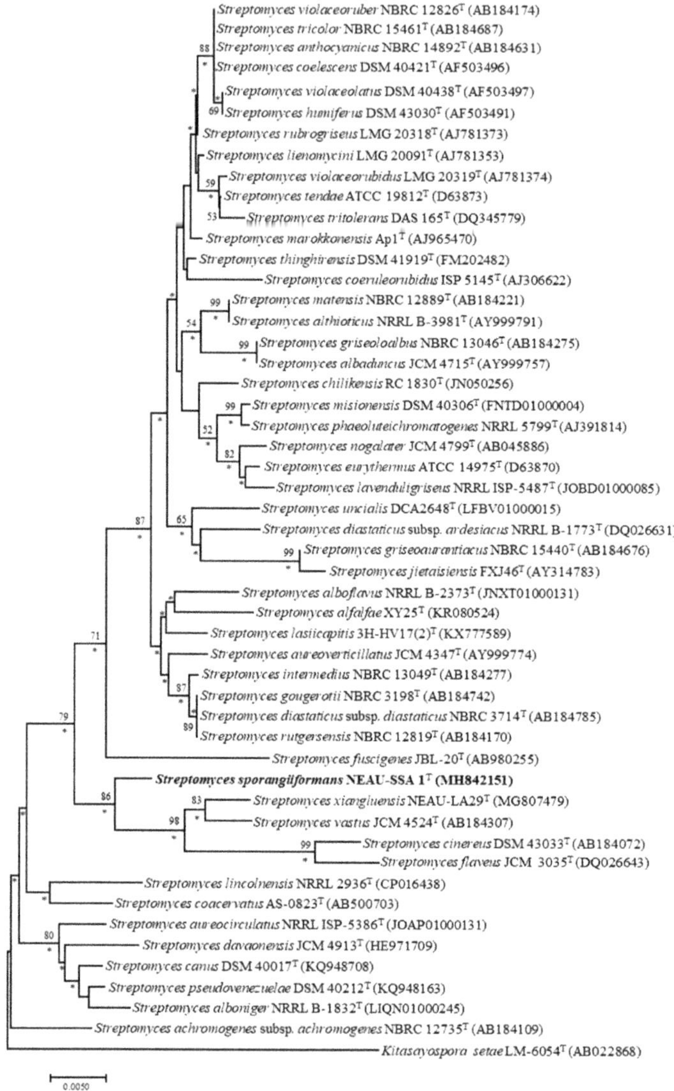

Figure 2. Neighbor-joining tree showing the phylogenetic position of strain NEAU-SSA 1^T (1412 bp) and the related species of the genus *Streptomyces* based on 16S rRNA gene sequences. The out-group used was *Kitasatospora setae* LM-6054^T. Only bootstrap values above 50% (percentages of 1000 replications) are indicated. Asterisks indicate branches also recovered in the maximum-likelihood tree. Scale bar represents 0.005 nucleotide substitutions per site.

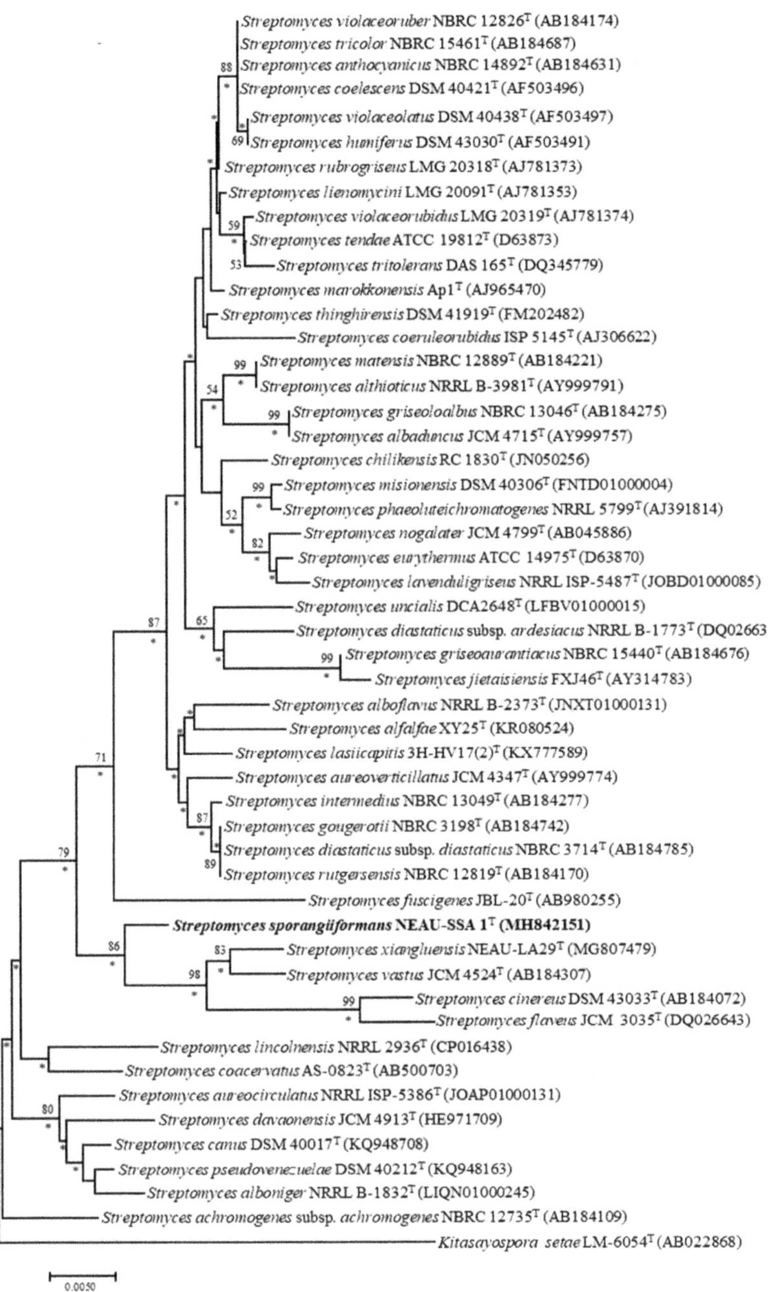

Figure 3. Neighbor-joining tree based on MLSA analysis of the concatenated partial sequences from five housekeeping genes (*atpD*, *gyrB*, *recA*, *rpoB*, and *trpB*) of isolate NEAU-SSA 1[T] (in bold) and related taxa. Only bootstrap values above 50% (percentages of 1000 replications) are indicated. *Kitasatospora setae* LM-6054[T] was used as an out-group. Asterisks indicate branches also recovered in the maximum-likelihood tree. Scale bar represents 0.02 nucleotide substitutions per site.

3.2. Description of Streptomyces sporangiiformans sp. nov.

Streptomyces sporangiiformans (spo.ran.gi.i.for'mans. N.L. neut. n. sporangium; L. pres. part. *formans* forming; N.L. part. adj. *sporangiiformans* forming sporangia).

Gram-stain-positive, aerobic actinomycete that formed well-developed, branched substrate hyphae and aerial mycelia. Sporangia consisted of cylindrical and rough surfaced spores (0.6–0.8 µm × 0.9–1.6 µm) were produced on aerial mycelia, but spore chains were not observed. Good growth on ISP 3, ISP 4, ISP 7, and Nutrient agar media; moderate growth on ISP 1, ISP 2, ISP 5, ISP 6, and Czapek's agar media; and poor growth on Bennett's agar medium. Growth occurred at pH values between 6.0 and 11.0, the optimum being pH 7.0. Tolerates up to 6.0% NaCl and grows optimally in 0–1% (*w/v*) NaCl. Growth was observed at temperatures between 15 and 45 °C, with an optimum temperature of 28 °C. Positive for decomposition of Tweens (40 and 80) and cellulose, hydrolysis of aesculin and starch, liquefaction of gelatin and production of urease; and negative for coagulation and peptonization of milk, hydrolysis of Tween 20, production of H_2S, and reduction of nitrate. D-fructose, D-galactose, D-glucose, inositol, lactose, D-maltose, D-mannose, D-raffinose, L-rhamnose, and D-sucrose were utilized as sole carbon sources, but not L-arabinose, dulcitol, D-ribose, D-sorbitol, or D-xylose. L-alanine, L-arginine, L-asparagine, L-aspartic acid, creatine, L-glutamic acid, L-glutamine, L-proline, L-serine, and L-threonine were utilized as sole nitrogen sources, but not glycine or L-tyrosine. Cell wall contained *LL*-diaminopimelic acid and the whole-cell hydrolysates were ribose, mannose, and galactose. The polar lipids contained diphosphatidylglycerol (DPG), phosphatidylethanolamine (PE), hydroxy-phosphatidylethanolamine (OH-PE), phosphatidylinositol (PI), two phosphatidylinositol mannosides (PIMs), and an unidentified phospholipid (PL). The menaquinones were MK-9(H_4), MK-9(H_6), and MK-9(H_8). Major fatty acids were *iso*-$C_{17:0}$, $C_{16:0}$, and $C_{17:1}\omega 9c$.

The type strain was NEAU-SSA 1^T (=CCTCC AA 2017028^T = DSM 105692^T), isolated from soil collected from Mount Song, Dengfeng, Henan Province, China. The DNA G+C content of the type strain was 69.9 mol %, calculated from the assembly for the draft genome sequence. The GenBank/EMBL/DDBJ accession number for the 16S rRNA gene sequence of strain NEAU-SSA 1^T is MH842151. This Whole Genome Shotgun project has been deposited at DDBJ/ENA/GenBank under the accession VCHX00000000. The version described in this paper is version VCHX00000000.2.

3.3. Antibacterial Activity of NEAU-SSA 1^T against Ralstonia solanacearum

Strain NEAU-SSA 1^T exhibited antibacterial activity against *Ralstonia solanacearum* with inhibitory zone diameters of 23 mm (Figure 4a). However, no inhibitory effect on the growth of *Micrococcus luteus* (Figure 4b) was observed. Comparison of the antibacterial activity of the extract of the supernatant with that of the cell pellet suggested that the antibacterial substances of strain NEAU-SSA 1^T were in both the supernatant and cell pellet since the extracts all showed inhibition of the growth of *Ralstonia solanacearum* with the inhibitory zone diameters of 31.5 and 26.4 mm, respectively (Figure 5a,b). The antibacterial substances in the supernatant were stable after they were placed in a water bath at 40 and 60 °C for 30 min, while they did not show antibacterial activity after 80 and 100 °C bath (Figure 6a), which indicated that they were sensitive to temperature. In contrast, the antibacterial substances in the cell pellet were insensitive to temperature (Figure 6b), which demonstrated that the antibacterial substances in the supernatant and cell pellet were different. The antiSMASH analysis led to the identification of 49 gene clusters, including 24 gene clusters that showed very low similarity to the known gene clusters of mediomycin A, cremimycin, primycin, ibomycin, naphthomycin, lasalocid, informatipeptin, polyoxypeptin, kutznerides, anisomycin, paulomycin, himastatin, desotamide, nystatin, tiacumicin B, oxazolomycin, and 4-Z-annimycin. Therefore, the relationships between the corresponding secondary metabolites produced by NEAU-SSA 1^T and the antibacterial activity are still ambiguous. *Streptomyces* are well known as important biological resources for their biologically active secondary metabolites, which play important roles in protecting plants against pathogens [50]. Strain NEAU-SSA 1^T, which shows a stronger antibacterial activity against *Ralstonia solanacearum*, is a novel species of the genus *Streptomyces*, and possesses 24 lower similarity gene clusters. Therefore, it is interesting and significant to isolate and identify the secondary metabolites of the strain in further studies.

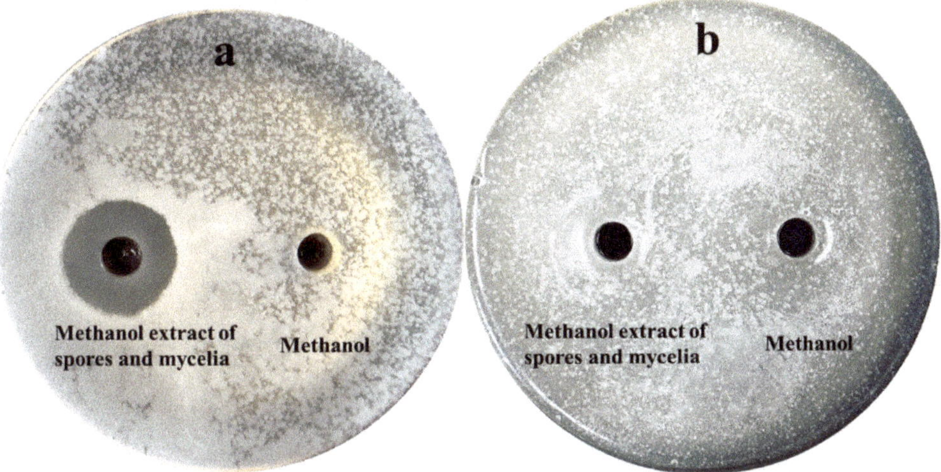

Figure 4. The antibacterial activity of strain NEAU-SSA 1^T against *Ralstonia solanacearum* (**a**) and *Micrococcus luteus* (**b**).

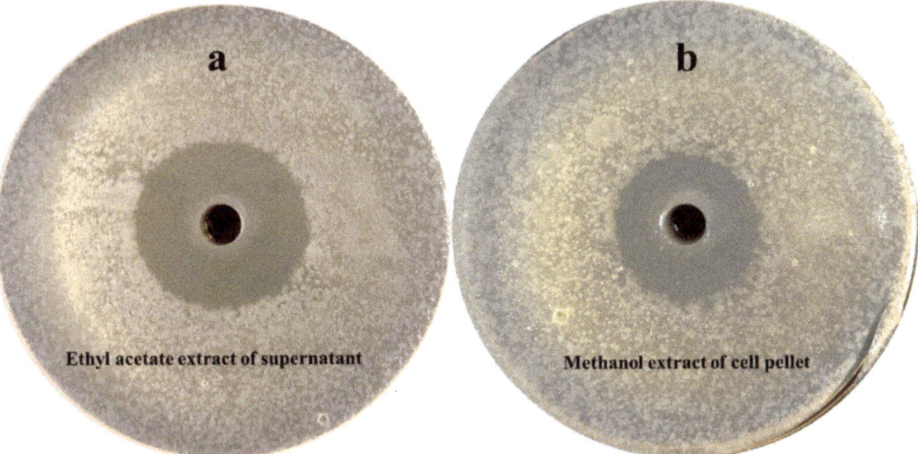

Figure 5. The antibacterial activity of the extract of the supernatant (**a**) and cell pellet of strain NEAU-SSA 1^T (**b**) against *Ralstonia solanacearum*.

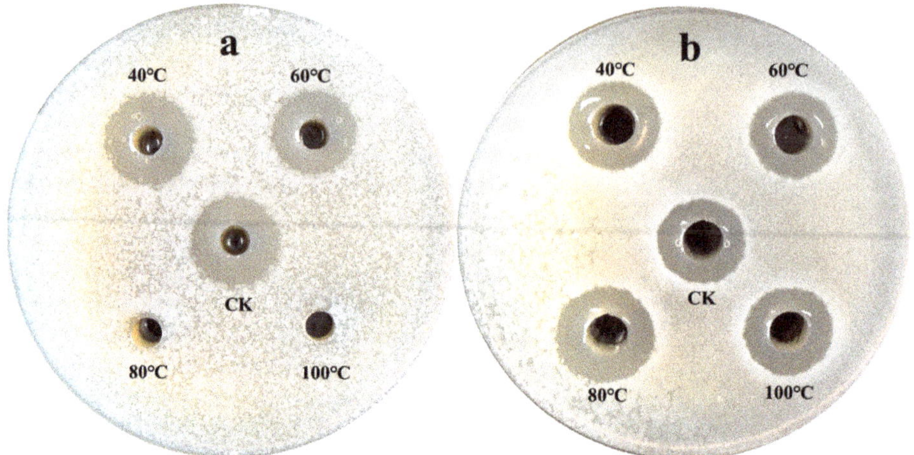

Figure 6. The effect of temperature on the antibacterial activity of the extract of the supernatant (**a**) and cell pellet of strain NEAU-SSA 1^T (**b**) against *Ralstonia solanacearum*.

4. Conclusions

A novel strain NEAU-SSA 1^T that exhibited antibacterial activity against *Ralstonia solanacearum* was isolated from a soil sample. Morphological features, phylogenetic analysis based on 16S rRNA gene sequences, and multilocus sequence analysis based on five other house-keeping genes (*atp*D, *gyr*B, *rec*A, *rpo*B, and *trp*B) suggested that strain NEAU-SSA 1^T belonged to the genus *Streptomyces*. Physiology and biochemical characteristics, together with DDH relatedness values and ANI values, clearly indicated that strain NEAU-SSA 1^T could be differentiated from the closely related strains *S. aureoverticillatus* JCM 4347^T, *S. vastus* JCM 4524^T, *S. cinereus* DSM 43033^T, *S. xiangluensis* NEAU-LA29^T, and *S. flaveus* JCM 3035^T. Based on the polyphasic analysis, it is proposed that strain NEAU-SSA 1^T should be classified as representatives of a novel species of the genus *Streptomyces*, for which the name *Streptomyces sporangiiformans* sp. nov. is proposed. The type strain is NEAU-SSA 1^T (=CCTCC AA 2017028^T = DSM 105692^T).

Supplementary Materials: The following are available online at http://www.mdpi.com/2076-2607/7/9/360/s1, Figure S1: Maximum-likelihood tree showing the phylogenetic position of strain NEAU-SSA 1^T (1412 bp) and the related species based on 16S rRNA gene sequences. The out-group used was *Kitasatospora setae* LM-6054^T. Only bootstrap values above 50% (percentages of 1000 replications) are indicated. Bar, 0.01 nucleotide substitutions per site. Figure S2: Maximum-likelihood tree based on MLSA analysis of the concatenated partial sequences from five housekeeping genes (*atp*D, *gyr*B, *rec*A, *rpo*B, and *trp*B) of isolate NEAU-SSA 1^T and related taxa. Only bootstrap values above 50% (percentages of 1000 replications) are indicated. *Kitasatospora setae* LM-6054^T was used as an out-group. Bar, 0.05 nucleotide substitutions per site. Table S1. Growth and cultural characteristics of strain NEAU-SSA 1^T. Table S2: MLAS distance values for selected strains in this study. Table S3: General features of the genome sequence of the type strain NEAU-SSA 1^T.

Author Contributions: J.Z. and L.H. performed the isolation and morphological and biochemical characterization of strain NEAU-SSA 1^T. M.Y. performed the antifungal test. P.C. analyzed DNA sequencing data and genomic sequencing data. D.L. performed chemotaxonomic analysis and phylogenetic analysis. X.G. prepared the figures and tables. Y.L. performed the morphological observation by transmission electron microscopy. X.W. and W.X. designed the experiments and edited the manuscript.

Funding: This work was supported in part by grants from the National Key Research and Development Program of China (No. 2017YFD0201606), the National Natural Youth Science Foundation of China (No. 31701858), the China Postdoctoral Science Foundation (2018M631907), the Heilongjiang Postdoctoral Fund (LBH-Z17015), the University Nursing Program for Young Scholars with Creative Talents in Heilongjiang Province (UNPYSCT-2017017), and the "Young Talents" Project of Northeast Agricultural University (17QC14).

Acknowledgments: The authors would like to thank Aharon Oren (Department of Plant and Environmental Sciences, the Alexander Silberman Institute of Life Sciences, the Hebrew University of Jerusalem) for helpful advice on the specific epithet.

Conflicts of Interest: The authors declare that there are no conflict of interest.

References

1. Mansfield, J.; Genin, S.; Magori, S.; Citovsky, V.; Sriariyanum, M.; Ronald, P.; Dow, M.; Verdier, V.; Beer, S.V.; Machado, M.A.; et al. Top 10 plant pathogenic bacteria in molecular plant pathology. *Mol. Plant Pathol.* **2012**, *13*, 614–629. [CrossRef] [PubMed]

2. Hayward, A.C. Biology and epidemiology of bacterial wilt caused by *Pseudomonas solanacearum*. *Annu. Rev. Phytopathol.* **1991**, *29*, 65–87. [CrossRef] [PubMed]

3. Jiang, G.F.; Wei, Z.; Xu, J.; Chen, H.L.; Zhang, Y.; She, X.M.; Macho, A.P.; Ding, W.; Liao, B.S. Bacterial wilt in china: history, current status, and future perspectives. *Front. Plant Sci.* **2017**, *8*, 1549. [CrossRef] [PubMed]

4. Labeda, D.P.; Goodfellow, M.; Brown, R.; Ward, A.C.; Lanoot, B.; Vanncanneyt, M.; Swings, J.; Kim, S.B.; Liu, Z.; Chun, J.; et al. Phylogenetic study of the species within the family *Streptomycetaceae*. *Antonie Van Leeuwenhoek* **2012**, *101*, 73–104. [CrossRef]

5. Bérdy, J. Bioactive microbial metabolites, a personal view. *J. Antibiot.* **2005**, *58*, 1–26. [CrossRef] [PubMed]

6. Li, C.; He, H.R.; Wang, J.B.; Liu, H.; Wang, H.Y.; Zhu, Y.J.; Wang, X.J.; Zhang, Y.Y.; Xiang, W.S. Characterization of a LAL-type regulator NemR in nemadectin biosynthesis and its application for increasing nemadectin production in *Streptomyces cyaneogriseus*. *Sci. China Life Sci.* **2019**, *62*, 394–405. [CrossRef]

7. Waksman, S.A.; Henrici, A.T. The nomenclature and classification of the actinomycetes. *J. Bacteriol.* **1943**, *46*, 337–341. [PubMed]

8. Berdy, J. Are actinomycetes exhausted as a source of secondary metabolites? *Biotechnologia* **1995**, 13–34.

9. Fiedler, H.P.; Bruntner, C.; Bull, A.T.; Ward, A.C.; Goodfellow, M.; Potterat, O.; Puder, C.; Mihm, G. Marine actinomycetes as a source of novel secondary metabolites. *Antonie Van Leeuwenhoek* **2005**, *87*, 37–42. [CrossRef]

10. Hayakawa, M.; Nonomura, H. Humic acid-vitamin agar, a new medium for the selective isolation of soil actinomycetes. *J. Ferment Technol.* **1987**, *65*, 501–509. [CrossRef]

11. Shirling, E.B.; Gottlieb, D. Methods for characterization of *Streptomyces* species. *Int. J. Syst. Bacteriol.* **1966**, *16*, 313–340. [CrossRef]

12. Jin, L.Y.; Zhao, Y.; Song, W.; Duan, L.P.; Jiang, S.W.; Wang, X.J.; Zhao, J.W.; Xiang, W.S. *Streptomyces inhibens* sp. nov., a novel actinomycete isolated from rhizosphere soil of wheat (*Triticum aestivum* L.). *Int. J. Syst. Evol. Microbiol.* **2019**, *69*, 688–695. [CrossRef] [PubMed]

13. Waksman, S.A. *The Actinomycetes. A Summary of Current Knowledge*; The Ronald Press Co.: New York, NY, USA, 1967; p. 286.

14. Jones, K.L. Fresh isolates of actinomycetes in which the presence of sporogenous aerial mycelia is a fluctuating characteristic. *J. Bacteriol.* **1949**, *57*, 141–145. [PubMed]

15. Waksman, S.A. *The Actinomycetes, Volume 2, Classification, Identification and Descriptions of Genera and Species*; Williams and Wilkins Company: Philadelphia, PA, USA, 1961; p. 363.

16. Kelly, K.L. Color-name charts illustrated with centroid colors. In *Inter-Society Color Council-National Bureau of Standards*; U.S. National Bureau of Standards: Washington, DC, USA, 1965.

17. Jia, F.Y.; Liu, C.X.; Wang, X.J.; Zhao, J.W.; Liu, Q.F.; Zhang, J.; Gao, R.X.; Xiang, W.S. *Wangella harbinensis* gen. nov., sp. nov., a new member of the family *Micromonosporaceae*. *Antonie Van Leeuwenhoek* **2013**, *103*, 399–408. [CrossRef] [PubMed]

18. Smibert, R.M.; Krieg, N.R. Phenotypic characterization. In *Methods for General and Molecular Bacteriology*; Gerhardt, P., Murray, R.G.E., Wood, W.A., Krieg, N.R., Eds.; American Society for Microbiology: Washington, DC, USA, 1994; pp. 607–654.

19. Gordon, R.E.; Barnett, D.A.; Handerhan, J.E.; Pang, C. *Nocardia coeliaca*, *Nocardia autotrophica*, and the nocardin strain. *Int. J. Syst. Bacteriol.* **1974**, *24*, 54–63. [CrossRef]

20. Yokota, A.; Tamura, T.; Hasegawa, T.; Huang, L.H. *Catenuloplanes japonicas* gen. nov., sp. nov., nom. rev., a new genus of the order *Actinomycetales*. *Int. J. Syst. Bacteriol.* **1993**, *43*, 805–812. [CrossRef]

21. McKerrow, J.; Vagg, S.; McKinney, T.; Seviour, E.M.; Maszenan, A.M.; Brooks, P.; Seviour, R.J. A simple HPLC method for analysing diaminopimelic acid diastereomers in cell walls of Gram-positive bacteria. *Lett. Appl. Microbiol.* **2000**, *30*, 178–182. [CrossRef] [PubMed]

22. Lechevalier, M.P.; Lechevalier, H.A. The chemotaxonomy of actinomycetes. In *Actinomycete Taxonomy*; Special Publication Volume 6; Dietz, A., Thayer, D.W., Eds.; Society of Industrial Microbiology: Arlington, TX, USA, 1980; pp. 227–291.

23. Minnikin, D.E.; O'Donnell, A.G.; Goodfellow, M.; Alderson, G.; Athalye, M.; Schaal, A.; Parlett, J.H. An integrated procedure for the extraction of bacterial isoprenoid quinones and polar lipids. *J. Microbiol. Methods* **1984**, *2*, 233–241. [CrossRef]

24. Collins, M.D. Isoprenoid quinone analyses in bacterial classification and identification. In *Chemical Methods in Bacterial Systematics*; Goodfellow, M., Minnikin, D.E., Eds.; Academic Press: London, UK, 1985; pp. 267–284.

25. Qu, Z.; Ruan, J.S.; Hong, K. Application of high performance liquid chromatography and gas chromatography in the identification of actinomyces. *Biotechnol. Bull.* **2009**, *s1*, 79–82.

26. Gao, R.X.; Liu, C.X.; Zhao, J.W.; Jia, F.Y.; Yu, C.; Yang, L.Y.; Wang, X.J.; Xiang, W.S. *Micromonospora jinlongensis* sp. nov., isolated from muddy soil in China and emended description of the genus *Micromonospora*. *Antonie Van Leeuwenhoek* **2014**, *105*, 307–315. [CrossRef] [PubMed]

27. Xiang, W.S.; Liu, C.X.; Wang, X.J.; Du, J.; Xi, L.J.; Huang, Y. *Actinoalloteichus nanshanensis* sp. nov., isolated from the rhizosphere of a fig tree (*Ficus religiosa*). *Int. J. Syst. Evol. Microbiol.* **2011**, *61*, 1165–1169. [CrossRef] [PubMed]

28. Zhou, J.Z.; Bruns, M.A.; Tiedje, J.M. DNA recovery from soils of diverse composition. *Appl. Environ. Microbiol.* **1996**, *62*, 316–322. [PubMed]

29. Woese, C.R.; Gutell, R.; Gupta, R.; Noller, H.F. Detailed analysis of the higher-order structure of 16S-like ribosomal ribonucleic acids. *Microbiol. Rev.* **1983**, *47*, 621–669. [PubMed]

30. Springer, N.; Ludwig, W.; Amann, R.; Schmidt, H.J.; Görtz, H.D.; Schleifer, K.H. Occurrence of fragmented 16S rRNA in an obligate bacterial endosymbiont of *Paramecium caudatum*. *Proc. Natl. Acad. Sci. USA* **1993**, *90*, 9892–9895. [CrossRef] [PubMed]

31. Saitou, N.; Nei, M. The neighbor-joining method: a new method for reconstructing phylogenetic trees. *Mol. Biol. Evol.* **1987**, *4*, 406–425.

32. Felsenstein, J. Evolutionary trees from DNA sequences: a maximum likelihood approach. *J. Mol. Evol.* **1981**, *17*, 368–376. [CrossRef] [PubMed]

33. Kumar, S.; Stecher, G.; Tamura, K. Mega7: molecular evolutionary genetics analysis version 7.0 for bigger datasets. *Mol. Biol. Evol.* **2016**, *33*, 1870–1874. [CrossRef]

34. Felsenstein, J. Confidence limits on phylogenies: an approach using the bootstrap. *Evolution* **1985**, *39*, 83–791. [CrossRef]

35. Kimura, M. A simple method for estimating evolutionary rates of base substitutions through comparative studies of nucleotide sequences. *J. Mol. Evol.* **1980**, *16*, 111–120. [CrossRef]

36. Yoon, S.H.; Ha, S.M.; Kwon, S.; Lim, J.; Kim, Y.; Seo, H.; Chun, J. Introducing EzBioCloud: A taxonomically united database of 16S rRNA and whole genome assemblies. *Int. J. Syst. Evol. Microbiol.* **2017**, *67*, 1613–1617.

37. Blin, K.; Wolf, T.; Chevrette, M.G.; Lu, X.; Schwalen, C.J.; Kautsar, S.A.; Suarez Duran, H.G.; de Los Santos, E.L.C.; Kim, H.U.; Nave, M.; et al. antiSMASH 4.0—Improvements in chemistry prediction and gene cluster boundary identification. *Nucleic Acids Res.* **2017**, *45*, W36–W41. [CrossRef]

38. Li, R.Q.; Zhu, H.M.; Ruan, J.; Qian, W.B.; Fang, X.D.; Shi, Z.B.; Li, Y.R.; Li, S.T.; Shan, G.; Kristiansen, K.; et al. De novo assembly of human genomes with massively parallel short read sequencing. *Genome Res.* **2010**, *20*, 265–272. [CrossRef]

39. Li, R.; Li, Y.; Kristiansen, K.; Wang, J. SOAP: short oligonucleotide alignment program. *Bioinformatics* **2008**, *24*, 713–714. [CrossRef]

40. De Ley, J.; Cattoir, H.; Reynaerts, A. The quantitative measurement of DNA hybridization from renaturation rates. *Eur. J. Biochem.* **1970**, *12*, 133–142. [CrossRef]

41. Huss, V.A.R.; Festl, H.; Schleifer, K.H. Studies on the spectrometric determination of DNA hybridisation from renaturation rates. *Syst. Appl. Microbiol.* **1983**, *4*, 184–192. [CrossRef]

42. Yoon, S.H.; Ha, S.M.; Lim, J.; Kwon, S.; Chun, J. A large-scale evaluation of algorithms to calculate average nucleotide identity. *Antonie Van Leeuwenhoek* **2017**, *110*, 1281–1286. [CrossRef]

43. Meier-Kolthoff, J.P.; Auch, A.F.; Klenk, H.P.; Goker, M. Genome sequence-based species delimitation with confidence intervals and improved distance functions. *Bmc Bioinform.* **2013**, *14*, 60. [CrossRef]

44. Boyanova, L.; Gergova, G.; Nikolov, R.; Derejian, S.; Lazarova, E.; Katsarov, N.; Mitov, I.; Krastev, Z. Activity of Bulgarian propolis against 94 *Helicobacter pylori* strains in vitro by agar-well diffusion, agar dilution and disc diffusion methods. *J. Med. Microbiol.* **2005**, *54*, 481–483. [CrossRef]

45. Goodfellow, M.; Kämpfer, P.; Busse, H.J.; Trujillo, M.E.; Suzuki, K.I.; Ludwig, W.; Whitman, W.B. *Bergey's Manual® of Systematic Bacteriology*; Springer: New York, NY, USA, 2012.

46. Rong, X.; Huang, Y. Taxonomic evaluation of the *Streptomyces hygroscopicus* clade using multilocus sequence analysis and DNA-DNA hybridization, validating the MLSA scheme for systematics of the whole genus. *Syst. Appl. Microbiol.* **2012**, *35*, 7–18. [CrossRef]

47. Wayne, L.G.; Brenner, D.J.; Colwell, R.R.; Grimont, P.A.D.; Kandler, O. International Committee on Systematic Bacteriology. Report of the ad hoc committee on reconciliation of approaches to bacterial systematics. *Int. J. Syst Bacteriol.* **1987**, *37*, 463–464. [CrossRef]

48. Richter, M.; Rossello-Mora, R. Shifting the genomic gold standard for the prokaryotic species definition. *Proc. Natl. Acad. Sci. USA* **2009**, *106*, 19126–19131. [CrossRef]

49. Chun, J.; Rainey, F.A. Integrating genomics into the taxonomy and systematics of the Bacteria and Archaea. *Int. J. Syst. Evol. Microbiol.* **2014**, *64*, 316–324. [CrossRef]

50. Ueno, M.; Quyet, N.T.; Shinzato, N.; Matsui, T. Antifungal activity of collected in subtropical region, Okinawa, against *Magnaporthe oryzae*. *Trop. Agricult. Dev.* **2016**, *60*, 48–52.

Article

Community Structures and Antifungal Activity of Root-Associated Endophytic Actinobacteria of Healthy and Diseased Soybean

Chongxi Liu [1,2], Xiaoxin Zhuang [1], Zhiyin Yu [1,2], Zhiyan Wang [2], Yongjiang Wang [2], Xiaowei Guo [1,2], Wensheng Xiang [1,3,*] and Shengxiong Huang [2,*]

[1] Heilongjiang Provincial Key Laboratory of Agricultural Microbiology, Northeast Agricultural University, Harbin 150030, China
[2] State Key Laboratory of Phytochemistry and Plant Resources in West China, Kunming Institute of Botany, Chinese Academy of Sciences, Kunming 650201, China
[3] State Key Laboratory for Biology of Plant Diseases and Insect Pests, Institute of Plant Protection, Chinese Academy of Agricultural Sciences, Beijing 100193, China
[*] Correspondence: xiangwensheng@neau.edu.cn (W.X.); sxhuang@mail.kib.ac.cn (S.H.)

Received: 24 July 2019; Accepted: 5 August 2019; Published: 7 August 2019

Abstract: The present study was conducted to examine the influence of a pathogen *Sclerotinia sclerotiorum* (Lib.) de Bary on the actinobacterial community associated with the soybean roots. A total of 70 endophytic actinobacteria were isolated from the surface-sterilized roots of either healthy or diseased soybeans, and they were distributed under 14 genera. Some rare genera, including *Rhodococcus*, *Kribbella*, *Glycomyces*, *Saccharothrix*, *Streptosporangium* and *Cellulosimicrobium*, were endemic to the diseased samples, and the actinobacterial community was more diverse in the diseased samples compared with that in the heathy samples. Culture-independent analysis of root-associated actinobacterial community using the high-throughput sequencing approach also showed similar results. Four *Streptomyces* strains that were significantly abundant in the diseased samples exhibited strong antagonistic activity with the inhibition percentage of 54.1–87.6%. A bioactivity-guided approach was then employed to isolate and determine the chemical identity of antifungal constituents derived from the four strains. One new maremycin analogue, together with eight known compounds, were detected. All compounds showed significantly antifungal activity against *S. sclerotiorum* with the 50% inhibition (EC_{50}) values of 49.14–0.21 mg/L. The higher actinobacterial diversity and more antifungal strains associated with roots of diseased plants indicate a possible role of the root-associated actinobacteria in natural defense against phytopathogens. Furthermore, these results also suggest that the root of diseased plant may be a potential reservoir of actinobacteria producing new agroactive compounds.

Keywords: *Sclerotinia sclerotiorum* (Lib.) de Bary; diseased soybean root; antifungal activity; actinobacterial community; new agroactive compounds

1. Introduction

Sclerotinia stem rot (SSR) caused by a fungus *Sclerotinia sclerotiorum* (Lib.) de Bary is a highly destructive disease leading to serious economic losses to crops throughout the world. This fungus can infect over 400 plant species, including many economically important crops and vegetables [1–3]. Generally, the development of resistant cultivars is a long-term approach for controlling the disease [2,4]. However, the disease has yet been difficult to control because of the limited resource of the resistant genes. Therefore, fungicides have been used as the auxiliary method for controlling SSR in practice [5]. The benzimidazole and dicarboximide fungicides were the most efficient fungicides in controlling

SSR [6]. However, the continuous use of these fungicides with high concentration can amplify the resistant level of phytopathogens [7–9]. Thus, development of new antifungal agents would be a constant need for controlling the disease.

Endophytic microorganisms residing inside plants have been found in majority of plant species [10]. A growing body of literature recognizes that some of these microorganisms are involved in plant defense against the phytopathogens through a range of mechanisms, including competition for an ecological niche or a substrate, secretion of antibiotics and lytic enzymes, and induction of systemic resistance (ISR) [11,12]. Recent studies on plant-microbe interactions reveal that plants can specifically attract bacteria for their ecological and evolutionary benefit by secreting root exudates [13–15]. It has even been postulated that plants can recruit beneficial microorganisms from soil to counteract pathogen assault [16,17]. For example, it has previously been observed that colonization of the roots of *Arabidopsis* by beneficial rhizobacteria *Bacillus subtilis* FB17 was greatly stimulated when leaves were infected by *Pseudomonas syringae* pv. *tomato* [18].

The phylum *Actinobacteria* consists of a wide range of Gram-positive bacteria with high guanine-plus-cytosine (G + C) content. Actinobacterial species are known to produce a vast diversity of active natural products including antibiotics, antitumor gents, enzymes and immunosuppressive agents, which have been widely used in pharmaceutical, agricultural and other industries [19,20]. Recently, endophytic actinobacteria have attracted significant interest for their capacity to produce abundant bioactive metabolites, which may contribute to their host plants by promoting growth and health [21,22]. A vast majority of endophytic actinobacteria have been isolated from a variety of plants including various crop plants, medicinal plants, and different woody tree species [23–28]. Further, recent cultivation-independent analysis using 16S rRNA gene-based methods revealed that actinobacteria can be specifically enriched in plant roots, and are more abundant in diseased plants than in healthy plants, which may provide probiotic functions for the host plants [29–31]. Thus, it is hypothesized that endophytic actinobacteria from disease plants may be a promising source for the discovery of new antifungal agents against *S. sclerotiorum*.

This prospective study was designed to test the above hypothesis by (i) using culture-independent and dependent methods to compare the generic diversity and antifungal activity of root-associated endophytic actinobacteria of field-growing healthy and diseased plants soybean plants and (ii) identifying antifungal metabolited produced by the outstanding actinobacteria isolated.

2. Materials and Methods

2.1. Plant Materials

Root samples were collected from soybean (cultivar: Hefeng-50) plants identified as SSR symptomatic (diseased) or asymptomatic (healthy) based on typical symptoms in a heavily infected soybean field in Suihua, Heilongjiang province, North China (46.63° N 126.98° E). The diseased samples showed lesions encircling up to 1/3 of stem diameter referring to a severity scale of three [32]. Each healthy plant and diseased plant located as close neighbors were determined as one group. The samples were brought to the lab in a cooler with ice in July 2017 and were processed immediately.

2.2. Isolation of Endophytic Actinobacteria

Three groups of root samples were used for isolation of endophytic actinobacteria. The root samples were air dried for 24 h at room temperature and then washed in water with an ultrasonic step (160 W, 15 min) (KH-160TDV, Hechuang, China) to remove the surface soils and adherent epiphytes completely. After drying, the sample was cut into pieces of 5–10 mm in length and then subjected to a seven-step surface sterilization procedure: A 60-sec wash in sterile tap water containing cycloheximide (100 mg/L) and nalidixic acid (20 mg/L), followed by a wash in sterile water, a 5 min wash in 5% (v/v) NaOCl, a 10 min wash in 2.5% (w/v) $Na_2S_2O_3$, a 5 min wash in 75% (v/v) ethanol, a wash in sterile water and a final rinse in 10% (w/v) $NaHCO_3$ for 10 min. After being thoroughly dried under sterile conditions, the surface-sterilized samples were subjected to continuous drying at 100 °C for 15 min. The sample was then cut up in a commercial blender and ground with a mortar and pestle, employing 1 mL of 0.5 M potassium phosphate buffer (pH 7.0) per 100 mg tissue. Tissue particles were allowed to settle down at 4 °C for 20–30 min, and an aliquot of 200 μL supernatants were spread on a series of isolation media and incubated at 28 °C for 2–3 weeks. Each isolation medium was supplemented with nalidixic acid (20 mg/L) and cycloheximide (50 mg/L) to inhibit the growth of Gram-negative bacteria and fungi. Five isolation media: Humic acid-vitamin (HV) agar [33], Gause's synthetic agar no. 1 [34], dulcitol-proline agar (DPA) [35], cellulose-proline agar [36], and amino acid agar (serine 0.05%, threonine 0.05%, alanine 0.05%, arginine 0.05%, agar powder 2%, pH 7.2–7.4) were selected for the isolation. After 14 days of aerobic incubation at 28 °C, the actinobacterial colonies were transferred onto oatmeal agar (International *Streptomyces* Project medium 3, ISP3) [37] and repeatedly re-cultured until pure cultures were obtained, and maintained as glycerol suspensions (20%, v/v) at −80 °C.

2.3. Phenotypic and Molecular Characterization of Actinobacterial Isolates

The purified colonies were cultivated on ISP 3 at 28 °C for two weeks, and then grouped according to their phenotypic characteristics, including the characteristics of colonies on plates, color of aerial and substrate mycelium, spore mass color, spore chain morphology, and production of diffusible pigment. Those colonies with the same characteristics were classified as one species. The number of species was counted to compare the diversity of root-associated endophytic actinobacteria from healthy and diseased soybean.

Different phenotypic isolates were further subjected to 16S rRNA gene sequence analysis for the genus and species identification. The total DNA was extracted using the lysozyme-sodium dodecyl sulfate-phenol/chloroform method [38]. The primers and procedure for PCR amplification were carried out as described by Kim et al. [39]. The PCR products were purified and ligated into the vector pMD19-T (Takara Biomedical Technology, Beijing, China) and sequenced by an Applied Biosystems DNA sequencer (model 3730XL). The almost full-length 16S rRNA gene sequences (~1500 bp) were obtained and aligned with multiple sequences obtained from the GenBank/EMBL/DDBJ databases using CLUSTAL X 1.83 software. Phylogenetic tree was constructed with neighbor-joining method [40] using Molecular Evolutionary Genetics Analysis (MEGA) software version 7.0 [41]. The bootstrap method with 1000 repetitions was using to assess the topology of the phylogenetic tree [42]. Phylogenetic distances were calculated according to the Kimura two-parameter model [43]. The 16S rRNA gene sequence similarities were determined using the EzBiocloud server [44]. The obtained gene sequences have been deposited in the GenBank database.

2.4. Screening for Antagonistic Actinobacteria

The phytopathogenic *S. sclerotiorum* strain used in this study was kindly provided by the Soybean Research Institute of Northeast Agricultural University (Harbin, China). Antagonistic activity of isolates were evaluated through the dual culture plate assay [45]. The isolates were point-inoculated at the margin of potato dextrose agar (PDA) [46] plates and incubated for three days at 28 °C, after which a fresh mycelial PDA agar plug of the fungus was transferred to the opposite margin of the corresponding plate. After additional days of incubation at 20 °C for seven days, inhibition of hyphal growth of the fungus was scored. The inhibition rates were calculated as follows:

$$\text{Inhibition rate (\%)} = Wi/W \times 100\% \tag{1}$$

where Wi is the width of inhibition and W is the width between the pathogen and actinobacteria. Each test was repeated three times and the average was calculated.

2.5. Isolation and Characterization of Antifungal Compounds

The antifungal compounds were isolated using an in vitro antifungal activity-guided method [47]. The active isolates were inoculated into 250 mL flask containing 50 mL of tryptone soy broth (TSB) [48] and cultivated for two days at 28 °C with shaking at 200 rpm. Then, 12.5 mL of the seed culture was transferred into 1 L Erlenmeyer flask containing 250 mL of the fermentation medium (soluble starch 1%, dextrose 2%, tryptone 0.5%, yeast extract 0.5%, NaCl 0.4%, K_2HPO_4 $3H_2O$ 0.05%, $MgSO_4$ $7H_2O$ 0.05%, $CaCO_3$ 0.5%, pH 7.2–7.4) and incubated at 28 °C for seven days with shaking at 200 rpm. The fermentation broth (20 L) was centrifuged (4000 rev/min, 20 min), and the supernatant and bacterial biomass were extracted with ethylacetate and methanol, respectively. Both extracts were concentrated by a rotary evaporator under reduced pressure until dry and mixed after dissolving their dried residues with methanol.

The crude extracts were divided into seven fractions using column fractionation packed with silica gel (200–300 mesh, Qingdao Marine Chemical Inc., Qingdao, China) eluting with petroleum ether/ethylacetate (20:1, 10:1, 5:1, 3:1, 2:1, 1:1 and 0:1). The bioactive fractions were then subjected to Sephadex LH-20 (Pharmacia, Uppsala, Sweden) and eluted with methanol to obtain several subfractions. The active subfractions were further separated by semipreparative HPLC (Hitachi-DAD, Tokyo, Japan) using a YMC-Triart C_{18} column (250 × 10 mm i.d., 5 μm) at a flow rate of 3.0 mL/min, and the potent active principles were finally isolated.

Structural determination of the active compounds were made according to spectroscopic analysis. NMR spectra were measured with a Bruker Avance III-600 spectrometer in $CDCl_3$ or CD_3OD using TMS as internal standard. The ESI-MS spectrum was taken on a Waters Xevo TQ-S ultrahigh pressure liquid chromatography triple quadrupole mass spectrometer. The HR-ESI-MS spectrum was acquired with an Agilent G6230 Q-TOF mass instrument. The UV spectrum was recorded in chloroform using a Shimadzu UV-2401PC UV-VIS spectrophotometer. The IR spectrum was obtained using a Bruker Tensor 27 FTIR. Optical rotation was determined in chloroform using a JASCO P-1020 polarimeter. The ECD spectrum was measured on a Chirascan circular dichroism spectrometer (Applied Photophysics Corporation Limited, Leatherhead, UK).

2.6. Antifungal Assay of Elucidated Bioactive Compounds

The active compounds were dissolved in methanol and diluted to different concentrations, which were then added to PDA medium. A fresh fungal plug of the fungus (5 mm in diameter) was placed in the center of the agar plate and incubated at 20 °C. Experiments were performed in triplicate, and the plate with the same amount of methanol was used as control. When the control plate was covered completely with fungal mass, the percentage of inhibition was calculated with the formula as follows:

$$\text{Inhibition (\%)} = (1 - D/D_c) \times 100 \tag{2}$$

where D is average diameter of the treatment and D_c is average diameter of the control. Data were subjected to linear regression analysis, and the effective concentrations required for 50% inhibition (EC_{50}) were calculated.

2.7. Culture-Independent Community Analysis

Total community DNA was extracted from three groups of surface-sterilized root samples using FastDNA®SPIN for soil kit (MP Biomedicals, Solon, CA, USA) according to the manufacturers' instructions. The purity and concentration of DNA were detected using NanoPhotometer spectrophotometer (Implen, München, Germany) and Qubit 2.0 Flurometer (Life Technologies, Carlsbad, CA, USA). Bacterial DNA pyrosequencing was based on ~460 bp amplicons generated by the PCR primers: 341F (5′-CCTACGGGNGGCWGCAG-3′) and 805R (5′-GACTACHVGGGTATCTAATCC-3′) with the barcode spanning the hypervariable regions V3-V4 of the 16S rRNA gene The PCR reaction was carried out in 30 μL reactions with 15 μL of Phusion High-Fidelity PCR Master Mix (New England

Biolabs, Ipswich, MA, USA), 0.2 µM of forward and reverse primers, and about 10 ng template DNA. Thermal cycling conditions were 95 °C for 3 min, followed by 25 cycles of 95 °C for 30 s, 55 °C for 30 s, and 72 °C for 30 s, with a final extension at 72 °C for 10 min. PCR products was mixed in equidensity ratios. Then, mixture PCR products was purified with GeneJET Gel Extraction Kit (Thermo Scientific, Fermentas, Germany).

Sequencing libraries were generated using NEB Next Ultra DNA Library Prep Kit for Illumina (New England Biolabs, Ipswich, Massachusetts, USA) following the manufacturer's recommendations and index codes were added. The library quality was assessed on the Qubit@ 2.0 Fluorometer and Agilent Bioanalyzer 2100 system (Agilent Technologies, Palo Alto, CA, USA). At last, the library was sequenced on an Illumina MiSeq platform and 250 bp paired-end reads.

When the sequencing was finished, we needed to filter the raw data to secure the quality, which mainly included the following steps: (1) Cut the polluted adapter, (2) remove low quality reads, specifically reads with average quality less than 19, based on the Phred algorithm, and (3) remove the reads with N base exceeding 5%.

According to overlap of the clean data, we spliced the paired reads by using the PEAR software [49] to merged sequences. The sequences were then removed Chimeras and clustered into operational taxonomic units (OTUs) by UCLUST [50] based on 97% pairwise identity. Taxonomic classification of the representative sequence for each OTU was done using a RDP classifier or QIIME's closed reference strategy against the 16S rRNA gene database [51].

2.8. Statistical Analysis

Statistical analyses were performed via SPSS software, version 18.0. Means were compared via analysis of variance one-way ANOVA using the least significant differences test (LSD, $p < 0.05$). The data were reported as means ± standard deviation.

3. Results

3.1. Isolation and Distribution of Endophytic Actinobacteria

A total of 1574 endophytic actinobacterial colonies were successfully isolated. Based on their phenotypic characteristics, the colonies were preliminary classified into 70 species. Among the 70 species, 15 species were from healthy soybean roots, 27 species were from diseased soybean roots, and 28 species were shared by healthy and diseased soybean roots. The diversity of actinobacteria from diseased soybean root were greater than those from healthy soybean root. 16S rRNA gene sequence analysis of the 70 isolates revealed that they were distributed under 14 genera: *Streptomyces, Micromonospora, Actinomadura, Nonomuraea, Microbacterium, Rhodococcus, Promicromonospora, Microbispora, Kribbella, Mycolicibacterium, Glycomyces, Saccharothrix, Streptosporangium* and *Cellulosimicrobium* within the class Actinobacteria. *Streptomyces* was the most frequently isolated genus (58%, 41 isolates), followed by *Micromonospora* (5 isolates) and *Nonomuraea* (4 isolates). Some rare genera, including *Rhodococcus, Kribbella, Glycomyces, Saccharothrix, Streptosporangium* and *Cellulosimicrobium*, were endemic to the diseased samples (Figure 1). The 16S rRNA gene sequences were deposited in GenBank with accession numbers: MH919371–MH919374 and MN058215–MH058280.

3.2. In Vitro Antagonism of S. sclerotiorum and Identification of Bioactive Strains

Strains DAAG3-11, DGS1-1, DDPA2-14 and DGS3-15 exhibited strong antagonistic activity against *S. sclerotiorum*, with inhibition activity rates ranging from 54.1% to 87.6% (Figure 2). Strains DGS1-1, DDPA2-14 and DGS3-15 were from diseased soybean roots. For strain DAAG3-11, 176 colonies were from diseased soybean roots, whereas 11 colonies were from healthy soybean roots. Based on the 16S rRNA gene sequences, strains DAAG3-11, DGS1-1, DDPA2-14 and DGS3-15 were closely related to *Streptomyces sporoclivatus, Streptomyces cavourensis, Streptomyces capitiformicae* and *Streptomyces pratensis*, respectively (Table 1).

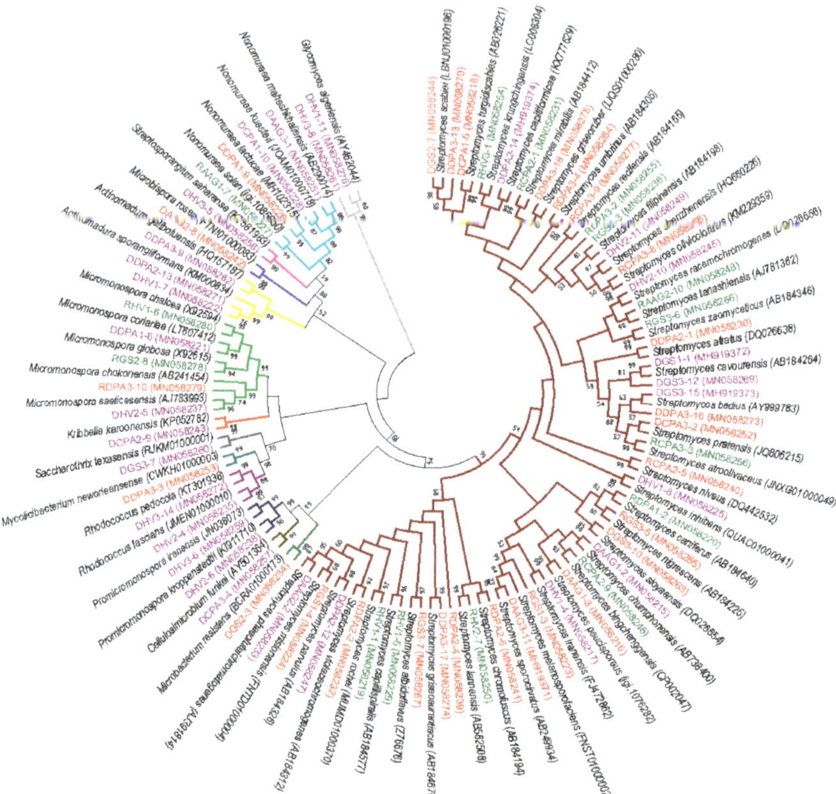

Figure 1. Neighbor-joining phylogenetic tree of 16S rRNA gene sequences from 70 endophytic actinobacteria in this study and their phylogenetic neighbors. Numbers at nodes are bootstrap values (percentages of 1000 replications); only values > 50% are shown. GenBank accession numbers of 16S rRNA gene sequences are shown next to isolate names. A branch indicated by the same color belongs to the same genus. Isolates indicated by green and purple are endemic to the healthy and diseased samples, respectively. Isolates shared by healthy and diseased samples are indicated with red.

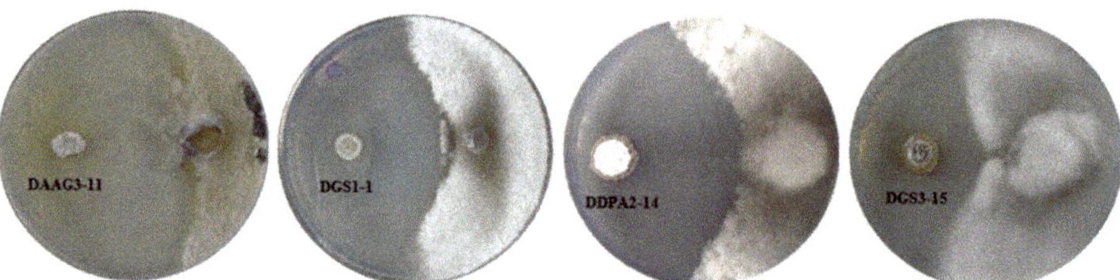

Figure 2. Dual culture plate assay between four endophytic actinobacteria against *S. sclerotiorum*.

Table 1. Antagonistic potential endophytic actinobacteria isolated from healthy and diseased soybean root, and similarity values for 16S rRNA gene sequences.

Isolate No. and NCBI Genbank Accesion No.	Closest Type Strain with Accession Number	Similarity	Isolated From	Colony Number	*S. sclerotiorum* Mycelial Growth Inhibition (%) *
DAAG3-11 (MH919371)	*Streptomyces sporoclivatus* (AB249934)	100%	Healthy soybean root	11	87.6 ± 1.8 a
			Diseased soybean root	176	
DGS1-1 (MH919372)	*Streptomyces cavourensis* (AB184264)	99.9%	Diseased soybean root	13	78.9 ± 1.9 b
DDPA2-14 (MH919374)	*Streptomyces capitiformicae* (KX777629)	100%	Diseased soybean root	9	68.6 ± 3.4 c
DGS3-15 (MH919373)	*Streptomyces pratensis* (JQ806215)	99.9%	Diseased soybean root	6	54.1 ± 2.2 d

* Values are the means ± SE ($n = 4$). Data within the same column followed by different letters are significantly different.

3.3. Identification and Activity Evaluation of the Antifungal Compounds

An antifungal activity-guided separation of the components of four active strains against *S. sclerotiorum*, using the in vitro antifungal assay, led to the isolation of nine compounds as their active principles (Figure 3). Out of the nine compounds, compounds **1** and **2** were from strain DAAG3-11, compound **3** was from strain DGS1-1, compounds **4–7** were from strain DGS3-15, and compounds **8** and **9** were from strain DDPA2-14. Compounds **1–8** are known compounds, which structures were elucidated as azalomycins F_{4a} (**1**) [52], azalomycins F_{5a} (**2**) [52], bafilomycin B_1 (**3**) [53], actinolactomycin (**4**) [54], dimeric dinactin (**5**) [55], tetranactin (**6**) [56], dinactin (**7**) [56] and maremycin G (**8**) [57] by analysis of their spectroscopic data and comparison with literature values (Figure S1). Compound **9** is a new maremycin analogue.

Compound **9** was obtained as a yellow, amorphous powder. Its molecular formula $C_{22}H_{27}N_3O_4S$ was determined by high resolution electrospray ionization mass spectrometry (HRESIMS) data (*m/z* 468.1555, $[M + Na]^+$, calculated for 468.1564), corresponding to 11 degrees of unsaturation. The IR spectrum indicated the presence of hydroxy (3424 cm^{-1}) and carbonyl (1720, 1682 cm^{-1}) groups. The ^{1}H NMR data (Table 2) of **9** suggested the presence of one 1, 2-disubstituted benzene system at δ_H 7.43 (1H, d, $J = 7.8$ Hz, H-4), 7.12 (1H, td, $J = 7.6$, 1.0 Hz, H-5), 7.37 (1H, td, $J = 7.7$, 1.2 Hz, H-6) and 6.89 (1H, d, $J = 7.9$ Hz, H-7). The ^{1}H NMR data of **9** also revealed the presence of two methyl signals at δ_H 2.17, (3H, s, H-23) and δ_H 3.23, (3H, s, H-25). ^{13}C NMR spectrum of **9** showed 11 sp^2-carbons including three carbonyls at δ_C 204.84 (C-21), 178.33 (C-2), 168.5 (C-13) and eight aromatic or olefinic carbons at δ_C 152.27 (C-16), 142.83 (C-9), 130.36 (C-7), 130.01 (C-4), 125.53 (C-5), 123.24 (C-6), 109.08 (C-8), 100.43 (C-17). In the sp^3-carbon region, the spectrum showed three methine at δ_C 42.84 (C-10), 52.94 (C-11), 52.8 (C-14), four methylene at δ_C 27.04 (C-18), 21.04 (C-19), 38.79(C-20), 38.75(C-22), and three methyl carbons at δ_C 16.36 (C-23), 8.92 (C-24), 26.65 (C-25).

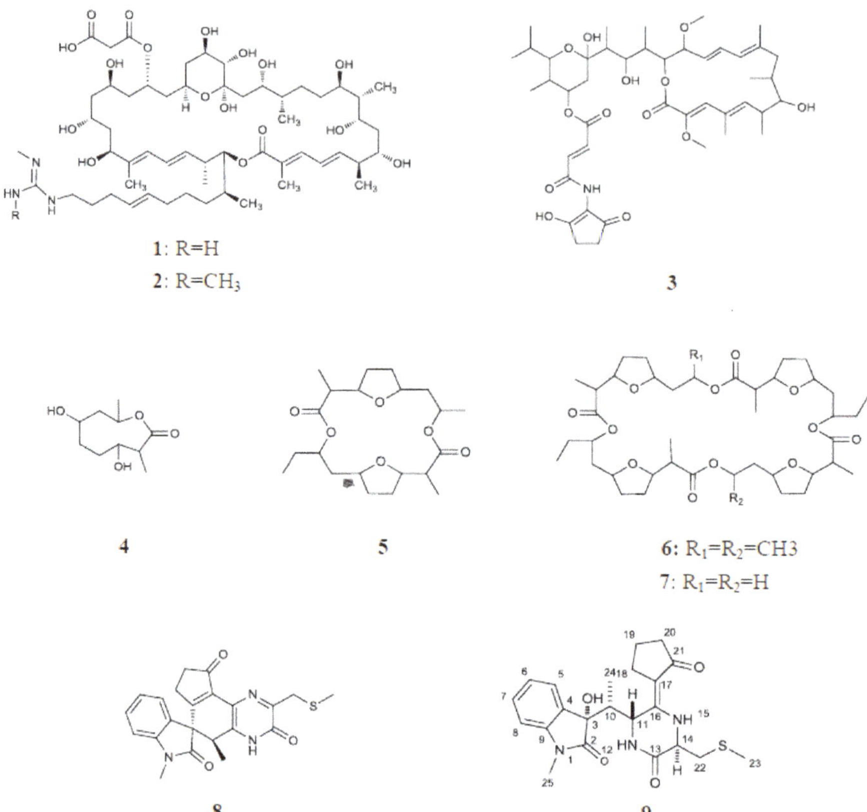

Figure 3. The structures of compounds **1–9**.

Comparison the NMR data of **9** with FR900452 [58], an indole diketopiperazine motif linked with a cyclopentenone moiety, which was isolated from the fermentation broth of *Streptomyces* sp. B9173, implied that **9** was identified as a reduced form of FR900452 in which the cyclopentenone moiety is hydrogenated to cyclopentanone. As shown in Figure 4, the accurate assignments of all protons and carbons for compound **9** were preformed through the correlations in 2D-NMR spectra (^{1}H–^{1}H COSY, HSQC and HMBC, Figure S2). The HMBC couplings Me-25/C-9/C-2, H-5/C-3, and H-10/C-2/C-3/C-4, along with ^{1}H–^{1}H COSY correlations of H-5/H-6/H7/H-8, revealed N-methyl-2-oxindole unit. In addition, ^{1}H–^{1}H COSY correlations of H-18/H-19/H-20, as well as the HMBC cross peaks H-18/C-17/C-16, H-20/C-21, demonstrated that oxopiperazinyl moiety was linked to C-16/C-17 on the cyclopentenone moiety. ^{1}H–^{1}H COSY correlations of Me-24/H-10/H-11, together with the HMBC correlations from Me-24/C-3/C-10/C-11, indicated that N-methyl-2-oxindole unit was linked to C-10/C-5 on the oxopiperazinyl moiety. The ^{1}H and ^{13}C NMR spectroscopic data of **9** were also indicative of methyl mercaptomethylene moieties [δ_C 38.75, $\delta_{H\text{-}CH2}$ 3.17 (1H, m), 2.83 (1H, dd, 14.0, 8.2), $\delta_{C\text{-}CH3}$ 16.36, $\delta_{H\text{-}S\text{-}CH3}$ 2.15 (3H, s)]. The HMBC cross peaks Me-23/C22, H-22/C-14/C-13, along with ^{1}H–^{1}H COSY correlations of H-22/H-14, evidenced that methyl mercaptomethylene moieties were linked to C-14 on the oxopiperazinyl moieties, respectively. Therefore, the planar structure of **9** was elucidated as a reduced form of FR900452, depicted in Figure 3.

Table 2. ^{1}H NMR and ^{13}C NMR data of compound **9** in CDCl$_3$.

Position	δ_C	δ_H (*J* in Hz)
2	178.33	
3	78.69	
4	130.01	
5	125.53	7.43 (1H, d, 7.8)
6	123.24	7.12 (1H, td, 7.6, 1.0)
7	130.36	7.37 (1H, td, 7.7, 1.2)
8	109.08	6.89 (1H, d, 7.9)
9	142.83	
10	42.84	2.06 (1H, m)
11	52.94	5.36 (1H, s)
12		11.01 (brs)
13	168.23	
14	52.8	4.14 (1H, dd, 8.2, 3.3)
15		7.09 (brs)
16	152.27	
17	100.43	
18	27.04	2.24 (1H, s)
		2.44 (1H, ddd, 13.9, 8.4, 3.8)
19	21.04	1.83 (2H, m)
20	38.79	2.28 (2H, m)
21	204.84	
22	38.75	3.17 (1H, m)
22		2.83 (1H, dd, 14.0, 8.2)
23	16.36	2.15 (3H, s)
24	8.92	1.19 (3H, d, 7.0)
25	26.65	3.23 (3H, s)

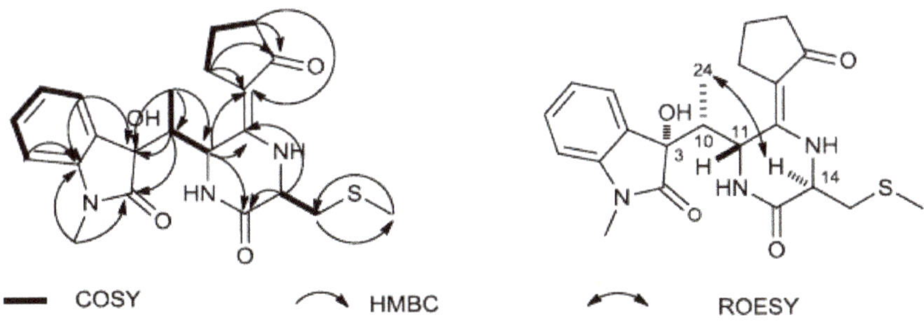

Figure 4. Key 1H–1H COSY, HMBC and ROESY correlations of compound **9**.

The relative configuration of compound **9** was determined by interpretation of its ROESY NMR spectrum. The correlations of H-14/Me-24, and H-10/H-11, indicated that H-14 and Me-24 were α-oriented, whereas H-10 and H-11 were β-oriented (Figure 4). Based on the close skeleton, the comparison of ECD spectra between **9** and N-demethylmaremycin B [57], and the largely consistent data supported that the configurations of 3-OH was α-oriented. Ultimately, the absolute configuration of was identified as 3S, 10R, 11R, 14R, resulting from the same trends of cotton effects (CEs) in the experimental ECD spectra of **9** and N-demethylmaremycin B.

The in vitro antifungal activity of compounds **1–9** against *S. sclerotiorum* was determined at various concentrations. All compounds showed significantly antifungal activity against *S. sclerotiorum* with the EC$_{50}$ values ranging from 49.14 mg/L to 0.21 mg/L (Table 3). Thus, it further confirmed that these compounds were the main antifungal constituents produced by the four active strains.

Table 3. EC₅₀ values of active compounds against S. sclerotiorum.

Compounds	1	2	3	4	5	6	7	8	9
EC_{50} (mg/L)	4.87 ± 0.16 a	4.96 ± 0.13 a	0.21 ± 0.02 b	49.14 ± 0.82 c	5.33 ± 0.15 ae	3.69 ± 0.05 d	5.60 ± 0.11 e	3.46 ± 0.12 d	3.70 ± 0.05 d

Values are the means ± SE ($n = 9$). Data within the same column followed by different letters are significantly different.

3.4. Culture-Independent Communities

A total of 4116 OTUs containing 745708 high-quality reads were detected in the soybean root microbiome. The raw sequencing reads for this project were submitted to the National Center for Biotechnology Information Short Read Archive under accession numbers SRR8056376–SRR8056381. The predominant bacterial phyla were *Proteobacteria*, *Bacteroidetes* and *Actinobacteria* in the soybean roots. To compare the microbial communities obtained in healthy and diseased root samples from each group, the relative abundance of order *Rhizobiales* that can improve rhizobial nodulation and nitrogen fixation was significantly greater in the healthy samples, whereas the order *Actinobacteria* were more abundant in the diseased samples (Figure 5).

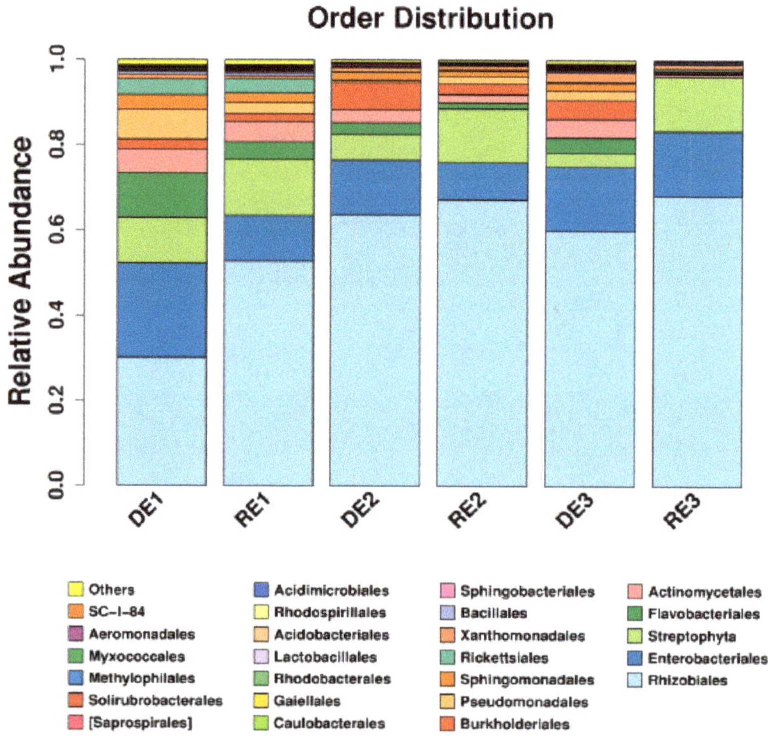

Figure 5. Analysis of culture-independent endophytic communities at order level in the soybean roots. DE, diseased sample; RE, healthy sample.

4. Discussion

The multifaceted approach adopted in this study, linking culture-independent and culture-dependent analysis, showed that actinobacteria were more abundant or diverse in the diseased soybean roots. This finding was in agreement with the previous study that the phylum *Actinobacteria* was higher in '*Candidatus* Liberibacter asiaticus'-infected citrus samples compared with that in healthy samples [31]. Another similar study also showed that potato plants infected with *Erwinia carotovora* subsp. *atroseptica* increased bacterial diversity [59]. The higher diversity of endophytic actinobacteria in diseased but healthy plants suggests that they may be involved in pathogen defense [60]. Indeed, extensive research has shown that endophytic actinobacteria has the capacity to control plant pathogens [22]. The in vitro antagonism assays demonstrated that four strains showed strong antifungal activity against *S. sclerotiorum*. Among the four antagonistic strains, all colonies of three strains were absolutely from diseased soybean roots, and another strain was also significantly enriched in diseased soybean roots compared to healthy soybean roots. A similar study has also shown that the rhizosphere soil of diseased tomato plant harbored a high percentage of antagonists [61]. Studies over the past few years have provided important information that plants possess a sophisticated defense mechanism by actively recruiting root-associated microbes from soil upon pathogen attack [18,62]. By adjusting the quantity and composition of its root secretion, plants can determine the composition of the root microbiome by affecting microbial diversity, density, and activity [63,64]. Our results seem to be consistent with previous observations. However, those strains with antagonistic activity in vitro may not be simply translated into biocontrol bacteria. Their biocontrol effects are influenced by various factors. For example, the antagonistic strains should reach a certain amount inside the plants to demonstrate a significant biocontrol effect [65,66]. Moreover, their secondary metabolite producing ability inside the plants may be influenced by plant physiological environment. Therefore, further work is required to assess the biocontrol efficiency in vivo and root-colonizing capacity of antagonistic strains by pathogen infection.

To learn more about the chemical nature of the antifungal activity, nine active compounds including six macrolides, two diketopiperazines and one 2-oxonanonoids, were finally obtained. Out of which, bafilomycin B_1 (**3**) showed strongest inhibitory activity against *S. sclerotiorum*. Bafilomycin B_1 has been reported to be produced by several *Streptomyces* strains and to show inhibitory activity against various fungi in vitro, such as *Rhizoctonia solani*, *Aspergillus fumigatus*, *Botrytis cinerea*, *Penicillium roqueforti*, and so on [67,68]. The antifungal activity of this compound against *S. sclerotiorum* was first reported in this paper. Azalomycins F_{4a} (**1**) and F_{5a} (**2**) were first isolated from the broth of *Streptomyces hygroscopicus* var. *azalomyceticus* [69]. Azalomycins F complex, including azalomycins F_{3a}, F_{4a} and F_{5a} showed remarkable antifungal activity against asparagus (*Asparagus officinalis*) pathogens *Fusarium moliniforme* and *Fusarium oxysporumas* as well as powdery mildew pathogen *Botrytis* spp. [70]. The antifungal activity of the pure compound was first demonstrated in our research. Azalomycins possess broad-spectrum antibacterial and antifungal activities, and almost all of them were produced by *Streptomyces*, which were isolated repeatedly from soil and plant roots in the field by our laboratory (data no shown). This emphasizes the possible importance of *Streptomyces* producing azalomycins to protect plants against phytopathogens. A mixture of dinactin (**7**), trinactin and the major component tetranactin (**6**) is known as commercial pesticides polynactin (liuyangmycin in China), which can effectively control spider mites under wet conditions [71]. In addition, tetranactin (**6**) also exhibited significant antifungal activity against plant pathogen *Botrytis cinerea* with a minimum inhibitory concentration (MIC) of 24 μg·mL^{-1} [72]. Besides dinactin (**7**) and tetranactin (**6**), the monomer (**4**) and dimer (**5**) of polynactin were also isolated in this study, all of which were active against *S. sclerotiorum*. The antifungal activities of actinolactomycin (**4**), dimeric dinactin (**5**) and dinactin (**7**) have not been reported as yet. The findings reported here shed new light on the application of polynactin. Natural indole diketopiperazines exhibited a wide range of biological activities including antitumor [73], antibacterial [74], antifungal [75] and antiviral activities [76]. FR900452 is sulfur-containing indole diketopiperazines that showed specific and potent inhibitory activity against the platelet aggregation induced by platelet-activating factor [77]. Maremycin G (**8**) and compound **9**, structurally related to FR900452, showed significant antifungal activity against *S. sclerotiorum*. To our knowledge, this is the first report of the antifungal property of maremycins. Further research is needed to confirm the efficacy of in vivo disease control provided by the nine active compounds under laboratory and field conditions.

5. Conclusions

In summary, we report that soybean infected by *S. sclerotiorum* (Lib.) de Bary had a higher populations of actinobacteria and enhanced root colonization of antagonistic populations. In addition, eight known compounds and one new compound that exerted significant antifungal activity against *S. sclerotiorum* were obtained. These findings suggest that diseased plant samples could be a potential source for screening novel agroactive compounds, which contribute to a better understanding of plant–microbe interactions and provide new strategies for the development of agricultural antibiotics.

Supplementary Materials: The following are available online at http://www.mdpi.com/2076-2607/7/8/243/s1.

Author Contributions: C.L., X.Z., Z.W. and X.G. performed the experiments. Z.Y. and Y.W. analyzed the data. C.L. wrote the paper. W.X. and S.H. designed the experiments and reviewed the manuscript. All authors read and approved the final manuscript.

Funding: This research was supported by the University Nursing Program for Young Scholars with Creative Talents in Heilongjiang Province (UNPYSCT-2016153), the Academic Backbone Project of Northeast Agricultural University (17XG17), and the National Natural Science Foundation of China (No. 31872037).

Conflicts of Interest: The authors declare that they have no conflicts of interest.

References

1. Boland, G.J.; Hall, R. Index of plant hosts of *Sclerotinia sclerotiorum*. *Can. J. Plant. Pathol.* **1994**, *16*, 93–108. [CrossRef]

2. Bardin, S.D.; Huang, H.C. Research on biology and control of *Sclerotinia* diseases in Canada. *Can. J. Plant. Pathol.* **2001**, *23*, 88–98. [CrossRef]

3. Firoz, M.J.; Xiao, X.; Zhu, F.X.; Fu, Y.P.; Jiang, D.H.; Schnabel, G.; Luo, C.X. Exploring mechanisms of resistance to dimethachlone in *Sclerotinia sclerotiorum*. *Pest. Manag. Sci.* **2016**, *72*, 770–779. [CrossRef] [PubMed]

4. Boland, G.J. Stability analysis for evaluating the influence of environment on chemical and biological control of white mold (*Sclerotinia sclerotiorum*) of bean. *Biol. Control.* **1997**, *9*, 7–14. [CrossRef]

5. Lee, Y.H.; Cho, Y.S.; Lee, S.W.; Hong, J.K. Chemical and biological controls of balloon flower stem rots caused by *Rhizoctonia solani* and *Sclerotinia sclerotiorum*. *Plant. Pathol. J.* **2012**, *28*, 156–163. [CrossRef]

6. Hu, S.; Zhang, J.; Zhang, Y.; He, S.; Zhu, F. Baseline sensitivity and toxic actions of boscalid against *Sclerotinia sclerotiorum*. *Crop. Prot.* **2018**, *110*, 83–90. [CrossRef]

7. Ma, H.X.; Feng, X.J.; Chen, Y.; Chen, C.J.; Zhou, M.G. Occurrence and characterization of dimethachlon insensitivity in *Sclerotinia sclerotiorum* in jiangsu province of China. *Plant. Dis.* **2009**, *93*, 36–42. [CrossRef] [PubMed]

8. Kuang, J.; Hou, Y.P.; Wang, J.X.; Zhou, M.G. Sensitivity of *Sclerotinia sclerotiorum* to fludioxonil: in vitro determination of baseline sensitivity and resistance risk. *Crop. Prot.* **2011**, *30*, 876–882. [CrossRef]

9. Zhou, F.; Zhang, X.L.; Li, J.L.; Zhu, F.X. Dimethachlon resistance in *Sclerotinia sclerotiorum* in China. *Plant. Dis.* **2014**, *98*, 1221–1226. [CrossRef]

10. Smith, S.A.; Tank, D.C.; Boulanger, L.A.; Bascom-Slack, C.A.; Eisenman, K.; Kingery, D. Bioactive endophytes warrant intensified exploration and conservation. *PLoS ONE* **2008**, *3*, e3052. [CrossRef]

11. Sturz, A.V.; Christie, B.R.; Nowak, J. Bacterial endophytes: potential role in developing sustainable systems of crop production. *Crit. Rev. Plant. Sci.* **2000**, *19*, 1–30. [CrossRef]

12. Lodewyckx, C.; Vangronsveld, J.; Porteous, F.; Moore, E.R.B.; Taghavi, S.; Mezgeay, M.; van der Lelie, D. Endophytic bacteria and their potential applications. *CRC. Crit. Rev. Plant. Sci.* **2002**, *21*, 583–606. [CrossRef]

13. Bais, H.P.; Weir, T.L.; Perry, L.G.; Gilroy, S.; Vivanco, J.M. The role of root exudates in rhizosphere interactions with plants and other organisms. *Annu. Rev. Plant. Biol.* **2006**, *57*, 233–266. [CrossRef]

14. Schulz, B.; Boyle, C. What are endophytes? In *Microbial Root Endophytes*; Schulz, B.J.E., Boyle, C.J.C., Sieber, T.N., Eds.; Springer: Berlin, Germany; pp. 1–13.

15. Sorensen, J.; Sessitsch, A. Plant-associated bacteria—Lifestyle and molecular interactions. In *Modern Soil Microbiology*; Van Elsas, J.D., Jansson, J.K., Trevors, J.T., Eds.; CRC Press: Boca Raton, FL, USA; pp. 211–236.

16. Mendes, R.; Kruijt, M.; de Bruijn, I.; Dekkers, E.; van der Voort, M.; Schneider, J.H.M.; Piceno, Y.M.; DeSantis, T.Z.; Andersen, G.L.; Bakker, P.A.H.M.; et al. Deciphering the rhizosphere microbiome for disease-suppressive bacteria. *Science* **2011**, *332*, 1097–1100. [CrossRef]

17. Liu, C.X.; Song, J.; Wang, X.J.; Xiang, W.S. Recruitment of defensive microbs and plant protection. *Sci. Sin. Vitae* **2016**, *46*, 1–8.

18. Rudrappa, T.; Czymmek, K.J.; Paré, P.W.; Bais, H.P. Root-secreted malic acid recruits beneficial soil bacteria. *Plant. Physiol.* **2008**, *148*, 1547–1556. [CrossRef]

19. Bèrdy, J. Bioactive microbial metabolites. *J. Antibiot.* **2005**, *58*, 1–26. [CrossRef]

20. Qin, S.; Li, W.J.; Dastager, S.G.; Hozzein, W.N. Editorial: Actinobacteria in special and extreme habitats: Diversity, function roles, and environmental adaptations. *Front. Microbiol.* **2016**, *7*, 1415. [CrossRef]

21. Singh, R.; Dubey, A.K. Diversity and applications of endophytic actinobacteria of plants in special and other ecological niches. *Front. Microbiol.* **2018**, *9*, 1767. [CrossRef]

22. Ek-Ramos, M.J.; Gomez-Flores, R.; Orozco-Flores, A.A.; Rodríguez-Padilla, C.; González-Ochoa, G.; Patricia Tamez-Guerra, P. Bioactive products from plant-endophytic Gram-positive bacteria. *Front. Microbiol.* **2019**, *10*, 463. [CrossRef]

23. Conn, V.M.; Franco, C.M. Isolation and identification of actinobacteria from surface-sterilized wheat roots. *Appl. Environ. Microbiol.* **2003**, *69*, 5603–5608.

24. Tian, X.; Cao, L.; Tan, H.; Han, W.; Chen, M.; Liu, Y.; Zhou, S. Diversity of cultivated and uncultivated actinobacterial endophytes in the stems and roots of rice. *Microb. Ecol.* **2007**, *53*, 700–707. [CrossRef]

25. Zhao, K.; Penttinen, P.; Guan, T.W.; Xiao, J.; Chen, Q.; Xu, J.; Lindström, K.; Zhang, L. The diversity and antimicrobial activity of endophytic actinomycetes isolated from medicinal plants in Panxi plateau, China. *Curr. Microbiol.* **2011**, *62*, 182–190. [CrossRef]

26. Li, J.; Zhao, G.Z.; Huang, H.Y.; Qin, S.; Zhu, W.Y.; Zhao, L.X.; Xu, L.H.; Zhang, S.; Li, W.J.; Strobel, G. Isolation and characterization of culturable endophytic actinobacteria associated with *Artemisia annua* L. *Antonie. Leeuwenhoek.* **2012**, *101*, 515–527. [CrossRef]

27. Miao, G.P.; Zhu, C.S.; Feng, J.T.; Han, L.R.; Zhang, X. Effects of plant stress signal molecules on the production of wilforgine in an endophytic actinomycete isolated from *Tripterygium wilfordii* Hook. f. *Curr. Microbiol.* **2015**, *70*, 571–579. [CrossRef]

28. Wei, W.; Zhou, Y.; Chen, F.; Yan, X.; Lai, Y.; Wei, C.; Chen, X.; Xu, J.; Wang, X. Isolation, diversity, and antimicrobial and immunomodulatory activities of endophytic actinobacteria from tea cultivars Zijuan and Yunkang-10 (*Camellia sinensis* var assamica). *Front. Microbiol.* **2018**, *9*, 1304. [CrossRef]

29. Bulgarelli, D.; Rott, M.; Schlaeppi, K.; Ver Loren van Themaat, E.; Nahal Ahmadinejad, N.; Assenza, F.; Rauf, P.; Huettel, B.; Reinhardt, R.; Schmelzer, E.; et al. Revealing structure and assembly cues for *Arabidopsis* root-inhabiting bacterial microbiota. *Nature* **2012**, *488*, 91–95. [CrossRef]

30. Lundberg, D.S.; Lebeis, S.L.; Paredes, S.H.; Yourstone, S.; Gehring, J.; Malfatti, S.; Tremblay, J.; Engelbrektson, A.; Kunin, V.; del Rio, T.G.; et al. Defining the core *Arabidopsis thaliana* root microbiome. *Nature* **2012**, *488*, 86–90. [CrossRef]

31. Trivedi, P.; He, Z.; Van Nostrand, J.D.; Albrigo, G.; Zhou, J.; Wang, N. Huanglongbing alters the structure and functional diversity of microbial communities associated with citrus rhizosphere. *ISME. J.* **2012**, *6*, 363–383. [CrossRef]

32. Boland, G.J.; Hall, R. Growth room evaluation of soybean cultivars for resistance to *Sclerotinia sclerotiorum*. *Can. J. Plant. Sci.* **1986**, *66*, 559–564. [CrossRef]

33. Hayakawa, M.; Nonomura, H. Humic acid-vitamin agar, a new medium for the selective isolation of soil actinomycetes. *J. Ferment. Technol.* **1987**, *65*, 501–509. [CrossRef]

34. Atlas, R.M. *Handbook of Microbiological Media*; Parks, L.C., Ed.; CRC Press: Boca Raton, FL, USA, 1993.

35. Guan, X.J.; Liu, C.X.; Fang, B.Z.; Zhao, J.W.; Jin, P.J.; Li, J.M. *Baia soyae* gen. nov., sp. nov., a mesophilic representative of the family *Thermoactinomycetaceae*, isolated from soybean root [*Glycine max* (L.) Merr]. *Int. J. Syst. Evol. Microbiol.* **2015**, *65*, 3241–3247. [CrossRef]

36. Qin, S.; Li, J.; Chen, H.H.; Zhao, G.Z.; Zhu, W.Y.; Jiang, C.L.; Xu, L.H.; Li, W.J. Isolation, diversity, and antimicrobial activity of rare actinobacteria from medicinal plants of tropical rain forests in Xishuangbanna, China. *Appl. Environ. Microbiol.* **2009**, *75*, 6176–6186. [CrossRef]

37. Shirling, E.B.; Gottlieb, D. Methods for characterization of *Streptomyces* species. *Int. J. Syst. Bacteriol.* **1966**, *16*, 313–340. [CrossRef]

38. Maniatis, T.; Fritsch, E.F.; Sambrook, J. *Molecular Cloning: A Laboratory Manual*, 2nd ed.; Cold Spring Harbor Laboratory: Cold Spring Harbor, NY, USA, 1982.

39. Kim, S.B.; Brown, R.; Oldfield, C.; Gilbert, S.C.; Iliarionov, S.; Goodfellow, M. *Gordonia amicalis* sp. nov., a novel dibenzothiophene-desulphurizing actinomycete. *Int. J. Syst. Evol. Microbiol.* **2000**, *50*, 2031–2036. [CrossRef]

40. Saitou, N.; Nei, M. The neighbor-joining method: A new method for reconstructing phylogenetic trees. *Mol. Biol. Evol.* **1987**, *4*, 406–425.

41. Kumar, S.; Stecher, G.; Tamura, K. MEGA7: molecular evolutionary genetics analysis version 7.0 for bigger datasets. *Mol. Biol. Evol.* **2016**, *33*, 1070. [CrossRef]

42. Felsenstein, J. Confidence limits on phylogenies: an approach using the bootstrap. *Evolution* **1985**, *39*, 783–791. [CrossRef]

43. Kimura, M. *The Neutral Theory of Molecular Evolution*; Cambridge Universiry Press: Cambridge, UK, 1983.

44. Yoon, S.H.; Ha, S.M.; Kwon, S.; Lim, J.; Kim, Y.; Seo, H.; Chun, J. Introducing EzBioCloud: A taxonomically united database of 16S rRNA gene sequences and whole-genome assemblies. *Int. J. Syst. Evol. Microbiol.* **2017**, *67*, 1613–1617.

45. Hamzah, T.N.T.H.; Lee, S.Y.; Hidayat, A.; Terhem, R.; Faridah-Hanum, I.; Mohamed, R. Diversity and characterization of endophytic fungi isolated from the tropical mangrove species, *Rhizophora mucronata*, and identification of potential antagonists against the soil-borne fungus, *Fusarium solani*. *Front. Microbiol.* **2018**, *9*, 1707. [CrossRef]

46. Zhang, J.; Wang, X.J.; Yan, Y.J.; Jiang, L.; Wang, J.D.; Li, B.J.; Xiang, W.S. Isolation and identification of 5-hydroxyl-5-methyl-2-hexenoic acid from *Actinoplanes* sp. HBDN08 with antifungal activity. *Bioresource. Technol.* **2010**, *101*, 8383–8388. [CrossRef]

47. Liu, C.X.; Zhang, J.; Wang, X.J.; Qian, P.T.; Wang, J.D.; Gao, Y.M.; Yan, Y.J.; Zhang, S.Z.; Xu, P.F.; Li, W.B.; et al. Antifungal activity of borrelidin produced by a *Streptomyces* strain isolated from soybean. *J. Agric. Food. Chem.* **2012**, *60*, 1251–1257. [CrossRef]

48. Zhang, Y.J.; Liu, C.X.; Zhang, J.; Shen, Y.; Li, C.; He, H.R.; Wang, X.J.; Xiang, W.S. *Actinomycetospora atypica* sp. nov., a novel soil actinomycete and emended description of the genus *Actinomycetospora*. *Antonie. Leeuwenhoek.* **2014**, *105*, 891–897. [CrossRef]

49. Zhang, J.; Kobert, K.; Flouri, T.; Stamatakis, A. PEAR: A fast and accurate Illumina Paired-End reAd mergeR. *Bioinformatics* **2014**, *30*, 614–620. [CrossRef]

50. Edgar, R.C. Search and clustering orders of magnitude faster than BLAST. *Bioinformatics* **2010**, *26*, 2460–2461. [CrossRef]

51. Caporaso, J.G.; Kuczynski, J.; Stombaugh, J.; Bittinger, K.; Bushman, F.D.; Costello, E.K.; Fierer, N.; Peña, A.G.; Goodrich, J.K.; Gordon, J.I.; et al. QIIME allows analysis of high-throughput community sequencing data. *Nat. Methods* **2010**, *7*, 335–336. [CrossRef]

52. Yuan, G.; Lin, H.; Wang, C.; Hong, K.; Liu, Y.; Li, J. ^{1}H and ^{13}C assignments of two new macrocyclic lactones isolated from *Streptomyces* sp. 211726 and revised assignments of azalomycins F_{3a}, F_{4a} and F_{5a}. *Magn. Reson. Chem.* **2011**, *49*, 30–37. [CrossRef]

53. Crevelin, E.J.; Canova, S.P.; Melo, I.S.; Zucchi, T.D.; da Silva, R.E.; Moraes, L.A. Isolation and characterization of phytotoxic compounds produced by *Streptomyces* sp. AMC 23 from red mangrove (*Rhizophora mangle*). *Appl. Biochem. Biotechnol.* **2013**, *171*, 1602–1616. [CrossRef]

54. Han, B.; Cui, C.B.; Cai, B.; Ji, X.F.; Yao, X.S. Actinolactomycin, a new 2-oxonanonoidal antitumor antibiotic produced by *Streptomyces flavoretus* 18522, and its inhibitory effect on the proliferation of human cancer cells. *Chin. Chem. Lett.* **2005**, *4*, 471–474.

55. Zhao, P.J.; Fan, L.M.; Li, G.H.; Zhu, N.; Shen, Y.M. Antibacterial and antitumor macrolides from *Streptomyces* sp. Is9131. *Arch. Pharm. Res.* **2005**, *28*, 1228–1232. [CrossRef]

56. Shishlyannikova, T.A.; Kuzmin, A.V.; Fedorova, G.A.; Shishlyannikov, S.M.; Lipko, I.A.; Sukhanova, E.V.; Belkova, N.L. Ionofore antibiotic polynactin produced by *Streptomyces* sp. 156A isolated from Lake Baikal. *Nat. Prod. Res.* **2017**, *31*, 639–644. [CrossRef]

57. Lan, Y.X.; Zou, Y.; Huang, T.T.; Wang, X.Z.; Brock, N.L.; Deng, Z.X.; Lin, S.J. Indole methylation protects diketopiperazine configuration in the maremycin biosynthetic pathway. *Sci. Chin. Chem.* **2016**, *59*, 1224–1228. [CrossRef]

58. Duan, Y.Y.; Liu, Y.Y.; Huang, T.; Zou, Y.; Huang, T.T.; Hu, K.F.; Deng, Z.X.; Lin, S.J. Divergent biosynthesis of indole alkaloids FR900452 and spiro-maremycins. *Org. Biomol. Chem.* **2018**, *16*, 5446–5451. [CrossRef]

59. Reiter, B.; Pfeifer, U.; Schwab, H.; Sessitsch, A. Response of endophytic bacterial communities in potato plants to infection with *Erwinia carotovora* subsp. *atroseptica*. *Appl. Environ. Microbiol.* **2002**, *68*, 2261–2268. [CrossRef]

60. Upreti, R.; Thomas, P. Root-associated bacterial endophytes from Ralstonia solanacearum resistant and susceptible tomato cultivars and their pathogen antagonistic effects. *Front. Microbiol.* **2015**, *6*, 255. [CrossRef]

61. Huang, J.F.; Wei, Z.; Tan, S.Y.; Mei, X.L.; Yin, S.X.; Shen, Q.R.; Xu, Y.C. The rhizosphere soil of diseased tomato plants as a source for novel microorganisms to control bacterial wilt. *Appl. Soil. Ecol.* **2013**, *72*, 79–84. [CrossRef]

62. Lakshmanan, V.; Kitto, S.L.; Caplan, J.L.; Hsueh, Y.H.; Kearns, D.B.; Wu, Y.S.; Bai, H.P. Microbe-associated molecular patterns-triggered root responses mediate beneficial rhizobacterial recruitment in *Arabidopsis*. *Plant. Physiol.* **2012**, *160*, 1642–1661. [CrossRef]

63. Barea, J.M.; Pozo, M.J.; Azcón, R.; Azcón-Aguilar, C. Microbial co-operation in the rhizosphere. *J. Exp. Bot.* **2005**, *56*, 1761–1778. [CrossRef]

64. Dennis, P.G.; Miller, A.J.; Hirsch, P.R. Are root exudates more important than other sources of rhizodeposits in structuring rhizosphere bacterial communities? *FEMS. Microbiol. Ecol.* **2010**, *72*, 313–327. [CrossRef]

65. Bull, C.T. Relationship between root colonization and suppression of *Gaeumannomyces graminis* var tritici by *Pseudomonas fluorescens* strain 2–79. *Phytopathology* **1991**, *81*, 954–959. [CrossRef]

66. Raaijmakers, J.M.; Leeman, M.; van Oorschot, M.M.P.; van der Sluis, I.; Schippers, B.; Bakker, P.A.H.M. Dose-response relationships in biological-control of Fusarium-wilt of radish by *Pseudomonas* spp. *Phytopathology* **1995**, *85*, 1075–1081. [CrossRef]

67. Werner, G.; Hagenmaier, H.; Drautz, H.; Baumgartner, A.; Zähner, H. Metabolic products of microorganisms. 224. Bafilomycins, a new group of macrolide antibiotics. Production, isolation, chemical structure and biological activity. *J. Antibiot.* **1984**, *37*, 110–117. [CrossRef]

68. Frändberg, E.; Petersson, C.; Lundgren, L.N.; Schnürer, J. *Streptomyces halstedii* K122 produces the antifungal compounds bafilomycin B_1 and C_1. *Can. J. Microbiol.* **2000**, *46*, 753–758. [CrossRef]

69. Arai, M. Azalomycins B and F, two new antibiotics. II. Properties of azalomycins B and F. *J. Antibiot.* **1960**, *13*, 51–56.

70. Chandra, A.; Nair, M.G. Azalomycin F complex from *Streptomyces hygroscopicus*, MSU/MN-4-75B. *J. Antibiot.* **1995**, *48*, 896–898. [CrossRef]

71. Copping, L.G.; Menn, J.J. Biopesticides: A review of their action, applications and efficacy. *Pest. Manage. Sci.* **2000**, *56*, 651–676. [CrossRef]

72. Silva, L.J.; Crevelin, E.J.; Souza, W.R.; Moraes, L.A.; Melo, I.S.; Zucchi, T.D. *Streptomyces araujoniae* produces a multiantibiotic complex with ionophoric properties to control *Botrytis cinerea*. *Phytopathology* **2014**, *104*, 1298–1305. [CrossRef]

73. Cui, C.B.; Kakeya, H.; Osada, H. Novel mammalian cell cycle inhibitors, spirotryprostatins A and B, produced by *Aspergillus fumigatus*, which inhibit mammalian cell cycle at G2/M phase. *Tetrahedron* **1996**, *52*, 12651–12666. [CrossRef]

74. Zheng, C.J.; Kim, Y.H.; Kim, W.G. Glioperazine B, as a new antimicrobial agent against *Staphylococcus aureus*, and glioperazine C: two new dioxopiperazines from *Bionectra byssicola*. *Biosci. Biotechnol. Biochem.* **2007**, *71*, 1979–1983. [CrossRef]

75. Byun, H.G.; Zhang, H.; Mochizuki, M.; Adachi, K.; Shizuri, Y.; Lee, W.J.; Kim, S.K. Novel antifungal diketopiperazine from marine fungus. *J. Antibiot.* **2003**, *56*, 102–106. [CrossRef]

76. Ma, Y.M.; Liang, X.A.; Kong, Y.; Jia, B. Structural diversity and biological activities of indole diketopiperazine alkaloids from fungi. *J. Agric. Food. Chem.* **2016**, *64*, 6659–6671. [CrossRef]

77. Takase, S.; Shigematsu, N.; Shima, I.; Uchida, I.; Hashimoto, M.; Tada, T. Structure of FR900452, a novel platelet-activating factor inhibitor from a *Streptomyces*. *J. Org. Chem.* **1987**, *52*, 3485–3487. [CrossRef]

Communication

Survey of Biosynthetic Gene Clusters from Sequenced Myxobacteria Reveals Unexplored Biosynthetic Potential

Katherine Gregory, Laura A. Salvador, Shukria Akbar, Barbara I. Adaikpoh and D. Cole Stevens *

Department of BioMolecular Sciences, School of Pharmacy, University of Mississippi, University, MS 38677, USA; kcgregor@go.olemiss.edu (K.G.); lasalvad@go.olemiss.edu (L.A.S.); sakbar@go.olemiss.edu (S.A.); biadaikp@go.olemiss.edu (B.I.A.)
* Correspondence: stevens@olemiss.edu; Tel.: +1-662-915-5730

Received: 24 May 2019; Accepted: 21 June 2019; Published: 24 June 2019

Abstract: Coinciding with the increase in sequenced bacteria, mining of bacterial genomes for biosynthetic gene clusters (BGCs) has become a critical component of natural product discovery. The order Myxococcales, a reputable source of biologically active secondary metabolites, spans three suborders which all include natural product producing representatives. Utilizing the BiG-SCAPE-CORASON platform to generate a sequence similarity network that contains 994 BGCs from 36 sequenced myxobacteria deposited in the antiSMASH database, a total of 843 BGCs with lower than 75% similarity scores to characterized clusters within the MIBiG database are presented. This survey provides the biosynthetic diversity of these BGCs and an assessment of the predicted chemical space yet to be discovered. Considering the mere snapshot of myxobacteria included in this analysis, these untapped BGCs exemplify the potential for natural product discovery from myxobacteria.

Keywords: myxobacteria; biosynthetic gene clusters; natural product discovery

---

1. Introduction

Ubiquitous to soils and marine sediments, bacteriovorous myxobacteria display organized social behaviors and predation strategies [1–4]. Perhaps intrinsic to their role as predators, myxobacteria are a critical source of diverse secondary metabolites that exhibit unique modes-of-action across a broad range of biological activities [5]. Distinct from other bacterial sources, the vast majority of the 60 species within the order Myxococcales produce natural products [5,6]. This gifted diversity of secondary metabolite producing representatives has established myxobacteria as a prolific resource for drug discovery efforts perhaps only second to Actinomycetales [7,8]. Bolstered by the observed lack of overlap between actinomycetal and myxobacterial drug-like metabolites, the potential to discover novel specialized metabolites from myxobacteria remains considerably high [7,8]. Herein, we report a survey of all myxobacterial natural product biosynthetic gene clusters (BGCs) deposited in the antiSMASH database and provide an account of all BGCs with and without characterization and assigned metabolites in an effort to observe the capacity for discovery from readily cultivable, sequenced myxobacteria [9,10]. Such analysis provides an assessment of the potential associated with the continued discovery efforts as well as development and application of methodologies to activate situational or cryptic secondary metabolism not functional during axenic cultivation [11,12]. A homology network of 994 BGCs from 36 sequenced myxobacterial genomes was constructed using the combined BiG-SCAPE-CORe Analysis of Syntenic Orthologues to prioritize Natural products biosynthetic gene clusters (CORASON) platform [13]. BiG-SCAPE facilitates the exploration of calculated BGC sequence similarity networks and provides the opportunity to visualize biosynthetic diversity across datasets [13]. Gene cluster families (GCFs) rendered by BiG-SCAPE are connected by edges that indicate shared domain types, sequence similarity, and similarity of domain pair-types amongst input BGCs [13]. Comparative analysis against the Minimum Information about a Biosynthetic Gene Cluster (MIBiG) repository (v1.4) indicates an untapped reservoir of BGCs that encompasses a broad range of biosynthetic diversity [14]. The 36 Myxococcales within the antiSMASH database currently span all 3 suborders with 26 Cystobacterineae, 7 Sorangineae, and 3 Nannocystineae included. Considering that the myxobacteria within the antiSMASH database minimally represent the breadth of the order Myxococcales, these observations not only support thorough investigation of identified myxobacteria and the presented biosynthetic space but also continued efforts for the identification and subsequent exploration of new myxobacteria [1,3].

2. Materials and Methods

Dataset. All BGCs associated with the order Myxococcales, a total of 994 BGCs from 36 myxobacteria, were downloaded as .gbk files from the publicly available antiSMASH database (https://antismash-db.secondarymetabolites.org) [9]. The original genome sequence data for all included myxobacteria are also publicly available and can be accessed at the National Center for Biotechnology Information, U.S. National Library of Medicine (https://www.ncbi.nlm.nih.gov/genome/browse#!/prokaryotes/myxobacteria).

BIG-SCAPE-CORASON analysis. BiG-SCAPE version 20181005 (available at: https://git.wageningenur.nl/medema-group/BiG-SCAPE) was utilized locally to analyse the 994 BGCs as individual .gbk files downloaded from the antiSMASH database (1/30/2019) [9,13]. BiG-SCAPE analysis was supplemented with Pfam database version 31 [15]. The singleton parameter in BiG-SCAPE was selected to ensure that BGCs with distances lower than the default cutoff distance of 0.3 were included in the corresponding output data. The MIBiG parameter in BiG-SCAPE was set to include the MIBiG repository version 1.4 of annotated BGCs [14]. The hybrids-off parameter was selected to prevent hybrid BGC redundancy. Generated network files separated by BiG-SCAPE class were combined for visualization using Cytoscape version 3.7.1; annotations associated with each BGC were included into Cytoscape networks by importing curated tables generated by BiG-SCAPE [16]. Phylogenetic trees provided by CORASON were generated during BiG-SCAPE analysis. Annotated network and table files including GCF associations are provided as Supplementary files. All BGCs with sequence similarities to deposited MIBiG clusters ≥75% were indicated and annotated using Cytoscape. An annotated .cys Cytoscape file is included as Supplementary Material. All associated .network and .tsv files are provided as Supplementary Materials. All histograms were generated GraphPad Prism version 7.0d for Mac OS X, GraphPad Software, San Diego, California, USA, www.graphpad.com.

3. Results

3.1. BiG-SCAPE Analysis of BGCs from Sequenced Myxobacteria

A sequence similarity network calculated using BiG-SCAPE consisted of 994 total BGCs as unique nodes from 36 myxobacteria and included 1035 edges (included self-looped nodes) representing homology across 753 GCFs (Figure 1). Of these 994 BGCs from the antiSMASH database, a total of 124 were determined to be located on contig edges by antiSMASH. Clusters determined to be on contig edges could contribute to redundancy within our analysis. While no 2 BGCs from an individual myxobacterium were found within a GCF, this does not preclude a single BGC split across multiple contigs from being included multiple times. A total of 613 singletons without homology using a similarity cutoff of 0.30 were also included in the network to appropriately depict all myxobacterial BGCs within the antiSMASH database [9,13]. Predicted BGC classes included 64 type I or modular polyketide synthases (t1PKS), 57 PKS categorized by antiSMASH as "PKSother" that includes all non-modular categories of PKSs, 125 nonribosomal peptide synthetases (NRPS), 166 hybrid PKS-NRPS, 245 ribosomally synthesized and post-translationally modified peptides (RiPPs), 149 terpene clusters, 3 saccharide clusters, and 185 clusters not belonging to any of the aforementioned classes that antiSMASH categorizes as "Others" clusters [9,10].

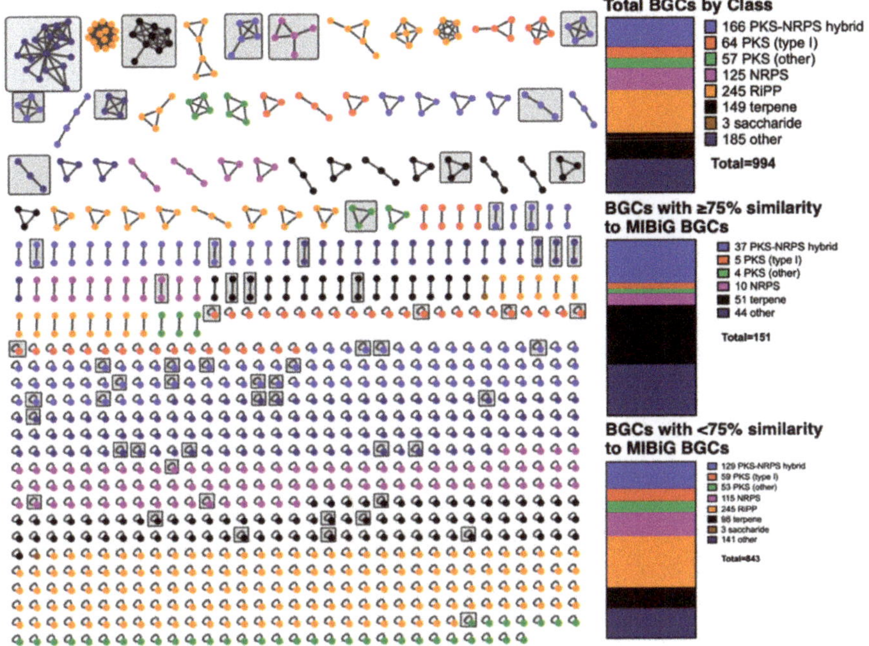

Figure 1. Sequence similarity network of 994 myxobacterial BGCs deposited in the antiSMASH database generated by BiG-SCAPE and rendered with Cytoscape [9,10,13,14,16]. All GCFs that include at least 1 BGC with sequence similarity greater than ≥75% to a characterized cluster deposited in the MIBiG repository are boxed in grey (excluding 25 geosmin BGCs) [9,14]. Totals for BGC class diversity and BGCs (including 25 geosmin BGCs identified as 22 Terpene and 3 Other clusters) with and without homology to MIBiG clusters as well as color reference provided (right).

While hybrid PKS-NRPS pathways that include both PKS and NRPS domains are organized into a specific separate grouping, all other hybrid pathways that include more than one BGC are categorized in the Others class [9,13]. The Others-associated BGCs included clusters with 133 predicted products as well as 52 hybrid BGCs (Figure 2). This breadth of biosynthetic diversity from just 36 myxobacteria includes 23 out of 52 BGC-types currently designated by antiSMASH [9,10].

3.2. Discovered Metabolites from Myxobacteria and Associated BGCs

Of the 994 BGCs analysed, 151 possess sequence similarities ≥75% with annotated BGCs in the MIBiG repository (v 1.4) [14]. Sequence similarities from the antiSMASH database are provided by KnownClusterBlast analysis of BGCs within the database against characterized pathways within the MIBiG repository [9,14,17,18]. As these BGCs produce characterized metabolites or potentially analogues thereof (Figure 3), a total of 85% of the BGCs within the network might produce yet to be discovered metabolites [19–43]. Considering the range in quality across the 36 total genomes and draft genomes incorporated in the antiSMASH database, we also considered additional BGCs with similarity scores lower than 75% that had similarities with MIBiG clusters reported from myxobacteria identified by antiSMASH. This analysis provided an additional 23 BGCs that might produce metabolites with overlapping chemical diversities to the products delineated within the MIBiG repository (Figure 4) [44–58]. Of these 23 BGCs omitted from our original analysis, only 10 would have been included if our sequence similarity cutoff had been lowered to 67% sequence similarity. Including this inference, 82% of the BGCs within the network lack any association with a reported myxobacterial metabolite. The biosynthetic diversity of these mapped BGCs includes 5 t1PKS, 10 NRPS, 37 hybrid PKS-NRPS, 4 PKSother, 51 terpene clusters, and 44 Others (Figure 1).

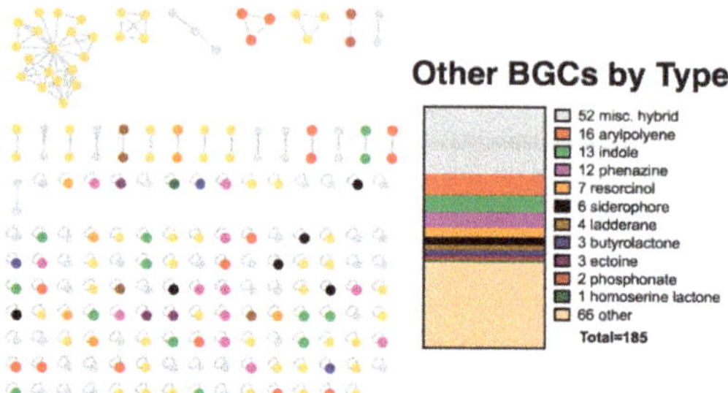

Figure 2. Sequence similarity network of myxobacterial BGCs classified as Others in the antiSMASH database with predicted product type and totals (right) [9].

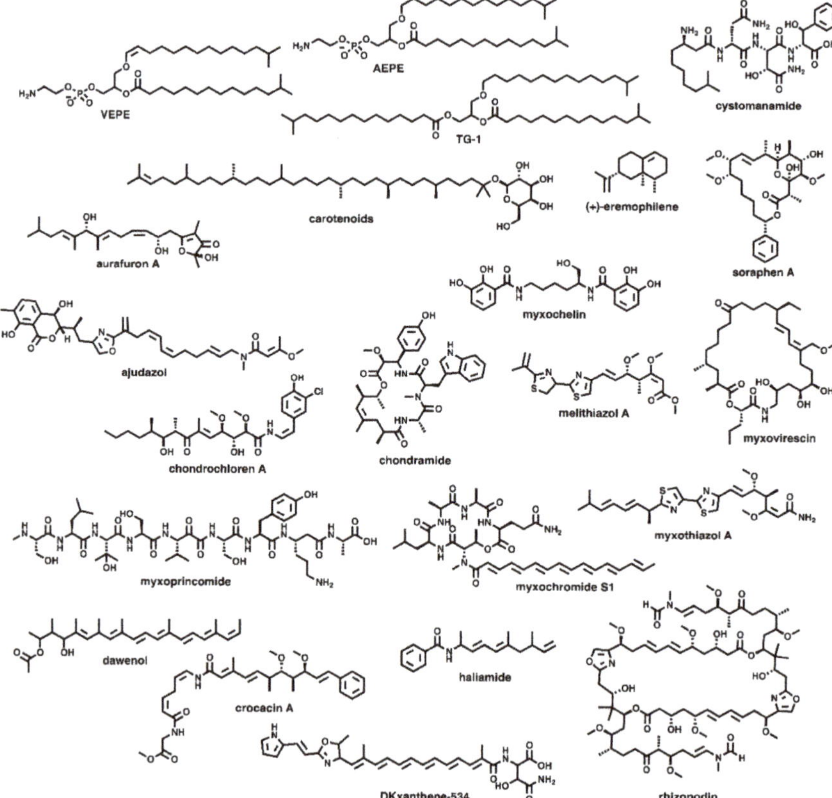

Figure 3. Secondary metabolites associated with BGCs determined to possess ≥75% sequence similarity to characterized clusters in MIBiG [17–41].

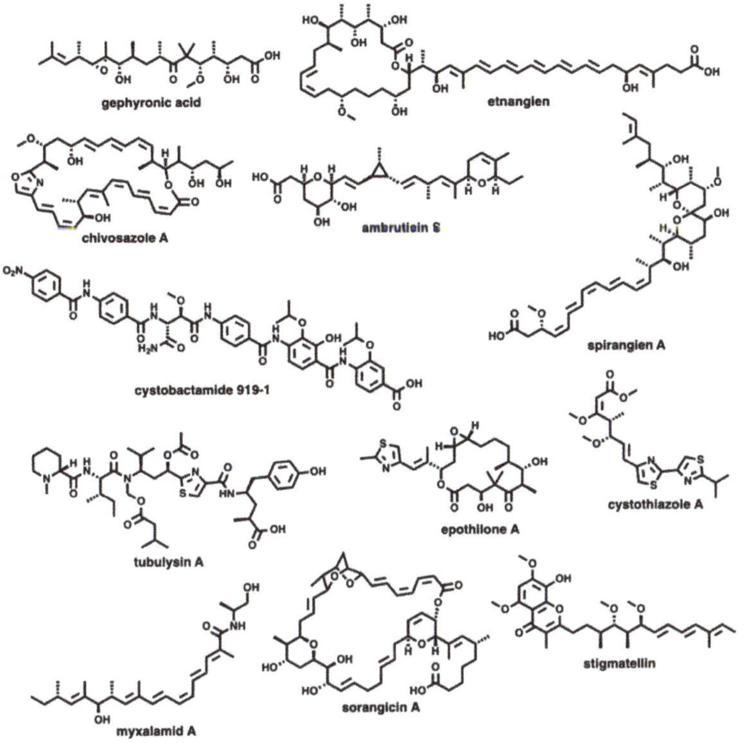

Figure 4. Secondary metabolites associated with known BGCs from myxobacteria with sequence similarity to BGCs included in the MIBiG dataset below the 75% similarity cutoff [42–56].

While the vast majority of BGCs were considered singletons or unclustered individual nodes without sequence similarity to other analysed BGCs, GCFs with more than 1 member BGC often shared sequence similarities with characterized MIBiG clusters. Interestingly, BGCs with high sequence similarity to specific MIBiG clusters were not always assigned the same cluster class nor were they included within an individual GCF. For example, 9 GCFs that include a single BGC with high homology to the myxochelin BGC were assigned as NRPS, hybrid PKS-NRPS, and Others type clusters [33,39]. Trees generated by CORASON provide the phylogenetic diversity associated with these myxochelin BGCs (Figure S1) [13]. Analysis of these trees indicated that such wholesale affiliation with each of these GCFs led to inclusion of BGCs that were in fact not related to the myxochelin BGC but instead shared proximal similarity to a BGC within the family that also included a neighbouring myxochelin-like BGC (Figure S1) [33,39]. While this omits unexplored BGCs and demonstrates the limitations of our totals, this only supports our conclusion that a vast wealth of biosynthetic space from myxobacteria remains unexplored. Other BGCs observed across multi-member GCFs included: 26 BGCs within 11 GCFs homologous to a carotenoid cluster from *Myxococcus xanthus*, 24 BGCs and 4 GCFs associated with the characterized VEPE/AEPE/TG-1 biosynthetic pathway from *M. xanthus* DK1622, and 11 BGCs across 5 GCFs with similarity to the hybrid PKS-NRPS DKxanthene cluster [19,20,24,30,31]. While all of the BGCs included in this charted biosynthetic space might not correlate to the corresponding metabolites associated with each MIBiG cluster, we consider this a rigorous assessment that provides a conservative estimate of uncharacterized BGCs and remaining opportunity for natural product discovery.

4. Discussion

This survey assesses the potential to discover novel metabolites from these myxobacteria and depicts unexplored biosynthetic space. Perhaps the most obvious absence in the 151 BGCs associated with characterized BGCs was that no RiPP clusters with sequence similarity to MIBiG clusters were observed [59–61]. However, there are no myxobacterial RiPP BGCs currently deposited in the MIBiG database, and crocagin A produced by *Chondromyces crocatus* is the only myxobacterial RiPP discovered to date [62]. Considering the 245 uncharacterized BGCs predicted to produce RiPPs within our network, myxobacteria are an excellent resource for the discovery of RiPPs. Also, with respect to notable outliers, no sequence similarities were observed for the 3 saccharide BGCs that include the aminoglycoside and aminocyclitol subtypes [63–65]. All other BGCs considered unexplored accounted for the vast majority of BGCs within each cluster class, including the following: 92% of t1PKS, 98% of PKSother, 92% of NRPS, 81% of terpene clusters, 78% of hybrid PKS-NRPS, and 76% of Others. Interestingly, within the BGCs assigned to the Others class, 3 butyrolactone and 1 homoserine lactone clusters were identified. Specialized metabolites belonging to these types of clusters are typically quorum-signaling molecules produced by Streptomyces and numerous non-myxobacterial Proteobacteria respectively [66–69]. Although putative quorum signal receptors are present within myxobacterial genomes and exogenous homoserine lactones increase the predatory behavior of *M. xanthus*, no metabolite associated with these quorum signaling systems has been reported from a myxobacteria [70,71].

5. Conclusions

The continued discovery of novel, biologically active bacterial metabolites is required to address the need for antimicrobials and anticancer therapeutics. Assessment of biosynthetic space within the growing amount of genome data from myxobacteria can provide insight to direct responsible discovery efforts [72–75]. This survey likely underestimates the unexplored biosynthetic space from myxobacteria. However, the vast discrepancies between BGCs with and without sequence similarity to characterized pathways suggests continued discovery of novel metabolites from this subset of 36 myxobacteria and exemplifies the outstanding potential associated with the Myxococcales at large.

Supplementary Materials: The following are available online at http://www.mdpi.com/2076-2607/7/6/181/s1. Supplemental Figure S1, annotated .cys file for all BGCs, and annotated .cys file for Other type BGCs.

Author Contributions: Conceptualization, supervision, and administration, D.C.S.; methodology, formal analysis, data curation, and validation, K.G., L.A.S., and D.C.S.; writing, K.G., L.A.S., S.A., B.I.A., and D.C.S.

Funding: This research was funded by the American Association of Colleges of Pharmacy New Investigator Award (D.C.S.), and salary support was provided for S.A. and D.C.S. by the National Cancer Institute (1R03CA219320-01A1).

Acknowledgments: The authors would like to acknowledge the University of Mississippi School of Pharmacy for startup support and the Sally Barksdale Honors College for encouraging undergraduate research.

Conflicts of Interest: The authors declare no conflict of interest. The funders had no role in the design of the study; in the collection, analyses, or interpretation of data; in the writing of the manuscript, or in the decision to publish the results.

References

1. Brinkhoff, T.; Fischer, D.; Vollmers, J.; Voget, S.; Beardsley, C.; Thole, S.; Mussmann, M.; Kunze, B.; Wagner-Dobler, I.; Daniel, R.; et al. Biogeography and phylogenetic diversity of a cluster of exclusively marine myxobacteria. *ISME J.* **2012**, *6*, 1260–1272. [CrossRef] [PubMed]

2. Cao, P.; Dey, A.; Vassallo, C.N.; Wall, D. How Myxobacteria Cooperate. *J. Mol. Biol.* **2015**, *427*, 3709–3721. [CrossRef] [PubMed]

3. Mohr, K.I. Diversity of Myxobacteria-We Only See the Tip of the Iceberg. *Microorganisms* **2018**, *6*, 84. [CrossRef] [PubMed]

4. Munoz-Dorado, J.; Marcos-Torres, F.J.; Garcia-Bravo, E.; Moraleda-Munoz, A.; Perez, J. Myxobacteria: Moving, Killing, Feeding, and Surviving Together. *Front. Microbiol.* **2016**, *7*, 781. [CrossRef] [PubMed]

5.   Herrmann, J.; Fayad, A.A.; Müller, R. Natural products from myxobacteria: Novel metabolites and bioactivities. *Nat. Prod. Rep.* **2017**, *34*, 135–160. [CrossRef] [PubMed]

6.   Landwehr, W.; Wolf, C.; Wink, J. Actinobacteria and Myxobacteria-Two of the Most Important Bacterial Resources for Novel Antibiotics. *Curr. Top. Microbiol. Immunol.* **2016**, *398*, 273–302. [CrossRef] [PubMed]

7.   Baltz, R.H. Natural product drug discovery in the genomic era: Realities, conjectures, misconceptions, and opportunities. *J. Ind. Microbiol. Biotechnol.* **2019**, *46*, 281–299. [CrossRef]

8.   Liu, R.; Deng, Z.; Liu, T. Streptomyces species: Ideal chassis for natural product discovery and overproduction. *Metab. Eng.* **2018**, *50*, 74–84. [CrossRef] [PubMed]

9.   Blin, K.; Pascal Andreu, V.; de Los Santos, E.L.C.; Del Carratore, F.; Lee, S.Y.; Medema, M.H.; Weber, T. The antiSMASH database version 2: A comprehensive resource on secondary metabolite biosynthetic gene clusters. *Nucleic Acids Res.* **2019**, *47*, D625–D630. [CrossRef]

10. Blin, K.; Shaw, S.; Steinke, K.; Villebro, R.; Ziemert, N.; Lee, S.Y.; Medema, M.H.; Weber, T. antiSMASH 5.0: Updates to the secondary metabolite genome mining pipeline. *Nucleic Acids Res.* **2019**. [CrossRef]

11. Baral, B.; Akhgari, A.; Metsa-Ketela, M. Activation of microbial secondary metabolic pathways: Avenues and challenges. *Synth. Syst. Biotechnol.* **2018**, *3*, 163–178. [CrossRef] [PubMed]

12. Mao, D.; Okada, B.K.; Wu, Y.; Xu, F.; Seyedsayamdost, M.R. Recent advances in activating silent biosynthetic gene clusters in bacteria. *Curr. Opin. Microbiol.* **2018**, *45*, 156–163. [CrossRef] [PubMed]

13. Navarro-Muñoz, J.C.; Selem-Mojica, N.; Mullowney, M.W.; Kautsar, S.; Tryon, J.H.; Parkinson, E.I.; de los Santos, E.L.C.; Yeong, M.; Cruz-Morales, P.; Abubucker, S.; et al. A computational framework for systematic exploration of biosynthetic diversity from large-scale genomic data. *Biorxiv* **2018**. [CrossRef]

14. Medema, M.H.; Kottmann, R.; Yilmaz, P.; Cummings, M.; Biggins, J.B.; Blin, K.; de Bruijn, I.; Chooi, Y.H.; Claesen, J.; Coates, R.C.; et al. Minimum Information about a Biosynthetic Gene cluster. *Nat. Chem. Biol.* **2015**, *11*, 625–631. [CrossRef] [PubMed]

15. Richardson, L.J.; Rawlings, N.D.; Salazar, G.A.; Almeida, A.; Haft, D.R.; Ducq, G.; Sutton, G.G.; Finn, R.D. Genome properties in 2019: A new companion database to InterPro for the inference of complete functional attributes. *Nucleic Acids Res.* **2019**, *47*, D564–D572. [CrossRef] [PubMed]

16. Shannon, P.; Markiel, A.; Ozier, O.; Baliga, N.S.; Wang, J.T.; Ramage, D.; Amin, N.; Schwikowski, B.; Ideker, T. Cytoscape: A software environment for integrated models of biomolecular interaction networks. *Genome Res.* **2003**, *13*, 2498–2504. [CrossRef] [PubMed]

17. Blin, K.; Wolf, T.; Chevrette, M.G.; Lu, X.; Schwalen, C.J.; Kautsar, S.A.; Suarez Duran, H.G.; de Los Santos, E.L.C.; Kim, H.U.; Nave, M.; et al. antiSMASH 4.0-improvements in chemistry prediction and gene cluster boundary identification. *Nucleic Acids Res.* **2017**, *45*, W36–W41. [CrossRef] [PubMed]

18. Weber, T.; Blin, K.; Duddela, S.; Krug, D.; Kim, H.U.; Bruccoleri, R.; Lee, S.Y.; Fischbach, M.A.; Müller, R.; Wohlleben, W.; et al. antiSMASH 3.0-a comprehensive resource for the genome mining of biosynthetic gene clusters. *Nucleic Acids Res.* **2015**, *43*, W237–W243. [CrossRef]

19. Bhat, S.; Ahrendt, T.; Dauth, C.; Bode, H.B.; Shimkets, L.J. Two lipid signals guide fruiting body development of Myxococcus xanthus. *MBio* **2014**, *5*, e00939-13. [CrossRef]

20. Botella, J.A.; Murillo, F.J.; Ruiz-Vazquez, R. A cluster of structural and regulatory genes for light-induced carotenogenesis in Myxococcus xanthus. *Eur. J. Biochem* **1995**, *233*, 238–248. [CrossRef]

21. Buntin, K.; Weissman, K.J.; Müller, R. An unusual thioesterase promotes isochromanone ring formation in ajudazol biosynthesis. *Chembiochem* **2010**, *11*, 1137–1146. [CrossRef] [PubMed]

22. Cervantes, M.; Murillo, F.J. Role for vitamin B(12) in light induction of gene expression in the bacterium Myxococcus xanthus. *J. Bacteriol.* **2002**, *184*, 2215–2224. [CrossRef] [PubMed]

23. Cortina, N.S.; Krug, D.; Plaza, A.; Revermann, O.; Müller, R. Myxoprincomide: A natural product from Myxococcus xanthus discovered by comprehensive analysis of the secondary metabolome. *Angew. Chem. Int. Ed. Engl.* **2012**, *51*, 811–816. [CrossRef] [PubMed]

24. Etzbach, L.; Plaza, A.; Garcia, R.; Baumann, S.; Müller, R. Cystomanamides: Structure and biosynthetic pathway of a family of glycosylated lipopeptides from myxobacteria. *Org. Lett.* **2014**, *16*, 2414–2417. [CrossRef] [PubMed]

25. Frank, B.; Wenzel, S.C.; Bode, H.B.; Scharfe, M.; Blocker, H.; Müller, R. From genetic diversity to metabolic unity: Studies on the biosynthesis of aurafurones and aurafuron-like structures in myxobacteria and streptomycetes. *J. Mol. Biol.* **2007**, *374*, 24–38. [CrossRef] [PubMed]

26. Gaitatzis, N.; Kunze, B.; Müller, R. In vitro reconstitution of the myxochelin biosynthetic machinery of Stigmatella aurantiaca Sg a15: Biochemical characterization of a reductive release mechanism from nonribosomal peptide synthetases. *Proc. Natl. Acad. Sci. USA* **2001**, *98*, 11136–11141. [CrossRef] [PubMed]

27. Li, Y.; Weissman, K.J.; Müller, R. Myxochelin biosynthesis: Direct evidence for two- and four-electron reduction of a carrier protein-bound thioester. *J. Am. Chem. Soc.* **2008**, *130*, 7554–7555. [CrossRef] [PubMed]

28. Ligon, J.; Hill, S.; Beck, J.; Zirkle, R.; Molnar, I.; Zawodny, J.; Money, S.; Schupp, T. Characterization of the biosynthetic gene cluster for the antifungal polyketide soraphen A from Sorangium cellulosum So ce26. *Gene* **2002**, *285*, 257–267. [CrossRef]

29. Lopez-Rubio, J.J.; Elias-Arnanz, M.; Padmanabhan, S.; Murillo, F.J. A repressor-antirepressor pair links two loci controlling light-induced carotenogenesis in Myxococcus xanthus. *J. Biol. Chem.* **2002**, *277*, 7262–7270. [CrossRef]

30. Lorenzen, W.; Ahrendt, T.; Bozhuyuk, K.A.; Bode, H.B. A multifunctional enzyme is involved in bacterial ether lipid biosynthesis. *Nat. Chem. Biol.* **2014**, *10*, 425–427. [CrossRef]

31. Meiser, P.; Weissman, K.J.; Bode, H.B.; Krug, D.; Dickschat, J.S.; Sandmann, A.; Müller, R. DKxanthene biosynthesis–understanding the basis for diversity-oriented synthesis in myxobacterial secondary metabolism. *Chem. Biol.* **2008**, *15*, 771–781. [CrossRef] [PubMed]

32. Muller, S.; Rachid, S.; Hoffmann, T.; Surup, F.; Volz, C.; Zaburannyi, N.; Müller, R. Biosynthesis of crocacin involves an unusual hydrolytic release domain showing similarity to condensation domains. *Chem. Biol.* **2014**, *21*, 855–865. [CrossRef] [PubMed]

33. Osswald, C.; Zaburannyi, N.; Burgard, C.; Hoffmann, T.; Wenzel, S.C.; Müller, R. A highly unusual polyketide synthase directs dawenol polyene biosynthesis in Stigmatella aurantiaca. *J. Biotechnol.* **2014**, *191*, 54–63. [CrossRef] [PubMed]

34. Perez-Marin, M.C.; Padmanabhan, S.; Polanco, M.C.; Murillo, F.J.; Elias-Arnanz, M. Vitamin B12 partners the CarH repressor to downregulate a photoinducible promoter in Myxococcus xanthus. *Mol. Microbiol.* **2008**, *67*, 804–819. [CrossRef] [PubMed]

35. Pistorius, D.; Müller, R. Discovery of the rhizopodin biosynthetic gene cluster in Stigmatella aurantiaca Sg a15 by genome mining. *Chembiochem* **2012**, *13*, 416–426. [CrossRef] [PubMed]

36. Rachid, S.; Scharfe, M.; Blocker, H.; Weissman, K.J.; Müller, R. Unusual chemistry in the biosynthesis of the antibiotic chondrochlorens. *Chem. Biol.* **2009**, *16*, 70–81. [CrossRef] [PubMed]

37. Schifrin, A.; Ly, T.T.; Gunnewich, N.; Zapp, J.; Thiel, V.; Schulz, S.; Hannemann, F.; Khatri, Y.; Bernhardt, R. Characterization of the gene cluster CYP264B1-geoA from Sorangium cellulosum So ce56: Biosynthesis of (+)-eremophilene and its hydroxylation. *Chembiochem* **2015**, *16*, 337–344. [CrossRef] [PubMed]

38. Silakowski, B.; Schairer, H.U.; Ehret, H.; Kunze, B.; Weinig, S.; Nordsiek, G.; Brandt, P.; Blocker, H.; Hofle, G.; Beyer, S.; et al. New lessons for combinatorial biosynthesis from myxobacteria. The myxothiazol biosynthetic gene cluster of Stigmatella aurantiaca DW4/3-1. *J. Biol. Chem.* **1999**, *274*, 37391–37399. [CrossRef]

39. Simunovic, V.; Zapp, J.; Rachid, S.; Krug, D.; Meiser, P.; Müller, R. Myxovirescin A biosynthesis is directed by hybrid polyketide synthases/nonribosomal peptide synthetase, 3-hydroxy-3-methylglutaryl-CoA synthases, and trans-acting acyltransferases. *Chembiochem* **2006**, *7*, 1206–1220. [CrossRef]

40. Sun, Y.; Tomura, T.; Sato, J.; Iizuka, T.; Fudou, R.; Ojika, M. Isolation and Biosynthetic Analysis of Haliamide, a New PKS-NRPS Hybrid Metabolite from the Marine Myxobacterium Haliangium ochraceum. *Molecules* **2016**, *21*, 59. [CrossRef]

41. Weinig, S.; Hecht, H.J.; Mahmud, T.; Müller, R. Melithiazol biosynthesis: Further insights into myxobacterial PKS/NRPS systems and evidence for a new subclass of methyl transferases. *Chem. Biol.* **2003**, *10*, 939–952. [CrossRef] [PubMed]

42. Wenzel, S.C.; Kunze, B.; Hofle, G.; Silakowski, B.; Scharfe, M.; Blocker, H.; Müller, R. Structure and biosynthesis of myxochromides S1-3 in Stigmatella aurantiaca: Evidence for an iterative bacterial type I polyketide synthase and for module skipping in nonribosomal peptide biosynthesis. *Chembiochem* **2005**, *6*, 375–385. [CrossRef] [PubMed]

43. Rachid, S.; Krug, D.; Kunze, B.; Kochems, I.; Scharfe, M.; Zabriskie, T.M.; Blocker, H.; Müller, R. Molecular and biochemical studies of chondramide formation-highly cytotoxic natural products from Chondromyces crocatus Cm c5. *Chem. Biol.* **2006**, *13*, 667–681. [CrossRef] [PubMed]

44. Baumann, S.; Herrmann, J.; Raju, R.; Steinmetz, H.; Mohr, K.I.; Huttel, S.; Harmrolfs, K.; Stadler, M.; Müller, R. Cystobactamids: Myxobacterial topoisomerase inhibitors exhibiting potent antibacterial activity. *Angew. Chem. Int. Ed. Engl.* **2014**, *53*, 14605–14609. [CrossRef] [PubMed]

45. Beyer, S.; Kunze, B.; Silakowski, B.; Müller, R. Metabolic diversity in myxobacteria: Identification of the myxalamid and the stigmatellin biosynthetic gene cluster of Stigmatella aurantiaca Sg a15 and a combined polyketide-(poly)peptide gene cluster from the epothilone producing strain Sorangium cellulosum So ce90. *Biochim. Biophys. Acta* **1999**, *1445*, 185–195. [PubMed]

46. Feng, Z.; Qi, J.; Tsuge, T.; Oba, Y.; Kobayashi, T.; Suzuki, Y.; Sakagami, Y.; Ojika, M. Construction of a bacterial artificial chromosome library for a myxobacterium of the genus Cystobacter and characterization of an antibiotic biosynthetic gene cluster. *Biosci. Biotechnol. Biochem.* **2005**, *69*, 1372–1380. [CrossRef] [PubMed]

47. Frank, B.; Knauber, J.; Steinmetz, H.; Scharfe, M.; Blocker, H.; Beyer, S.; Müller, R. Spiroketal polyketide formation in Sorangium: Identification and analysis of the biosynthetic gene cluster for the highly cytotoxic spirangienes. *Chem. Biol.* **2007**, *14*, 221–233. [CrossRef]

48. Irschik, H.; Kopp, M.; Weissman, K.J.; Buntin, K.; Piel, J.; Müller, R. Analysis of the sorangicin gene cluster reinforces the utility of a combined phylogenetic/retrobiosynthetic analysis for deciphering natural product assembly by trans-AT PKS. *Chembiochem* **2010**, *11*, 1840–1849. [CrossRef]

49. Julien, B.; Shah, S.; Ziermann, R.; Goldman, R.; Katz, L.; Khosla, C. Isolation and characterization of the epothilone biosynthetic gene cluster from Sorangium cellulosum. *Gene* **2000**, *249*, 153–160. [CrossRef]

50. Julien, B.; Tian, Z.Q.; Reid, R.; Reeves, C.D. Analysis of the ambruticin and jerangolid gene clusters of Sorangium cellulosum reveals unusual mechanisms of polyketide biosynthesis. *Chem. Biol.* **2006**, *13*, 1277–1286. [CrossRef]

51. Menche, D.; Arikan, F.; Perlova, O.; Horstmann, N.; Ahlbrecht, W.; Wenzel, S.C.; Jansen, R.; Irschik, H.; Müller, R. Stereochemical determination and complex biosynthetic assembly of etnangien, a highly potent RNA polymerase inhibitor from the myxobacterium Sorangium cellulosum. *J. Am. Chem. Soc.* **2008**, *130*, 14234–14243. [CrossRef] [PubMed]

52. Molnar, I.; Schupp, T.; Ono, M.; Zirkle, R.; Milnamow, M.; Nowak-Thompson, B.; Engel, N.; Toupet, C.; Stratmann, A.; Cyr, D.D.; et al. The biosynthetic gene cluster for the microtubule-stabilizing agents epothilones A and B from Sorangium cellulosum So ce90. *Chem. Biol.* **2000**, *7*, 97–109. [CrossRef]

53. Perlova, O.; Gerth, K.; Kaiser, O.; Hans, A.; Müller, R. Identification and analysis of the chivosazol biosynthetic gene cluster from the myxobacterial model strain Sorangium cellulosum So ce56. *J. Biotechnol.* **2006**, *121*, 174–191. [CrossRef] [PubMed]

54. Sandmann, A.; Sasse, F.; Müller, R. Identification and analysis of the core biosynthetic machinery of tubulysin, a potent cytotoxin with potential anticancer activity. *Chem. Biol.* **2004**, *11*, 1071–1079. [CrossRef] [PubMed]

55. Silakowski, B.; Nordsiek, G.; Kunze, B.; Blocker, H.; Müller, R. Novel features in a combined polyketide synthase/non-ribosomal peptide synthetase: The myxalamid biosynthetic gene cluster of the myxobacterium Stigmatella aurantiaca Sga15. *Chem. Biol.* **2001**, *8*, 59–69. [CrossRef]

56. Tang, L.; Shah, S.; Chung, L.; Carney, J.; Katz, L.; Khosla, C.; Julien, B. Cloning and heterologous expression of the epothilone gene cluster. *Science* **2000**, *287*, 640–642. [CrossRef] [PubMed]

57. Young, J.; Stevens, D.C.; Carmichael, R.; Tan, J.; Rachid, S.; Boddy, C.N.; Müller, R.; Taylor, R.E. Elucidation of gephyronic acid biosynthetic pathway revealed unexpected SAM-dependent methylations. *J. Nat. Prod.* **2013**, *76*, 2269–2276. [CrossRef] [PubMed]

58. Zhu, L.P.; Li, Z.F.; Sun, X.; Li, S.G.; Li, Y.Z. Characteristics and activity analysis of epothilone operon promoters from Sorangium cellulosum strains in Escherichia coli. *Appl. Microbiol. Biotechnol.* **2013**, *97*, 6857–6866. [CrossRef]

59. Hetrick, K.J.; van der Donk, W.A. Ribosomally synthesized and post-translationally modified peptide natural product discovery in the genomic era. *Curr. Opin. Chem. Biol.* **2017**, *38*, 36–44. [CrossRef]

60. Hudson, G.A.; Mitchell, D.A. RiPP antibiotics: Biosynthesis and engineering potential. *Curr. Opin. Microbiol.* **2018**, *45*, 61–69. [CrossRef]

61. Ortega, M.A.; van der Donk, W.A. New Insights into the Biosynthetic Logic of Ribosomally Synthesized and Post-translationally Modified Peptide Natural Products. *Cell Chem. Biol.* **2016**, *23*, 31–44. [CrossRef] [PubMed]

62. Viehrig, K.; Surup, F.; Volz, C.; Herrmann, J.; Abou Fayad, A.; Adam, S.; Kohnke, J.; Trauner, D.; Müller, R. Structure and Biosynthesis of Crocagins: Polycyclic Posttranslationally Modified Ribosomal Peptides from Chondromyces crocatus. *Angew. Chem. Int. Ed. Engl.* **2017**, *56*, 7407–7410. [CrossRef] [PubMed]
63. Flatt, P.M.; Mahmud, T. Biosynthesis of aminocyclitol-aminoglycoside antibiotics and related compounds. *Nat. Prod. Rep.* **2007**, *24*, 358–392. [CrossRef] [PubMed]
64. Kudo, F.; Eguchi, T. Aminoglycoside Antibiotics: New Insights into the Biosynthetic Machinery of Old Drugs. *Chem. Rec.* **2016**, *16*, 4–18. [CrossRef] [PubMed]
65. Yu, Y.; Zhang, Q.; Deng, Z. Parallel pathways in the biosynthesis of aminoglycoside antibiotics. *F1000Res* **2017**, *6*. [CrossRef]
66. Biarnes-Carrera, M.; Breitling, R.; Takano, E. Butyrolactone signalling circuits for synthetic biology. *Curr. Opin. Chem. Biol.* **2015**, *28*, 91–98. [CrossRef] [PubMed]
67. Camilli, A.; Bassler, B.L. Bacterial small-molecule signaling pathways. *Science* **2006**, *311*, 1113–1116. [CrossRef]
68. Papenfort, K.; Bassler, B.L. Quorum sensing signal-response systems in Gram-negative bacteria. *Nat. Rev. Microbiol.* **2016**, *14*, 576–588. [CrossRef]
69. Polkade, A.V.; Mantri, S.S.; Patwekar, U.J.; Jangid, K. Quorum Sensing: An Under-Explored Phenomenon in the Phylum Actinobacteria. *Front. Microbiol.* **2016**, *7*, 131. [CrossRef]
70. Brotherton, C.A.; Medema, M.H.; Greenberg, E.P. luxR Homolog-Linked Biosynthetic Gene Clusters in Proteobacteria. *mSystems* **2018**, *3*. [CrossRef]
71. Lloyd, D.G.; Whitworth, D.E. The Myxobacterium Myxococcus xanthus Can Sense and Respond to the Quorum Signals Secreted by Potential Prey Organisms. *Front. Microbiol.* **2017**, *8*, 439. [CrossRef] [PubMed]
72. Amiri Moghaddam, J.; Crusemann, M.; Alanjary, M.; Harms, H.; Davila-Cespedes, A.; Blom, J.; Poehlein, A.; Ziemert, N.; Konig, G.M.; Schaberle, T.F. Analysis of the Genome and Metabolome of Marine Myxobacteria Reveals High Potential for Biosynthesis of Novel Specialized Metabolites. *Sci. Rep.* **2018**, *8*, 16600. [CrossRef] [PubMed]
73. Bouhired, S.; Rupp, O.; Blom, J.; Schaberle, T.F.; Schiefer, A.; Kehraus, S.; Pfarr, K.; Goesmann, A.; Hoerauf, A.; Konig, G. Complete Genome Sequence of the Corallopyronin A-Producing Myxobacterium Corallococcus coralloides B035. *Microbiol. Resour. Announc.* **2019**, *8*. [CrossRef] [PubMed]
74. Garcia, R.; Müller, R. Simulacricoccus ruber gen. nov., sp. nov., a microaerotolerant, non-fruiting, myxospore-forming soil myxobacterium and emended description of the family Myxococcaceae. *Int. J. Syst. Evol. Microbiol.* **2018**, *68*, 3101–3110. [CrossRef] [PubMed]
75. Livingstone, P.G.; Morphew, R.M.; Whitworth, D.E. Genome Sequencing and Pan-Genome Analysis of 23 Corallococcus spp. Strains Reveal Unexpected Diversity, With Particular Plasticity of Predatory Gene Sets. *Front. Microbiol.* **2018**, *9*, 3187. [CrossRef] [PubMed]

Article

Evaluation of Antimicrobial, Enzyme Inhibitory, Antioxidant and Cytotoxic Activities of Partially Purified Volatile Metabolites of Marine *Streptomyces* sp.S2A

Saket Siddharth and Ravishankar Rai Vittal *

Department of Studies in Microbiology, University of Mysore, Manasagangotri, Mysore 570006, India; saketsiddharth@gmail.com
* Correspondence: raivittal@gmail.com; Tel.: +91-821-241-9441

Received: 30 April 2018; Accepted: 13 July 2018; Published: 18 July 2018

Abstract: In the present study, marine actinobacteria *Streptomyces* sp.S2A was isolated from the Gulf of Mannar, India. Identification was carried out by 16S rRNA analysis. Bioactive metabolites were extracted by solvent extraction method. The metabolites were assayed for antagonistic activity against bacterial and fungal pathogens, inhibition of α-glucosidase and α-amylase enzymes, antioxidant activity and cytotoxic activity against various cell lines. The actinobacterial extract showed significant antagonistic activity against four gram-positive and two gram-negative pathogens. Excellent reduction in the growth of fungal pathogens was also observed. The minimum inhibitory concentration of the partially purified extract (PPE) was determined as 31.25 µg/mL against *Klebsiella pneumoniae*, 15.62 µg/mL against *Staphylococcus epidermidis, Staphylococcus aureus* and *Bacillus cereus*. The lowest MIC was observed against *Micrococcus luteus* as 7.8 µg/mL. MIC against fungal pathogens was determined as 62.5 µg/mL against *Bipolaris maydis* and 15.62 µg/mL against *Fusarium moniliforme*. The α-glucosidase and α-amylase inhibitory potential of the fractions were carried out by microtiter plate method. IC$_{50}$ value of active fraction for α-glucosidase and α-amylase inhibition was found to be 21.17 µg/mL and 20.46 µg/mL respectively. The antioxidant activity of partially purified extract (PPE) (DPPH, ABTS, FRAP and Metal chelating activity) were observed and were also found to have significant cytotoxic activity against HT-29, MDA and U-87MG cell lines. The compound analysis was performed using gas chromatography-mass spectrometry (GC-MS) and resulted in three constituents; pyrrolo[1–a]pyrazine-1,4-dione,hexahydro-3-(2-methylpropyl)-, being the main component (80%). Overall, the strain possesses a wide spectrum of antimicrobial, enzyme inhibitory, antioxidant and cytotoxic activities which affords the production of significant bioactive metabolites as potential pharmacological agents.

Keywords: marine actinobacteria; *Streptomyces* sp.; enzyme inhibition; antimicrobial; antioxidant; cytotoxicity; GC-MS; pyrrolopyrazines

---

1. Introduction

The microbial natural products are a source of several important drugs of high therapeutic value. Going back to the history of drugs of the first choice, it suggests that novel chemical moieties forming the backbone of bioactive compounds are primarily obtained from natural sources [1]. The microbial natural products are a source of several important drugs of high therapeutic value, namely antitumor agents [2], antibiotics [3], immunosuppressive agents [4], and enzyme inhibitors [5]. The majority of commercially available pharmaceutical products are secondary metabolites or their derivatives produced by bacteria, fungi and actinobacteria [6]. Among producers of important metabolites, actinobacteria have proven to be most prolific source accounting for more than two-third of available clinical products of several medical uses [7]. Actinobacteria are filamentous gram-positive bacteria with high G + C content [8]. They are characterized by complex morphological differentiation and are considered as an intermediate group of bacteria and fungi [9]. Their presence in various ecological habitats and marine environments has enabled research communities to exploit their tremendous potential as the richest source of pharmaceutical and biologically active products [10]. Therefore, they are contemplated as the most economical and biotechnologically beneficial prokaryotes.

Secondary metabolites are organic compounds having no direct role in the vegetative growth and the development of the organism. About 40–45% of active metabolites produced by the microorganisms are contributed by various genera of actinobacteria and are currently in clinical use [11]. Over the last few decades, actinobacterial metabolites have been used as a template for the development of anticancer agents, antibiotics, enzyme inhibitors, immunomodulators and plant growth hormones [12]. Among important genera of actinobacteria, *Streptomyces* is the most dominant and prolific source of bioactive metabolites with the broad spectrum of activity. Of 10,000 known compounds, genus *Streptomyces* alone accounts for nearly 7500 compounds, while the rare actinobacterial genera including *Nocardia, Micromonospora, Streptosporangium, Actinomadura, Saccharopolyspora* and *Actinoplanes* represent 2500 compounds [13]. Although the majority of the actinobacterial bioactive metabolites come from terrestrial habitats, recent studies on actinobacteria from diverse habitats have suggested new chemical entities and bioactive compounds [14]. Moreover, the possibility of finding a novel bioactive molecule from the terrestrial habitat has diminished over the years [15]. The marine ecosystem is an untapped and underexploited source for the discovery of novel metabolites. Species isolated from marine environments have found to be different in physiological, biochemical and molecular characteristics from their terrestrial counterparts and therefore might produce novel metabolites [16]. With the increase in resistance among pathogens and unavailability of novel metabolites from terrestrial sources, marine-derived drugs could be of great importance. However, the distribution of actinobacteria in the marine ecosystem has not been explored much and the knowledge about the marine-derived metabolites remains elusive. But, recent outbreaks about the marine actinobacterial-derived bioactive metabolites with distinct lead molecules have made a significant contribution in drug discovery and may lead to the development of new drugs in future.

The present work therefore aimed to investigate the potential of secondary metabolites produced by marine actinobacteria *Streptomyces* sp.S2A and their characterization.

2. Materials and Methods

2.1. Sample Collection

Marine sediment samples were collected from Gulf of Mannar Marine National Park (Latitude 9.127823° N, Longitude 79.466155° E), Rameshwaram, India. The collected sediment samples were brought to the laboratory in sterile zip-lock plastic bags and stored at 4 °C until further use. The sediments were pre-treated with $CaCO_3$ and kept in hot air oven at 55 °C for 20 min [17].

2.2. Actinobacterial Isolates

Isolation of the actinobacterial strain was determined by serial dilution method on starch casein agar (Himedia, New Delhi, India) supplemented with nalidixic acid (25 µg/mL) and nystatin (50 µg/mL). The plates were incubated at 28 °C for 7 days. After incubation, individual colonies were maintained on ISP-2 slants and stored at 4 °C for further use. The ornamentation of the spore chain was analyzed by SEM.

2.3. Molecular Identification of Actinobacteria

Genomic DNA extraction of the strain was done using the phenol-chloroform method. The selected colony was grown in ISP-2 (International Streptomyces Project) broth on the rotary shaker

(140 rpm, Hahn-Shin, Bucheon, South Korea) at 28 °C, pH 7.2 for 14 days. The cells were harvested by centrifugation at 8000 rpm for 10 min and the pellet was washed twice with normal saline. Washed pellet was suspended in 10 mM Tris-HCl (pH 8, Merck, Burlington, VT, USA) and lysozyme (2.5 mg/Ml, Sigma, Burlington, VT, USA), incubated at 37 °C for 1 h and was re-suspended in lysis buffer (50 mM Tris, 10 mM ethylenediaminetetraacetic acid (EDTA, Qualigens Fine Chemicals Pvt. Ltd., San Diego, CA, USA), 1% sodium dodecyl sulfate (SDS, Sigma)) and proteinase K (1 mg/mL, Sigma) and incubated for 1 h at 50 °C. 400 μL of phenol (Tris-saturated, Himedia, New Delhi, India) was added and mixed vigorously for 2 min. After centrifugation, the upper aqueous layer was transferred to the fresh tube (Tarsons, Kolkata, India), followed by the addition of $CHCl_3$ (Sigma) and isoamyl alcohol (Merck) (24:1) and centrifuged at 1000 rpm for 15 min (4 °C). To the supernatant 50 μL of NaCl (5M) and twice the volume of absolute alcohol (Himedia) was added and kept for overnight incubation. Again, it was centrifuged at 14,000 rpm for 15 min and the pellet was washed with 70% alcohol. Pellet was air dried to remove traces of ethanol (EtOH, Himedia) and was suspended in 30 μL of Tris-EDTA (TE) buffer (Himedia). DNA was analyzed by 1% agarose gel electrophoresis (Bio-Rad, Hercules, CA, USA).

16S rRNA gene amplification was carried out using universal primer set, 27F (5′-AGAGTTTGA TCCTGGCTCAG-3′) and 1492R (5′-ACGGCTACCTTGTTACGACTT-3′). The PCR conditions were programmed as follows: Initial denaturation at 95 °C for 5 min; followed by 35 cycles at 95 °C for 1 min, primer annealing at 54 °C for 1 min, extension at 72 °C for 1 min. Final extension was done at 72 °C for 10 min and was kept for cooling at 10 °C. The amplified products were determined at 1.8% agarose gel electrophoresis. The sequence was compared with similar 16S rRNA sequences obtained from BLAST search in National Center for Biotechnology Information (NCBI) database and the phylogenetic tree was constructed by the neighbor joining tree algorithm using MEGA 7.0 software (Mega, Raynham, MA, USA).

2.4. Fermentation

Isolate *Streptomyces* sp.S2A was inoculated in ISP-2 broth (Himedia) and kept for incubation at the rotary shaker (140 rpm, 28 °C) for 14 days. The culture broth obtained was extracted thrice with ethyl acetate (EA, Fisher Scientific, Madison, WI, USA) and concentrated under the rotary evaporator at 50 °C.

2.5. Antimicrobial Assays

2.5.1. Disc Diffusion Method

Antimicrobial activity of active fraction was assessed by disk diffusion method against *Staphylococcus epidermidis* (MTCC 435), *Staphylococcus aureus* (MTCC 740), *Bacillus cereus* (MTCC 1272), *Escherichia coli* (MTCC 40), *Klebsiella pneumoniae* (MTCC 661), *Micrococcus luteus* (MTCC 7950) *Aspergillus flavus* (MTCC 2590), *Fusarium moniliforme* (MTCC 6576), *Bipolaris maydis* and *Alternaria alternata* (MTCC 1362). The sterile discs (6 mm, Himedia) were impregnated with 30 μL of crude extract. The pathogens were inoculated in Mueller-Hinton broth (24 h for bacteria, Himedia) and Sabouraud Dextrose both (72 h for fungi, Himedia). The well-grown bacterial and fungal cultures were plated on Mueller-Hinton agar and Potato Dextrose agar respectively (Himedia). Sterile discs loaded with extract were placed on the plate. Chloramphenicol discs (Himedia) were used as positive control for antibacterial assay, while nystatin discs (Himedia) were used for the antifungal assay. Discs impregnated with dimethyl sulfoxide (DMSO, Himedia) were used as the solvent control. The plates were incubated at 37 °C and room temperature (for test bacteria and fungi respectively) and the zone of inhibition was measured.

2.5.2. Determination of Minimum Inhibitory Concentration

The minimum inhibitory concentration (MIC) value of the partially purified extract (PPE) was determined by micro dilution method. Bacterial and fungal pathogens were grown in sterile broth and

10 μL of log phase culture was added into 96 well micro titre plates. Partially purified fractions were dissolved in 1% DMSO and serially diluted to give required concentrations (1 mg/mL–3.9 μg/mL). Diluted fractions and sterile broth were added into pre-coated microbial cultures, making up a total of volume of 200 μL. The plate was incubated at 37 °C and room temperature (for test bacteria and fungi respectively).

2.6. Antioxidant Assays

2.6.1. 2,2 diphenyl 1 picrylhydrazyl Radical Scavenging Activity (DPPH)

DPPH free radicals are highly stable and widely used to evaluate the radical scavenging activity of the antioxidants. Scavenging activity is based upon the reduction of DPPH radicals by hydrogen donating antioxidant compounds by forming DPPH-H. Radical scavenging activity of the ethyl acetate extract of the strain S2A was examined based on the previously described method by Ser et al. with minor changes [18]. Varying concentration of S2A extract was dissolved in methanol and reacted with freshly prepared DPPH solution (60 Mm, Sigma). The reaction mixture was incubated for 30 min in the dark. The absorbance was measured at 520 nm. Decreasing absorbance of DPPH solution indicates an increase in radical scavenging activity. The scavenging activity (%) was calculated using the following equation:

$$\text{DPPH scavenging activity (\%)} = [(A_o - A_1)/A_o] \times 100,$$

where A_o is the absorbance of control (blank) and A_1 is the absorbance of the sample. Methanol was used as a blank whereas trolox (Sigma) was used as the reference compound [19].

2.6.2. Metal Chelating Activity

The metal chelating activity was examined by measuring the ability of the compound to compete with ferrozine for Fe^{2+}, complex of which can be quantified spectrophotometrically. Metal chelating activity was measured by the method previously described by Adjimani and Asare with minor modifications [20]. Assay measures the reduction in the color intensity as a result of disruption of ferrous ion and ferrozine complexes. Briefly, varying concentration of extract was added to 0.15 mL of 2 mM $FeCl_2$. The reaction was initiated with the addition of 5 mM ferrozine (Sigma), followed by the incubation at the room temperature for 10 min. The absorbance was measured at 562 nm. The percentage of inhibition was calculated using the following equation:

$$\text{Metal chelating activity (\%)} = [(A_o - A_1)/A_o] \times 100,$$

where A_o is the absorbance of control and A_1 is the absorbance of the sample. EDTA was used as a positive control.

2.6.3. 2,2′-Azino-bis(3-ethylbenzothiazoline-6-sulfonic acid) Radical Scavenging Activity (ABTS)

ABTS (Sigma) scavenging activity is based upon the reduction of ABTS* radicals by compounds having lower redox potential than that of ABTS. The 2,2′-azino-bis(3-ethylbenzothiazoline-6-sulfonic acid) (ABTS) radical scavenging assay was carried out according to the method developed by Ser et al. [21]. Initially, ABTS stock solution (7 mM) was mixed with potassium persulfate (2.45 Mm, Himedia) to form ABTS cation complex for 12 h. The ABTS complex solution was added to varying concentrations of the extract preloaded in a 96-well microplate. The reaction was kept for incubation at room temperature for 20 min and the absorbance was measured at 734 nm. The percentage scavenging activity was calculated using the following formula:

$$\text{ABTS radical scavenging activity (\%)} = [(A_o - A_1)/A_o] \times 100,$$

where A_o is the absorbance of control and A_1 is the absorbance of the sample. Trolox (Sigma) was used as a positive control.

2.6.4. Ferric Reducing Antioxidant Power (FRAP) Assay

This assay determined the reduction of ferric ions to ferrous ions which was monitored spectrophotometrically at 593 nm. The FRAP assay was performed according to the method previously described by Benzy and Strain with minor modification [22], based on the reduction of ferric complex to ferrous complex by the antioxidants. Initially, FRAP reagent was prepared by adding acetate buffer (pH 3.6), 10 mM TPTZ (Sigma) and 20 mM $FeCl_3$ (Himedia) at a ratio of 10:1:1. The reaction was started with the addition of varying concentration of extracts to the FRAP reagent. The mixture was then incubated at 37 °C for 10 min and absorbance was measured at 593 nm. Trolox was used as the positive control. The final FRAP values were expressed as Trolox equivalent antioxidant capacity (μM TE/g sample).

2.7. Enzyme Inhibitory Activities

2.7.1. Inhibition Assays for α-glucosidase Activity

The α-glucosidase inhibition was determined by the 96-well microtiter plate method based on the calorimetric assay as previously described by Vinholes et al. [23]. α-glucosidase enzyme solution (2U mL^{-1}, Sigma) was prepared in 100 mM phosphate buffer (pH 7.0). Ethyl acetate extracts were used in concentrations ranging from 10–100 μg mL^{-1}. 2 mM of *para*-nitrophenyl-α-D-glucopyranoside (Sigma) was prepared in 50 mM phosphate buffer (pH 7.0). 50 μL of the partially purified fraction was pre-incubated with an equal volume of yeast enzyme at 37 °C for 5 min, followed by the addition of 30 μL of pNPG and further incubation for 30 min. After incubation, 100 μL of stopping reagent (0.1 M Na_2CO_3) was added to cease the reaction. Color produced was quantified by UV spectrophotometer (Shimadzu, Kyoto, Japan) at 405 nm. Each experiment was performed in triplicate. Acarbose (Sigma) was used as a positive control, whereas purified fraction was replaced by phosphate buffer in control. Reaction mixture without enzyme was taken as blank. The percentage inhibition (%) was determined by the formula:

$$\text{percentage inhibition } (\%) = \frac{\text{absorbance of control} - \text{absorbance of sample}}{\text{absorbance of control}} \times 100$$

2.7.2. Inhibition Assays for α-amylase Activity

The α-amylase inhibition was determined by 96-well microtiter plate method based on calorimetric assay as previously described by Balasubramaniam et al. [24]. Equal volume of test samples (5 mg mL^{-1}) and α-amylase solution (0.5 mg mL^{-1}, Sigma) prepared in 30 mM phosphate buffer (pH 7.0) was pre- incubated at 37 °C for 10 min. 50 μL of 0.5% starch solution was added and incubated for 10 min at 37 °C. 120 μL of DNS reagent (Sigma) was added to stop the reaction. The reaction mixture was incubated at 95 °C for 5 min, cooled to room temperature. Absorbance was measured at 540 nm in a microplate reader. Acarbose at the concentration 2 mg mL^{-1} was taken as positive control. The inhibition percentage of amylase was determined by the formula reported in the previous paragraph.

2.8. Cytotoxicity Assay

The human cell lines HT-29 (Colon cancer), MDA (Breast cancer) and U-87 MG (Brain cancer) were procured from National Centre for Cell Science, Pune, India. The cell lines were cultured and maintained in Dulbecco's modified Eagle's medium (DMEM, Sigma) in T-flasks in the incubator at 37 °C and internal atmosphere of 95% air and 5% CO_2. The cytotoxicity was determined by standard MTT dye assay, according to the method described by Carmichael et al. [25]. Briefly, varying concentration of extracts were dissolved in 1% DMSO and treated to cells seeded in 96 well tissue culture plates. The plates were kept for incubation at 37 °C for 24 h, MTT solution (Sigma) was added and incubated for 4 h at 37 °C. The amount of purple formazan crystals resulting from the reduction of MTT dye by succinic dehydrogenase in mitochondria of the viable cells was determined by measuring OD at 570 nm. The IC_{50} value was calculated using graph pad prism. Each assay was performed in triplicate.

2.9. Gas Chromatography-Mass Spectrometry (GC-MS)

The analysis of the volatile constituents in extracts was determined by GC-MS technique (Perkin Elmer Clarus, USA). Perkin Elmer Clarus 680 employed a fused silica column, packed with Elite-5MS and the compounds were separated using helium as a carrier gas at a constant flow of 1 mL/min. The injector temperature was kept at 260 °C. Oven temperature was set as follows: 60 °C (2 min); followed by 300 °C at the rate of 10 °C min^{-1}. The spectrum thus obtained was compared with the database of the already known spectrum of components stored in GC-MS NIST library. The infrared spectrum of the extract was analyzed by FT-IR spectrophotometer in the range of 400–4000 cm^{-1}.

3. Results

3.1. Isolation and Molecular Identification of the Strain

Marine sediment from Gulf of Mannar was pre-treated with physical and chemical methods. Grown on SCA and ISP-2 medium, the cultural characteristics were identical on either of them. The aerial hyphae were white in color and substrate mycelium was colorless. SCA and ISP-2 culture plates did not show any pigment diffusion. Micromorphological studies of strain using SEM showed smooth spore ornamentation and rectiflexibilis spore morphology (Figure 1). The genomic DNA of the strain was isolated using the phenol-chloroform method and examined for 16S r-RNA sequence. The amplified sequences were subjected to BLAST analysis using the megablast tool of Genebank at NCBI under the accession number (KU921225). The BLAST search revealed that the strain belonged to *Streptomyces* sp. The highest similarity value index was found between the sequences of *Streptomyces* sp.S2A and *Streptomyces griesoruber* (100%). The neighbor-joining phylogenetic tree was drawn using MEGA 7.0 (Figure 2).

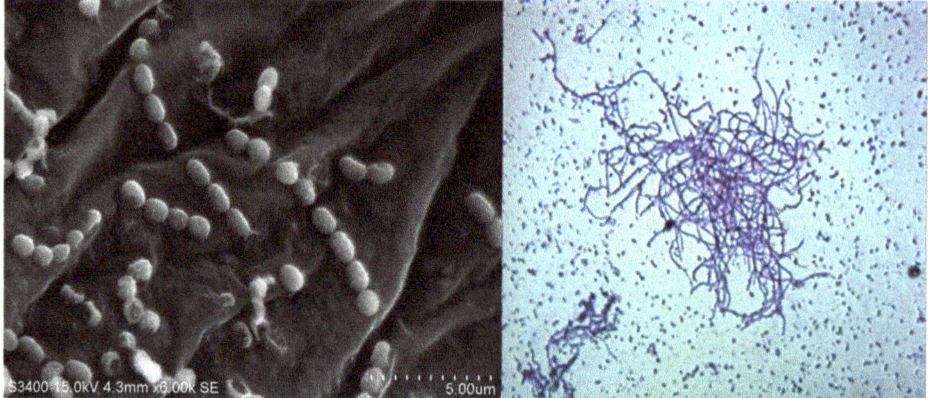

Figure 1. (**A**) Scanning electron micrograph showing spore ornamentation in *Streptomyces* sp.S2A; (**B**) Microscopic image of *Streptomyces* sp.S2A under 100×.

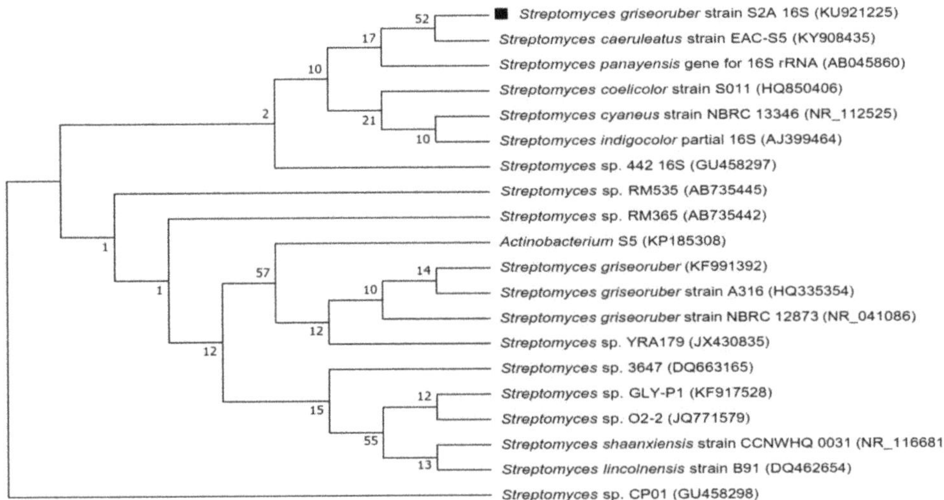

Figure 2. Phylogenetic tree of *Streptomyces* sp.S2A and the relationships with the closest species based on 16S rRNA gene sequencing using the neighbor-joining method.

3.2. Antimicrobial Assays

3.2.1. Disc Diffusion Method

Antagonistic characteristics of the bioactive extract of *Streptomyces* sp.S2A showed potent antagonistic activity against bacterial and fungal pathogens (Table 1). Of six bacterial pathogens, the highest inhibition activity was manifested against *Micrococcus luteus* and *Staphylococcus epidermidis* (16 mm). Susceptibility of *Bacillus cereus*, *Klebsiella pneumoniae* and *Staphylococcus aureus* to bioactive compounds was highly noticeable (14 mm). *Escherichia coli* was less susceptible to the compound (10 mm). Among fungal pathogens, reduction in mycelial growth was not seen against *Aspergillus flavus* and *Alternaria alternata* whereas inhibitory activity was significantly observed against *Fusarium moniliforme* and *Bipolaris maydis* (See Supplementary Figure S1).

Table 1. Antimicrobial activity and MIC (μg/mL) of *Streptomyces* sp.S2A by broth dilution method.

Test Microorganisms	Zone of Inhibition (mm)		MIC (μg/mL)
Bacteria	**Extract**	**Antibiotics (Chloramphenicol)**	
Klebsiella pneumoniae MTCC 661	14 ± 0.4	30 ± 1.1	31.25
Micrococcus luteus MTCC 7950	16 ± 0.8	28 ± 1.6	7.81
Escherichia coli MTCC 40	10 ± 0.8	22 ± 1.9	15.62
Bacillus cereus MTCC 1272	14 ± 1.2	25 ± 1.1	15.62
Staphylococcus epidermidis MTCC 435	16 ± 0.4	23 ± 1.8	15.62
Staphylococcus aureus MTCC 740	14 ± 0.8	24 ± 0.8	15.62
Fungi		**(Nystatin)**	
Aspergillus flavus MTCC 2590	-	-	-
Bipolaris maydis	14 ± 1.2	20±1.2	31.25
Alternaria alternata MTCC 1362	-	-	-
Fusarium moniliforme MTCC 6576	18 ± 1.2	22±1.0	7.81

3.2.2. Determination of Minimum Inhibitory Concentration (MIC)

The minimum inhibitory concentration of the extract was determined as 31.25 μg/mL against *Klebsiella pneumoniae*, 15.62 μg/mL against *Staphylococcus epidermidis*, *Staphylococcus aureus*,

Bacillus cereus and *Escherichia coli*. Lowest MIC was observed against *Micrococcus luteus* as 7.8 µg/mL. The solvent DMSO (1%) had no significant inhibitory activity against pathogens. MIC against fungal pathogens was determined as 62.5 µg/mL against *Bipolaris maydis* and 15.62 µg/mL against *Fusarium moniliforme*. (Table 1).

3.3. Antioxidant Assays

3.3.1. DPPH Radical Scavenging Activity

The highest inhibition concentration of radical scavenging activity of the extract was found to be 56.55 ± 3.1%, as compared to the trolox that was found to be 74.73 ± 1.13%. The IC_{50} value for DPPH radical scavenging activity of the extract was 0.86 mg (Table 2).

Table 2. Radical scavenging activity of ethyl acetate extract of *Streptomyces* sp.S2A.

Antioxidant Assays	Concentration of Extract (mg/mL)	% Inhibition	Absorbance	IC_{50} (mg/mL)
DPPH	1.0	56.55 ± 3.1	-	0.86
	0.50	32.33 ± 1.4	-	
	0.25	17.29 ± 1.6	-	
Metal chelating	2.0	59.98 ± 2.12	-	1.56
	1.0	37.50 ± 2.36	-	
	0.50	24.90 ± 2.11	-	
	0.25	18.40 ± 1.4	-	
ABTS	0.10	42.48 ± 3.1	-	0.011
	0.05	30.24 ± 3.74	-	
	0.02	7.29 ± 3.62	-	
FRAP	0.1	-	0.248	-
	0.08	-	0.202	
	0.06	-	0.145	
	0.04	-	0.060	
	0.02	-	0.028	

3.3.2. Metal Chelating Activity

The study showed the decrease in the formation of ferrozine-Fe^{2+} complex with increase in the concentration of the extract. It showed the significant chelating activity measuring from 18.40 ± 1.4% to 59.98 ± 2.12% at the concentration ranging from 0.25–2 mg/mL (Table 2).

3.3.3. ABTS Radical Scavenging Activity

This assay showed the significant increase in the scavenging activity with increase in the concentration of the extract, thus decolorized the blue-green color of ABTS* back into ABTS, which is colorless. The IC_{50} value for ABTS radical scavenging activity of the extract was 11.77 µg (Table 2).

3.3.4. Ferric Reducing Antioxidant Power (FRAP) Assay

The increase in absorbance was observed with the increase in the concentration of the extract suggesting the significant antioxidant activity (Table 2).

3.4. In Vitro Enzyme Inhibition Assay

The EA extract exhibited α-amylase and α-glucosidase inhibitory activity in a dose-dependent manner. Acarbose was used as a standard. IC_{50} value of EA extract for α-glucosidase and α-amylase inhibition was found to be 21.17 and 20.46 respectively, whereas that of acarbose was 15.47 and 18.15 µg/mL respectively (Tables 3 and 4).

Table 3. α-glucosidase inhibition and IC_{50} values of ethyl acetate extract of *Streptomyces* sp.S2A.

Concentration (µg/mL)	Inhibition % (EA Extract)	IC_{50} (µg/mL) (EA Extract)	Inhibition % (Acarbose)	IC_{50} (µg/mL) (Acarbose)
6.25	29.12 ± 0.33		36.44 ± 0.58	
12.5	38.54 ± 0.77		45.27 ± 0.34	
25	55.1 ± 1.16	21.17	62.19 ± 1.10	15.47
50	68.4 ± 1.55		78.52 ± 1.99	
100	72.31 ± 1.01		86.83 ± 2.01	
200	81.74 ± 2.65		94.22 ± 2.33	

Table 4. α-amylase inhibition and IC_{50} values of ethyl acetate extract of *Streptomyces* sp.S2A.

Concentration (µg/mL)	Inhibition % (EA Extract)	IC_{50} (µg/mL) (EA Extract)	Inhibition % (Acarbose)	IC_{50} (µg/mL) (Acarbose)
6.25	16.44 ± 0.21		20.19 ± 0.78	
12.5	34.77 ± 0.44		40.05 ± 0.10	
25	59.29 ± 1.15	20.46	64.44 ± 1.45	18.15
50	74.32 ± 1.09		87.57 ± 1.33	
100	81.13 ± 1.34		97.03 ± 1.10	
200	88.67 ± 1.93		97.84 ± 1.78	

3.5. Cytotoxicity Assay

The tested results of the extract against cell lines were shown in (See Supplementary Figure S2). The results revealed that the extract showed varying efficacy against cell lines. The highest activity against U-87 at 100 µg/mL was found to be 59.63 ± 1.9%. It also showed significant activity against MDA and HT-29 with cell inhibition was found to be 55.23 ± 1.09% and 52.31 ± 2.4% respectively at 100 µg/mL (Table 5). Overall, the results suggested the potential cytotoxic activity against various cell lines.

Table 5. Cytotoxic activity of extract of *Streptomyces* sp.S2A against HT-29, MDA and U-87 MG.

Concentration (µg/mL)	Inhibition %		
	U-87 MG	**MDA**	**HT-29**
5	13.76 ± 1.81	3.57 ± 1.76	18.51 ± 3.89
10	16.51 ± 2.01	10.71 ± 3.75	21.76 ± 2.32
20	19.26 ± 3.79	15.0 ± 4.10	23.15 ± 1.96
50	36.19 ± 2.11	30.95 ± 2.87	35.31 ± 2.77
100	59.63 ± 1.90	55.23 ± 1.09	52.31 ± 2.40
IC_{50} (µg/mL)	93.32	80.02	88.68

3.6. Gas Chromatography-Mass Spectrometry (GC-MS)

Analysis of components of the active fraction with the highest activity by GC-MS analysis implied nine peaks at the retention time of (i) 17.239; (ii) 17.309; (iii) 20.811; (iv) 21.311; (v) 21.406; (vi) 21.586; (vii) 22.071; (viii) 22.126; (ix) 24.257. Further examination of MS peaks revealed m/z at 168, 259, 210 and 350. According to NIST library search, peak retentions at 21.311, 21.406 and 21.586 correspond to single compound i.e., pyrrolo[1–a]pyrazine-1,4-dione,hexahydro-3-(2-methylpropyl) (Figure 3). Other tentatively identified compounds were diphenylmethane, 2-Isopropyl-1-Phenyl-3-Pyrrolidin-1-yl Propane-1,3-Dione and Benzene, 1′1-tetradecyclidenebis (The GC-MS spectrum indicated the ions at 70 and 154 corresponded to molecule $C_7H_{10}NO_2$ and C_4H_8 ions (See Supplementary Figures S3 and S4). The spectrum was similar to that of pyrrolo[1–a]pyrazine-1,4-dione,hexahydro-3-(2-methylpropyl) spectra in the GC-MS library and a study reported by Yang et al. [26]. The FT-IR spectrum of the partially purified metabolite showed the characteristic functional groups such as NH stretching peak of a primary amine at 3313.20 cm^{-1}. The functional group at 2923.91 cm^{-1} corresponded to strong C-H stretching in alkanes. The peak at 1646.10 cm^{-1} was assigned to C-N stretch in primary amine. The absorption peak at 1516.05 cm^{-1} was assigned to C=C stretch in an alkene. The peak at 1454.69 cm^{-1} was assigned to C-H bend in alkanes. The peak at 1240.27 cm^{-1} was assigned to C-O stretch. The absorption peak at 1033.54 cm^{-1} was assigned to the ether. The peak ranging from 605.8 cm^{-1} to 701.09 cm^{-1} was assigned to strong C-H bend in alkenes (Figure 4).

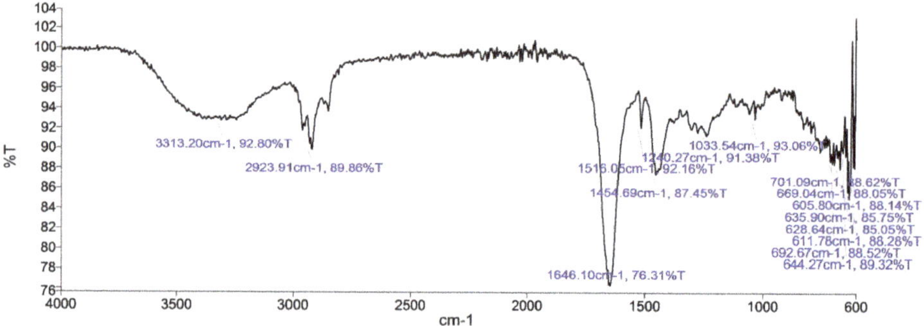

Figure 3. Chemical structure of the compound pyrrolo[1–a]pyrazine-1,4-dione,hexahydro-3-(2-methylpropyl).

Figure 4. FT-IR spectrum of the active extract of *Streptomyces* sp.S2A.

4. Discussion

With this outlook, the present investigation was carried out to identify bioactive compound from marine actinobacteria exhibiting antagonistic activity against bacterial and fungal pathogens, enzyme inhibition activity, antioxidant and cytotoxic activity. *Streptomyces* sp.S2A isolated from marine sediment of Gulf of Mannar produced white aerial mycelium to colorless substrate mycelium on SCA medium. Shirling and Gottileb [27] reported that the pigmentation paradigm could be used for the classification and identification. However, there was no pigment pattern observed with the isolate. Extraction of metabolites with ethyl acetate yielded the dark color residue. Purification by silica gel column chromatography resulted in six fractions, of which one fraction exhibited significant activity. The antagonistic activity of the bioactive compound showed the high zone of inhibition against *Micrococcus luteus* and *Staphylococcus epidermidis* (16 mm). Moderate activity was observed against *Bacillus cereus*, *Klebsiella pneumoniae* and *Staphylococcus aureus* (14 mm). Susceptibility of *Escherichia coli* to the compound was found to be weaker (10 mm). Inhibitory potential of the bioactive compound against fungal pathogens showed good activity against *Fusarium moniliforme* and *Bipolaris maydis*, whereas no zone of inhibition was observed against *Aspergillus flavus* and *Alternaria aleternata*. MIC was used to determine the efficacy of the compound at different concentration. The reduction in the growth of bacterial pathogens was observed with the concentration ranging from 7.8–31.25 µg/mL. MIC of the compound showed excellent antifungal activity against *Fusarium moniliforme* and *Bipolaris maydis*. Ethyl acetate extract of *Streptomyces* sp.S2A also showed significant α-glucosidase and α-amylase inhibition activity, though IC$_{50}$ of the extract were less than that of acarbose. Antioxidant and cytotoxic activities of the extract was determined by the DPPH, ABTS, FRAP and metal chelating assays and against HT-29, MDA and U-87 MG cell lines. The results thus obtained showed the presence of potent antioxidant and anticancer agents.

The partial chemical composition relates to the metabolite was detected by GC-MS. The chromatogram of fraction A43 showed a total of nine peaks. Of all, the major constituent was pyrrolo[1–a]pyrazine-1,4-dione,hexahydro-3-(2-methylpropyl)-, constituted 80.7% and this may be the active principle compound. The other chemical compound was identified as diphenylmethane (6%), 2-Isopropyl-1-Phenyl-3-Pyrrolidin-1-yl-Propane-1,3-Dione (2%) and Benzene, 1'1-tetradecyclidenebis (2%). Pyrrolo[1–a]pyrazine-1,4-dione,hexahydro-3-(2-methylpropyl is a peptide derivative of diketopiperazine with the molecular weight as 210 and empirical formula as $C_{11}H_{18}N_2O_2$. All the bioassays mentioned in the above paragraph were confirmed with the commercially available purified compound: pyrrolo[1–a]pyrazine-1,4-dione,hexahydro-3-(2-methylpropyl). Antimicrobial and the cytotoxic activity of the purified compound were found to be significantly higher the partially purified compound, whereas the enzyme inhibition potential of the partially purified compound against α-glucosidase and α-amylase were better in comparison with commercial compound. The mass spectrum of the commercial compound corresponds to 70.0315 *m/z* and 154.0152, which is same as shown by the compound present in partially purified extract (See Supplementary Figures S5 and S6, Tables S1–S3).

Pyrrolopyrazines are known for their wide range of biological activities such as antioxidant, anti-angiogenesis, anti-tumor and antimicrobial [28]. Manimaram et al. reported the presence of antibacterial metabolite, pyrrolo[1–a]pyrazine-1,4-dione,hexahydro-3-(2-methylpropyl) in the crude extract of *Streptomyces* sp. VITMK1 isolated from mangrove soil [29]. Antifouling potential of pyrrolo[1–a]pyrazine-1,4-dione,hexahydro-3-(2-methylpropyl) against *Vibrio halioticoli* and *Loktanella honkongensis* was studied by Dash et al. [30]. Sponge-derived marine bacteria significantly inhibited the larval settlement of *Balanus amphitrite* and *Hydroides elegans*. Another marine bacteria isolated from the sponge, *Spongia officinalis* showed the potent antibacterial and antifungal activity of pyrrolo[1–a]pyrazine-1,4-dione,hexahydro-3-(2-methylpropyl) [31]. Diketopiperzines derivatives present in marine *Streptomyces* sp. had shown good anti-H1N1 activity [32]. Mithun and Rao also reported the presence of pyrrolopyrazines in *Micrococcus luteus* with anti-cancer activity against HCT-15 cell line [33]. Presence of pyrrolo[1–a]pyrazine-1,4-dione,hexahydro-3-(2-methylpropyl) was detected in *Streptomyces* sp. MUM 256 isolated from the mangrove forest in Malaysia. This compound was reported to possess antioxidant and anticancer activities [34]. Anti-cancer metabolites were also reported from *Streptomyces malaysiense* sp.MUSC 136 isolated from the mangrove ecosystem. The bioactive metabolite exhibited strong antioxidant activity and high cytotoxic activity against HCT-116 cells [35]. The first report on marine *Staphylococcus* sp. derived pyrrolo[1–a]pyrazine-1,4-dione,hexahydro-3-(2-methylpropyl) was reported by Lalitha et al. [36]. Purified metabolite was potentially active against lung (A549) and cervical (HeLa) cancer cells in a dose-dependent manner. Thus, the present study suggested that the pyrrolopyrazines derivative pyrrolo[1–a]pyrazine-1,4-dione,hexahydro-3-(2-methylpropyl) may account for the observed antagonistic, antioxidant, and cytotoxic activities in marine actinobacteria, *Streptomyces* sp.S2A. The results obtained in the current study demonstrate that bioactive metabolites produced by marine actinobacteria have tremendous potential for pharmaceutical product and are a subject of future investigation.

Supplementary Materials: The following are available online at http://www.mdpi.com/2076-2607/6/3/72/s1.

Author Contributions: Conceptualization, S.S. and R.R.V.; Methodology, S.S. and R.R.V.; Software, S.S. and R.R.V.; Validation, S.S. and R.R.V.; Formal Analysis, S.S. and R.R.V.; Investigation, S.S.; Resources, R.R.V.; Data curation, Writing—Original Draft Preparation, S.S. and R.R.V.; Writing—Review & Editing, S.S. and R.R.V.; Visualization, S.S. and R.R.V.; Supervision, R.R.V.; Project Administration, R.R.V.; Funding Acquisition, R.R.V.

Funding: This research was funded by DST-PURSE program, DST, New Delhi, grant number DST-PURSE/Phase-II/2017-2018." The APC was funded by DST-PURSE program.

Acknowledgments: The authors would like to thank SIF, VIT University, Vellore for carrying out GC-MS analysis. The authors would also like to acknowledge the facilities provided by DST-PURSE program, DST, New Delhi to the University of Mysore, Mysuru.

Conflicts of Interest: The authors confirm that this article content has no conflict of interest.

References

1. Ganesan, A. The impact of natural products upon drug discovery. *Curr. Opin. Chem. Biol.* **2008**, *12*, 306–317. [CrossRef] [PubMed]
2. Chin, Y.-W.; Balunas, M.J.; Chai, H.B.; Kinghorn, A.D. Drug discovery from natural sources. *AAPS J.* **2006**, *8*, 239–253. [CrossRef]
3. Berdy, J. Bioactive microbial metabolites. *J. Antibiot.* **2005**, *58*, 1–26. [CrossRef] [PubMed]
4. Mann, J. Natural products as immunosuppressive agents. *Nat. Prod. Rep.* **2001**, *18*, 417–430. [CrossRef] [PubMed]
5. Imada, C. Enzyme inhibitors and other bioactive compounds from marine actinomycetes. *Antonie Leeuwenhoek* **2005**, *87*, 59–63. [CrossRef] [PubMed]
6. Naine, J.; Srinivasan, M.V.; Devi, S.C. Novel anticancer compounds from marine actinomycetes: A review. *J. Pharm. Res.* **2011**, *4*, 1285–1287.
7. Williams, S.T.; Goodfellow, M.; Wellington, E.M.H.; Vicker, J.C.; Alderson, G.; Sneath, P.H.A.; Sackin, M.J.; Mortimer, A.M. A probability matrix for identification of some streptomycetes. *Microbiology* **1983**, *129*, 1815–1830. [CrossRef] [PubMed]
8. Ventura, M.; Canchaya, C.; Tauch, A.; Chandra, G.; Fitzgerald, G.F.; Chater, K.F.; Van Sinderen, D. Genomics of *Actinobateria*: Tracing the evolutionary history of an ancient phylum. *Microbiol. Mol. Biol. Rev.* **2007**, *71*, 495–548. [CrossRef] [PubMed]
9. Goodfellow, M.; Williams, S.T. Ecology of actinomycetes. *Annu. Rev. Microbiol.* **1983**, *37*, 189–216. [CrossRef] [PubMed]
10. Bull, A.T.; Stach, J.E. Marine actinobacteria: New opportunities for natural product search and discovery. *Trends Microbiol.* **2007**, *15*, 491–499. [CrossRef] [PubMed]
11. Fiedler, H.P.; Bruntner, C.; Bull, A.T.; Ward, A.C.; Goodfellow, M.; Potterat, O. Marine actinomycetes as a source of novel secondary metabolites. *Antonie Leeuwenhoek* **2004**, *87*, 37–42. [CrossRef] [PubMed]
12. Shivlata, L.; Satyanarayana, T. Thermophilic and alkaliphilic *Actinobacteria*: Biology and potential applications. *Front. Microbiol.* **2015**, *6*, 1014. [CrossRef] [PubMed]
13. Miao, V.; Davies, J. Actinobacteria: The good, the bad and the ugly. *Antonie Leeuwenhoek* **2010**, *98*, 143–150. [CrossRef] [PubMed]
14. Jensen, P.R.; Mincer, T.J.; Williams, P.G.; Fenical, W. Marine actinomycete diversity and natural product discovery. *Antonie Leeuwenhoek* **2005**, *87*, 43–48. [CrossRef] [PubMed]
15. Subramani, R.; Aalbersberg, W. Marine actinomycetes: An ongoing source of novel bioactive metabolites. *Microbiol. Res.* **2012**, *167*, 571–580. [CrossRef] [PubMed]
16. Sharma, S.R.; Shah, G.S. Isolation and screening of actinomycetes for bioactive compounds from the marine coast of South-Gujarat Region. *Int. J. Res. Sci. Innov.* **2014**, *1*, 345–349.
17. Saadoun, I.; Hameed, K.M.; Moussauui, A. Characterization and analysis of antibiotic activity of some aquatic actinomycetes. *Microbios* **1999**, *99*, 173–179. [PubMed]
18. Ser, H.-L.; Palanisamy, U.D.; Yin, W.-F.; Malek, A.; Nurestri, S.; Chan, K.-G. Presence of antioxidative agent, Pyrrolo[1,2a] pyrazine-1,4-dione, hexahydro-in newly isolated *Streptomyces mangrovisoli* sp. nov. *Front. Microbiol.* **2015**, *6*, 854. [CrossRef] [PubMed]
19. Mišan, A.; Mimica-Dukić, N.; Sakač, M.; Mandić, A.; Sedej, I.; Šimurina, O.; Tumbas, V. Antioxidant activity of medicinal plant extracts in cookies. *J. Food Sci.* **2011**, *76*, 1239–1244. [CrossRef] [PubMed]

20. Adjimani, J.P.; Asare, P. Antioxidant and free radical scavenging activity of iron chelators. *Toxicol. Rep.* **2015**, *2*, 721–728. [CrossRef] [PubMed]

21. Ser, H.-L.; Tan, L.T.-H.; Palanisamy, U.D.; Abd Malek, S.N.; Yin, W.-F.; Chan, K.G. *Streptomyces antioxidans* sp. nov., a novel mangrove soil actinobacterium with antioxidative and neuroprotective potentials. *Front. Microbiol.* **2016**, *7*, 899. [CrossRef] [PubMed]

22. Benzie, I.F.F.; Strain, J.J. The ferric reducing ability of plasma (FRAP) as a measure of antioxidant power: The FRAP assay. *Anal. Biochem.* **1996**, *239*, 70–76. [CrossRef] [PubMed]

23. Vinholes, J.; Grosso, C.; Andrade, P.B.; Gil-Izquierdo, A.; Valentao, P.; de Pinho, P.G.; Ferreres, F. In vitro studies to assess the antidiabetic: Anti-cholinesterase and antioxidant potential of *Spergularia rubra*. *Food Chem.* **2011**, *129*, 454–462. [CrossRef]

24. Balasubramaniam, V.; Mustar, S.; Khalid, N.M.; Rashed, A.A.; Noh, M.F.M.; Wilcox, M.D.; Chater, P.I.; Brownlee, I.A.; Pearson, J.P. Inhibitory activities of three Malaysian edible seaweeds on lipase and alpha-amylase. *J. Appl. Phycol.* **2013**, *25*, 1405–1412. [CrossRef]

25. Carmichael, J.; DeGraff, W.G.; Gazdar, A.F.; Minna, J.D.; Mitchell, J.B. Evaluation of a tetrazolium-based semiautomated colorimetric assay, assessment of chemosensitivity testing. *Cancer Res.* **1987**, *47*, 936–942. [PubMed]

26. Guo, X.; Liu, X.; Yang, H. Synergistic algicidal effect and mechanism of two diketopiperazines produced by *Chryseobacterium* sp. strain GLY-1106 on the harmful bloom-florming *Microcystis aeruginosa*. *Sci. Rep.* **2015**, *5*, 14720. [CrossRef] [PubMed]

27. Shirling, E.B.; Gottileb, D. Methods for characterization of *Streptomyces* species. *Int. J. Syst. Bactriol.* **1966**, *16*, 312–340. [CrossRef]

28. Wang, C. Antifungal activity of volatile organic compounds from *Streptomyces alboflavus* TD-1. *FEMS Microbiol. Lett.* **2013**, *341*, 45–51. [CrossRef] [PubMed]

29. Manimaran, M.; Gopal, J.V.; Kannabiran, K. Antibacterial activity of *Streptomyces* sp. VITMK1 isolated from mangrove soil of Pichavaram, Tamil Nadu, India. *Proc. Natl. Acad. Sci. USA India Sect. B Biol. Sci.* **2015**, *87*, 499–506. [CrossRef]

30. Dash, S.; Jin, C.; Lee, O.O.; Xu, Y.; Qian, P. Antibacterial and antilarval-settlement potential and metabolite profiles of novel sponge-associated marine bacteria. *J. Ind. Microbiol. Biotechnol.* **2009**, *36*, 1047–1056. [CrossRef] [PubMed]

31. Sathiyanarayanan, G.; Gandhimathi, R.; Sabarathnam, B.; Kiran, G.S.; Selvin, J. Optimization and production of pyrrolidone antimicrobial agent from marine sponge-associated *Streptomyces* sp. MAPS15. *Bioprocess Biosyst. Eng.* **2014**, *37*, 561–573. [CrossRef] [PubMed]

32. Wang, P.; Xi, L.; Liu, P.; Wang, Y.; Wang, W.; Huang, Y.; Zhu, W. Diketopiperazine derivatives from the marine-derived actinomycete *Streptomyces* sp. FXJ7.328. *Mar. Drugs* **2013**, *11*, 1035–1049. [CrossRef] [PubMed]

33. Mithun, V.S.L.; Rao, C.S.V. Isolation and molecular characterization of anti-cancerous compound producing marine bacteria by using 16S rRNA sequencing and GC-MS techniques. *IJMER* **2012**, *2*, 4510–4515.

34. Tan, L.T.H.; Ser, H.L.; Yin, W.F.; Chan, K.G.; Lee, L.H.; Goh, B.H. Investigation of antioxidative and anticancer potentials of *Streptomyces* sp. MUM256 isolated from Malaysia mangrove soil. *Front. Microbiol.* **2015**, *6*, 1316. [CrossRef] [PubMed]

35. Ser, H.L.; Palanisamy, U.D.; Yin, W.F.; Chan, K.G.; Goh, B.H.; Lee, L.H. *Streptomyces malaysiense* sp. nov.: A novel Malaysian mangrove soil actinobacterium with antioxidative activity and cytotoxic potential against human cancer cell lines. *Sci. Rep.* **2016**, *6*, 24247. [CrossRef] [PubMed]

36. Lalitha, P.; Veena, V.; Vidhyapriya, P.; Lakshmi, P.; Krishna, R.; Sakthivel, N. Anticancer potential of pyrrole (1, 2, a) pyrazine 1, 4, dione, hexahydro 3-(2-methyl propyl) (PPDHMP) extracted from a new marine bacterium, *Staphylococcus* sp. strain MB30. *Apoptosis* **2016**, *21*, 566–577. [CrossRef] [PubMed]

 *microorganisms*

Review

Biosynthesis of Polyketides in *Streptomyces*

Chandra Risdian [1,2,*], **Tjandrawati Mozef** [3] **and Joachim Wink** [1,*]

1   Microbial Strain Collection (MISG), Helmholtz Centre for Infection Research (HZI), 38124 Braunschweig, Germany
2   Research Unit for Clean Technology, Indonesian Institute of Sciences (LIPI), Bandung 40135, Indonesia
3   Research Center for Chemistry, Indonesian Institute of Sciences (LIPI), Serpong 15314, Indonesia; tjandrawm@yahoo.com
*   Correspondence: chandra.risdian@gmail.com or chandra.risdian@helmholtz-hzi.de (C.R.); Joachim.Wink@helmholtz-hzi.de (J.W.); Tel.: +49-531-6181-4222 (C.R.); +49-531-6181-4223 (J.W.)

Received: 18 March 2019; Accepted: 27 April 2019; Published: 9 May 2019

Abstract: Polyketides are a large group of secondary metabolites that have notable variety in their structure and function. Polyketides exhibit a wide range of bioactivities such as antibacterial, antifungal, anticancer, antiviral, immune-suppressing, anti-cholesterol, and anti-inflammatory activity. Naturally, they are found in bacteria, fungi, plants, protists, insects, mollusks, and sponges. *Streptomyces* is a genus of Gram-positive bacteria that has a filamentous form like fungi. This genus is best known as one of the polyketides producers. Some examples of polyketides produced by *Streptomyces* are rapamycin, oleandomycin, actinorhodin, daunorubicin, and caprazamycin. Biosynthesis of polyketides involves a group of enzyme activities called polyketide synthases (PKSs). There are three types of PKSs (type I, type II, and type III) in *Streptomyces* responsible for producing polyketides. This paper focuses on the biosynthesis of polyketides in *Streptomyces* with three structurally-different types of PKSs.

Keywords: *Streptomyces*; polyketides; secondary metabolite; polyketide synthases (PKSs)

---

1. Introduction

Polyketides, a large group of secondary metabolites, are known to possess remarkable variety, not only in their structure, and but also in their function [1,2]. Polyketides exhibit a wide range of bioactivities such as antibacterial (e.g., tetracycline), antifungal (e.g., amphotericin B), anticancer (e.g., doxorubicin), antiviral (e.g., balticolid), immune-suppressing (e.g., rapamycin), anti-cholesterol (e.g., lovastatin), and anti-inflammatory activity (e.g., flavonoids) [3–9]. Some organisms can produce polyketides such as bacteria (e.g., tetracycline from *Streptomyces aureofaciens*) [10], fungi (e.g., lovastatin from *Phomopsis vexans*) [11], plants (e.g., emodin from *Rheum palmatum*) [12], protists (e.g., maitotoxin-1 from *Gambierdiscus australes*) [13], insects (e.g., stegobinone from *Stegobium paniceum*) [14], and mollusks (e.g., elysione from *Elysia viridis*) [15]. These organisms could use the polyketides they produce as protective compounds and for pheromonal communication in the case for insects.

Since the beginning of the 1940s, the history of antibiotics has greatly related to microorganisms. One of the groups of bacteria that produce many important antibiotics is Actinobacteria. Actinobacteria are Gram-positive, have high GC content, and comprise various genera known for their secondary metabolite production, such as *Streptomyces, Micromonospora, Kitasatospora, Nocardiopsis, Pseudonocardia, Nocardia, Actinoplanes, Saccharopolyspora*, and *Amycolatopsis* [16,17]. Their most important genus is *Streptomyces*, which has a filamentous form like fungi and has become a source of around two-thirds of all known natural antibiotics [18]. Among the antibiotics produced by *Streptomyces*, polyketides are one group of the very important compounds. Some examples of polyketides produced by *Streptomyces* are rapamycin (produced by *Streptomyces hygroscopicus*), oleandomycin (produced by *Streptomyces antibioticus*), actinorhodin (produced by *Streptomyces coelicolor* A3(2)), daunorubicin (produced by *Streptomyces peucetius*) and caprazamycin (produced *by Streptomyces* sp. MK730-62F2) [19–23].

Biosynthesis of polyketides is very complex because the process involves multifunctional enzymes called polyketide synthases (PKSs). The mechanism of PKS is similar to fatty acid synthase (FAS). The process includes many enzymatic reactions with different enzymes such as acyltransferase (AT), which has a role in catalyzing the attachment of the substrate (e.g., acetyl or malonyl) to the acyl carrier protein (ACP), and ketosynthase (KS), which catalyzes the condensation of substrates attached in ACP. After condensation of the substrates, the reaction continues by incorporating ketoreductase (KR), which reduces keto ester, dehydratase (DH), which dehydrates the compound, and enoylreductase (ER), which reduces the carbon-carbon double bond in the molecule (Figure 1). Unlike in FAS, the process catalyzed by KR, DH, and ER is optional in PKSs, which can give the various structures of polyketides with keto groups, hydroxy groups, and/or double bonds in different locations of the molecule [24–26]. In *Streptomyces*, there are three types of PKSs (type I, type II, and type III) [27–29]. This review describes the biosynthesis of polyketides in *Streptomyces* with three distinct types of PKSs. The focus is only on the *Streptomyces* genus because it is one of the most important producers of bioactive compounds and one of the most well-studied microbes in terms of polyketide biosynthesis. To the best of our knowledge, this is the first review that describes the three types of PKSs that are involved in the biosynthesis of polyketides in *Streptomyces*.

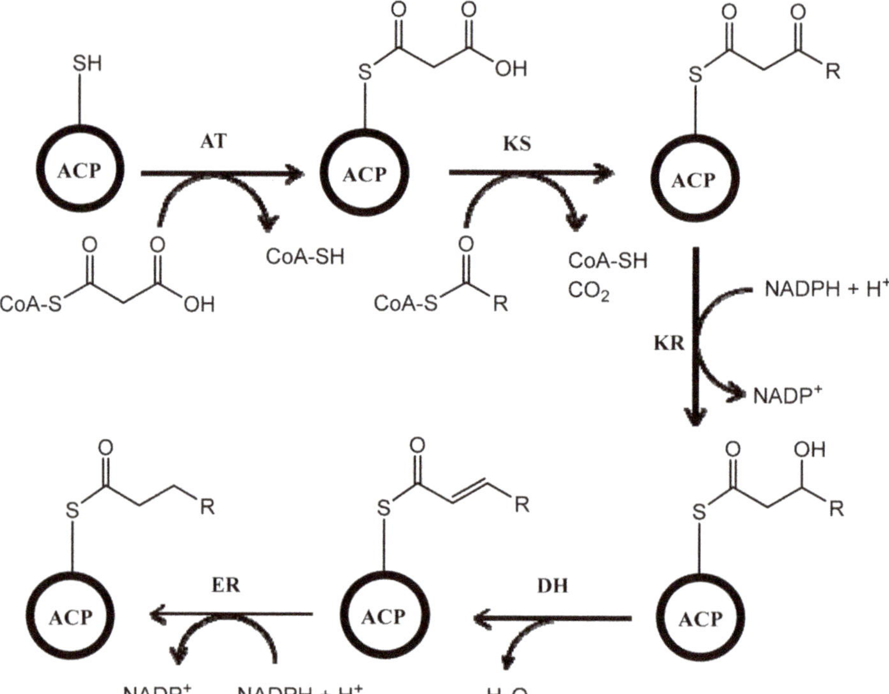

Figure 1. Scheme of the reaction occurring in polyketide synthases (PKSs). ACP, acyl carrier protein; AT, acyltransferase; KS, ketosynthase; KR, ketoreductase; DH, dehydratase; ER, enoylreductase. Adapted with permission of Portland Press, from Vance, S.; Tkachenko, O.; Thomas, B.; Bassuni, M.; Hong, H.; Nietlispach, D.; Broadhurst, W. Sticky swinging arm dynamics: studies of an acyl carrier protein domain from the mycolactone polyketide synthase. *Biochem. J.* **2016**, *473*, 1097–1110 [30].

2. Polyketide Synthases Type I

The type I polyketide synthases (type I PKSs) involve huge multifunctional proteins that have many modules containing some domains, in which a particular enzymatic reaction occurs (Figure 2). Each module has the responsibility of performing one condensation cycle in a non-iterative way. Because this system works with some modules, it is also called modular PKS. The essential domains existing in each module are acyltransferase (AT), keto synthase (KS), and acyl carrier protein (ACP) which collaborate to produce β-keto ester intermediate. In addition, the other domains that may be present in the module are β-ketoreductase (KR), dehydratase (DH), and enoyl reductase (ER), which are responsible for keto group modification. In the process of producing polyketide, the expanding polyketide chain is transferred from one module to another module until the completed molecule is liberated from the last module by a special enzyme [2,26,31].

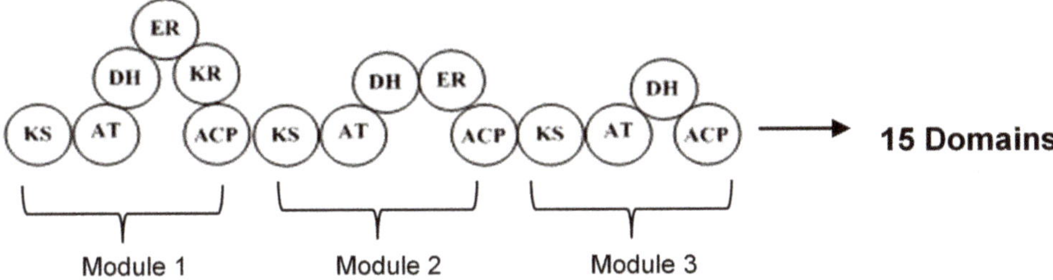

Figure 2. Structure of type I PKSs with three modules and 15 domains. ACP, acyl carrier protein; AT, acyltransferase; KS, ketosynthase; KR, ketoreductase; DH, dehydratase; ER, enoylreductase.

Furthermore, type I PKSs are generally responsible for producing macrocyclic polyketides (macrolides), although there was also a study reporting that type I PKSs are also involved in the biosynthesis of linear polyketide tautomycetin [32]. Macrolide belongs to a polyketide compound characterized by a macrocyclic lactone ring, which has various bioactivities such as antibacterial, antifungal, immunosuppressing, and anticancer. As an antibacterial agent, macrolide works by inhibiting protein synthesis by binding to the 50S ribosomal subunit and blocking the translocation steps of protein synthesis [8,27,33]. Some examples of macrolides produced by *Streptomyces* are rapamycin, FK506, spiramycin, avermectin, methymycin, narbomycin, and pikromycin, as shown in Figure 3 [34–37]. These compounds were produced by multifunctional polypeptides encoded by a biosynthetic gene cluster. The list of some polyketides produced by *Streptomyces* with their huge multifunctional proteins can be seen in Table 1.

2.1. Biosynthesis of Rapamycin

Rapamycin is a 31-membered ring macrolide produced by *Streptomyces hygroscopicus* isolated firstly from the soil of Easter Island (Chile) in the South Pacific Ocean. It is a hydrophobic compound and known as an antifungal compound against *Candida albicans*, *Cryptococcus neoformans*, *Aspergillus fumigatus*, *Fusarium oxysporum*, and some pathogenic species from the genus *Penicillium*. The antifungal mechanism of this compound has been described by diffusing into the cell and binding to intracellular receptor immunophilin FKB12. The FKBP12-rapamycin complexes inhibit enzymes required for signal transduction and cell growth. These enzymes are TOR (target of rapamycin) kinases that are conserved and very important for cell cycle progression. Interestingly, it was also reported that rapamycin has not only antifungal activity, but also anticancer and immunosuppressant activity [8,27,38,39].

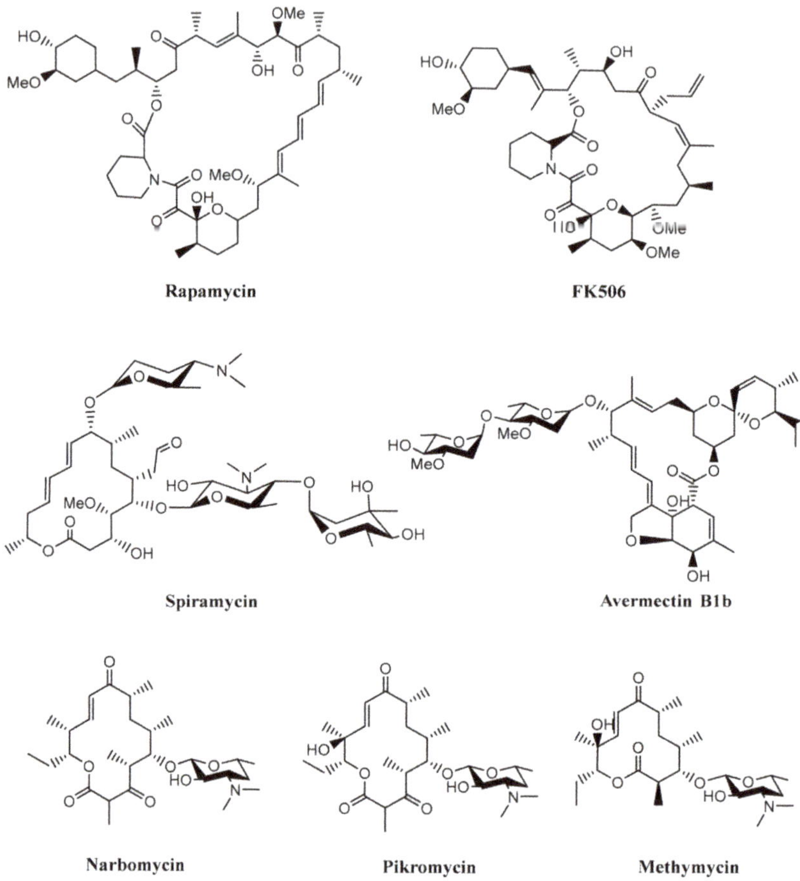

Figure 3. Some of the macrolides produced by *Streptomyces*.

Rapamycin is synthesized by type I PKS rapamycin synthase (RAPS) [40]. The rapamycin-PKS gene cluster (*rapPKS*) is 107.3 kb in size and has three remarkably large ORFs (open reading frames), *rapA*, *rapB*, and *rapC* which encode multifunctional protein RAPS1 (~900 kDa), RAPS2 (~1.07 MDa), and RAPS3 (~660 kDa), respectively. Protein RAPS1 comprises four modules for polyketide chain extension; protein RAPS2 contains six modules responsible for continuing the process of polyketide chain elongation until C-16; and RAPS3 possesses four modules that have a role in completing the polyketide fraction of the rapamycin molecule. Overall, these three giant proteins encompass 70 domains or enzymatic functions, and because of this, rapamycin PKSs are considered as the most complex multienzyme system discovered so far [26,27,34].

In rapamycin PKSs, there is a loading domain (LD) before module 1. In LD, there are three domains, i.e., coenzyme A ligase (CL), enoylreductase (ER), and acyl carrier protein (ACP) domain, which are considered to play a role in activating, reducing a free shikimic-acid-derived moiety starter unit, and finally passing it to the ketosynthase (KS) domain of the first module, respectively. The extender units required for producing rapamycin are malonyl-CoA and methylmalonyl-CoA. The mechanism of transferring from the last domain in rapamycin PKSs and cyclisation of polyketide molecule is assisted by pipecolate-incorporating enzyme (PIE), as depicted in Figure 4. This enzyme (170 kDa) is encoded by gene *rapP*, which is also located in the *rapPKS* gene cluster [26,27,34].

Table 1. Some polyketides produced by *Streptomyces* and their type I PKSs.

Polyketide	Structure	Producer	Type I PKSs	Ref.
Avermectin	16-membered ring macrolide	*Streptomyces avermitilis*	AVES1-4	[41]
Chalcomycin	16-membered ring macrolide	*Streptomyces bikiniensis*	ChmGI-V	[42]
Candicidin	38-membered ring polyene macrolide	*Streptomyces griseus* IMRU 3570	CanP1-3, and CanPF	[43,44]
FK506 (Tacrolimus)	23-membered ring macrolide	*Streptomyces tsukubaensis, Streptomyces* sp. *MA6858*	FkbABC	[35,45]
FK520 (Ascomycin)	23-membered ring macrolide	*Streptomyces hygroscopicus* var. *ascomyceticus*	FkbABC	[46]
Methymycin, Neomethymycin, Narbomycin, Pikromycin	12-membered ring macrolide, 12-membered ring macrolide, 14-membered ring macrolide, 14- membered ring macrolide	*Streptomyces venezuelae*	PikAI-IV	[37]
Pimaricin	26-membered ring polyene macrolide	*Streptomyces natalensis*	PIMS0 and PIMS1	[47]
Rapamycin	31- membered ring macrolide	*Streptomyces hygroscopicus*	RAPS1-3	[34]
Spiramycin	16- membered ring macrolide	*Streptomyces ambofaciens*	SrmGI-V	[36]
Tautomycetin	Linear	*Streptomyces* sp. *CK4412*	TmcA and TmcB	[32]
Tylosin	16- membered ring macrolide	*Streptomyces fradiae*	TYLGI-V	[48]

2.2. Biosynthesis of Avermectin

Avermectin is a 16-membered ring macrolide and one of the notable anthelmintic compounds produced by *Streptomyces avermitilis* [41,49]. The biosynthesis of avermectin involves type I PKSs (AVES1, AVES2, AVES3, and AVES4). AVES1 (414 kDa) contains one loading domain and two modules; AVES2 (666 kDa) consists of four modules; AVES 3 (575 kDa) comprises three modules; and AVES4 (510 kDa) has three modules. The process of avermectin biosynthesis includes assembling of the polyketide-derived initial aglycon (6, 8a-seco-6, 8a-deoxy-5-oxoavermectin aglycons) by AVES1–4, alteration of the initial aglycon to avermectin aglycons, and, as the last step, the glycosylation of avermectin aglycons to produce avermectins. The starter unit for avermectin biosynthesis is isobutyryl-CoA (derived from valine) or 2-methylbutyryl-CoA (derived from isoleucine), whereas the extender units involved in the production of avermectin are seven malonyl-CoAs (for acetate units) and five methylmalonyl-CoAs (for propionate units). The nucleotide sequence of the avermectin biosynthetic gene cluster comprises 18 ORFs spanning a distance of 82 kb, in which four large ORFs encode the avermectin polyketide synthase (AVES1, AVES2, AVES3, and AVES4) and some of the 14 ORFs encode polypeptides having important roles in avermectin biosynthesis [41].

2.3. Biosynthesis of Candicidin

Candicidin is a 38-membered ring polyene macrolide produced by *Streptomyces griseus* IMRU 3570 that has antifungal activity. Like the other polyene compounds, the antifungal mechanism of candicidin is also by disrupting the fungal cell membrane. Candicidin has both the amino sugar mycosamine and the aromatic component p-aminoacetophenone in its macrolide structure [43,50].

The candicidin biosynthetic gene cluster (<205 kb) was cloned and partially sequenced. Four genes, canP1, canP3, canP2 (incomplete), and canPF (incomplete), were determined as genes encoding parts of type I PKSs (CanP1, CanP2, CanP3, and CanPF). CanP1 contains one loading domain and one module; CanP2 consists of three modules; and CanP2 comprises six modules. CanPF hypothetically serves as one end of the PKS gene cluster. The starter unit is PABA (p-aminobenzoic acid), and the extender units are four methylmalonyl-CoAs and 17 malonyl-CoAs. At the end of the process in PKS, the molecule is released by thioesterase (CanT). In the next step, the compound is cyclized to become candicidin aglycone, oxidized by P450 monooxygenase (CanC) with aid from ferredoxin (CanF). The last step is glycosylation by adding mycosamine to the structure [43,44].

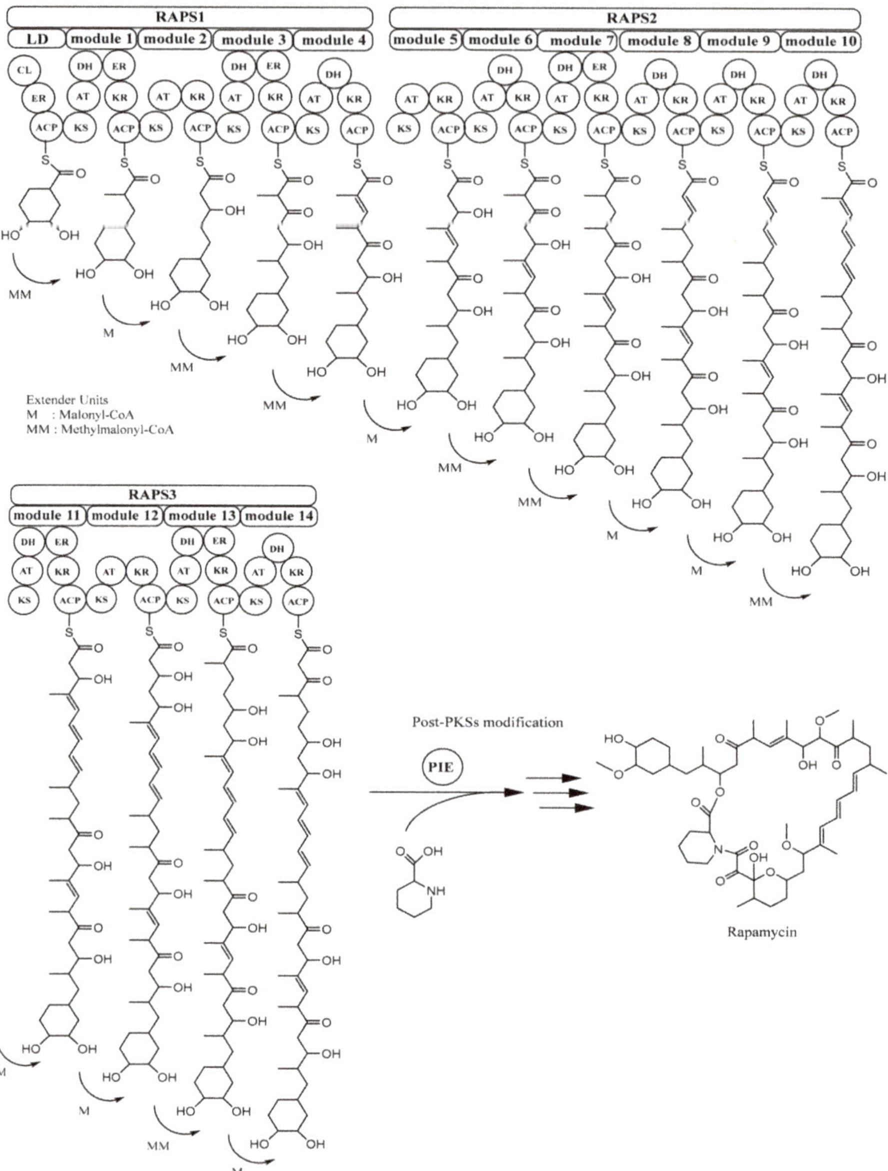

Figure 4. Biosynthesis of rapamycin. ACP, acyl carrier protein; AT, acyltransferase; KS, ketosynthase; KR, ketoreductase; DH, dehydratase; ER, enoylreductase; PIE, pipecolate-incorporating enzyme. Adapted with permission from Schwecke, T.; Aparicio, J.F.; Molnár, I.; König, A; Khaw, L.E.; Haydock, S.F.; Oliynyk, M.; Caffrey, P.; Cortés, J.; Lester, J.B. The biosynthetic gene cluster for the polyketide immunosuppressant rapamycin. Proc. Natl. Acad. Sci. USA 1995, 92, 7839–7843, doi:10.1073/pnas.92.17.7839 [34]. Copyright (1995) National Academy of Sciences, U.S.A. Adapted with permission of The Royal Society of Chemistry 2001, from Staunton, J.; Weissman, K.J. Polyketide biosynthesis: A millennium review. *Nat. Prod. Rep.* **2001**, *18*, 380–416 [26]; permission conveyed through Copyright Clearance Center, Inc.

2.4. Biosynthesis of Tautomycetin

Tautomycetin, firstly isolated from *Streptomyces griseochromogenes* and then from *Streptomyces* sp. CK4412, is an antifungal compound and an activated T cell-specific immunosuppressive compound. The inhibition of T-cells' proliferation is by the apoptosis mechanism. Unlike the other type I polyketide-derived compounds, tautomycetin has a linear structure [32,51].

The tautomycetin (TMC) biosynthetic gene cluster (~70 kb) has two ORFs that encode type I PKSs (Tmc A and TmcB). TmcA has six modules including the loading module, and TmcB has four modules, the TE (thioesterase) domain of which is located in TmcB. TE domain is responsible for releasing the intermediate chain of the compound from the PKS. The biosynthesis of TMC requires malonyl-CoA as a starter unit and the extender units such as 4 malonyl-CoAs, 4 methylmalonyl-CoAs, and 1 ethylmalonyl-CoA. After being released from PKS, the intermediate compound is modified by post-PKS mechanisms such as hydroxylation, decarboxylation, dehydration, and esterification with the cyclic C8 dialkylmaleic anhydride moiety [32].

3. Polyketide Synthases Type II

The type II polyketide synthases (type II PKSs) are responsible for producing aromatic polyketide. Based on the polyphenolic ring system and their biosynthetic pathways, the aromatic polyketides produced by type II PKSs generally are classified into seven groups, i.e., anthracyclines, angucyclines, aureolic acids, tetracyclines, tetracenomycins, pradimicin-type polyphenols, and benzoisochromanequinones [52].

Anthracyclines consists of a linear tetracyclic ring system with quinone–hydroquinone groups in rings B and C. Angucyclines have an angular tetracyclic ring system. The aureolic acids have a tricyclic chromophore. Tetracyclines contain a linear tetracyclic ring system without quinone–hydroquinone groups in rings B and C. Tetracenomycins have a linear tetracyclic ring system with the quinone group in ring B. Pradimicin-type polyphenols are considered as extended angucyclines. Benzoisochromanequinones contain a quinone derivative from the isochroman structure [52]. Some examples of aromatic polyketide produced by *Streptomyces* are actinorhodin (benzoisochromanequinones), doxorubicin (anthracyclines), jadomycin B (angucyclines), oxytetracycline (tetracyclines), mithramycin (aureolic acids), tetracenomycin C (tetracenomycins), and benastatin A (pradimicin-type polyphenols) (Figure 5) [28,52–57].

Unlike type I PKSs that involve huge multifunctional proteins that have many modules containing domains and perform the enzymatic reaction in a non-iterative way, the type II PKSs have monofunctional polypeptides and work iteratively to produce aromatic polyketide. However, like the type I PKS, the type II PKSs also comprise the acyl carrier protein (ACP) that functions as an anchor for the nascent polyketide chain. In addition to possessing ACP, the type II PKSs also consists of two ketosynthases units (KS_α and KS_β) that work cooperatively to produce the poly-β-keto chain. The KS_α unit catalyzes the condensation of the precursors; on the other hand, the role of KS_β in the type II PKSs is as a chain length-determining factor. The three major systems (ACP, KS_α, and KS_β) are called "minimal PKS" that work iteratively to produce aromatic polyketide. The other additional enzymes such as ketoreductases, cyclases, and aromatases cooperate to transform the poly-β-keto chain into the aromatic compound core. Furthermore, the post-tailoring process is conducted by oxygenases and glycosyl and methyl transferases [52,58–60]. The list of some aromatic polyketides produced by *Streptomyces* with their type II PKSs can be seen in Table 2.

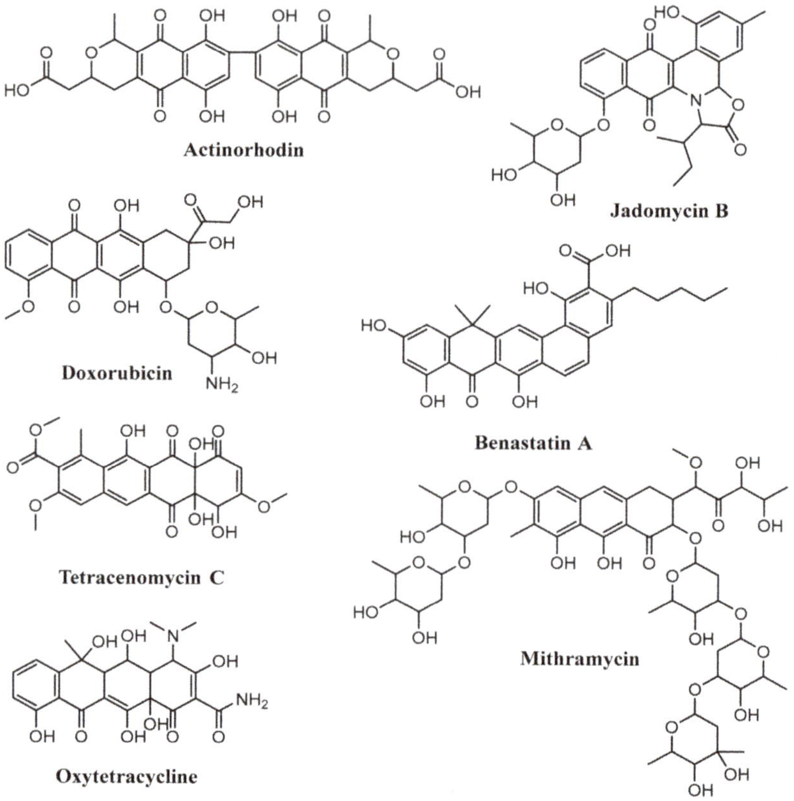

Figure 5. Some aromatic polyketides produced by *Streptomyces*.

3.1. Biosynthesis of Doxorubicin

Doxorubicin was isolated from *Streptomyces peucetius* in the early of 1960s. It belongs to anthracyclines that have a tetracyclic ring containing quinone and a hydroquinone group in their structure. Doxorubicin is one of the important drugs for the treatment of cancer such as breast cancer, childhood solid tumors, soft tissue sarcomas, and aggressive lymphomas. There are some proposed mechanisms for how doxorubicin kills the cancer cells: (i) intercalation of DNA and interference of topoisomerase-II-mediated DNA repair; and (ii) formation of free radicals and their deterioration of cell components such as cellular membranes, DNA, and proteins [61–63].

Daunorubicin (DNR)-doxorubicin (DXR) type II PKSs, encoded by *dps* genes in *Streptomyces peucetius*, are involved in the formation of doxorubicin. The biosynthesis of doxorubicin requires one propionyl-CoA as the starter unit and nine malonyl-CoAs as the extender units. The process involves two "minimal PKSs" (DpsC-DpsD-DpsG and DpsA-DpsB-DpsG) to produce a 21-carbon decaketide as an intermediate compound. The repetitive process is conducted by KS_α (DpsA), KS_β (DpsB), and ACP (DpsG). The next process employs several enzymes such as ketoreductase (DpsE), cyclases (DpsF, DpsY, and DnrD), oxygenase (DnrG and DnrF), and methyl transferase (DnrC) to produce ε-rhodomycinone, an important intermediate of doxorubicin biosynthesis. The remaining steps to synthesize doxorubicin utilize glycosyltransferase (DnrS) with the thymidine-diphospho (TDP) derivative of L-daunosamine, methyl esterase (DnrP), oxygenase (DoxA), and methyl transferase (DnrK) (Figure 6) [60,64–68].

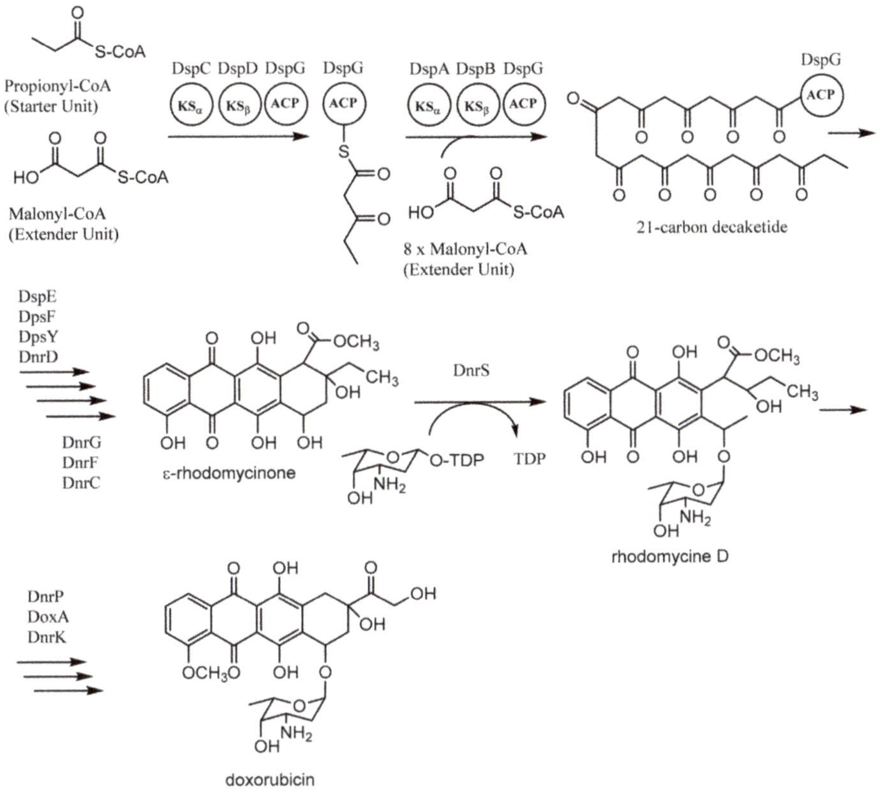

Figure 6. Biosynthesis of doxorubicin. Adapted with permission of The Royal Society of Chemistry 2009, from Chan, Y.A.; Podevels, A.M.; Kevany, B.M.; Thomas, M.G. Biosynthesis of polyketide synthase extender units. *Nat. Prod. Rep.* **2009**, *26*, 90–114 [60]; permission conveyed through Copyright Clearance Center, Inc.

Table 2. Some polyketides produced by *Streptomyces* and their minimal type II PKSs.

Polyketide	Intermediate Backbone Structure	Producer	Minimal Type II PKSs	Ref.
Medermycin	octaketide	*Streptomyces* sp. *K73*	Med-1,2,23	[69,70]
Doxorubicin	decaketide	*Streptomyces peucetius*	DpsABCDG	[60,65]
Oxytetracycline	decaketide	*Streptomyces rimosus*	OxyABCD	[71]
Gilvocarcin	decaketide	*Streptomyces griseoflavus* Gö 3592	GilABC	[72]
Oviedomycin	decaketide	*Streptomyces antibioticus*	OvmPKS	[73]
Chartreusin	decaketide	*Streptomyces chartreusis*	ChaABC	[74]
Cervimycin	decaketide	*Streptomyces tendae* HKI-179	CerABC	[75,76]
Resistomycin	decaketide	*Streptomyces resistomycificus*	RemABC	[77]
Chromomycin	decaketide	*Streptomyces griseus* subsp. *griseus*	CmmPKS	[78,79]
Hedamycin	dodecaketide	*Streptomyces griseoruber*	HedCDE	[80,81]
Fredericamycin	pentadecaketide	*Streptomyces griseus* ATCC 49344	FdmFSGH	[82,83]

3.2. Biosynthesis of Medermycin

Medermycin is a benzoisochromanequinone (BIQ) antibiotic, isolated from *Streptomyces* sp. K73. It has high activity against some Gram-positive bacteria such as *Staphylococcus aureus*, *Staphylococcus epidermidis*, *Sarcina lutea*, *Bacillus subtilis*, and *Bacillus cereus*. Besides antibiotic activity, medermycin also has potent activity as a platelet aggregation inhibitor. Because of its unique ability to give different colors in acidic and alkaline aqueous solution, medermycin is considered as an indicator type antibiotic [69,70,84].

Biosynthesis of medermycin requires eight malonyl-CoAs and a sugar molecule, angolosamine, which is derived from the deoxyhexose (DOH) pathway. In the first step of biosynthesis, the minimal PKS that consists of ACP (encoded by the *med*-ORF23 gene), KS_α (encoded by *med*-ORF1), and KS_β (encoded by *med*-ORF2) forms an octaketide moiety. The next process employs several enzymes such as keto reductase, aromatase, cyclase, enoyl reductase, and oxygenase/hydroxylase to produce the aglycone compound dihydrokalafungin. The aglycone structure then is combined by C-glycosyl transferase with an angolosamine structure to yield the final structure medermycin [69].

3.3. Biosynthesis of Hedamycin

Hedamycin is a pluramycin antitumor antibiotic, produced by *Streptomyces griseoruber*. This aromatic polyketide has a planar anthrapyrantrione chromophore, two amino sugars in its structure (α-L-*N*,*N*-dimethylvancosamine and β-D-angolosamine), and a bisepoxide-containing a side chain. The compound could inhibit 50% of human cancer cell growth at a subnanomolar concentration in three days. It is a monofunctional DNA alkylating agent, and because of its low therapeutic index, hedamycin is not clinically used [81,85].

Biosynthesis of hedamycin uses twelve malonyl-CoAs and two amino sugars, vancosamine and an angolosamine moiety. The minimal type II PKSs of hedamycin biosynthesis consist of HedC (KS_α), HedD (CLF), and HedE (ACP). Uniquely, the initial process involves type I PKSs (HedT and HedU proteins) that produce the 2,4-hexadienyl primer unit from three malonyl-CoAs, and then, it is transferred to the minimal type II PKSs of hedamycin biosynthesis. After that, a dodecaketide structure is formed by processing nine malonyl-CoAs. The structure then is modified with keto reductase, aromatase/cyclase, and oxygenase into the aglycone compound. In the last step, two glycosyltransferases are used for incorporating two amino sugars to produce hedamycin [80,81].

3.4. Biosynthesis of Fredericamycin

Fredericamycin, isolated from *Streptomyces griseus* ATCC 49344, is an aromatic polyketide that contains a spirocyclic structure. It has moderate antitumor and cytotoxic activity in various cell lines. These bioactivities are suggested because of the blockage of topoisomerases I and II or the peptidyl-prolyl cis-trans isomerase Pin1 [83].

The biosynthesis of fredericamycin employs the minimal type II PKSs that contains KS_α (FdmF and FdmS), KS_β (FdmG), and ACP (FdmH). There are two alternative mechanisms for chain initiation in the biosynthesis of fredericamycin. The first one requires acetyl-CoA and two malonyl-CoAs to produce the hexadienyl-priming unit. The second mechanism is by utilizing butyryl- or crotonyl-CoA and one malonyl-CoA to yield the hexadienyl-priming unit. The next step is carried out by processing twelve malonyl-CoAs as extender units to give the pentadecaketide intermediate, and then, the cyclases and oxygenases modify the intermediate compound into the final product [82,83,86].

4. Polyketide Synthases Type III

Unlike the type I and type II PKSs, the type III PKSs do not utilize ACP as an anchor for the production of polyketide metabolite. In this case, acyl-CoAs are used directly as substrates for generating polyketide compounds. In order to create polyketides, this system contains enzymes that construct homodimers and catalyzes many reactions such as priming, extension, and cyclization in an iterative way. With this fact, the type III PKSs are the simplest structures among the other types of PKSs. The type III PKSs found in bacteria were reported the first time in1999, and before that time, the type III PKSs were known only to be detected in plants [87–89].

Some studies previously revealed that type III PKSs could also be identified in the *Streptomyces* such as RppA, found in *Streptomyces griseus*, which is responsible in the synthesis of

1,3,6,8-tetrahydroxynaphthalene (THN) [90]. Gcs, identified in *Streptomyces coelicolor* A3(2), is reported to have an important role in the biosynthesis of germicidin [91]. SrsA, encoded by the *srsA* gene and isolated from *Streptomyces griseus*, is known to have an important role in the biosynthesis of phenolic lipids, i.e., alkylresorcinols and alkylpyrones [29].

The type III PKS Ken2, isolated from *Streptomyces violaceoruber*, was suggested to be involved in the production of 3,5-dihydroxyphenylglycine (3,5-DHPG). This compound is a nonproteinogenic amino acid needed for the formation of kendomycin and several other glycopeptide antibiotics such as balhimycin, chloroeremomycin, and also vancomycin [92]. Cpz6, encoded by the *cpz6* gene and isolated from *Streptomyces* sp. MK730–62F2, was reported to be engaged in the biosynthesis of caprazamycins by producing a group of new triketidepyrenes (presulficidins) [93]. Moreover, another finding also suggested that DpyA catalyzes the formation of alkyldihydropyrones in *Streptomyces reveromyceticus* (Figure 7) [94].

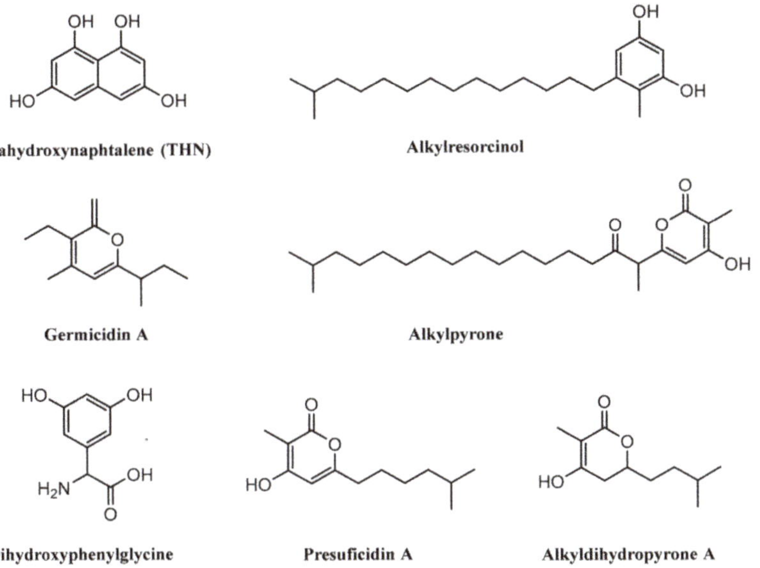

Tetrahydroxynaphtalene (THN)

Alkylresorcinol

Germicidin A

Alkylpyrone

Dihydroxyphenylglycine

Presuficidin A

Alkyldihydropyrone A

Figure 7. Some compounds produced by type III PKSs.

4.1. Biosynthesis of Germicidin

Germicidin, a pyrone-derived polyketide, is produced by a type III PKS germicidin synthase (Gcs) and is known to inhibit spore germination. Germicidin A, produced by *Streptomyces viridochromogenes* and *Streptomyces coelicolor*, prevents the spore germination reversibly at a very low concentration (40 pg/mL). The mechanism of inhibition is suggested by affecting the sporal respiratory chain and blocking Ca^{2+}-activated ATPase, thus resulting in inadequate energy for spore germination. Furthermore, germicidin A also has antibacterial properties against various Gram-positive bacteria [95,96].

Although many bacterial type III PKSs use only malonyl-CoA as both starter and extender units, the type III PKS Gcs, which is responsible for germicidin biosynthesis, is suggested to have the ability to utilize either acyl-ACP or acyl-CoA such as medium-chain acyl-CoAs (C4–C8) as starter units and malonyl-CoA, methylmalonyl-CoA, and ethylmalonyl-CoA as extender units [97,98]. In the first step, the starter unit is transacylated onto the cysteine residue of Gcs, and then, Gcs catalyzes the condensation reaction between the starter unit and extender unit concomitantly with the decarboxylation process, resulting in β-ketoacyl-CoA. The process continues with β-ketoacyl-CoA, which transacylates back onto the cysteine residue of Gcs (repetitive process) and subsequently undergoes a condensation reaction with either methylmalonyl-CoA or ethylmalonyl-CoA simultaneously with decarboxylation to formulate β,δ-diketothioester of CoA or a triketide intermediate. In the end of the reaction, cyclization of the β,δ-diketothioester of CoA is carried out to produce various types of germicidins (Figure 8) [91].

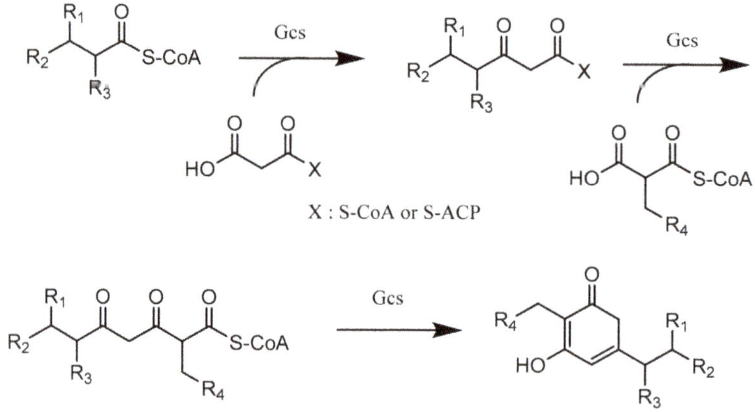

X : S-CoA or S-ACP

Germicidins

Germicidin A   : $R_1 = H$, $R_2 = R_3 = R_4 = CH_3$

Isogermicidin A : $R_1 = R_2 = R_4 = CH_3$, $R_3 = H$

Germicidin B   : $R_1 = R_2 = H$, $R_3 = R_4 = CH_3$

Isogermicidin B : $R_1 = R_3 = H$, $R_2 = R_4 = CH_3$

Germicidin C   : $R_1 = R_4 = H$, $R_2 = R_3 = CH_3$

Figure 8. Biosynthesis of germicidins. Gcs: germicidin synthase. Adapted with permission from Song, L.; Barona-Gomez, F.; Corre, C.; Xiang, L.; Udwary, D.W.; Austin, M.B.; Noel, J.P.; Moore, B.S.; Challis, G.L. Type III polyketide synthase β-ketoacyl-ACP starter unit and ethylmalonyl-coA extender unit selectivity discovered by Streptomyces coelicolor genome mining. *J. Am. Chem. Soc.* **2006**, *128*, 14754–14755 [91]. Copyright 2006 American Chemical Society.

4.2. Biosynthesis of Tetrahydroxynaphthalene

Tetrahydroxynaphthalene or THN is a small aromatic compound that is produced by utilizing type III PKSs (RppA). The biosynthesis process of THN requires five molecules of malonyl-CoA to form a pentaketide intermediate structure, and then, it is cyclized and aromatized to yield THN product. Spontaneous oxidation of THN may result flaviolin (red pigment) [90,99,100].

4.3. Biosynthesis of Dihydroxyphenylglycine

In order to synthesize 3,5-dihydroxyphenylglycine (3,5-DHPG), four malonyl-CoAs are needed, and the process is catalyzed by type III PKS (Ken2 or DpgA), which leads to the formation of the intermediate tetraketide compound. The tetraketide further is modified by hydratase/dehydratase, and oxidase/thioesterase to form 3,5-dihydroxyphenylacetic acid. The final step involves transaminase and tyrosine, as the amino group donor, to yield 3,5-DHPG, which is known as a nonproteinogenic amino acid [92,100,101].

4.4. Biosynthesis of Alkylresorcinol

The alkylresorcinol biosynthesis in *Streptomyces griseus* is catalyzed by SrsA. The reaction needs fatty acid (starter unit), one methylmalonyl-CoA, and two malonyl-CoAs (extender unit), and the intermediate structure is tetraketide. The tetraketide structure then transforms into the aromatic compound nonenzymatically (alkylresorcinol). This reaction may occur because of the nucleophilic attack on the thioester group by the methine carbon of the intermediate tetraketide compound [29].

5. Conclusions

Streptomyces has various systems in order to produce polyketides with different structures and functions. Knowing the polyketide structures, activities, producing enzymes, starter units, extender units, and the structural genes are very important in the development of new drugs. Some mechanisms of polyketide biosynthesis in *Streptomyces* that have been reported previously could provide strong basic knowledge not only for the biosynthesis investigation of the new polyketides, but also engineering the producing system in the future.

Acknowledgments: The authors gratefully acknowledge support from German Federal Ministry of Education and Research (BMBF) under the German-Indonesian anti-infective cooperation (GINAICO) project, a fellowship awarded by the German Academic Exchange Service (German: Deutscher Akademischer Austauschdienst or DAAD), and The President's Initiative and Networking Funds of the Helmholtz Association of German Research Centres (German: Helmholtz Gemeinschaft Deutscher Forschungszentren or HGF) under Contract Number VH-GS-202.

Conflicts of Interest: The authors declare no conflict of interest.

References

1. Moore, B.S.; Hopke, J.N. Discovery of a new bacterial polyketide biosynthetic pathway. *ChemBioChem* **2001**, *2*, 35–38. [CrossRef]
2. Rokem, J.S.; Lantz, A.E.; Nielsen, J. Systems biology of antibiotic production by microorganisms. *Nat. Prod. Rep.* **2007**, *24*, 1262. [CrossRef]
3. Katsuyama, Y.; Funa, N.; Miyahisa, I.; Horinouchi, S. Synthesis of unnatural flavonoids and stilbenes by exploiting the plant biosynthetic pathway in *Escherichia coli*. *Chem. Biol.* **2007**, *14*, 613–621. [CrossRef]
4. Chopra, I.; Roberts, M. Tetracycline antibiotics: Mode of action, applications, molecular biology, and epidemiology of bacterial resistance tetracycline antibiotics: Mode of action, applications, molecular biology, and epidemiology of bacterial resistance. *Microbiol. Mol. Biol. Rev.* **2001**, *65*, 232–260. [CrossRef] [PubMed]
5. Ghannoum, M.A.; Rice, L.B. Antifungal agents: Mode of action, mechanisms of resistance, and correlation of these mechanisms with bacterial resistance. *Clin. Microbiol. Rev.* **1999**, *12*, 501–517. [CrossRef]
6. Tacar, O.; Sriamornsak, P.; Dass, C.R. Doxorubicin: An update on anticancer molecular action, toxicity and novel drug delivery systems. *J. Pharm. Pharmacol.* **2013**, *65*, 157–170. [CrossRef]
7. Shushni, M.A.M.; Singh, R.; Mentel, R.; Lindequist, U. Balticolid: A new 12-membered macrolide with antiviral activity from an *Ascomycetous* fungus of marine origin. *Mar. Drugs* **2011**, *9*, 844–851. [CrossRef]
8. Li, J.; Kim, S.G.; Blenis, J. Rapamycin: One drug, many effects. *Cell Metab.* **2014**, *19*, 373–379. [CrossRef] [PubMed]
9. Van de Donk, N.W.C.J.; Kamphuis, M.M.J.; Lokhorst, H.M.; Bloem, A.C. The cholesterol lowering drug lovastatin induces cell death in myeloma plasma cells. *Leukemia* **2002**, *16*, 1362–1371. [CrossRef]
10. Hleba, L.; Charousova, I.; Cisarova, M.; Kovacik, A.; Kormanec, J.; Medo, J.; Bozik, M.; Javorekova, S. Rapid identification of *Streptomyces* tetracycline producers by MALDI-TOF mass spectrometry. *J. Environ. Sci. Heal. Part A* **2018**. [CrossRef] [PubMed]
11. Parthasarathy, R.; Sathiyabama, M. Lovastatin-producing endophytic fungus isolated from a medicinal plant *Solanum xanthocarpum*. *Nat. Prod. Res.* **2015**, *29*, 2282–2286. [CrossRef]
12. Huang, Q.; Lu, G.; Shen, H.-M.; Chung, M.C.M.; Ong, C.N. Anti-cancer properties of anthraquinones from rhubarb. *Med. Res. Rev.* **2006**, *27*, 609–630. [CrossRef]
13. Kohli, G.S.; John, U.; Figueroa, R.I.; Rhodes, L.L.; Harwood, D.T.; Groth, M.; Bolch, C.J.S.; Murray, S.A. Polyketide synthesis genes associated with toxin production in two species of *Gambierdiscus* (Dinophyceae). *BMC Genomics* **2015**, *16*, 410. [CrossRef]
14. Florian, P.; Monika, H. Polyketides in insects: Ecological role of these widespread chemicals and evolutionary aspects of their biogenesis. *Biol. Rev.* **2008**, *83*, 209–226.
15. Adele, C.; Guido, C.; Guido, V.; Angelo, F. Shaping the polypropionate biosynthesis in the solar-powered mollusc *Elysia viridis*. *ChemBioChem* **2008**, *10*, 315–322.
16. Müller, R.; Wink, J. Future potential for anti-infectives from bacteria—How to exploit biodiversity and genomic potential. *Int. J. Med. Microbiol.* **2014**, *304*, 3–13. [CrossRef]

17. Widyastuti, Y.; Lisdiyanti, P.; Ratnakomala, S.; Kartina, G.; Ridwan, R.; Rohmatussolihat, R.; Rosalinda Prayitno, N.; Triana, E.; Widhyastuti, N.; et al. Genus diversity of Actinomycetes in Cibinong Science Center, West Java, Indonesia. *Microbiol. Indones.* **2012**, *6*, 165–172. [CrossRef]

18. Lucas, X.; Senger, C.; Erxleben, A.; Grüning, B.A.; Döring, K.; Mosch, J.; Flemming, S.; Günther, S. StreptomeDB: A resource for natural compounds isolated from *Streptomyces* species. *Nucleic Acids Res.* **2013**, *41*, D1130–D1136. [CrossRef]

19. Igarashi, M.; Takahashi, Y.; Shitara, T.; Nakamura, H.; Naganawa, H.; Miyake, T.; Akamatsu, Y. Caprazamycins, novel lipo-nucleoside antibiotics, from *Streptomyces* sp. *J. Antibiot.* **2005**, *58*, 327–337. [CrossRef]

20. Dutta, S.; Basak, B.; Bhunia, B.; Chakraborty, S.; Dey, A. Kinetics of rapamycin production by *Streptomyces hygroscopicus* MTCC 4003. *3 Biotech.* **2014**, *4*, 523–531. [CrossRef]

21. Rodríguez, L.; Rodríguez, D.; Olano, C.; Braña, A.F.; Méndez, C.; Salas, J.A. Functional analysis of OleY L-oleandrosyl 3-O-methyltransferase of the oleandomycin biosynthetic pathway in *Streptomyces antibioticus*. *J. Bacteriol.* **2001**, *183*, 5358–5363. [CrossRef] [PubMed]

22. Elibol, M. Optimization of medium composition for actinorhodin production by *Streptomyces coelicolor* A3(2) with response surface methodology. *Process Biochem.* **2004**, *39*, 1057–1062. [CrossRef]

23. Pokhrel, A.R.; Chaudhary, A.K.; Nguyen, H.T.; Dhakal, D.; Le, T.T.; Shrestha, A.; Liou, K.; Sohng, J.K. Overexpression of a pathway specific negative regulator enhances production of daunorubicin in *bldA* deficient *Streptomyces peucetius* ATCC 27952. *Microbiol. Res.* **2016**, *192*, 96–102. [CrossRef]

24. Shelest, E.; Heimerl, N.; Fichtner, M.; Sasso, S. Multimodular type I polyketide synthases in algae evolve by module duplications and displacement of AT domains *in trans*. *BMC Genomics* **2015**, *16*, 1–15. [CrossRef]

25. Hopwood, D.A. Cracking the polyketide code. *PLoS Biol.* **2004**, *2*, 166–169. [CrossRef]

26. Staunton, J.; Weissman, K.J. Polyketide biosynthesis: A millennium review. *Nat. Prod. Rep.* **2001**, *18*, 380–416. [CrossRef]

27. Lal, R.; Kumari, R.; Kaur, H.; Khanna, R.; Dhingra, N.; Tuteja, D. Regulation and manipulation of the gene clusters encoding type I PKSs. *Trends Biotechnol.* **2000**, *18*, 264–274. [CrossRef]

28. Okamoto, S.; Taguchi, T.; Ochi, K.; Ichinose, K. Biosynthesis of actinorhodin and related antibiotics: Discovery of alternative routes for quinone formation encoded in the act gene cluster. *Chem. Biol.* **2009**, *16*, 226–236. [CrossRef] [PubMed]

29. Funabashi, M.; Funa, N.; Horinouchi, S. Phenolic lipids synthesized by type III polyketide synthase confer penicillin resistance on *Streptomyces griseus*. *J. Biol. Chem.* **2008**, *283*, 13983–13991. [CrossRef]

30. Vance, S.; Tkachenko, O.; Thomas, B.; Bassuni, M.; Hong, H.; Nietlispach, D.; Broadhurst, W. Sticky swinging arm dynamics: Studies of an acyl carrier protein domain from the mycolactone polyketide synthase. *Biochem. J.* **2016**, *473*, 1097–1110. [CrossRef] [PubMed]

31. Ruan, X.; Stassi, D.; Lax, S.A.; Katz, L. A second type I PKS gene cluster isolated from *Streptomyces hygroscopicus* ATCC 29253, a rapamycin-producing strain. *Gene* **1997**, *203*, 1–9. [CrossRef]

32. Choi, S.S.; Hur, Y.A.; Sherman, D.H.; Kim, E.S. Isolation of the biosynthetic gene cluster for tautomycetin, a linear polyketide T cell-specific immunomodulator from *Streptomyces* sp. CK4412. *Microbiology* **2007**, *153*, 1095–1102. [CrossRef] [PubMed]

33. Mazzei, T.; Mini, E.; Noveffi, A.; Periti, P. Chemistry and mode of action of macrolides. *J. Antimicrob. Chemother.* **1993**, *31*, 1–9. [CrossRef]

34. Schwecke, T.; Aparicio, J.F.; Molnár, I.; König, A.; Khaw, L.E.; Haydock, S.F.; Oliynyk, M.; Caffrey, P.; Cortés, J.; Lester, J.B. The biosynthetic gene cluster for the polyketide immunosuppressant rapamycin. *Proc. Natl. Acad. Sci. USA* **1995**, *92*, 7839–7843. [CrossRef] [PubMed]

35. Motamedi, H.; Shafiee, A. The biosynthetic gene cluster for the macrolactone ring of the immunosuppressant FK506. *Eur. J. Biochem.* **1998**, *256*, 528–534. [CrossRef]

36. Karray, F.; Darbon, E.; Oestreicher, N.; Dominguez, H.; Tuphile, K.; Gagnat, J.; Blondelet-Rouault, M.H.; Gerbaud, C.; Pernodet, J.L. Organization of the biosynthetic gene cluster for the macrolide antibiotic spiramycin in *Streptomyces ambofaciens*. *Microbiology* **2007**, *153*, 4111–4122. [CrossRef] [PubMed]

37. Xue, Y.; Zhao, L.; Liu, H.W.; Sherman, D.H. A gene cluster for macrolide antibiotic biosynthesis in *Streptomyces venezuelae*: Architecture of metabolic diversity. *Proc. Natl. Acad. Sci. USA* **1998**, *95*, 12111–12116. [CrossRef]

38. Cruz, M.C.; Cavallo, L.M.; Görlach, J.M.; Cox, G.; Perfect, J.R.; Cardenas, M.E.; Heitman, J. Rapamycin antifungal action is mediated via conserved complexes with FKBP12 and TOR kinase homologs in *Cryptococcus neoformans*. *Mol. Cell Biol.* **1999**, *19*, 4101–4112. [CrossRef]

39. Bastidas, R.J.; Shertz, C.A.; Lee, S.C.; Heitman, J.; Cardenas, M.E. Rapamycin exerts antifungal *activity in vitro* and *in vivo* against *Mucor circinelloides* via FKBP12-dependent inhibition of tor. *Eukaryot. Cell* **2012**, *11*, 270–281. [CrossRef]

40. Kwan, D.H.; Schulz, F. The stereochemistry of complex polyketide biosynthesis by modular polyketide synthases. *Molecules* **2011**, *16*, 6092–6115. [CrossRef]

41. Ikeda, H.; Nonomiya, T.; Usami, M.; Ohta, T.; Omura, S. Organization of the biosynthetic gene cluster for the polyketide anthelmintic macrolide avermectin in *Streptomyces avermitilis*. *Proc. Natl. Acad. Sci. USA* **1999**, *96*, 9509–9514. [CrossRef] [PubMed]

42. Ward, S.L.; Hu, Z.; Schirmer, A.; Reid, R.; Revill, W.P.; Reeves, C.D.; Petrakovsky, O.V.; Dong, S.D.; Katz, L. Chalcomycin biosynthesis gene cluster from *Streptomyces bikiniensis*: Novel features of an unusual ketolide produced through expression of the *chm* polyketide synthase in *Streptomyces fradiae*. *Antimicrob. Agents Chemother.* **2004**, *48*, 4703–4712. [CrossRef]

43. Gil, J.A.; Campelo-Diez, A.B. Candicidin biosynthesis in *Streptomyces griseus*. *Appl. Microbiol. Biotechnol.* **2003**, *60*, 633–642. [CrossRef]

44. Campelo, A.B.; Gil, J.A. The candicidin gene cluster from *Streptomyces griseus* IMRU 3570. *Microbiology* **2002**, *148*, 51–59. [CrossRef]

45. Kino, T.; Hatanaka, H.; Hashimoto, M.; Nishiyama, M.; Goto, T.; Okuhara, M.; Kohsaka, M.; Aoki, H.; Imanaka, H. FK-506, a novel immunosuppressant isolated from a *Streptomyces*. I. Fermentation, isolation, and physico-chemical and biological characteristics. *J. Antibiot. (Tokyo)* **1987**, *40*, 1249–1255. [CrossRef]

46. Wu, K.; Chung, L.; Revill, W.P.; Katz, L.; Reeves, C.D. The FK520 gene cluster of *Streptomyces hygroscopicus* var. *ascomyceticus* (ATCC 14891) contains genes for biosynthesis of unusual polyketide extender units. *Gene* **2000**, *251*, 81–90. [CrossRef]

47. Aparicio, J.; Colina, A.; Ceballos, E.; Martin, J. The biosynthetic gene cluster for the 26-membered ring polyene macrolide pimaricin. A new polyketide synthase organization encoded by two subclusters separated by functionalization genes. *J. Biol. Chem.* **1999**, *274*, 10133–10139. [CrossRef]

48. Fiers, W.D.; Dodge, G.J.; Li, Y.; Smith, J.L.; Fecik, R.A.; Aldrich, C.C. Tylosin polyketide synthase module 3: Stereospecificity, stereoselectivity and steady-state kinetic analysis of β-processing domains *via* diffusible, synthetic substrates. *Chem. Sci.* **2015**, *6*, 5027–5033. [CrossRef]

49. Zhang, X.; Chen, Z.; Li, M.; Wen, Y.; Song, Y.; Li, J. Construction of ivermectin producer by domain swaps of avermectin polyketide synthase in *Streptomyces avermitilis*. *Appl. Microbiol. Biotechnol.* **2006**, *72*, 986–994. [CrossRef]

50. Hammond, S.M.; Kliger, B.N. Mode of action of the polyene antibiotic candicidin: Binding factors in the wall of *Candida albicans*. *Antimicrob. Agents Chemother.* **1976**, *9*, 561–568. [CrossRef]

51. Cheng, X.-C.; Kihara, T.; Ying, X.; Uramoto, M.; Osada, H.; Kusakabe, H.; Wang, B.-N.; Kobayashi, Y.; Ko, K.; Yamaguchi, I.; Shen, Y.-C.; Isono, K. A new antibiotic, tautomycetin. *J. Antibiot.* **1989**, *42*, 141–144.

52. Hertweck, C.; Luzhetskyy, A.; Rebets, Y.; Bechthold, A. Type II polyketide synthases: Gaining a deeper insight into enzymatic teamwork. *Nat. Prod. Rep.* **2007**, *24*, 162–190. [CrossRef]

53. Malla, S.; Prasad Niraula, N.; Singh, B.; Liou, K.; Kyung Sohng, J. Limitations in doxorubicin production from *Streptomyces peucetius*. *Microbiol. Res.* **2010**, *165*, 427–435. [CrossRef]

54. Jakeman, D.L.; Bandi, S.; Graham, C.L.; Reid, T.R.; Wentzell, J.R.; Douglas, S.E. Antimicrobial activities of jadomycin B and structurally related analogues. *Antimicrob. Agents Chemother.* **2009**, *53*, 1245–1247. [CrossRef]

55. Lombó, F.; Menéndez, N.; Salas, J.A.; Méndez, C. The aureolic acid family of antitumor compounds: Structure, mode of action, biosynthesis, and novel derivatives. *Appl. Microbiol. Biotechnol.* **2006**, *73*, 1–14. [CrossRef]

56. Petkovic, H.; Cullum, J.; Hranueli, D.; Hunter, I.S.; Peric-Concha, N.; Pigac, J.; Thamchaipenet, A.; Vujaklija, D.; Long, P.F. Genetics of *Streptomyces rimosus*, the oxytetracycline producer. *Microbiol. Mol. Biol. Rev.* **2006**, *70*, 704–728. [CrossRef] [PubMed]

57. Shen, B.; Hutchinson, C.R. Triple hydroxylation of tetracenomycin A2 to tetracenomycin C in *Streptomyces glaucescens*. Overexpression of the *tcmG* gene in *Streptomyces lividans* and characterization of the tetracenomycin A2 oxygenase. *J. Biol. Chem.* **1994**, *1326*, 30726–30733.

58. Zhang, Z.; Pan, H.-X.; Tang, G.-L. New insights into bacterial type II polyketide biosynthesis. *F1000Research* **2017**, *6*, 172. [CrossRef] [PubMed]

59. Komaki, H.; Harayama, S. Sequence diversity of type II polyketide synthase genes in *Streptomyces*. *Actinomycetologica* **2006**, *20*, 42–48. [CrossRef]

60. Chan, Y.A.; Podevels, A.M.; Kevany, B.M.; Thomas, M.G. Biosynthesis of polyketide synthase extender units. *Nat. Prod. Rep.* **2009**, *26*, 90–114. [CrossRef] [PubMed]

61. Minotti, G. Anthracyclines: Molecular advances and pharmacologic developments in antitumor activity and cardiotoxicity. *Pharmacol. Rev.* **2004**, *56*, 185–229. [CrossRef]

62. Yang, F.; Teves, S.S.; Kemp, C.J.; Henikoff, S. Doxorubicin, DNA torsion, and chromatin dynamics. *Biochim. Biophys. Acta* **2014**, *1845*, 84–89. [CrossRef] [PubMed]

63. Thorn, C.F.; Oshiro, C.; Marsh, S.; Hernandez-Boussard, T., McLeod, H., Klein, T.E., Altman, R.B. Doxorubicin pathways: Pharmacodynamics and adverse effects. *Pharmacogenet. Genomics* **2011**, *21*, 440–446. [CrossRef] [PubMed]

64. Bao, W.; Sheldon, P.J.; Wendt-Pienkowski, E.; Richard Hutchinson, C. The *Streptomyces peucetius dpsC* gene determines the choice of starter unit in biosynthesis of the daunorubicin polyketide. *J. Bacteriol.* **1999**, *181*, 4690–4695.

65. Grimm, A.; Madduri, K.; Ali, A.; Hutchinson, C.R. Characterization of the *Streptomyces peucetius* ATCC 29050 genes encoding doxorubicin polyketide synthase. *Gene* **1994**, *151*, 1–10. [CrossRef]

66. Rajgarhia, V.B.; Strohl, W.R. Minimal *Streptomyces* sp. strain C5 daunorubicin polyketide biosynthesis genes required for aklanonic acid biosynthesis. *J. Bacteriol.* **1997**, *179*, 2690–2696. [CrossRef]

67. Hutchinson, C.R. Biosynthetic studies of daunorubicin and tetracenomycin C. *Chem. Rev.* **1997**, *97*, 2525–2536. [CrossRef]

68. Lomovskaya, N.; Doi-Katayama, Y.; Filippini, S.; Nastro, C.; Fonstein, L.; Gallo, M.; Colombo, A.L.; Hutchinson, C.R. The *Streptomyces peucetius dpsY* and *dnrX* genes govern early and late steps of daunorubicin and doxorubicin biosynthesis. *J. Bacteriol.* **1998**, *180*, 2379–2386.

69. Ichinose, K.; Ozawa, M.; Itou, K.; Kunieda, K.; Ebizuka, Y. Cloning, sequencing and heterologous expression of the medermycin biosynthetic gene cluster of *Streptomyces* sp. AM-7161: Towards comparative analysis of the benzoisochromanequinone gene clusters. *Microbiology* **2003**, *149*, 1633–1645. [CrossRef] [PubMed]

70. Takano, S.; Hasuda, K.; Ito, A.; Koide, Y.; Ishii, F. A new antibiotic, medermycin. *J. Antibiot. (Tokyo)* **1976**, *29*, 765–768. [CrossRef]

71. Pickens, L.B.; Tang, Y. Oxytetracycline biosynthesis. *J. Biol. Chem.* **2010**, *285*, 27509–27515. [CrossRef]

72. Fischer, C.; Lipata, F.; Rohr, J. The complete gene cluster of the antitumor agent gilvocarcin V and its implication for the biosynthesis of the gilvocarcins. *J. Am. Chem. Soc.* **2003**, *125*, 7818–7819. [CrossRef] [PubMed]

73. Lombó, F.; Abdelfattah, M.S.; Braça, A.F.; Salas, J.A. Elucidation of oxygenation steps during oviedomycin biosynthesis and generation of derivatives with increased antitumor activity. *ChemBioChem* **2009**, *10*, 296–303. [CrossRef]

74. Xu, Z.; Jakobi, K.; Welzel, K.; Hertweck, C. Biosynthesis of the antitumor agent chartreusin involves the oxidative rearrangement of an anthracyclic polyketide. *Chem. Biol.* **2005**, *12*, 579–588. [CrossRef] [PubMed]

75. Bretschneider, T.; Zocher, G.; Unger, M.; Scherlach, K.; Stehle, T.; Hertweck, C. A ketosynthase homolog uses malonyl units to form esters in cervimycin biosynthesis. *Nat. Chem. Biol.* **2011**, *8*, 154–161. [CrossRef] [PubMed]

76. Herold, K.; Xu, Z.; Gollmick, F.A.; Gräfe, U.; Hertweck, C. Biosynthesis of cervimycin C, an aromatic polyketide antibiotic bearing an unusual dimethylmalonyl moiety. *Organ. Biomol. Chem.* **2004**, *2*, 2411–2414. [CrossRef]

77. Jakobi, K.; Hertweck, C. A gene cluster encoding resistomycin biosynthesis in *Streptomyces resistomycificus*; exploring polyketide cyclization beyond linear and angucyclic patterns. *J. Am. Chem. Soc.* **2004**, *126*, 2298–2299. [CrossRef] [PubMed]

78. Menendez, N.; Nur-e-Alam, M.; Bran, A.F.; Rohr, J.; Salas, J.A.; Mendez, C. Biosynthesis of the antitumor chromomycin A3 in *Streptomyces griseus*: Analysis of the gene cluster and rational design of novel chromomycin analogs. *Chem. Biol.* **2004**, *11*, 21–32.

79. Sun, L.; Zeng, J.; Cui, P.; Wang, W.; Yu, D.; Zhan, J. Manipulation of two regulatory genes for efficient production of chromomycins in *Streptomyces reseiscleroticus*. *J. Biol. Eng.* **2018**, *12*, 1–11. [CrossRef]

80. Das, A.; Khosla, C. *In vivo* and *in vitro* analysis of the hedamycin polyketide synthase. *Chem. Biol.* **2009**, *16*, 1197–1207. [CrossRef]

81. Bililign, T.; Hyun, C.-G.; Williams, J.S.; Czisny, A.M.; Thorson, J.S. The hedamycin locus implicates a novel aromatic PKS priming mechanism. *Chem. Biol.* **2004**, *11*, 959–969. [CrossRef]

82. Das, A.; Szu, P.; Fitzgerald, J.T.; Khosla, C. Mechanism and engineering of polyketide chain initiation in fredericamycin biosynthesis. *J. Am. Chem. Soc.* **2010**, *132*, 8831–8833. [CrossRef]

83. Chen, Y.; Wendt-Pienkowski, E.; Shen, B. Identification and utility of FdmR1 as a *Streptomyces* antibiotic regulatory protein activator for fredericamycin production in *Streptomyces griseus* ATCC 49344 and heterologous hosts. *J. Bacteriol.* **2008**, *190*, 5587–5596. [CrossRef] [PubMed]

84. Nakagawa, A.; Fukamachi, N.; Yamaki, K.; Hayashi, M.; Oh-ishi, S.; Kobayashi, B.; Omura, S. Inhibition of platelet aggregation by medermycin and it's related isochromanequinone antibiotics. *J. Antibiot.* **1987**, *40*, 1075–1076. [CrossRef]

85. Tu, L.C.; Melendy, T.; Beerman, T.A. DNA damage responses triggered by a highly cytotoxic monofunctional DNA alkylator, hedamycin, a pluramycin antitumor antibiotic. *Mol. Cancer Ther.* **2004**, *3*, 577–585. [PubMed]

86. Szu, P.-H.; Govindarajan, S.; Meehan, M.J.; Das, A.; Nguyen, D.D.; Dorrestein, P.C.; Minshull, J.; Khosla, C. Analysis of the ketosynthase-chain length factor heterodimer from the fredericamycin polyketide synthase. *Acta Neurobiol. Exp. (Wars)* **2011**, *18*, 1021–1031. [CrossRef]

87. Shen, B. Polyketide biosynthesis beyond the type I, II and III polyketide synthase paradigms. *Curr. Opin. Chem. Biol.* **2003**, *7*, 285–295. [CrossRef]

88. Yu, D.; Xu, F.; Zeng, J.; Zhan, J. Type III polyketide synthases in natural product biosynthesis. *IUBMB Life* **2012**, *64*, 285–295. [CrossRef] [PubMed]

89. Nakano, C.; Ozawa, H.; Akanuma, G.; Funa, N.; Horinouchi, S. Biosynthesis of aliphatic polyketides by type III polyketide synthase and methyltransferase in *Bacillus subtilis*. *J. Bacteriol.* **2009**, *191*, 4916–4923. [CrossRef] [PubMed]

90. Funa, N.; Funabashi, M.; Ohnishi, Y.; Horinouchi, S. Biosynthesis of hexahydroxyperylenequinone melanin via oxidative aryl coupling by cytochrome P-450 in *Streptomyces griseus*. *J. Bacteriol.* **2005**, *187*, 8149–8155. [CrossRef]

91. Song, L.; Barona-Gomez, F.; Corre, C.; Xiang, L.; Udwary, D.W.; Austin, M.B.; Noel, J.P.; Moore, B.S.; Challis, G.L. Type III polyketide synthase β-ketoacyl-ACP starter unit and ethylmalonyl-coA extender unit selectivity discovered by *Streptomyces coelicolor* genome mining. *J. Am. Chem. Soc.* **2006**, *128*, 14754–14755. [CrossRef]

92. Wenzel, S.C.; Bode, H.B.; Kochems, I.; Müller, R. A type I/type III polyketide synthase hybrid biosynthetic pathway for the structurally unique *ansa* compound kendomycin. *ChemBioChem* **2008**, *9*, 2711–2721. [CrossRef]

93. Tang, X.; Eitel, K.; Kaysser, L.; Kulik, A.; Grond, S.; Gust, B. A two-step sulfation in antibiotic biosynthesis requires a type III polyketide synthase. *Nat. Chem. Biol.* **2013**, *9*, 610–615. [CrossRef]

94. Aizawa, T.; Kim, S.Y.; Takahashi, S.; Koshita, M.; Tani, M.; Futamura, Y.; Osada, H.; Funa, N. Alkyldihydropyrones, new polyketides synthesized by a type III polyketide synthase from *Streptomyces reveromyceticus*. *J. Antibiot. (Tokyo)* **2014**, *67*, 819–823. [CrossRef] [PubMed]

95. Čihák, M.; Kameník, Z.; Šmídová, K.; Bergman, N.; Benada, O.; Kofroňová, O.; Petříčková, K.; Bobek, J. Secondary metabolites produced during the germination of *Streptomyces coelicolor*. *Front. Microbiol.* **2017**, *8*, 1–13. [CrossRef]

96. Petersen, F.; Zähner, H.; Metzger, J.W.; Freund, S.; Hummel, R.P. Germicidin, an autoregulative germination inhibitor of *Streptomyces viridochromogenes* NRRL B-1551. *J. Antibiot. (Tokyo)* **1993**, *46*, 1126–1138. [CrossRef] [PubMed]

97. Lim, Y.P.; Go, M.K.; Yew, W.S. Exploiting the biosynthetic potential of type III polyketide synthases. *Molecules* **2016**, *21*, 1–37. [CrossRef] [PubMed]

98. Chemler, J.A.; Buchholz, T.J.; Geders, T.W.; Akey, D.L.; Rath, C.M.; Chlipala, G.E.; Smith, J.L.; Sherman, D.H. Biochemical and structural characterization of germicidin synthase: Analysis of a type III polyketide synthase that employs acyl-ACP as a starter unit donor. *J. Am. Chem. Soc.* **2012**, *134*, 7359–7366. [CrossRef]

99. Izumikawa, M.; Shipley, P.R.; Hopke, J.N.; O'Hare, T.; Xiang, L.; Noel, J.P.; Moore, B.S. Expression and characterization of the type III polyketide synthase 1,3,6,8-tetrahydroxynaphthalene synthase from *Streptomyces coelicolor* A3(2). *J. Ind. Microbiol. Biotechnol.* **2003**, *30*, 510–515. [CrossRef]

100. Chen, H.; Tseng, C.C.; Hubbard, B.K.; Walsh, C.T. Glycopeptide antibiotic biosynthesis: Enzymatic assembly of the dedicated amino acid monomer (S)-3,5-dihydroxyphenylglycine. *Proc. Natl. Acad. Sci. USA* **2001**, *98*, 14901–14906. [CrossRef] [PubMed]
101. Pfeifer, V.; Nicholson, G.J.; Ries, J.; Recktenwald, J.; Schefer, A.B.; Shawky, R.M.; Schröder, J.; Wohlleben, W.; Pelzer, S. A polyketide synthase in glycopeptide biosynthesis: The biosynthesis of the non-proteinogenic amino acid (S)-3,5-dihydroxyphenylglycine. *J. Biol. Chem.* **2001**, *276*, 38370–38377. [CrossRef] [PubMed]

Review

Diversity of Myxobacteria—We Only See the Tip of the Iceberg

Kathrin I. Mohr

Microbial Drugs (MWIS), Helmholtz Centre for Infection Research (HZI), 38124 Braunschweig, Germany; kathrin.mohr@helmholtz-hzi.de; Tel.: +49-531-6181-4251

Received: 6 June 2018; Accepted: 8 August 2018; Published: 11 August 2018

Abstract: The discovery of new antibiotics is mandatory with regard to the increasing number of resistant pathogens. One approach is the search for new antibiotic producers in nature. Among actinomycetes, *Bacillus* species, and fungi, myxobacteria have been a rich source for bioactive secondary metabolites for decades. To date, about 600 substances could be described, many of them with antibacterial, antifungal, or cytostatic activity. But, recent cultivation-independent studies on marine, terrestrial, or uncommon habitats unequivocally demonstrate that the number of uncultured myxobacteria is much higher than would be expected from the number of cultivated strains. Although several highly promising myxobacterial taxa have been identified recently, this so-called Great Plate Count Anomaly must be overcome to get broader access to new secondary metabolite producers. In the last years it turned out that especially new species, genera, and families of myxobacteria are promising sources for new bioactive metabolites. Therefore, the cultivation of the hitherto uncultivable ones is our biggest challenge.

Keywords: myxobacteria; diversity; uncultured; secondary metabolites; new antibiotics

1. Introduction

We know little about the real diversity of myxobacteria in the environment. How many myxobacteria are cultivable under standard laboratory conditions? How many resist these cultivation efforts and lie undiscovered in the ground? After an introduction to myxobacteria, the status of antibiotics, myxobacterial secondary metabolites, the Great Plate Count Anomaly phenomenon, and microbial biogeography, the diversity of cultivable and uncultivated myxobacteria in different habitats is presented. Therefore, numerous sequences from the NCBI database were analysed. The intent of this review is to draw attention to the high amount of undiscovered myxobacteria and encourage further discovery and isolation of these hidden treasures with regard to their potential as new antibiotic producers.

2. Biology and Phylogeny of Myxobacteria

Myxobacteria are soil dwelling deltaproteobacteria and are distributed all over the world. Temperate zones, tropical rain forests, arctic tundra, deserts, acidic soils [1–3], marine and other saline environments [4–7], and even caves [8], for example, are appropriate habitats. Myxobacteria can be isolated from various natural sources as soil, bark, rotting wood, leaves of trees, compost [9], or dung of herbivores [1,10]. They live aerobically, except the only described facultative anaerobic genus and species, *Anaeromyxobacter dehalogenans* [11]. Nevertheless, it is highly likely that further facultative or even strictly anaerobic myxobacteria exist, which hitherto withstand the common isolation efforts. Currently the monophyletic order Myxococcales comprises 3 suborders, 10 families, 29 genera, and 58 species (Figure 1).

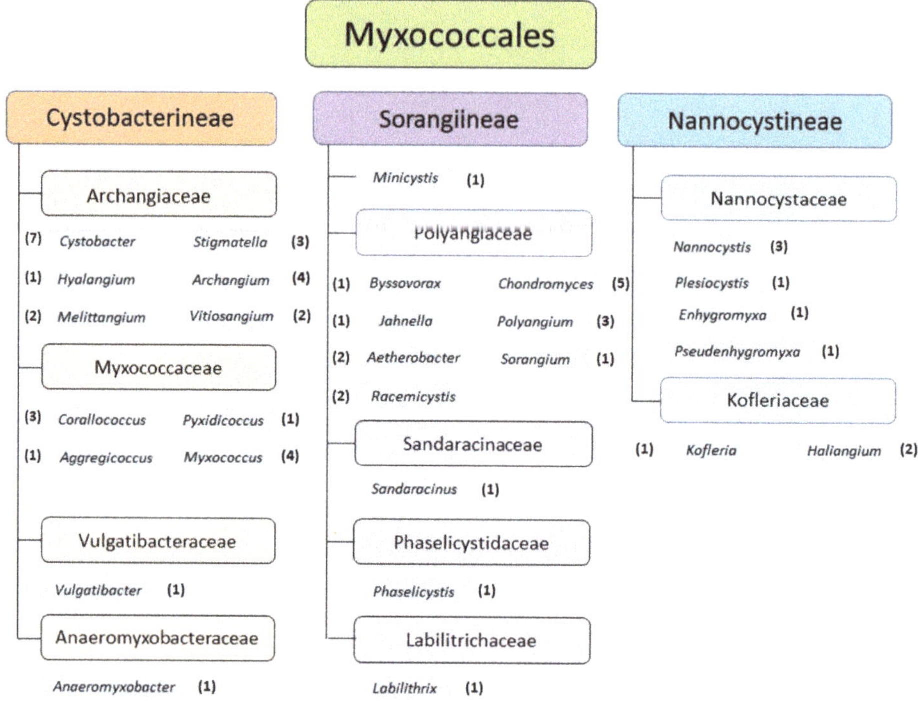

Figure 1. Monophyletic order Myxococcales (delta-proteobacteria), suborders, families, and genera of myxobacteria (status May 2018). The number of species within the genera is mentioned in brackets (original graphic from Corinna Wolf, modified by K. I. Mohr).

Myxobacteria are fascinating because of their extraordinary social lifestyle, which is unique in the bacterial domain. Under appropriate environmental conditions, vegetative cells move in swarms by gliding over solid surfaces [12]. Myxobacteria do not have flagella, but two motility systems, used for locomotion, and well studied in *Myxococcus xanthus*, are known: social (S-) motility, powered by retraction of type IV pili is responsible for the movement of cells which travel in groups [13]. In addition, extracellular matrix polysaccharide (EPS), also referred to as fibrils, are used. Therefore, the intrinsic polarity of rod-shaped cells lays the foundation, and each cell uses two polar engines for gliding on surfaces. It sprouts retractile type IV pili from the leading cell pole and secretes capsular polysaccharide through nozzles from the trailing pole [14]. On the other hand, slime secretion enables cell movements, when cells were isolated from the group (adventurous A-motility) [15]. For a detailed description of myxobacterial gliding mechanisms see Nan et al. [13] and Faure et al. [16].

Due to their nutritional behavior and based on their specialization in degradation of biomacromolecules, members of the order Myxococcales can be divided into two groups: predators (the majority), which are able to lyse whole living cells of other microorganisms by exhausting lytic enzymes, and cellulose-decomposers, the latter are represented by the genera *Sorangium* and *Byssovorax* [12]. But, as mentioned for the (facultative) anaerobic myxobacteria, it is also highly likely that further cellulose-degrading genera exist, which successfully resisted standard cultivation attempts.

If nutrients become rare, the cells undergo an impressive process of cooperative morphogenesis. Cells agglomerate and form species-specific fruiting bodies by directed cell movement [12]. These fruiting bodies consist of one to several sporangioles [1]. The architecture of these fruiting bodies ranges from simple, single sporangioles (*M. xanthus*, *Cystobacter* spp.), stalked sporangioles (*M. stipitatus*), or even delicate tree-like structures of high complexity (*Chondromyces* spp.; *Stigmatella* spp.) [10]. Colors of cells/fruiting bodies vary from milky, yellow, orange, red, brown to even black (Figure 2) [1].

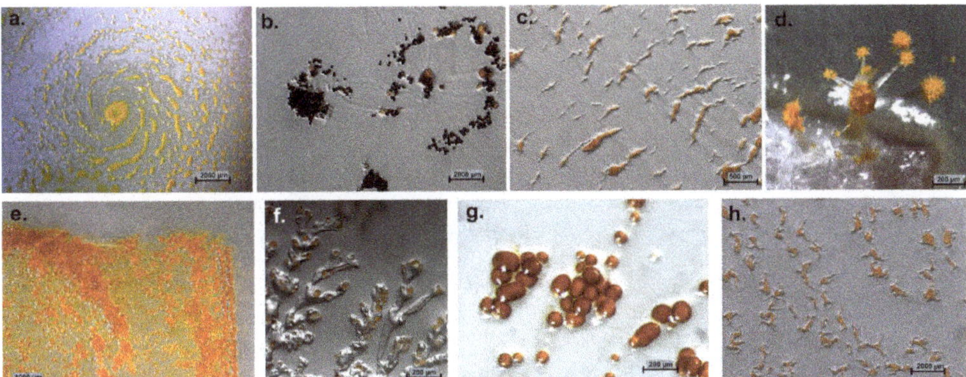

Figure 2. Variation of myxobacterial fruiting bodies. Genus/species, strain designation, (agar medium) are mentioned. (**a**) *Myxococcus xanthus* Mxx42 (P); (**b**) *Cystobacter ferrugineus* Cbfe48 (VY/2); (**c**) *Archangium* sp. Ar7747 (VY/2); (**d**) *Chondromyces* sp. (Stan 21 with filter); (**e**) *Sorangium* sp. Soce 1462 degrading filter paper on Stan 21 agar; (**f**) *Polyangium* sp. Pl3323 (VY/2); (**g**) *Cystobacter fuscus* Cbf18 (VY/2); (**h**) *Corallococcus coralloides* Ccc379 (VY/2).

A known function of these mainly carotenoid or melanoid pigments is to provide protection against photo-oxidation [17]. Within the fruiting bodies, most of the vegetative cells die and serve as food for the remaining cells, which convert into short and hardy myxospores, especially resistant to desiccation [18,19]. These spores are not as heat-resistant as *Bacillus* spores, but they can survive in the environment and are able to germinate under appropriate conditions even after decades of resting [1]. Therefore, it is possible to isolate myxobacteria from dried environmental samples, which were stored for several years at room temperature [10]. A fruiting body consists of 10^5–10^6 cells [18]. This ensures that the new cycle starts with a sufficient amount of cells, necessary for the typical collaborative feeding [20]. A further very interesting feature of myxobacteria is their ability to produce a large number and variety of secondary metabolites, as described in the next section.

In 1892, Thaxter was the first who described myxobacteria in literature [19]. He found out that *Chondromyces crocatus* was a bacterium and he had discovered its unicellular vegetative stage. This was spectacular, because until such time, *C. crocatus* had been considered a slime mold for more than 20 years [14]. Studies by Bauer [21], Kofler [20], Jahn [22,23], and Kühlwein [24] followed in the early 20th century. Myxobacteria have always fascinated scientists due to their social behavior, including cooperative swarming, group predation, and multicellular fruiting body formation. *Myxococcus xanthus* for example has become one of the model systems for the study of prokaryotic development [25]. Today, beside their capabilities to produce promising bioactive secondary metabolites, myxobacteria are of utmost importance in elucidating multicellular behavior in bacteria, as well as working out social evolution theory.

3. Current Status of Antibiotics and Myxobacterial Secondary Metabolites

Before the first antibiotics were commercially available in the early 20th century, people were delivered helplessly to various kinds of infections like pest, cholera, and tuberculosis, which often reached epidemic proportions and have cost the lives of millions of people [26]. In 1940, quinine was used against malaria, the arsenic derivative arsphenamine, Salvarsan, was used against syphilis, and sulfa drugs like Prontosil were used against mainly Gram-positive cocci infections. However, most agents of infectious diseases were still untreatable. The situation improved radically with the detection of the first beta-lactam antibiotic, penicillin, produced by the mold *Penicillium rubens* [27]. Henceforth, soil organisms like fungi [28] and bacteria [29] as producers of secondary metabolites with bioactive properties moved into the focus of research. The Golden Age of Antibiotics started. Aminoglycosides [30], tetracyclines [31], and macrolides [32] are only some examples of important antibiotic classes, discovered in those days. Numerous pharmaceutical companies participated on large-scale screening activities of antibiotic producing organisms, mainly actinobacteria [33]. However, in most cases, it took only a few years from the launch of a new antibiotic to the detection of the first resistant germs [34]. Incorrect use in human medicine, incorrectly prescribed antibiotics, extensive agricultural use and fast spread of resistant bacteria caused by increasing mobility led to substantial problems with multi-drug resistant bacteria. Some of the most problematic germs belong to the so-called ESKAPE-panel: *Enterococcus faecium*, *Staphylococcus aureus*, *Klebsiella pneumoniae*, *Acinetobacter baumannii*, *Pseudomonas aeruginosa*, and *Enterobacter* spp, are mainly responsible for nosocomial infections. Since the 1960s more and more companies retracted from the time- and cost-consuming screening procedures. Of the 18 largest pharmaceutical companies, 15 abandoned the antibiotic field [35]. Indeed, from the late 1960s through the early 1980s, the pharmaceutical industry introduced many new antibiotics to solve the resistance problem. After that the antibiotic pipeline began to dry up and fewer new drugs were brought to market [36]. This led to a dangerous bottleneck of currently available reserve-antibiotics and a widely held concern over the lack of innovation and productivity in the research and development of novel bioactive substances [37]. Eligible countermeasures include the development of synthetic and semi-synthetic drugs, evaluation of rediscovered drugs and the classical screen of natural secondary metabolite producers. Here, especially new genera and species are of great interest [38]. But for natural production of secondary metabolites in large-scale fermentation processes the corresponding producer strains have to be isolated from nature. Maintenance, cultivation, and upscale are challenging. Beside the appropriate expertise and equipment for fermentation and isolation of substances from the fermenter broth, for every producer strain the specific biotic and abiotic conditions need to be determined. Myxobacteria for instance are one of the most promising natural product producers, but demanding with regard to isolation and large-scale cultivation. Successful handling of these organisms places special challenges to microbiologists and biotechnologists in equal measure.

Myxobacteria are among the best natural product producers, together with actinomycetes [39], *Bacillus* species [40], and fungi [31]. Even shortly after their discovery, scientists described predatory and cellulolytic action of myxobacteria. Already in 1947, Singh complained that many antibiotics were isolated from various groups of microorganisms, except myxobacteria [41]. He observed that some species of the Myxococcaceae lyse living bacteria, including Gram-negatives such as *Pseudomonas fluorescens* and *Bacterium* (*Escherichia*) *coli*, and concluded that a detailed study of myxobacteria may be profitable in discovering new antibiotics. In 1955, Mathews and Dudani investigated the lysis of human pathogenic bacteria by myxobacteria [42] and in 1962, Noren and Raper described the antibiotic activity of myxobacteria in relation to their bacteriolytic capacity [43]. But, it took another 15 years until the first antifungal metabolite, ambruticin, was isolated from a *Sorangium* strain (Figure 3) [44].

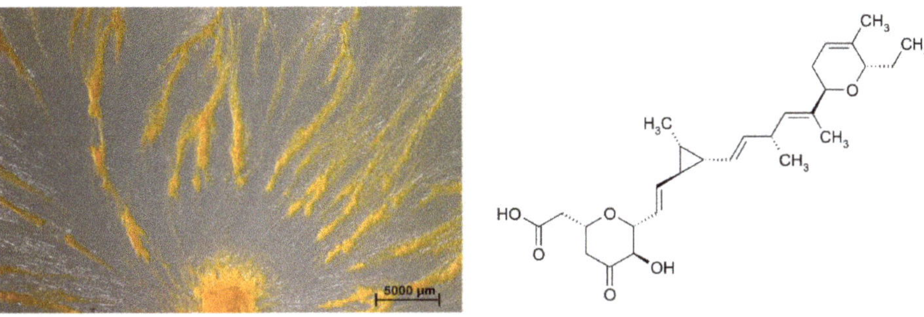

Figure 3. *Sorangium* sp. strain Soce 1014, an ambruticin-producer, swarming on VY/2-agar and the structure of ambruticin A, the first secondary metabolite which was isolated and described from myxobacteria.

The majority of myxobacterial compounds are polyketides, non-ribosomal polypeptides, and their hybrids, terpenoids, phenyl-propanoids, and alkaloids [45]. Many of these substances show promising activities against bacteria [46,47], viruses [48], fungi [49], cancer cells [50] immune cells [51], and malaria [52], respectively, as well as unusual modes of action [53]. Many strains produce metabolites belonging to multiple structural classes, as well as a number of chemical variants on each basic scaffold [48]. Whole-genome sequencing of several myxobacterial strains like *Sorangium cellulosum* [54] and *Myxococcus xanthus* [55] has revealed that the secondary metabolite potential is far greater than that suggested by fermentation under standard laboratory conditions.

It is, of course, possible to isolate new substances from known (myxobacterial) species [51,52]. But again: the low hanging fruit have long been harvested and it is more likely to find new substances in new families, genera and species [53,56–61]. The study of Hoffmann et al. confirmed this [41]. The authors found a correlation between taxonomic distance and the production of distinct secondary metabolite families, and supported the idea that the chances of discovering novel metabolites are greater by examining strains from new genera rather than additional representatives within the same genus. For comprehensive overviews about secondary metabolites produced by myxobacteria and their mode of action, I recommend Weissman and Müller [48] and Herrmann et al. [49].

4. The Great Plate Count Anomaly and Microbial Biogeography

Based on cultivation, approximately only 1% of the naturally occurring bacterial community is known and characterized so far [62]. Most bacterial groups remain uncultured and uncharacterized, because appropriate culture conditions are lacking [63]. This Great Plate Count Anomaly is the oldest unresolved microbiological challenge. The Austrian microbiologist Heinrich Winterberg was the first who described this phenomenon in 1898 [64]. Winterberg observed that the number of microbial cells in his samples did not match the number of colonies formed on nutrient media. Since Winterberg, numerous authors who investigated bacterial communities in different habitats confirmed this phenomenon. The establishment of culture independent analytical methods in the early 1990s greatly expanded the dimension of knowledge about the bacterial diversity again [65]. Estimations, that about 80% of bacteria resist standard laboratory cultivation approaches were obsolete after publication of the first culture-independent analyses of bacterial communities, which were based on 16S rRNA-coding genes. Now, the estimated amount of uncultivable species has increased to 90–99% and it can be assumed that many of these uncultured bacteria could be probably a source for new antibiotics [66].

Notwithstanding the frequent discovery and description of new species/genera, the real number of myxobacteria is unknown. The current knowledge about the diversity of organisms is always just a snapshot. However, several (NCBI) 16S rRNA-sequences of cultures belong to the order Myxococcales, but are only distantly related to valid type strains (up to 12% distance) and therefore probably belong to new species, genera, or even families. One example: "*Anaeromyxobacter dehalogenans*" strain WY75 (Acc. no. KC921178) was isolated from ginger foundation soil and shows highest similarity (87.4%) to the type strain of *Sandaracinus amylolyticus*. It is therefore at least a representative of a new myxobacterial family (Figure 4). Nevertheless, as long as a valid publication of such strains in taxonomic journals as for example *IJSEM* or *Antonie van Leeuwenhoek* is absent, even the current diversity of cultivable myxobacteria is not fully reflected.

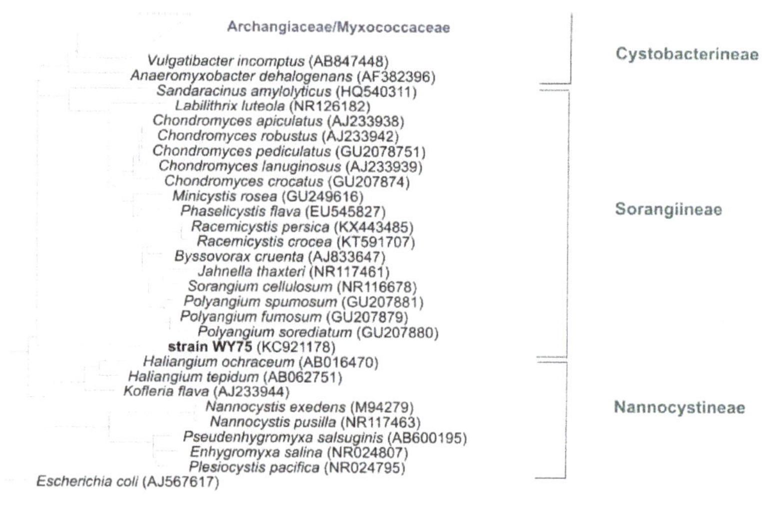

Figure 4. Neighbour joining tree with myxobacterial type strains shows the phylogenetic position of strain WY75, cultivated from ginger foundation soil, within the Sorangiineae suborder. Comparison of 16S rRNA sequences revealed only 87.4% similarity to the next myxobacterial type strain *S. amylolyticus*. Accession numbers are in brackets. Bar, 0.1 substitutions per nucleotide position.

Although there are numerous reports about cultivable myxobacteria in soils and other habitats [1], it has to be considered that myxospores may tolerate considerable environmental extremes. Most isolation techniques involve the cultivation of extensively dried samples [10]. Species which are present in the sample as vegetative cells will probably not survive this process and therefore will not grow on the isolation plates. Also, and irrespective of the detection method used, it is difficult to determine whether myxobacteria were present as dormant spores or metabolically active vegetative cells in the environmental sample taken [12]. The standard procedure to isolate myxobacteria is drying the sample (soil, plant material, etc.) at 30 °C to reduce growth of undesired bacteria and fungi, and subsequent placement on water agar with *E. coli*-bait (to attract predators) and on Stan 21 agar with filter paper (for cellulose decomposers), respectively (Figure 5). As the degradation of biomacromolecules like microbial cells (*E. coli* bait) or cellulose requires a sufficient amount of viable myxobacterial cells in the sample, underrepresented species will probably not be able to start growing.

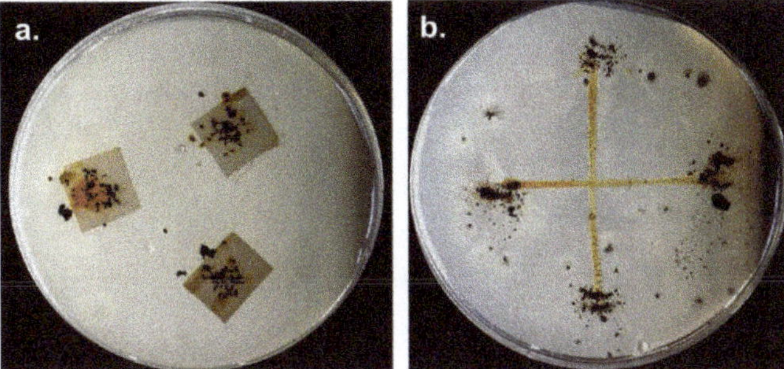

Figure 5. Common isolation procedure for myxobacteria: Soil/environmental sample is placed on **a.** Stan 21 with filter paper to enrich cellulose decomposing strains and **b.** on water agar with *E. coli* bait (cross) for predators. Numerous transfers of fruiting bodies/swarm edge material to fresh plates are necessary to purify myxobacteria.

As was mentioned at the beginning, myxobacteria live in various habitats. It is recommended, but not mandatory, to investigate uncommon habitats from different geographic regions with regard to new secondary metabolite producers. But, already within microscale areas of environmental samples, different strains of one myxobacterial species show surprising genetic differences, as biogeographical studies of myxobacteria revealed. Biogeography is the study of the distribution of organisms across space and time [67]. As mentioned by Ramette and Tiedje, prokaryotic biogeography is "the science that documents the spatial distribution of prokaryotic taxa in the environment at local, regional, and continental scales" [68]. Hanson et al. propose that four processes, selection, drift, dispersal, and mutation, create and maintain microbial biogeographic patterns on inseparable ecological and evolutionary scales [69]. For example, Bacteria and Archaea are globally distributed [70]. At the class level, the β-proteobacteria, cyanobacteria, actinobacteria, and flavobacteria have been shown to display worldwide distribution in marine or terrestrial ecosystems [71–73]. According to Hedlund and Staley, at the genus level, many prokaryotes have a cosmopolitan distribution in their respective habitats [74]. Recent global surveys indicate that most bacteria are restricted to broad habitat types, as there is little overlap among bacterial taxa found in soils, sediments, freshwater, and seawater [75,76]. Dawid gave a comprehensive overview about the ecology and global distribution of myxobacteria in the macroscale range [1]. The study was based on data given in the literature as well as on his own analyses of almost 1400 soil samples from 64 countries and all continents. The study found that an exceptionally high average species number was determined for soils from countries that belong to the winter rain climates of the Mediterranean type, the permanent wet rain forest climates and the tropical semi-desert climates. However, soils of countries with cold temperate coniferous forest climates and cool temperate intermediate climates with peat mosses and coniferous forests harbor a low average number of species. Jiang et al. determined biogeographic patterns of myxobacterial taxa in deep-sea sediments [77]. They screened DNA from four different depths for myxobacteria-like 16S rRNA genes and provided the first evidence, that marine myxobacteria are phylogenetically distinct from terrestrial species. Brinkhoff et al. studied the biogeography and phylogenetic diversity of marine myxobacteria and found a deep-branching monophyletic cluster of exclusively marine myxobacteria within the Myxococcales [78]. Wielgoss et al. sequenced the genomes of 22 *Myxococcus xanthus* isolates from a 16 × 16-cm-scale patch of soil. They found out "that two closely related *M. xanthus* clades inhabiting the same centimeter-scale patch of soil, display strong sexual isolation, with homologous recombination occurring frequently between members within each clade, but with almost no detectable levels of genetic exchange occurring across clades" [79]. Kraemer et al. resolved the micro biogeography of social identity and genetic relatedness in local populations of *M. xanthus* at small spatial scales [80]. The study comprises samples taken from fruiting bodies, neighboring fruiting bodies separated by millimeters, neighborhoods of fruiting bodies separated by centimeters and finally soil patches separated by meters and kilometers. They found out that "relatedness decreases greatly with spatial distance even across the smallest scale transition and that both, social relatedness and genetic relatedness are maximal within individual fruiting bodies at the micrometer scale but are much lower already across adjacent fruiting bodies at the millimeter scale." What will this mean with regard to myxobacteria and natural product research? The cellulose degrading genus/species *Sorangium cellulosum* serves as an example: already in 2003, the myxobacterial strain collection of the HZI (former GBF) comprises 7000 strains from which 23.2% belong to *S. cellulosum*. On the other side, *S. cellulosum* strains produced 48.4% of all known secondary metabolites described so far from myxobacteria [81]. This means that closely related strains also have huge potential to produce different chemical and biological bioactive metabolites [48,49] and that the search for new antibiotic producers can be successful in both, small and large scale. For a comprehensive overview about biogeographic patterns of myxobacteria, I refer to Velicer et al. [82]. For a deeper insight to prokaryotic biogeography, see the study of Ramette and Tiedje [73] and the review of Hanson et al. [74].

With regard to numerous studies based on cultivation-dependent approaches, the number of publications that focus on the non-cultivable myxobacteria is comparatively small. Nevertheless, there are about 4000 (often unpublished) 16S rRNA sequences deposited at the NCBI database which are mentioned to be "uncultured Myxococcales". Under consideration of further myxobacteria-related sequences which are just deposited as "uncultured (delta) proteobacterium" [83] or even "uncultured bacterium", the true extent of uncultivated myxobacteria can just be surmised. Most of the deposited sequences are "by-products" from cultivation-independent studies of bacterial communities in general, without special focus on Myxococcales.

To give an impression about the diversity of cultivable and uncultivable myxobacteria in different habitats, published and unpublished 16S rRNA sequences from NCBI are compared with each other and the results are summarised subsequently.

5. Distribution of Myxobacteria in Different Habitats

5.1. Terrestrial Habitats

Myxobacteria are optimally adapted to terrestrial habitats, which manifests as a wide range of different phenotypes, such as social swarming and gliding, fruiting-body formation, resting myxospores, excretion of secondary metabolites with antibiotic or antifungal activity into the environment, as well as predation or cellulose decomposition. It is therefore not surprising that the majority of known species (and secondary metabolite producers) was isolated or detected from soil samples. In 1947, Singh investigated myxobacteria in soils and composts, their distribution, number, and lytic action on bacteria [44]. From soils of Great Britain, he isolated species of *Myxococcus*, *Chondrococcus* (later renamed to *Corallococcus*) and *Archangium* and estimated that the numbers of myxobacteria ranged from 2000 to 76,400/gram in soil. In an actively decomposing compost of sludge and straw, the number of *Myxococcus fulvus* was more than 500,000/g. Singh was also the first who detected the potential of myxobacteria to produce antibiotics.

In 2005, Wu et al. were the first to explore the diversity of myxobacteria (in a soil niche) by cultivation-independent methods with myxobacteria-specific primers and probes [84]. Moreover, in the latter study members of *Myxococcus*, *Corallococcus*, *Cystobacter*, and *Nannocystis* were cultivated. Nevertheless, screening a special library using Cystobacterineae- and Sorangiineae-specific probes and subsequent sequence analyses revealed a somewhat higher number of myxobacteria within the sample, from which many show only minor similarity to known species. Therefore, even in this first cultivation-independent study about Myxococcales, the authors suggested that myxobacteria in nature are much more diverse than were ever known, even in a single soil sample.

Jiang et al. investigated fruiting and non-fruiting myxobacteria and gave a phylogenetic perspective of cultured and uncultured members of this group [85]. The authors analysed the diversity of myxobacteria in campus garden soil and found out that many undescribed relatives exist in nature and concluded that there are two forms: the fruiting and the non-fruiting types. They postulated that most of the uncultured myxobacteria might represent taxa, which rarely form fruiting bodies, or may lack some or all of the developmental genes needed for fruiting body formation. The majority of sequences from the cultivation-independent approach are only distantly related to known genera and species. As myxobacteria are widespread in terrestrial habitats, consequently, they are frequently detected in those cultivation-independent studies on microbial diversity. Even in uncommon habitats like adult worker ants myxobacteria were detected [86].

In our study about myxobacteria in two geographically distant locations, namely sand from Kiritimati Island and German compost, we also compared the diversity of cultivable myxobacteria to those from cultivation-independent clone libraries [9]. The study revealed an overrepresentation of the genera *Myxococcus* and *Corallococcus* with standard cultivation methods (Figure 6).

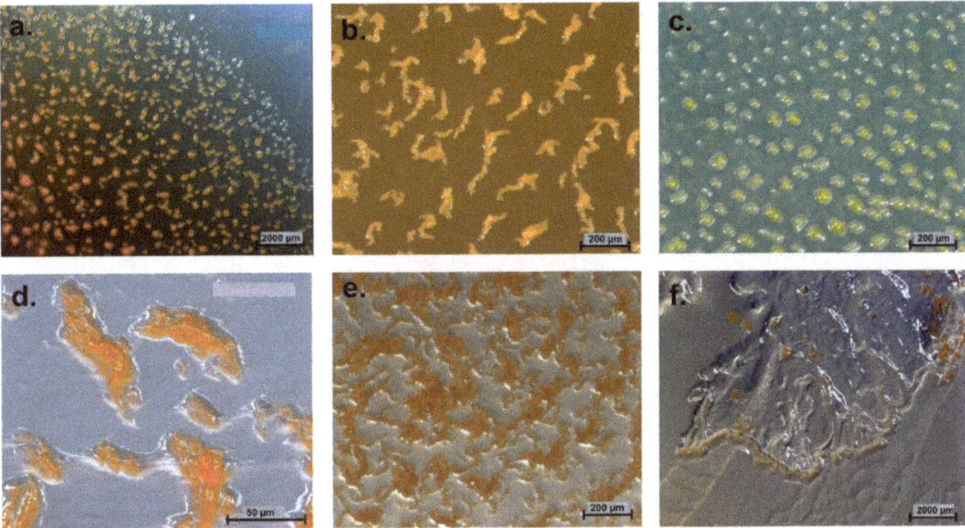

Figure 6. Myxobacterial cultures isolated from Kiritimati sand (**a,d**) and German compost (**b–f**) modified from Mohr et al. [9]. (**a**) *Corallococcus* (*Myxococcus*) *macrosporus*, (**b**) *Corallococcus* sp., (**c**) *Myxococcus* sp. (**d**) *Archangium gephyra*, (**e**) *Corallococcus* sp., (**f**) *Polyangium fumosum*.

However, phylogenetic analyses of the 16S rRNA gene sequences from clones revealed a great potential of undescribed myxobacteria in both sampling sites. Several OTUs (operational taxonomic units; groups of sequences with ≥97% similarity) represented unknown taxa exclusively detected by cultivation-independent analyses, but not by cultivation. Furthermore, clone library analyses indicated that the myxobacterial community of the investigated samples is predominantly indigenous.

Most of the known myxobacterial secondary metabolites were previously isolated from terrestrial myxobacteria, because the majority of strains was isolated from (moderate) terrestrial habitats. However, myxobacteria are extremely adaptable and can also be found in demanding environments like acidic soils, fresh water, oceans and salines, anaerobic/microaerophilic, and extreme habitats, respectively.

5.2. Acidic and Alkaline Habitats

Acidic wetlands have a major impact on the global carbon and water cycles. With high acidity (pH 3.5 to 5.0), low temperatures, and extremely low concentrations of mineral nutrients (5 to 50 mg per liter), wetlands are moderate to extreme habitats. Their microbial diversity remains poorly understood, because only microbial populations involved in CH_4 cycling, i.e., methanotrophic bacteria and methanogenic archaea, have attracted considerable research interest. Other members of the microbial communities in acidic *Sphagnum* peatlands remain largely unknown [87].

The pH range for growth of the majority of myxobacteria is rather narrow, approximately 6.5–8.5. Therefore, they are common in soils of pH 6–8 (neutral to slightly alkaline pH). However, acidic or alkaline habitats also seem to be suitable for myxobacteria [10]. Even in 1977, Hook isolated ten species from waters of an alkaline bog and adjacent soils [88] like *Archangium*, *Corallococcus*, *Melittangium*, *Myxococcus*, and *Sorangium* (former *Polyangium*). *Corallococcus coralloides*, (formerly *Myxococcus coralloides*) was dominant in the terrestrial samples. With pH between 6.0 and 8.7, the investigated habitats were between slightly acidic and slightly alkaline. In 1979, Rückert also described *C. coralloides* as the predominant species in soils of pH 4.1–4.9 and as dominant as *M. fulvus* in soils of pH 3.0–3.5 [89]. In alpine acidic soils *C. coralloides* was the third-most dominant species behind two *Myxococcus* species. But, Rückert also noted that the overall myxobacterial diversity in acidic soils (pH 3.5–4.9) was less than in slightly acidic or neutral environments (pH 5.0–7.8). In 1984, Dawid isolated *Myxococcus xanthus*, *M. virescens* and *Polyangium* sp., but no cellulolytic species from undisturbed *Sphagnum* bogs of the Hohen Venn, Belgium [3].

Mohr et al. studied myxobacteria in peat bog and fen with cultivation and cultivation-independent methods [2]. Therefore, 38 moor samples of soil, water, plant residues, mud, and feces-material (Figure 7a–c) were screened using standard as well as moor-adjusted cultivation conditions (low pH, low temperature, moor-water for preparation of agar plates), screening numerous replicates over several years. The pH of moor samples analyzed in this study was between 4.0 and 7.0 and therefore comparable to those from the other studies. But almost exclusively species of the genus *Corallococcus* could be isolated from acidic soils of the Harz-region (Figure 7d,e). A *Sorangium* strain was detected on a raw culture plate with filter (Figure 7f), but could not be purified. In addition, the community composition of acidic high moor and fen revealed by cultivation-independent 16S rRNA clone library analysis gave a rather different picture of the myxobacterial diversity.

Figure 7. (**a**) Brockenfeld high moor; (**b**) fen Am Sandbeek; (**c**) Brockenfeld high moor scarp. Isolated *Corallococcus* sp. strains from moors (**d**) strain B19, (**e**) strain B2t-1. (**f**) *Sorangium cellulosum* (orange) on a raw culture plate (Stan 21 with filter) inoculated with soil material from moor. Pictures are from Mohr et al. (2017) and modified [2].

Phylogenetic analyses of clone sequences revealed a high diversity of undescribed myxobacteria in high moor and fen. Many sequences represent totally unknown taxa. However, numerous clones were closely related to sequences from other cultivation-independent studies of eubacterial diversity in which samples from peat swamp, wetlands peat bog, *Sphagnum* moss, pine forest, acidic fen soil, and forest soil were analysed (Figure 8). As mentioned above, cultivation exclusively revealed strains from the genus *Corallococcus* (but from almost all analysed samples).

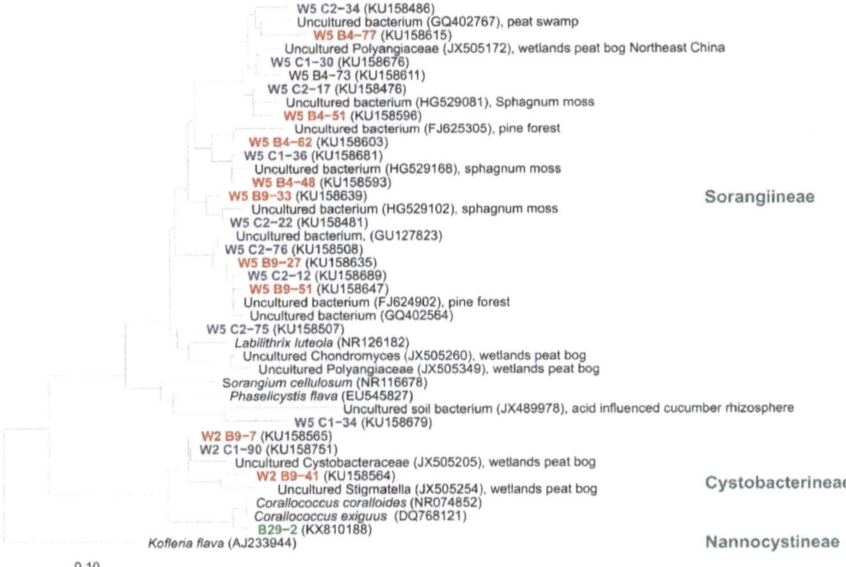

Figure 8. Part of a distance tree showing some myxobacterial type strains, some representative clone sequences and one representative culture sequence from our study [2] as well as sequences from uncultivated myxobacteria from other studies. Red: clones from Brockenfeld high moor; blue: clones from fen Am Sandbeek; green: representative culture from Brockenfeld high moor. Accession numbers are given in brackets. Origin of sequences from uncultured myxobacteria are mentioned. Bar, 0.1 substitutions per nucleotide position.

To my knowledge, no publications about (bioactive) secondary metabolites from myxobacteria isolated from acidic or alkaline habitats are available. However, we screened 21 *Corallococcus* spp.-strains from the moor study for production of bioactive metabolites. Raw extracts of all strains showed high activity against Gram positives (*Micrococcus luteus*, *Staphylococcus aureus*, *Bacillus subtilis*, and *Mycobacterium* sp.), the yeasts *Saccharomyces pombe* and *Rhodotorula glutinis*, as well as against the filamentous fungi *Mucor hiemalis*, but no activity against Gram negatives. HPLC analyses of the raw extracts revealed three dominant peaks. By HPLC-fractionation of bioactive extracts and subsequent HPLC-MS analyses the already known substances dibenzylpyrazine, myxothiazol A, and myxothiazol Z/A-methylester, respectively, were identified (data not published). In summary, the moor habitat is a promising source and of high interest with regard to the cultivation of prospective new bioactive secondary metabolite-producing myxobacteria.

5.3. Freshwater Habitats

In natural aquatic environments, microbial cells often build complex, surface-attached biofilm communities. Within the water body or pelagic zone of unpolluted freshwaters, the number and diversity of bacteria is normally lower than on the available substrates. Myxobacteria glide in swarms over solid surfaces. If it is possible, they prefer attached in contrast to planktonic living.

Only very few studies about myxobacteria in fresh water habitats are published. In the 1960th/1970th several studies dealt with nonpathogenic or pathogenic non-fruiting "myxobacteria" as colonizers of freshwater fish [90]. However, these publications deal with strains of the Cytophaga-group, which do not belong to the Myxococcales, but to the Cytophaga—Flavobacterium—Bacteroides group. No myxobacterial pathogens are published. Reichenbach mentioned that myxobacteria can also be isolated from fresh water, but explained these findings with soil organisms notoriously exchange into water bodies, being regularly washed or blown in and often surviving there periodically or permanently [12].

In 2012, Li and co-workers investigated the myxobacterial community in freshwater lake mud using high-throughput 454 pyrosequencing and myxobacteria-enriched libraries with Cystobacterineae- and Sorangiineae-specific primer pairs, respectively, and reported that myxobacteria were one of the major bacterial groups in the lake mud [91]. Phylogenetic analysis showed that the limnetic myxobacteria exhibit closer relationships to their soil than to their marine relatives, but there are also exclusive taxa of limnetic myxobacteria. The major conclusion was that the unclassified Myxococcales in the lake mud comprise a large portion of the microbiota and exhibit high species diversity. Kou et al. analysed bacterial communities in sediments of freshwater (Poyang Lake) in China. There, *Anaeromyxobacter dehalogenans* turned out to be a main part of the bacterial community composition (1–14.6%) [92]. In another study about methanogenic microbial communities in sediments of Amazonian lakes using terminal restriction fragment length polymorphism (T-RFLP) and pyrosequencing, the proteobacteria revealed as the most abundant phylum in all lake sediments. Delta-proteobacteria (mainly Myxococcales, Syntrophobacteriales and sulfate/sulfur-reducing bacteria) dominated this habitat [93]. In 2014, Kandel et al. investigated the abundance, diversity, and seasonal dynamics of predatory bacteria in aquaculture zero discharge systems by cultivation-independent analyses and found out that in addition to the detected *Bdellovibrio* and similar organisms, other potential predators were highly abundant, especially from the Myxococcales [94].

In the absence of cultures which are verifiable natural fresh water inhabitants, up to now, no (bioactive) metabolites have been isolated from limnic strains. However, the above-mentioned detection of exclusively limnic taxa [95] suggest that also the habitat fresh water could be a promising source for the cultivation of new secondary metabolite producing myxobacteria.

5.4. Marine/Saline Environments

Covering more than roughly 78% of the earth's surface, water is the most prevalent natural substance, of which approximately 97.5% is salt water in the world's oceans [96]. The salt tolerance of myxobacteria is low in general. It was assumed for a long time that myxobacteria exclusively live in terrestrial habitats. Indeed, even in 1963, Brockman observed fruiting myxobacteria in sand samples from an ocean beach in South Carolina [97]. Species of the already known terrestrial genera *Archangium*, *Chondrococcus* (*Corallococcus*), *Chondromyces*, *Myxococcus*, and *Polyangium*, could be cultivated. As late as 2002 with *Haliangium ochraceum* and *H. tepidum*, the first myxobacterial genus was isolated and described from coastal salt marshes. The strains differ from known terrestrial myxobacteria with regard to salt requirements (2–3% NaCl) and the presence of anteiso-branched fatty acids [4]. Other genera, exclusively detected in marine habitats like *Plesiocystis* [5], *Enhygromyxa* [6], and *Pseudenhygromyxa* [7] (all Nannocystineae-suborder) followed (Figure 9).

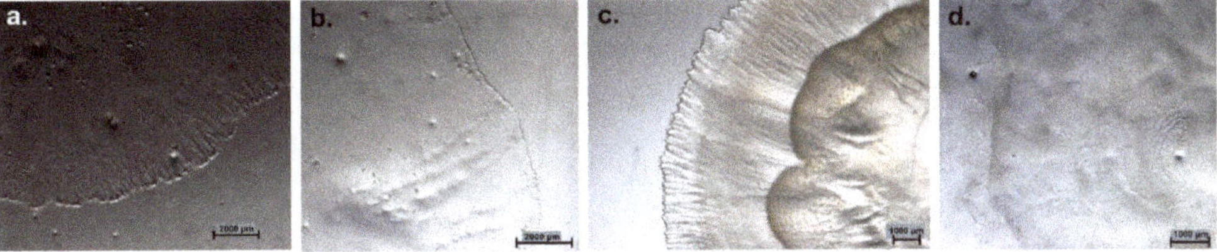

Figure 9. Marine myxobacterial type strains on agar plates. (**a**) *Haliangium tepidum* (DSM 14436T) on VY/2SWS, (**b**) *Enhygromyxa salina* (DSM 15217T) on VY/4SWS, (**c**) *Pseudenhygromyxa salsuginis* (DSM 21377T) on 1102, (**d**) *Plesiocystis pacifica* (DSM 14875T) on VY/2SWS.

In 2010, Jiang et al. investigated the diversity of marine myxobacteria in comparison to terrestrial soil myxobacteria [82]. Therefore, they established myxobacteria enriched libraries of 16S rRNA gene sequences from four deep-sea sediments and a hydrothermal vent and identified 68 different myxobacteria related sequences from randomly sequenced clones of these libraries. The authors concluded that the myxobacterial sequences were diverse but phylogenetically similar at different locations and depths. However, they separate from terrestrial myxobacteria at high levels of classification.

In 2012, the study of Brinkhoff and co-workers gave an impressive insight to the marine myxobacterial community [83]. They detected a cluster of exclusively marine myxobacteria (marine myxobacteria cluster, MMC) in sediments of the North Sea, but not in the limnetic section of the Weser estuary and other freshwater habitats. In a quantitative real-time PCR approach, the authors found out that the MMC constituted up to 13% of total bacterial 16S rRNA genes in surface sediments of the North Sea. In addition, in a global survey including sediments from the Mediterranean Sea, the Atlantic, Pacific, and Indian Oceans, and various climatic regions, the MMC appears in most samples and to a water depth of 4300 m, but there was no synteny to other myxobacterial genomes. The study of Brinkhoff et al. showed that the MMC is an important and widely distributed but largely unknown component of marine sediment-associated bacterial communities. Figure 10 shows some representative clones from different marine habitats mentioned in the Brinkhoff study. The 17 clones show 95.1–100% similarity to each other and 88–91% to the next type strain *Sandaracinus amylolyticus*. This implies that members of the MMC cluster do at least belong to new genera if not even families, but presumably belong to the Sorangiineae suborder. However, the genus *Sandaracinus* was published shortly after the Brinkhoff study, so this relative was not mentioned there.

In 2013, Zhang et al. isolated fifty-eight terrestrial and salt-tolerant myxobacteria from the saline-alkaline soils collected from Xinjiang, China [95]. Based on morphology and 16S rRNA gene sequences, the authors identified species of *Myxococcus*, *Cystobacter*, *Corallococcus*, *Sorangium*, *Nannocystis*, and *Polyangium*. They reported that all the strains grew better with 1% NaCl than without salt; some *Myxococcus* strains even grow with 2% NaCl.

Li et al. (2014) chose a cultivation-independent approach to analyze the diversity of myxobacteria from saline-alkaline soils of Xinjiang, China, too. A semi-nested PCR-denaturing gradient gel electrophoresis (DGGE) based on the taxon-specific gene mglA (a key gene involved in gliding motility) was used [98]. In accordance to previous studies, Li et al. also suggested that there are still many viable, but under standard laboratory conditions uncultured myxobacterial strains in the investigated saline-alkaline habitat. Natural product classes discovered from marine Myxococcales strains include polyketides, hybrid polyketide-nonribosomal peptides, degraded sterols, diterpenes, cyclic depsipeptides, and alkylidenebutenolides [99]. Four genera of marine/saline origin are known so far (Figure 10) and from two, *Haliangium* and *Enhygromyxa*, numerous (bioactive) secondary metabolites could be isolated [100]: Haliangicin [101], salimabromide [102], salimyxins, enhygrolides [103], and haliamide [104]. The above-mentioned data reveal that marine/saline environments as oceans harbor an enormous potential of new myxobacteria. These organisms are an unexplored resource of novel antibiotics of novel chemical scaffolds, as mentioned by Albataineh and Stevens, who highlighted the need for continued discovery and exploration of marine myxobacteria as producers of novel natural products [104].

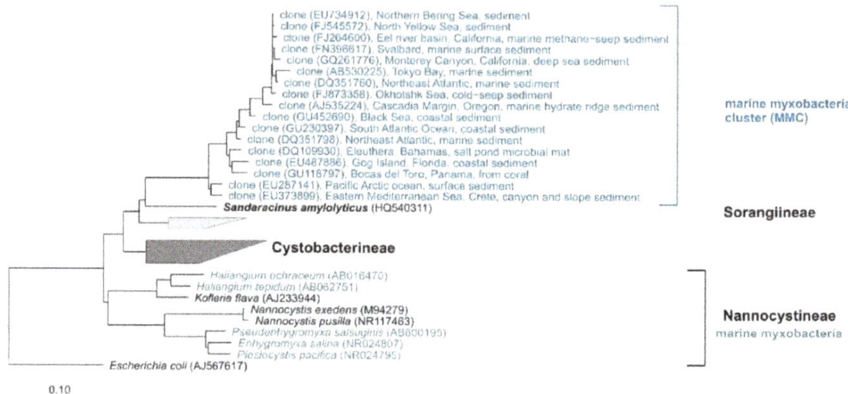

Figure 10. Neighbour joining tree of myxobacterial type strains (16S rRNA-genes), some representative clones from the MMC-cluster (blue) and the next cultivated relative *Sandaracinus amylolyticus*. Genera isolated from marine environment are in green blue. Suborders of the order Myxococcales, origin of clones, and accession numbers are shown. Bar, 0.1 substitutions per nucleotide position.

5.5. Facultative or Strictly Anaerobic Myxobacteria

All known myxobacteria live aerobically, with one exception: The facultative-anaerobic genus *Anaeromyxobacter* comprises one species, *A. dehalogenans*. The type strain was isolated from stream sediment and grows with acetate as electron donor and 2-chlorophenol (2-CPh) as electron acceptor [11]. Since 2002, several strains of *Anaeromyxobacter* were isolated from various habitats. Flooded rice field soil [105], uranium contaminated surface environment [106], corrosion material of drinking water pipelines [107], arsenic-contaminated soils [108], or chemically and electro-chemically enriched sodic-saline soil (unpublished) served as sources for the cultivation of *Anaeromyxobacter*-strains. A total of 23 sequences designated as *Anaeromyxobacter* (sequence lengths > 1000 bp) are available from the NCBI database (FJ90053–FJ90062, FJ90048, FJ90049, FJ90051, EF067314, AJ504438, KF952446, AF382397, AF382399, AF382400, FJ939131, KF952441, KF952438, KC921178) and were added to a phylogenetic tree of myxobacterial type strains. A similarity matrix calculated with arb (www.arb-home.de) revealed 98.4–100% similarity (on basis of 16S rRNA gene) for 20 of these strains to the type strain of *A. dehalogenans*. Assuming that the standard value for the definition of a new species is 98.65% [109] and 94% for a new genus, respectively, the above-mentioned 20 cultures probably do belong to *A. dehalogenans*. However, strains OnlyC-B2 (KF952441) and SSS-B8 (KF952438) show only 96.3% and 96.0% similarity to the corresponding type strain (but 99.4% to each other) and putatively represent a new species. One culture, isolated from ginger foundation soil, and also designated as *A. dehalogenans* (KC921178), shows only 86% similarity to the next cultivated (myxobacterial) type strain *Vulgatibacter incomptus*. This culture definitely represents at least a new family if not even a new suborder of myxobacteria (unpublished).

In summary, there are currently two cultivated species of *Anaeromyxobacter*, but only one is validly described. But what about further facultative or even strictly anaerobic myxobacteria?

The NCBI search for 16S rRNA gene sequences of "uncultured *Anaeromyxobacter*" revealed more than 1200 hits. Nevertheless, not all sequences designated as "Uncultured *Anaeromyxobacter*" are close relatives of *Anaeromyxobacter*, as a revision of randomly selected sequences revealed. For example: the sequence GU271851, mentioned as "Uncultured *Anaeromyxobacter*", shows 91% similarity to the next type strain *Haliangium tepidum*, but only 87% to the type strain of *A. dehalogenans* [110]. Clone GU271788, also mentioned as *A. dehalogenans*, shows 92% to the next type strain, *Sorangium cellulosum*, but only 86% to *Anaeromyxobacter* [111]. On the other hand, there are probably numerous sequences deposited at NCBI which are close relatives of *Anaeromyxobacter*. However, these sequences are just mentioned as "uncultivated Myxococcales", "uncultivated (delta) proteobacteria" or "uncultured bacterium clone" such as clone EUB_19 (FJ189540), which shows 98.7% [112] or clone A_Ac-2_16 (EU307085), which shows 97.8% similarity to the next type strain: *A. dehalogenans* [113]. In 2009, Thomas et al. analysed the diversity and distribution of *Anaeromyxobacter* strains in a uranium-contaminated environment by mainly cultivation-independent methods. Phylogenetic analyses of the clone and culture sequences revealed that there are at least three distinct *Anaeromyxobacter* clusters at the IFC (Integrated Field-Scale Subsurface Research Challenge) site near Oak Ridge, whereby two sides are exclusively represented by clones. As mentioned above, quantitative PCR assay and pyrosequencing analysis of 16S rRNA genes also revealed *A. dehalogenans* as a part of the microbial community in the sediment of Poyang Lake, the largest freshwater lake in China [95].

To get an impression about the diversity of uncultured *Anaeromyxobacter*, I added about 80 clone sequences from NCBI with corresponding designation to a phylogenetic tree of myxobacterial type strains (data not shown). The clones revealed 90.0–100% similarity on the basis of 16S rRNA gene to the type strain of *A. dehalogenans* (AF382396). Figure 11 shows the affiliation of some representative clones. These clones represent at least several new genera, if not even families of myxobacteria, which are probably also facultative or strictly anaerobic and which could not be cultivated so far.

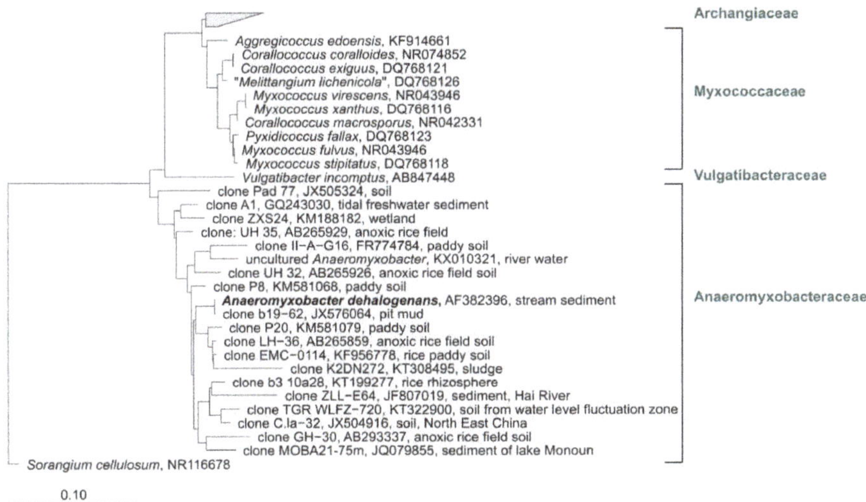

Figure 11. Neighbour joining tree of some myxobacterial type strains (16S rRNA-genes) and some representative clones. All clones show highest relationship to the next cultivated relative *A. dehalogenans*. All families of the Cystobacterineae suborder, origin of samples, and accession numbers of clones/cultures are shown. Bar, 0.1 substitutions per nucleotide position.

Although the 16S affiliation of clones does not give any information about metabolism of the corresponding organism, high similarities to aerobic or anaerobic cultures indicate similar metabolic capabilities. However, no bioactive secondary metabolites have been described so far from *Anaeromyxobacter* strains, which is certainly because anaerobic isolation, cultivation, and large scale fermentation requires special efforts regarding equipment and microbiological skills and experience.

5.6. Moderate to Extreme Hot or Cold Environments

Myxobacteria are mesophilic and grow well at 30 °C, although their temperature range is much wider. For most myxobacterial strains, the growth temperatures is between 4 °C and 44 °C. Usually vegetative cells cannot survive temperatures above 45 °C, but myxospores suspended in water tolerate 58–60 °C. This property can be used to be purify myxobacteria from mesophilic accompaniment organisms [12]. A moderate terrestrial habitat was investigated by Brockman in 1976 [114], who isolated strains of *Archangium*, *Chondromyces*, *Cystobacter*, *Myxococcus*, *Polyangium*, and *Stigmatella* from arid Mexican soils. He reported a greater species diversity from regions with higher annual rainfall (400–800 mm compared to 200–400 mm). Moderate thermophilic myxobacteria of Cystobacterineae and Sorangiineae-suborders, which grew very fast at temperatures of 42 °C–44 °C, were isolated from soil samples of semiarid and warm climates by Gerth and Müller [115] (Figure 12). One strain even grew at 48 °C, whereas the majority of the described species grows best at 30 °C.

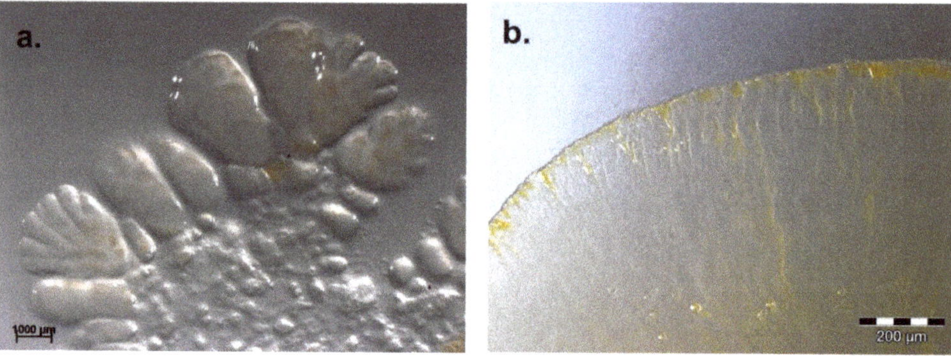

Figure 12. Moderately thermophilic strains of *Sorangium* on VY/2 agar. (**a**) GT47 and (**b**) GT 41, isolated by Dr. K. Gerth.

Recently, a new *Nannocystis* species, *N. konarekensis*, was isolated from an Iranian desert [111]. The strain shows an optimal growth temperature at 37 °C, in contrast to the other known *Nannocystis* species *N. pusilla* and *N. exedens*, which show optimal growth at 30 °C. Iizuka et al. reported about enrichment and phylogenetic analysis of moderately thermophilic myxobacteria. During their search for thermophilic myxobacteria in geothermal environments, four strains that grew at temperatures up to 50 °C (optimum 45 °C–49 °C) could be isolated from various hot springs in Japan [116]. Three of the cultures were from fresh water hot springs and one was from a coastal saline spring. Even after repeated enrichment procedures, other thin film-like spreading bacteria accompanied the strains. PCR, cloning, and sequencing of 16S genes revealed that all cultivated bacteria belong to the order Myxococcales and showed between 89–99% homology to strains of myxobacteria. Therefore, some of these cultures represent new undescribed but cultivable species, genera, and perhaps even families (Figure 13).

Although numerous (cultivation-independent) studies about bacterial diversity of hot springs/geothermal sources are published, the NCBI search for sequences of uncultured thermophilic myxobacteria or myxobacteria from hot springs revealed only very few matches. Hot springs are probably not the most suitable habitat for the mainly mesophilic myxobacteria. But, based on the cultivation success mentioned by Iizuka et al. [116], it is certainly worth investigating these habitats more precisely to isolate new myxobacteria.

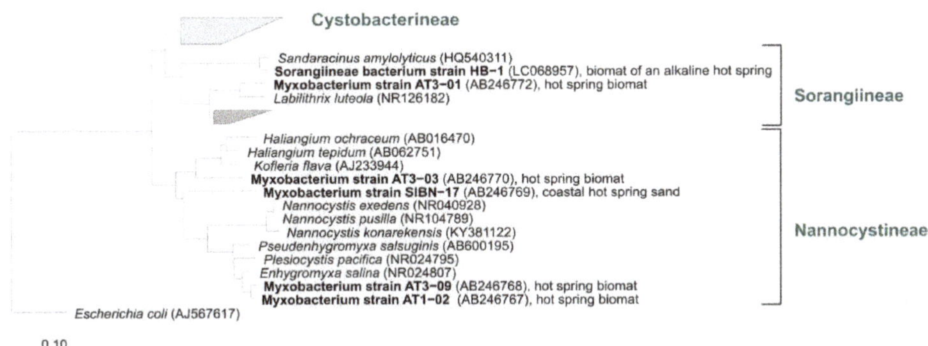

Figure 13. Neighbour joining tree of some myxobacterial type strains (16S rRNA-genes) and cultures (in bold) from hot springs (AB246767-AB246770, AB246772) [117] and alkaline hot spring, all from Japan. Suborders of the order Myxococcales, origin of samples, and accession numbers are shown. Bar, 0.1 substitutions per nucleotide position.

Some publications deal with myxobacteria from cold environments like Arctic soils. However, in the study of Brockman who tried to isolate myxobacteria from Alaskan and Canadian Arctic soils, myxobacterial growth was only observed when soil plates were incubated at 24 °C–26 °C, but not at 6 °C–8 °C [118]. In contrast, Dawid described psychrophilic myxobacteria which grow at 4 °C but not under mesophilic conditions between 18 °C and 30 °C (after 7–9 month of incubation) on samples of Antarctic soils [117].

Due to long incubation times of psychrophilic strains, their biotechnological use in large scale fermentation would be expensive and time consuming and would only be worthwhile if a highly promising antibiotic was detected in such a psychrophilic strain.

6. Conclusion

In summary, myxobacteria are highly adaptable cosmopolitans. They can grow/survive in various kind of habitats and areas of different, even extreme climatic conditions. In 1993, only 2 suborders, 4 families, 12 genera, and 38 species were assigned to the order Myxococcales [28], but in 2018, already 3 suborders, 10 families, 29 genera, and 58 species are described. Although the number of species grows every year, consideration of data from cultivation-independent studies reveals that we only see the tip of the diversity iceberg.

In 2010, 67 distinct core structures and about 500 derivatives were known from approximately 7500 myxobacterial strains [48]. Only seven years later, Herrmann et al. could refer to five natural product classes produced by myxobacteria [49]. These new molecules show such promising activity that several of them may serve as early lead structures for drug development. This shows the enormous potential of myxobacteria as producers of new, bioactive secondary metabolites. As mentioned by Müller and Wink, three of the most promising approaches toward finding novel anti-infectives from microorganisms are the use of biodiversity to find novel producers, the variation of culture conditions and induction of silent genes, and the exploitation of the genomic potential of producers via "genome mining" [119]. With focus on novel producers, the biggest challenge for microbiologists is to get access to the so far uncultivated bacteria.

Acknowledgments: I would like to thank Klaus Peter Conrad, Diana Telkemeyer and Birte Trunkwalter for taking the nice pictures of myxobacteria, Wera Collisi and Steffi Schulz for technical assistance, and Aileen Gollasch, and Sabrina Karwehl for HPLC-MS analyses.

Conflicts of Interest: The authors declare no conflict of interest.

References

1. Dawid, W. Biology and global distribution of myxobacteria in soils. *FEMS Microbiol. Rev.* **2000**, *24*, 403–427. [CrossRef] [PubMed]
2. Mohr, K.I.; Zindler, T.; Wink, J.; Wilharm, E.; Stadler, M. Myxobacteria in high moor and fen: An astonishing diversity in a neglected extreme habitat. *MicrobiologyOpen* **2017**, *6*. [CrossRef] [PubMed]
3. Dawid, W. Myxobakterien in ungestörten Hochmooren des Hohen Venn (Hautes Fagnes, Belgien). *Syst. Appl. Microbiol.* **1984**, *5*, 555–563. [CrossRef]
4. Fudou, R.; Jojima, Y.; Iizuka, T.; Yamanaka, S. *Haliangium ochraceum* gen. nov., sp. nov. and *Haliangium tepidum* sp. nov.: Novel moderately halophilic myxobacteria isolated from coastal saline environments. *J. Gen. Appl. Microbiol.* **2002**, *48*, 109–116. [CrossRef] [PubMed]
5. Iizuka, T.; Jojima, Y.; Fudou, R.; Hiraishi, A.; Ahn, J.W.; Yamanaka, S. *Plesiocystis pacifica* gen. nov., sp. nov., a marine myxobacterium that contains dihydro-genated menaquinone, isolated from the Pacific coasts of Japan. *IJSEM* **2003**, *53*, 189–195. [CrossRef] [PubMed]
6. Iizuka, T.; Jojima, Y.; Fudou, R.; Tokura, M.; Hiraishi, A.; Yamanaka, S. *Enhygromyxa salina* gen. nov., sp. nov., a slightly halophilic myxobacterium isolated from the coastal areas of Japan. *Syst. Appl. Microbiol.* **2003**, *26*, 189–196. [CrossRef] [PubMed]
7. Iizuka, T.; Jojima, Y.; Hayakawa, A.; Fujii, T.; Yamanaka, S.; Fudou, R. *Pseudenhygromyxa salsuginis* gen. nov., sp. nov., a myxobacterium isolated from an estuarine marsh. *IJSEM* **2013**, *63*, 1360–1369. [CrossRef] [PubMed]
8. Menne, B.; Rückert, G. Myxobakterien (Myxobacterales) in Höhlensedimenten des Hagengebirges (Nördliche Kalkalpen). Die Höhle. *Z Karst Höhlenkd* **1988**, *39*, 120–131.
9. Mohr, K.I.; Stechling, M.; Wink, J.; Wilharm, E.; Stadler, M. Comparison of Myxobacterial Diversity and Evaluation of Isolation Success in two niches: Kiritimati Island and German Compost. *MicrobiologyOpen* **2016**, *5*, 268–278. [CrossRef] [PubMed]
10. Shimkets, L.J.; Dworkin, M.; Reichenbach, H. The Myxobacteria. In *The Prokaryotes*; Dworkin, M., Falkow, S., Rosenberg, E., Schleifer, K.H., Stackebrandt, E., Eds.; Springer: New York, NY, USA, 2006; pp. 31–115.
11. Sanford, R.A.; Cole, J.R.; Tiedje, J.M. Characterization and Description of *Anaeromyxobacter dehalogenans* gen. nov., sp. nov., an Aryl-Halorespiring Facultative Anaerobic Myxobacterium. *AEM* **2002**, *68*, 893–900. [CrossRef]
12. Reichenbach, H. The ecology of the myxobacteria. *Environ. Microbiol.* **1999**, *1*, 15–21. [CrossRef] [PubMed]
13. Nan, B.; Chen, J.; Neu, J.C.; Berry, R.M.; Oster, G.; Zusman, D.R. Myxobacteria gliding motility requires cytoskeleton rotation powered by proton motive force. *Proc. Natl. Acad. Sci. USA* **2011**, *108*, 2498–2503. [CrossRef] [PubMed]
14. Kaiser, D.; Robinson, M.; Kroos, L. Myxobacteria, Polarity, and Multicellular Morphogenesis. *Cold Spring Harb. Perspect. Biol.* **2010**, *2*. [CrossRef] [PubMed]
15. Mauriello, E.M.F.; Mignot, T.; Yang, Z.; Zusman, D.R. Gliding Motility Revisited: How Do the Myxobacteria move without Flagella? *Microbiol. Mol. Biol. Rev.* **2010**, *74*, 229–249. [CrossRef] [PubMed]
16. Faure, L.M.; Fiche, J.B.; Espinosa, L.; Ducret, A.; Anantharaman, V.; Luciano, J.; Lhospice, S.; Islam, S.T.; Tréguier, J.; Sotes, M.; et al. The mechanism of force transmission at bacterial focal adhesion complexes. *Nature* **2016**, *539*, 530–535. [CrossRef] [PubMed]
17. Burchard, R.P.; Dworkin, M. Light-induced lysis and carotenogenesis in *Myxococcus xanthus*. *J. Bacteriol.* **1966**, *91*, 535–545. [PubMed]
18. Zusman, D.R.; Scott, A.E.; Yang, Z.; Kirby, J.R. Chemosensory pathways, motility and development in *Myxococcus xanthus*. *Nat. Rev. Microbiol.* **2007**, *5*, 862–872. [CrossRef] [PubMed]
19. Thaxter, R. On the Myxobacteriaceae, a new order of the Schizomycetes. *Bot. Gaz.* **1892**, *17*, 389. [CrossRef]
20. Kofler, L. Die Myxobakterien der Umgebung von Wien. *Sitzungsberichte der Akademie der Wissenschaften mathematisch-naturwissenschaftliche Klasse* **1913**, *122*, 845–876.
21. Baur, E. Myxobakterienstudien. *Arch. Protistenkunde* **1904**, *5*, 42.
22. Jahn, E. *Beiträge zur botanischen Protistologie*; Gebrüder Borntraeger: Leipzig, Germany, 1924.
23. Jahn, E. *Kulturmethoden und Stoffwechseluntersuchungen bei Myxobakterien (Polyangiden)*; Urban and Schwarzenberg: Berlin, Germany, 1936.

24. Kühlwein, H. Beiträge zur Biologie und Entwicklungsgeschichte der Myxobakterien. *Arch. Mikrobiol.* **1950**, *14*, 678–704. [CrossRef]

25. Reichenbach, H.; Höfle, G. Biologically active secondary metabolites from myxobacteria. *Biotechnol. Adv.* **1993**, *11*, 219–277. [CrossRef]

26. Mohr, K.I. History of antibiotics research. In *How to Overcome the Antibiotic Crisis—Facts, Challenges, Technologies & Future Perspective*; Stadler, M., Dersch, P., Eds.; Springer: Berlin, Germany, 2017; Volume 398, pp. 237–272.

27. Houbraken, J.; Frisvad, J.C.; Samson, R.A. Fleming's penicillin producing strain is not *Penicillium chrysogenum* but *P. rubens*. *IMA Fungus* **2011**, *1*, 87–95. [CrossRef] [PubMed]

28. Karwehl, S.; Stadler, M. Exploitation of fungal biodiversity for discovery of novel antibiotics. In *How to Overcome the Antibiotic Crisis—Facts, Challenges, Technologies & Future Perspective*; Stadler, M., Dersch, P., Eds.; Springer: Berlin, Germany, 2017; Volume 398, pp. 303–338.

29. Waksman, S.A.; Woodruff, H.B. Bacteriostatic and bacteriocidal substances produced by soil actinomycetes. *Proc. Soc. Exp. Biol.* **1940**, *45*, 609–614. [CrossRef]

30. Schatz, A.; Bugie, E.; Waksman, S. Streptomycin: A substance exhibiting antibiotic activity against gram positive and gram negative bacteria. *Proc. Exp. Biol. Med.* **1944**, *55*, 66–69. [CrossRef]

31. Duggar, B.M. Aureomycin: A product of the continuing search for new antibiotics. *Ann. N. Y. Acad. Sci.* **1948**, *30*, 177–181. [CrossRef]

32. McGuire, J.M.; Bunch, R.L.; Anderson, R.C.; Boaz, H.E.; Flynn, E.H.; Powell, H.M.; Smith, J.W. Ilotycin, a new antibiotic. *Antibiot. Chemother.* **1952**, *2*, 281–283.

33. Drews, J. Drug discovery: A historical perspective. *Science* **2000**, *287*, 1960–1964. [CrossRef] [PubMed]

34. Aminov, R.I. The role of antibiotics and antibiotic resistance in nature. *Environ. Microbiol.* **2009**, *11*, 2970–2988. [CrossRef] [PubMed]

35. Bartlett, J.G.; Gilbert, D.N.; Spellberg, B. Seven ways to preserve the miracle of antibiotics. *Clin. Infect. Dis.* **2013**, *56*, 1445–1450. [CrossRef] [PubMed]

36. Ventola, C.L. The Antibiotic Resistance Crisis: Part 1: Causes and Threats. *Pharm. Ther.* **2015**, *40*, 277–283.

37. Hesterkamp, T. Antibiotics Clinical Development and Pipeline. In *How to Overcome the Antibiotic Crisis—Facts, Challenges, Technologies & Future Perspective*; Stadler, M., Dersch, P., Eds.; Springer: Berlin, Germany, 2016; Volume 398, pp. 447–474.

38. Hoffmann, T.; Krug, D.; Bozkurt, N.; Duddela, S.; Jansen, R.; Garcia, R.; Gerth, K.; Steinmetz, H.; Müller, R. Correlating chemical diversity with taxonomic distance for discovery of natural products in myxobacteria. *Nat. Commun.* **2018**, *9*, 803. [CrossRef] [PubMed]

39. Landwehr, W.; Wolf, C.; Wink, J. Actinobacteria and Myxobacteria—Two of the Most Important Bacterial Resources for Novel Antibiotics. In *How to Overcome the Antibiotic Crisis—Facts, Challenges, Technologies & Future Perspective*; Stadler, M., Dersch, P., Eds.; Spinger: Berlin, Germany, 2016.

40. Sansinenea, E.; Ortiz, A. Secondary metabolites of soil *Bacillus* spp. *Biotechnol. Lett.* **2011**, *33*, 1523–1538. [CrossRef] [PubMed]

41. Singh, B.N. Myxobacteria in Soils and Composts; their Distribution, Number and Lytic Action on Bacteria. *Microbiology* **1947**, *1*, 1–10. [CrossRef] [PubMed]

42. Mathews, S.; Dudani, A. Lysis of human pathogenic bacteria by myxobacteria. *Nature* **1955**, *15*, 125. [CrossRef]

43. Noren, B.; Raper, K.B. Antibiotic activity of myxobacteria in relation to their bacteriolytic capacity. *J. Bacteriol.* **1962**, *84*, 157–162. [PubMed]

44. Ringel, S.M.; Greenough, R.C.; Roemer, S.; Connor, D.; Gutt, A.L.; Blair, B.; Kanter, G.; von Strandtmann, M. Ambruticin (W7783), a new antifungal antibiotic. *J. Antibiot.* **1977**, *30*, 371–375. [CrossRef] [PubMed]

45. Nett, M.; König, G.M. The chemistry of gliding bacteria. *Nat. Prod. Rep.* **2007**, *24*, 1245–1261. [CrossRef] [PubMed]

46. Baumann, S.; Herrmann, J.; Raju, R.; Steinmetz, H.; Mohr, K.I.; Hüttel, S.; Harmrolfs, K.; Stadler, M.; Müller, R. Cystobactamids: Myxobacterial Topoisomerase Inhibitors Exhibiting Potent Antibacterial Activity. *Angew. Chem. Int. Ed.* **2014**, *53*, 14605–14609. [CrossRef] [PubMed]

47. Surup, F.; Viehrig, K.; Mohr, K.I.; Jansen, R.; Herrmann, J.; Müller, R. Disciformycins A and B, unprecedented 12-membered Macrolide-Glycoside Antibiotics from the Myxobacterium Pyxidicoccus fallax active against multiresistant Staphylococci. *Angew. Chem. Int. Ed.* **2014**, *53*, 13588–13591. [CrossRef] [PubMed]

48. Plaza, A.; Garcia, R.; Bifulco, G.; Martinez, J.P.; Hüttel, S.; Sasse, F.; Meyerhans, A.; Stadler, M.; Müller, R. Aetheramides A and B, Potent HIV-Inhibitory Depsipeptides from a Myxobacterium of the New Genus "*Aetherobacter*". *Org. Lett.* **2012**, *14*, 2854–2857. [CrossRef] [PubMed]

49. Gerth, K.; Bedorf, N.; Irschik, H.; Höfle, G.; Reichenbach, H. The soraphens: A family of novel antifungal compounds from Sorangium cellulosum (Myxobacteria). I. Soraphen A1 alpha: Fermentation, isolation, biological properties. *J. Antibiot.* **1994**, *47*, 23–31. [CrossRef] [PubMed]

50. Gerth, K.; Bedorf, N.; Höfle, G.; Irschik, H.; Reichenbach, H. Epothilons A and B: Antifungal and cytotoxic compounds from *Sorangium cellulosum* (Myxobacteria). Production, physico-chemical and biological properties. *J. Antibiot.* **1996**, *49*, 560–563. [CrossRef] [PubMed]

51. Sasse, F.; Steinmetz, H.; Schupp, T.; Petersen, F.; Memmert, K.; Hofmann, H.; Heusser, C.; Brinkmann, V.; von Matt, P.; Höfle, G.; et al. Argyrins, immunosuppressive cyclic peptides from myxobacteria. I. Production, isolation, physico-chemical and biological properties. *J. Antibiot.* **2002**, *55*, 543–551. [CrossRef] [PubMed]

52. Held, J.; Gebru, T.; Kalesse, M.; Jansen, R.; Gerth, K.; Müller, R.; Mordmüller, B. Antimalarial activity of the myxobacterial macrolide chlorotonil a. *Antimicrob. Agents Chemother.* **2014**, *58*, 6378–6384. [CrossRef] [PubMed]

53. Wenzel, S.C.; Müller, R. The impact of genomics on the exploitation of the myxobacterial secondary metabolome. *Nat. Prod. Rep.* **2009**, *26*, 1385–1407. [CrossRef] [PubMed]

54. Schneiker, S.; Perlova, O.; Kaiser, O.; Gerth, K.; Alici, A.; Altmeyer, M.O.; Bartels, D.; Bekel, T.; Beyer, S.; Bode, E.; et al. Complete genome sequence of the myxobacterium *Sorangium cellulosum. Nat. Biotechnol.* **2007**, *25*, 1281–1289. [CrossRef] [PubMed]

55. Goldman, B.S.; Nierman, W.C.; Kaiser, D.; Slater, S.C.; Durkin, A.S.; Eisen, J.A.; Ronning, C.M.; Barbazuk, W.B.; Blanchard, M.; Field, C.; et al. Evolution of sensory complexity recorded in a myxobacterial genome. *Proc. Natl. Acad. Sci. USA* **2006**, *10*, 15200–15205. [CrossRef] [PubMed]

56. Steinmetz, H.; Mohr, K.I.; Zander, W.; Jansen, R.; Gerth, K.; Müller, R. Indiacens A and B: Prenyl Indoles from the Myxobacterium *Sandaracinus amylolyticus. J. Nat. Prod.* **2012**, *75*, 1803–1805. [CrossRef] [PubMed]

57. Mohr, K.I.; Garcia, R.O.; Gerth, K.; Irschik, H.; Müller, R. *Sandaracinus amylolyticus* gen. nov., sp. nov., a starch-degrading soil myxobacterium, and description of Sandaracinaceae fam. nov. *IJSEM* **2012**, *62*, 1191–1198. [CrossRef] [PubMed]

58. Garcia, R.; Stadler, M.; Gemperlein, K.; Müller, R. *Aetherobacter fasciculatus* gen. nov., sp. nov. and *Aetherobacter rufus* gen. nov., sp. nov., two novel myxobacteria with promising biotechnological applications. *IJSEM* **2015**, *66*, 928–938. [CrossRef] [PubMed]

59. Jansen, R.; Mohr, K.I.; Bernecker, S.; Stadler, M.; Müller, R. Indothiazinone, an indolyl-thiazolyl-ketone from a novel myxobacterium belonging to the Sorangiineae. *J. Nat. Prod.* **2014**, *25*, 1054–1060. [CrossRef] [PubMed]

60. Sood, S.; Awal, R.P.; Wink, J.; Mohr, K.I.; Rohde, M.; Stadler, M.; Kämpfer, P.; Glaeser, S.; Schumann, P.; Garcia, R.; et al. *Aggregicoccus edonensis* gen. nov., sp. nov., an unusually aggregating myxobacterium isolated from a soil sample. *IJSEM* **2014**, *65*, 745–753. [CrossRef] [PubMed]

61. Karwehl, S.; Mohr, K.I.; Jansen, R.; Sood, S.; Bernecker, S.; Stadler, M. Edonamides, the first secondary metabolites from the recently described Myxobacterium *Aggregicoccus edonensis. Tetrahedron Lett.* **2015**, *56*, 6402–6404. [CrossRef]

62. Muyzer, G. Genetic fingerprinting of microbial communities: Present status and future perspective. In *Microbial Biosystems: New Frontiers, Proceedings of the 8th International Symposium, Halifax, Canada, 9–14 August 1998*; Bell, C.R., Brylinsky, M., Johnson-Green, P., Eds.; Microbial Ecology Atlantic Canada Society for Microbial Ecology: Halifax, Canada, 2000.

63. Vaz-Moreira, I.; Silva, M.E.; Manaia, C.M.; Nunes, O.C. Diversity of bacterial isolates from commercial and homemade composts. *Microbiol. Ecol.* **2008**, *55*, 714–722. [CrossRef] [PubMed]

64. Winterberg, H. Zur Methodik der Bakterienzahlung. *Z. Hyg.* **1898**, *29*, 75–93. [CrossRef]

65. Ward, D.M.; Weller, R.; Bateson, M.M. 16S rRNA sequences reveal numerous uncultured microorganisms in a natural community. *Nature* **1990**, *345*, 63–65. [CrossRef] [PubMed]

66. Lewis, K. Platforms for antibiotic discovery. *Nat. Rev. Drug Disc.* **2013**, *12*, 371–387. [CrossRef] [PubMed]

67. Lomolino, M.V.; Riddle, B.R.; Whittaker, R.; Brown, J.H. *Biogeography*; Sinauer Associates: Sunderland, MA, USA, 2010.

68. Ramette, A.; Tiedje, J.M. Biogeography: An Emerging Cornerstone for Understanding Prokaryotic Diversity, Ecology, and Evolution. *Microb. Ecol.* **2005**, *53*, 192–207. [CrossRef] [PubMed]

69. Hanson, C.A.; Fuhrman, J.A.; Horner-Devine, M.-C.; Martiny, J.B.H. Beyond biogeographic patterns: Processes shaping the microbial landscape. *Nat. Rev. Microbiol.* **2012**, *10*, 497–506. [CrossRef] [PubMed]

70. DeLong, E.E.; Pace, N.R. Environmental diversity of Bacteria and Archaea. *Syst. Biol.* **2001**, *50*, 470–478. [CrossRef] [PubMed]

71. Bano, N.; Ruffin, S.; Ransom, B.; Hollibaugh, J.T. Phylogenetic composition of Arctic Ocean archaeal assemblages and comparison with Antarctic assemblages. *Appl. Environ. Microbiol.* **2004**, *70*, 781–789. [CrossRef] [PubMed]

72. Glöckner, F.O.; Zaichikov, E.; Belkova, N.; Denissova, L.; Pernthaler, J.; Pernthaler, A.; Amann, R. Comparative 16S rRNA analysis of lake bacterioplankton reveals globally distributed phylogenetic clusters including an abundant group of actinobacteria. *Appl. Environ. Microbiol.* **2000**, *66*, 5053–5065. [CrossRef] [PubMed]

73. Rejmánková, E.; Komárek, J.; Komárková, J. Cyanobacteria—A neglected component of biodiversity: Patterns of species diversity in inland marshes of northern Belize (Central America). *Divers. Distrib.* **2004**, *10*, 189–199. [CrossRef]

74. Hedlund, B.P.; Staley, J.T. Microbial endemism and biogeography. In *Microbial Diversity and Bioprospecting*; Bull, A.T., Ed.; ASM Press: Washington, DC, USA, 2003; pp. 225–231.

75. Nemergut, D.R.; Costello, E.K.; Hamady, M.; Lozupone, C.; Jiang, L.; Schmidt, S.K.; Fierer, N.; Townsend, A.R.; Cleveland, C.C.; Stanish, L.; et al. Global patterns in the biogeography of bacterial taxa. *Environ. Microbiol.* **2011**, *13*, 135–144. [CrossRef] [PubMed]

76. Lozupone, C.A.; Knight, R. Global patterns in bacterial diversity. *Proc. Natl. Acad. Sci. USA* **2007**, *104*, 11436–11440. [CrossRef] [PubMed]

77. Jiang, D.M.; Kato, C.; Zhou, X.W.; Wu, Z.H.; Sato, T.; Li, Y.Z. Phylogeographic separation of marine and soil myxobacteria at high levels of classification. *ISME J.* **2010**, *4*, 1520–1530. [CrossRef] [PubMed]

78. Brinkhoff, T.; Fischer, D.; Vollmers, J.; Voget, S.; Beardsley, C.; Thole, S.; Mussmann, M.; Kunze, B.; Wagner-Döbler, I.; Daniel, R.; et al. Biogeography and phylogenetic diversity of a cluster of exclusively marine myxobacteria. *IJSEM* **2012**, *6*, 1260–1272. [CrossRef] [PubMed]

79. Wielgoss, S.; Didelot, X.; Chaudhuri, R.R.; Liu, X.; Weedall, G.D.; Velicer, G.J.; Vos, M. A barrier to homologous recombination between sympatric strains of the cooperative soil bacterium *Myxococcus xanthus*. *ISME J.* **2016**, *10*, 2468–2477. [CrossRef] [PubMed]

80. Kraemer, S.A.; Wielgoss, S.; Fiegna, F.; Velicer, G.J. The biogeography of kin discrimination across microbial neighbourhoods. *Mol. Ecol.* **2016**, *25*, 4875–4888. [CrossRef] [PubMed]

81. Gerth, K.; Pradella, S.; Perlova, O.; Beyer, S.; Müller, R. Myxobacteria: Proficient producers of novel natural products with various biological activities—Past and future biotechnological aspects with the focus on the genus Sorangium. *J. Biotechnol.* **2003**, *106*, 233–253. [CrossRef] [PubMed]

82. Velicer, G.J.; Mendes-Soares, H.; Wielgoss, S. Whence Comes Social Diversity? Ecological and Evolutionary Analysis of the Myxobacteria. In *Myxobacteria: Genomics, Cellular and Molecular Biology*; Yang, Z., Higgs, P.I., Eds.; Caister Academic Press: Poole, UK, 2014.

83. Tian, F.; Yong, Y.; Chen, B.; Li, H.; Yao, Y.-F.; Guo, X.-K. Bacterial, archaeal and eukaryotic diversity in Arctic sediment as revealed by 16S rRNA and 18S rRNA gene clone libraries analysis. *Polar Biol.* **2009**, *32*, 93–103. [CrossRef]

84. Wu, Z.H.; Jiang, D.M.; Li, P.; Li, Y.Z. Exploring the diversity of myxobacteria in a soil niche by myxobacteria-specific primers and probes. *Environ. Microbiol.* **2005**, *7*, 1602–1610. [CrossRef] [PubMed]

85. Jiang, D.M.; Wu, Z.H.; Zhao, J.Y.; Li, Y.Z. Fruiting and non-fruiting myxobacteria: A phylogenetic perspective of cultured and uncultured members of this group. *Mol. Phylogenet. Evol.* **2007**, *44*, 545–552. [CrossRef] [PubMed]

86. Anderson, K.E.; Russell, J.A.; Moreau, C.S.; Kautz, S.; Sullam, K.E.; Hu, Y.; Basinger, U.; Mott, B.M.; Buck, N.; Wheeler, D.E. Highly similar microbial communities are shared among related and trophically similar ant species. *Mol. Ecol.* **2012**, *21*, 2282–2296. [CrossRef] [PubMed]

87. Dedysh, S.N.; Pankratov, T.A.; Belova, S.E.; Kulichevskaya, I.S.; Liesack, W. Phylogenetic Analysis and In Situ Identification of Bacteria Community Composition in an Acidic Sphagnum Peat Bog. *Appl. Environ. Microbiol.* **2005**, *72*, 2110–2117. [CrossRef] [PubMed]

88. Hook, L.A. Distribution of Myxobacters in Aquatic Habitats of an Alkaline Bog. *AEM* **1977**, *34*, 333–335.

89. Rückert, G. Myxobakterien-Artenspektren von Boden in Abhängigkeit von bodenbildenden Faktoren unter besonderer Berücksichtigung der Bodenreaktion. *Z. Pflanzenernaehr. Bodenkd.* **1979**, *142*, 330–343. [CrossRef]

90. Pacha, R.E.; Porter, S. Characteristics of Myxobacteria Isolated from the Surface of Freshwater Fish. *Appl. Microbiol.* **1968**, *16*, 1901–1906. [PubMed]

91. Li, S.G.; Zhou, X.W.; Li, P.F.; Han, K.; Li, W.; Li, Z.F.; Wu, Z.H.; Li, Y.Z. The existence and diversity of myxobacteria in lake mud—A previously unexplored myxobacteria habitat. *Environ. Microbiol. Rep.* **2012**, *4*, 587–595. [CrossRef] [PubMed]

92. Kou, W.; Zhang, J.; Lu, X.; Ma, Y.; Mou, X.; Wu, L. Identification of bacterial communities in sediments of Poyang Lake, the largest freshwater lake in China. *SpringerPlus* **2016**, *1*, 401. [CrossRef] [PubMed]

93. Ji, Y.; Angel, R.; Klose, M.; Claus, P.; Marotta, H.; Pinho, L.; Enrich-Prast, A.; Conrad, R. Structure and function of methanogenic microbial communities in sediments of Amazonian lakes with different water types. *Environ. Microbiol.* **2016**, *18*, 5082–5100. [CrossRef] [PubMed]

94. Kandel, P.P.; Pasternak, Z.; van Rijn, J.; Nahum, O.; Jurkevitch, E. Abundance, diversity and seasonal dynamics of predatory bacteria in aquaculture zero discharge systems. *FEMS Microbiol. Ecol.* **2014**, *89*, 149–161. [CrossRef] [PubMed]

95. Zhang, X.; Yao, Q.; Cai, Z.; Xie, X.; Zhu, H. Isolation and Identification of Myxobacteria from Saline-Alkaline Soils in Xinjiang, China. *PLoS ONE* **2013**, *8*, e70466. [CrossRef] [PubMed]

96. Kaushal, S.S.; Likens, G.E.; Pace, M.L.; Utz, R.M.; Haq, S.; Gorman, J.; Gresea, M. Freshwater salinization syndrome on a continental scale. *Proc. Natl. Acad. Sci. USA* **2018**, *115*, 574–583. [CrossRef] [PubMed]

97. Brockman, E.R. Fruiting myxobacteria from the South Carolina coast. *J. Bacteriol.* **1963**, *94*, 1253–1254.

98. Li, B.; Yao, Q.; Zhu, H. Approach to analyze the diversity of myxobacteria in soil by semi-nested PCR-denaturing gradient gel electrophoresis (DGGE) based on taxon-specific gene. *PLoS ONE* **2014**. [CrossRef] [PubMed]

99. Albataineh, H.D.; Stevens, D.C. Marine Myxobacteria: A Few Good Halophiles. *Mar. Drugs* **2018**, *16*, 209. [CrossRef] [PubMed]

100. Schäberle, T.F.; Goralski, E.; Neu, E.; Erol, O.; Hölzl, G.; Dörmann, P.; Bierbaum, G.; König, G.M. Marine myxobacteria as a source of antibiotics–comparison of physiology, polyketide-type genes and antibiotic production of three new isolates of *Enhygromyxa salina*. *Mar. Drugs* **2010**, *8*, 2466–2479. [CrossRef] [PubMed]

101. Fudou, R.; Iizuka, T.; Sato, S.; Ando, T.; Shimba, N.; Yamanaka, S. Haliangicin, a novel antifungal metabolite produced by a marine myxobacterium. 2. Isolation and structural elucidation. *J. Antibiot.* **2001**, *54*, 153–156. [CrossRef] [PubMed]

102. Felder, S.; Dreisigacker, S.; Kehraus, S.; Neu, E.; Bierbaum, G.; Wright, P.R.; Menche, D.; Schäberle, T.F.; König, G.M. Salimabromide: Unexpected chemistry from the obligate marine myxobacterium *Enhygromxya salina*. *Chemistry* **2013**, *19*, 9319–9324. [CrossRef] [PubMed]

103. Felder, S.; Kehraus, S.; Neu, E.; Bierbaum, G.; Schäberle, T.F.; König, G.M. Salimyxins and enhygrolides: Antibiotic, sponge-related metabolites from the obligate marine myxobacterium *Enhygromyxa salina*. *ChemBioChem* **2013**, *14*, 1363–1371. [CrossRef] [PubMed]

104. Sun, Y.; Tomura, T.; Sato, J.; Iizuka, T.; Fudou, R.; Ojika, M. Isolation and Biosynthetic Analysis of Haliamide, a New PKS-NRPS Hybrid Metabolite from the Marine Myxobacterium *Haliangium ochraceum*. *Molecules* **2016**, *21*, 59. [CrossRef] [PubMed]

105. Treude, N.; Rosencrantz, D.; Liesack, W.; Schnell, S. Strain FAc12, a dissimilatory iron-reducing member of the *Anaeromyxobacter* subgroup of Myxococcales. *FEMS Microbiol. Ecol.* **2003**, *44*, 261–269. [CrossRef]

106. Thomas, S.H.; Padilla-Crespo, E.; Jardine, P.M.; Sanford, R.A.; Löffler, F.E. Diversity and distribution of anaeromyxobacter strains in a uranium-contaminated subsurface environment with a nonuniform groundwater flow. *AEM* **2009**, *75*, 3679–3687. [CrossRef] [PubMed]

107. Lin, J.; Ratering, S.; Schnell, S. Microbial iron cylce in corrosion material of drinking water pipelines. *Ann. Agrar. Sci.* **2011**, *9*, 18–25.

108. Kudo, K.; Yamaguchi, N.; Makino, T.; Ohtsuka, T.; Kimura, K.; Dong, D.T.; Amachi, S. Release of arsenic from soil by a novel dissimilatory arsenatereducing bacterium, *Anaeromyxobacter* sp. strain PSR-1. *AEM* **2013**, *79*, 4635–4642. [CrossRef] [PubMed]

109. Kim, M.; Oh, H.S.; Park, S.C.; Chun, J. Towards a taxonomic coherence between average nucleotide identity and 16S rRNA gene sequence similarity for species demarcation of prokaryotes. *IJSEM* **2014**, *64*, 346–351. [CrossRef] [PubMed]

110. Zerzghi, H.; Brooks, J.P.; Gerba, C.P. Pepper IL. Influence of long-term land application of Class B biosolids on soil bacterial diversity. *J. Appl. Microbiol.* **2010**, *109*, 698–706. [CrossRef] [PubMed]

111. Mohr, K.I.; Moradi, A.; Glaeser, S.P.; Kämpfer, P.; Gemperlein, K.; Nübel, U.; Schumann, P.; Müller, R.; Wink, J. *Nannocystis konarekensis* sp. nov., a novel myxobacterium from an Iranian desert. *IJSEM* **2018**, *68*, 721–729. [CrossRef] [PubMed]
112. Xu, K.; Liu, H.; Li, X.; Chen, J.; Wang, A. Typical methanogenic inhibitors can considerably alter bacterial populations and affect the interaction between fatty acid degraders and homoacetogens. *Appl. Microbiol. Biotechnol.* **2010**, *87*, 2267–2279. [CrossRef] [PubMed]
113. Hansel, C.M.; Fendorf, S.; Jardine, P.M.; Francis, C.A. Changes in bacterial and archaeal community structure and functional diversity along a geochemically variable soil profile. *Appl. Environ. Microbiol.* **2008**, *74*, 1620–1633. [CrossRef] [PubMed]
114. Brockman, E.R. Myxobacters from Arid Mexican Soil. *AEM* **1976**, *32*, 642–644.
115. Gerth, K.; Müller, R. Moderately thermophilic Myxobacteria: Novel potential for the production of natural products isolation and characterization. *Environ. Microbiol.* **2005**, *7*, 874–880. [CrossRef] [PubMed]
116. Iizuka, T.; Tokura, M.; Jojima, Y.; Hiraishi, A.; Yamanaka, S.; Fudou, R. Enrichment and Phylogenetic Analysis of Moderately Thermophilic Myxobacteria from Hot Springs in Japan. *Microbes Environ.* **2006**, *21*, 189–199. [CrossRef]
117. Dawid, W.; Gallikowski, C.A.; Hirsch, P. Psychrophilic myxobacteria from Antarctic soils. *Polarforschung* **1988**, *58*, 271–278.
118. Brockman, E.R.; Boyd, W.L. Myxobacteria from soils of the Alaskan and Canadian arctic. *J. Bacteriol* **1963**, *86*, 605–606.
119. Müller, R.; Wink, J. Future potential for anti-infectives from bacteria—How to exploit biodiversity and genomic potential. *Int. J. Med. Microbiol.* **2014**, *304*, 3–13. [CrossRef] [PubMed]

Review

Antiviral Compounds from Myxobacteria

Lucky S. Mulwa [1,2] and Marc Stadler [1,*]

[1] Department of Microbial Drugs, Helmholtz Centre for Infection Research and German Centre for Infectio Research (DZIF), Partner Site Hannover/Braunschweig, Inhoffenstrasse 7, 38124 Braunschweig, Germany; luckymulwa@gmail.com

[2] Department of Microbial Strain Collection (MISG), Helmholtz Centre for Infection Research (HZI), Inhoffenstrasse 7, 38124 Braunschweig, Germany

* Correspondence: marc.stadler@helmholtz-hzi.de; Tel.: +49-531-6181-4240; Fax: +49-531-6181-9499

Received: 25 May 2018; Accepted: 17 July 2018; Published: 19 July 2018

Abstract: Viral infections including human immunodeficiency virus (HIV), cytomegalovirus (CMV), hepatitis B virus (HBV), and hepatitis C virus (HCV) pose an ongoing threat to human health due to the lack of effective therapeutic agents. The re-emergence of old viral diseases such as the recent Ebola outbreaks in West Africa represents a global public health issue. Drug resistance and toxicity to target cells are the major challenges for the current antiviral agents. Therefore, there is a need for identifying agents with novel modes of action and improved efficacy. Viral-based illnesses are further aggravated by co-infections, such as an HIV patient co-infected with HBV or HCV. The drugs used to treat or manage HIV tend to increase the pathogenesis of HBV and HCV. Hence, novel antiviral drug candidates should ideally have broad-spectrum activity and no negative drug-drug interactions. Myxobacteria are in the focus of this review since they produce numerous structurally and functionally unique bioactive compounds, which have only recently been screened for antiviral effects. This research has already led to some interesting findings, including the discovery of several candidate compounds with broad-spectrum antiviral activity. The present review looks at myxobacteria-derived antiviral secondary metabolites.

Keywords: myxobacteria; antivirals; secondary metabolites; HIV; Ebola; hepatitis viruses

---

1. Introduction

Antimicrobial resistance (AMR) threatens the effective treatment, control and management of an increasing range of infections caused by viruses, bacteria, parasites and fungal pathogens [1]. Development of effective therapies to suppress human immunodeficiency virus (HIV) and hepatitis B virus (HBV) has been achieved with great success [2]. However, the medicines are not curative, and therefore more efforts in HIV and HBV drug discovery are directed toward longer-acting therapies or compounds with new mechanisms of action that could potentially lead to a cure or complete eradication of the viruses [2]. In 2010, an estimated 7–20% of people starting antiretroviral therapy (ART) globally had drug-resistant HIV [3]. Some countries have recently reported levels of 15% amongst those starting HIV treatment, and up to 40% among people re-starting therapy [3]. Of great concern are the high levels of viral resistance towards nucleoside reverse transcriptase inhibitors (NRTIs) as recently found in Kenyan children [4]. In 2015, the World Health Organization (WHO) recommended that everyone living with HIV should start on antiretroviral treatment. Hence, increased ART resistance is expected as more people start ART [3]. The United States Food and Drug Administration (FDA) approved anti-HIV drugs are classified into eight classes according to modes of action [5]. The common side effects of Abacavir® and other NRTIs are that they can cause life-threatening side effects, including a serious allergic reaction, a build-up of lactic acid in the blood, and hepatotoxicity [6,7]. In fact, each drug class of the FDA approved anti-HIV drugs has side effects, and some are contra-indicated for co-infected patients with HIV and HBC or HCV [8]. On the other hand, all influenza A viruses circulating in humans were reported to be resistant to Amantadine® and Rimantadine®, two essential antivirals for treatment of epidemic and pandemic influenza A. However, the frequency of resistance to Oseltamivir® another antiviral with different mode of action for treating influenza A remains low at 1–2% [3]. Treatment failure of antivirals has been suggested to be caused by the emergence of recombinant viruses, drug resistance, and cell toxicity [9,10]. Compounds with a different mode of action can play an essential role in overcoming AMR. Viral disease such as influenza spreads fast and knows no borders, with the vast masses of people travelling all over the globe due to efficient transport systems. Hence, there is an urgent need for international collaboration to identify new antiviral agents with new modes of action and better efficacy.

Myxobacteria are well-known to be producers of biologically active secondary metabolites with novel carbon skeletons and new modes of action [11–13]. Many compounds isolated from myxobacteria have recently been found to have impressive antiviral activity. More so, some have been found to have an unusual broad-spectrum antiviral activity [11,12,14].

2. Myxobacteria

Myxobacteria are δ-proteobacteria belonging to the order *Myxococcales*. They are rod-shaped, Gram-negative bacteria that exhibit gliding motility and swarm on solid surfaces. Under nutrient-limiting conditions, they form species-specific fruiting bodies (Figure 1) [15]. Within these fruiting bodies, some vegetative cells convert to myxospores, which are desiccation-resistant and can survive over decades. Under appropriate conditions the spores germinate [16]. These soil-dwelling microorganisms have also been isolated from other habitats such as the bark of trees, oceans, freshwater lakes, and herbivore dung [15,17]. Myxobacteria have also been isolated from extreme environments such as desert soils [18]. Numerous unique classes of secondary metabolites have been isolated from myxobacteria, the majority of which are biogenetically derived from polyketide synthases (PKSs) and non-ribosomal peptide synthetases (NPRSs) or a hybrid of PKSs and NPRSs [12,15,19,20]. PKSs and NPRSs are enzymatic "assembly lines" of complex multi-step biosynthetic pathways for making compounds by catalysing the stepwise condensation of a starter unit with small monomeric building blocks [19]. The ability to produce unique metabolites is conferred by the creative biosynthetic pathways and the large genome of 9–14 Mb [21], consistent with the strengthening correlation between genome size and the extent of secondary metabolites produced [21–25]. In fact, the sequenced genomes of myxobacteria are the largest yet known from any bacterium [20–23,25]. In the last 35 years, over 100 new carbon skeleton secondary metabolites, with over 600 analogues, have been isolated from over 9000 strains of myxobacteria [12]. The metabolites exhibited antifungal, antibacterial, antimalarial, antitumor, and anti-immunomodulatory properties some with novel modes of action and have been reviewed extensively [12,20,26]. Microorganisms are valuable as producers of bioactive metabolites because they can be cultivated in bioreactors from as little as below 5 mL to large scales of over 100,000 L, making the production of natural products independent of season, locality, or climate [15]. Furthermore, conditions in a bioreactor are controllable to optimise production of the desired outcome. Particularly important as illustrated by Zeeck et al., 2002 in the 'OSMAC' (One Strain-Many Compounds) approach, which revealed that microorganisms do not exhaust their potential for producing metabolites under standard laboratory conditions [27].

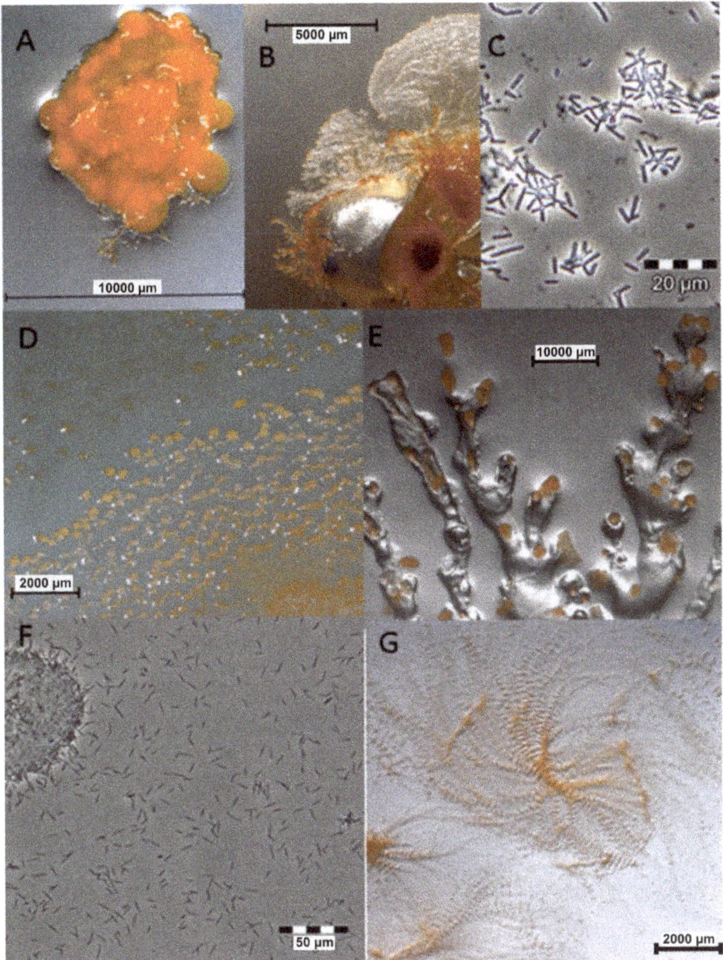

Figure 1. Images of myxobacteria. (**A–C**): *Sorangium cellulosum*; (**A**): Fruiting bodies; (**B**): Swarming on agar plate; (**C**): Cells from the liquid medium under the light microscope. (**D–E**): Images of the producers ofthiangazole (**7**), phenalamide A_1 (**8**) and phenoxan (**9**), from agar plates; (**D**): *Myxococcus stipitatus*; (**E**): *Polyangium* species; (**F–G**): *Angiococcus disciformis* (strain An d30) producer of myxochelins; (**F**): culture under the light microscope from liquid media; (**G**): culture on agar plate. Images provided by Joachim Wink (HZI Braunschweig).

3. Secondary Metabolites from Myxobacteria with Antiviral Activity

3.1. Human Immunodeficiency Virus (HIV)

HIV is a lentivirus of the Retroviridae family. HIV targets immune cells, and reverse transcribes its single-stranded RNA (ssRNA) genome, integrating into the host chromosomal DNA. The virus uses high antigenic diversity and multiple mechanisms to avert recognition by the human immune system thus posing a challenge to host defences and treatment [28].

Various assays have been developed and used to identify molecules with anti-HIV activity. Some of the assays includes structure-based design of a small molecule CD4-antagonist with broad spectrum anti-HIV-1 activity [29], structure-based identification of small molecule antiviral compounds targeted to the gp41 core structure of the human immunodeficiency virus type 1 [30], and identification of HIV inhibitors by high-throughput (HTP) two-step infectivity assay [31]. The HTP assay has been used on myxobacterial-derived molecules with success due to the ability to screen a large number of molecules in a short period.

Sulfangolids, the first sulfate esters containing a series of secondary metabolites produced by several strains of *Sorangium cellulosum*, were isolated together with the structurally related macrolide kulkenon (**5**) [32]. Sulfangolid C (**1**), soraphen F (**2**), epothilon D (**3**), and spirangien B (**4**), showed impressive activity, with EC_{50} values in the nM range with a selectivity index value greater than 15 (SI > 15) in the high-throughput two-step infectivity assay [31]. Despite the impressive antiviral activity of **5**, the SI is low because of toxicity. A search in the SciFinder database revealed dozens of analogues of soraphen and epothilone [33]. It may be promising to screen the large number of analogues in the soraphen and epothilone families to attempt to encounter more potent representatives with better SI values. The soraphens have been reported as acetyl-CoA carboxylate transferase inhibitors [34], while the epothilones stabilise microtubuli of macrophages in a similar manner as Taxol® without showing taxol-like endotoxin activity [35,36]. In fact, the FDA-approved anticancer drug, Ixabepilone®, is an epothilone B derivative [37]. Metabolites **3** and **4** are reported to decelerate the phosphorylation and degradation of inhibitor of kappa Bα (IkBα) [36,38]. The compounds identified as preliminary hits for anti-HIV included **1–5** (Table 1, Figure 2) [31]. Rhizopodin (**6**), from *Myxococcus stipitatus* was identified as interesting in the two step HTP assay, likely because of its mode of action [31]. HIV cell-to-cell transmission is the primary route of HIV infection in naive cells in vivo. Actin filaments are known to be essential for virological synapse formation, therefore, virus synapses are interfered by **6**, which is a known actin inhibitor. Disorazol, tubulysin and stigmatellin variants were also reported to have mild anti-HIV activity [31]. Thiangazole (**7**), phenalamide A_1 (**8**), and phenoxan (**9**) isolated from two strains of *Polyangium* sp. and *Myxococcus stipitatus* strain Mx s40 were reported to have anti-HIV activity (Figure 2) [39]. They all revealed high activity by suppressing HIV-1-mediated cell death in the MT-4 cell assay with EC_{50} values of **9** and **8** in the nanomolar range, whereas thiangazole (**7**) had an impressive EC_{50} value in the picomolar range, making it a possible lead compound for anti-HIV therapy (Table 2, Figure 2) [39]. In another assay involving measuring ATP levels as a parameter of cell viability of TZM-bl cells aetheramide A (**10a**) and aetheramide B (**10b**) isolated from the recently described genus *Aetherobacter*, inhibited HIV-1 infection with IC_{50} value of 0.015 and 0.018 μM, respectively [40–42]. Concurrently, the aetheramides were reported to be moderately antifungal and cytotoxic [41]. The chemical structures of **10a** and **10b** are rare, containing a polyketide moiety and two amino acid residues, thus forming a new class of antivirals [40–42]. This discovery of new antivirals from the recently described myxobacteria genus, *Aetherobacter*, represents an example of the enormous biosynthetic capabilities of myxobacteria and their importance to drug discovery efforts [42]. Ratjadon A (**11**), an α-pyrone metabolite isolated from *Sorangium cellulosum* (strain Soce 360), was reported to inhibit HIV infection by blocking the Rev/CRM1-mediated nuclear export pathway [43,44]. The CRM1-Rev complex is an attractive target for the development of new antiviral drugs because the nuclear export of unspliced and partially spliced HIV-1 mRNA is mediated by the recognition of a leucine-rich nuclear export signal (NES) in the HIV Rev protein by the host protein CRM1/Exportin1 [44]. Despite **11** being reported to exhibit a strong anti-HIV activity, it has a low selectivity due to toxic effect. The low SI value limits the potential use of **11** as a therapeutic drug. More studies with derivatives of **11** need to be done. It is important to observe that different assays were used to screen for anti-HIV compounds. There is a need therefore for a standardised method to be able to adequately compare the anti-HIV activity of those compounds that have been identified as preliminary hits. Equally important is an evaluation of the mechanism of action on viruses in comparison to the mechanism of action on bacteria or fungi. Investigation for synergism between the identified anti-HIV compounds for possible use at a lower concentration to improve the selectivity index of the metabolites should be looked into in the future. Even the active metabolites that cannot realistically be further developed as drug candidates because they are too toxic, could be used as biochemical tools to attain a better understanding of the invasion mechanism of HIV, or for development of synthetic analogues that mimic these compounds without causing toxicity.

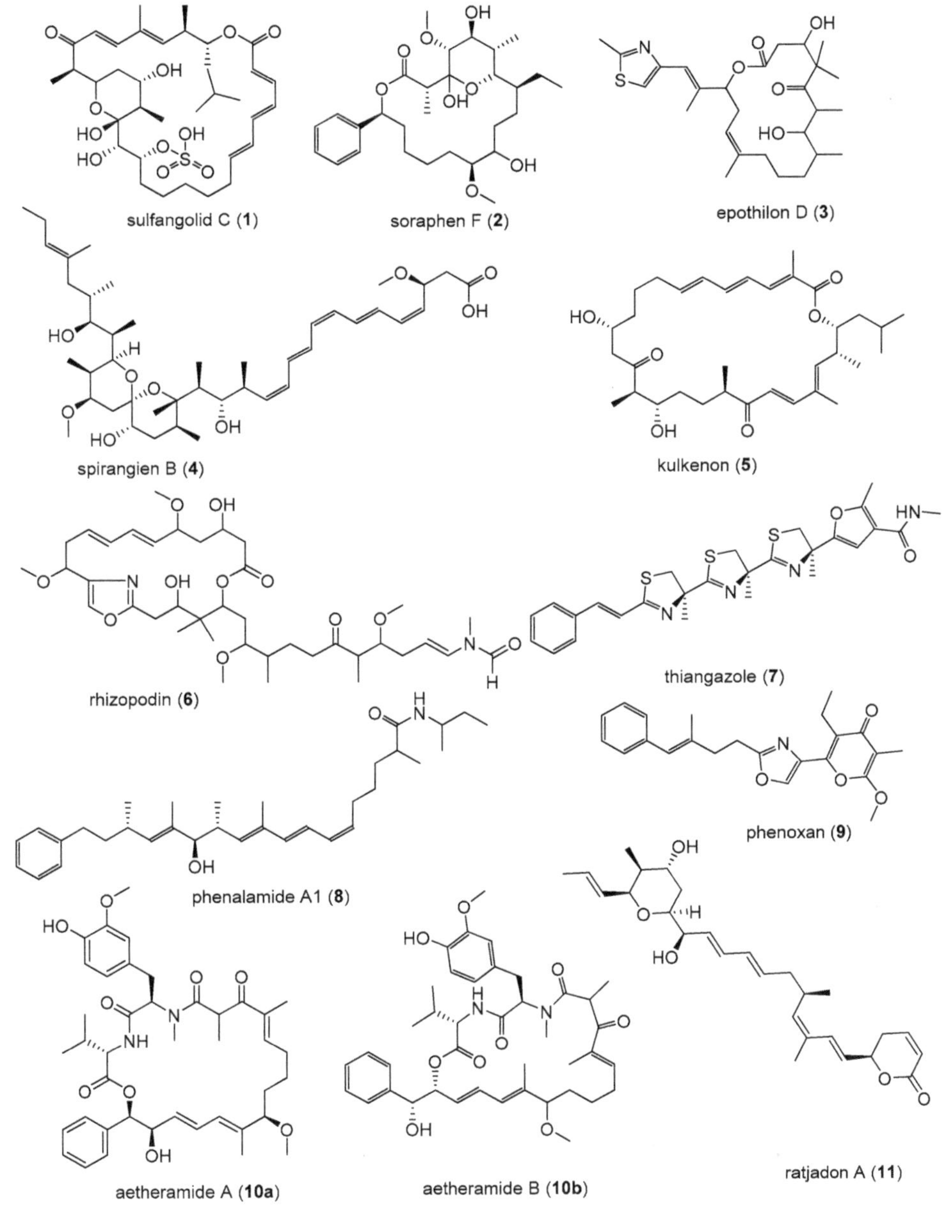

Figure 2. Myxobacterial-derived compounds with activity against human immunodeficiency virus (HIV).

Table 1. Preliminary anti-HIV hits from a high-throughput two-step infectivity assay [31].

Compound	MW [1]	EC$_{50}$ (μM)	CC$_{50}$ (μM)	SI **
Nevirapine *	266	0.07	81.8	>10^3
sulfangolid C (**1**)	682	0.41	8.18	20.2
soraphen F (**2**)	522	0.30	5.02	16.5
epothilon D (**3**)	491	0.0005	0.012	24.4
spirangien B (**4**)	717	0.007	0.35	52
kulkenon (**5**)	734	0.07	0.36	5.3

* Control, [1] Molecular weight, EC$_{50}$: effective concentration; CC$_{50}$: cytotoxic concentration; ** Selectivity Index = CC$_{50}$/EC$_{50}$. EC$_{50}$ is the concentration of a drug or metabolite which induces a response halfway between the baseline and maximum after a specified exposure time or gives the desired effect to 50% of test subjects. While SI is a comparison of the amount of a drug or metabolite that causes the desired effect to the amount that causes death or toxicity. Metabolites with a low EC$_{50}$ and a high SI values are good drug candidates.

Table 2. Anti-HIV-1 activities of compounds derived from myxobacteria by MT-4 cell assay [40,41].

Compound	MW [1]	Tx [2] (nM)	AE [3] (μM)	SI **
AZT *	267	250,000	25	10^4
thiangazole (**7**)	539	>4700	0.0047	>10^6
phenalamide A$_1$ (**8**)	491	102,000	1.02	10^5
Phenoxan (**9**)	379	>6600	6.6	>10^3

* azidothymidine Control, [1] Molecular weight, [2] Toxicity (Tx) is the lowest toxic concentration (nM) of the compound in the MT-4 cell assay, [3] Antiviral efficacy (AE) is given as the lowest effective concentration (μM) of the compound at which 100% prevention of the virus-mediated cytopathogenicity was observed in the MT-4 cell assay, ** Selectivity Index = Tx (nM)/AE (μM).

3.2. Human Cytomegalovirus (HCMV)

HCMV belongs to the β-herpesvirus family, with a high prevalence, infecting up to 80% of the general population usually asymptomatic in healthy people [45]. Diseases associated with HCMV include glandular fever and pneumonia. HCMV is also an important pathogen in organ transplant patients responsible for significant morbidity and mortality in organ transplant recipients, and a major cause of disease in patients with HIV infection [46]. HCMV infections in newborns may result in hearing loss, mental retardation and palsy [47]. The available FDA-approved therapeutic options for HCMV infection include ganciclovir, foscarnet, cidoforvir, and fomivirsen [48]. These drugs have different mechanisms of actions or applications, and represent the successes that had been made against the challenges of HCMV [48–50]. Several anti-HCMV drugs were reported to have low potency, poor oral bioavailability, and adverse side effects [50]. Moreover, drug resistance strains were reported to emerge [51]. Hence, there has been a renewed interest in search of new inhibitors of HCVM [50,51]. Of greater concern is the increase in the number of people living with transplanted organs, and the increase in HIV infected people [52]. Technological advancements have enabled organ transplants to be more accessible while the increase in HIV-infected individuals is due to new retroviral therapies that have converted HIV infection to chronic disease as infected people live longer, leading to increased cases of HCMV infection [52]. In 2011, the first case of HCMV-treated with AIC246, a novel anti-CMV compound that targets the viral terminase complex and remains active against virus resistant to DNA polymerase inheritors was reported [53], which represents a good example of the renewed interest for HCMV inhibitors [53]. Almost 60 patents claiming novel agents for the treatment of HCMV were launched from January 1996 to 2000, but so far none of these projects has led to the approval of an anti-HCMV drug [49]. However, the recent FDA approval of letermovir (Figure 3), providing a long-awaited alternative for preventing cytomegalovirus infection in allogeneic hematopoietic stem cell transplant recipients [54] is very encouraging.

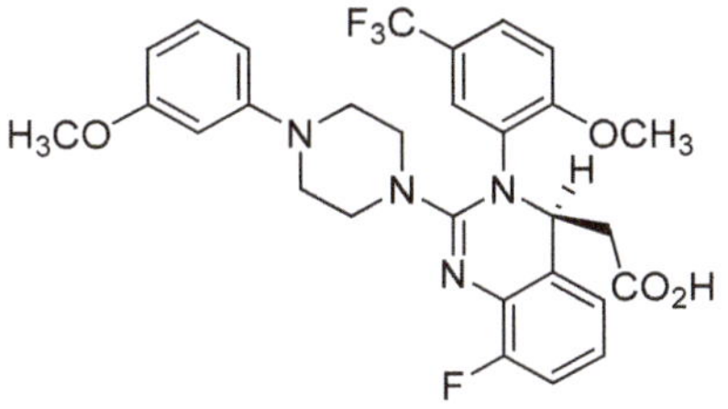

Figure 3. PREVYMIS™ (letermovir) a recently (2017) FDA-approved drug for the prevention of *Human cytomegalovirus* (HCMV) infection and disease in organ transplant patients.

A class of myxobacterial compounds, myxochelin, belonging to a larger group of natural products, siderophores, were isolated from several strains of myxobacteria [53,55]. Siderophores are secondary metabolites produced by some microorganisms under iron-limiting conditions, and enhance the uptake of iron [56]. Other siderophores isolated from myxobacteria includes nannochelins and hylachelins. Various studies have revealed myxochelins to be potent antitumour agents [55,57–59]. The antitumour activity was demonstrated to be caused by inhibition of human 5-lipoxygenase (5-LO) [58]. Surprisingly the inhibition of 5-LO by myxochelins was found not correlating with the iron affinities [58]. The enzyme 5-LO is responsible for the catalysis of two initial steps in the biosynthesis of leukotriens, starting from arachidonic acid [58]. Leukotriens are well-known mediators of a variety of allergic reactions such as inflammatory, rheumatic arthritis, allergic rhinitis and cardiovascular diseases [58]. Importantly, 5-LO pathways were associated with cancer proliferation, hence explaining the observed strong anticancer activity of myxochelin [58,60]. Nannochelins are reported to have no significant antimicrobial activity [24].

Myxochelin A (**12a**) was initially isolated from the culture broth of *Angiococcus disciformis* (strain An d30). Later on, myxochelins B (**12b**), C (**12c**), D (**12d**), E (**12e**), and F (**12f**) were isolated and also synthesised (Figure 4) [60,61]. The corresponding biosynthetic gene clusters have been identified in *Stigmatella aurantiaca* (Sga 15), and *Sorangium cellulosum* (Soce 56) [62]. Additional siderophores have been isolated from *Nannocystis exedens* (**21a–21c**) and *Hyalangium minutum* (**20a–20c**) [63,64]. Myxochelin C (**12c**) inhibited HCMV with an IC_{50} value of 0.7 µg/mL [46,53]. It could in future become feasible to test others among the over 500 different siderophores that are known to science [65]. In particular, the known myxobacterial-derived siderophores, such as nannochelins (**21a–21c**), hylachelins (**20a–20c**), and all the other myxochelin analogues (Figure 4) should be screened for various antiviral activities, especially anti-HCMV, and should be studied for structure activity relationship for possible discovery of more potent antivirals.

3.3. Ebola Virus Disease (EVD)

Ebola haemorrhagic fever is caused by the Ebola virus (EBOV), a single stranded RNA enveloped virus belonging to the family *Filoviridae*. EVD first appeared in 1976 in two simultaneous outbreaks, one in Nzara, South Sudan, and the other in Yambuku, the Democratic Republic of Congo. The latter occurred in a village near the Ebola River, from which the disease takes its name [65,66]. EVD case fatality rate is around 50%, with different cases from 25% to 90% fatality in past outbreaks reported [65]. Furthermore, EBOV is known to persist in immune-privileged sites, such as testicles, inside of the eye, and central nervous system, and in some people who have recovered from EVD [65]. The effect of the persistence is yet to be known [66].

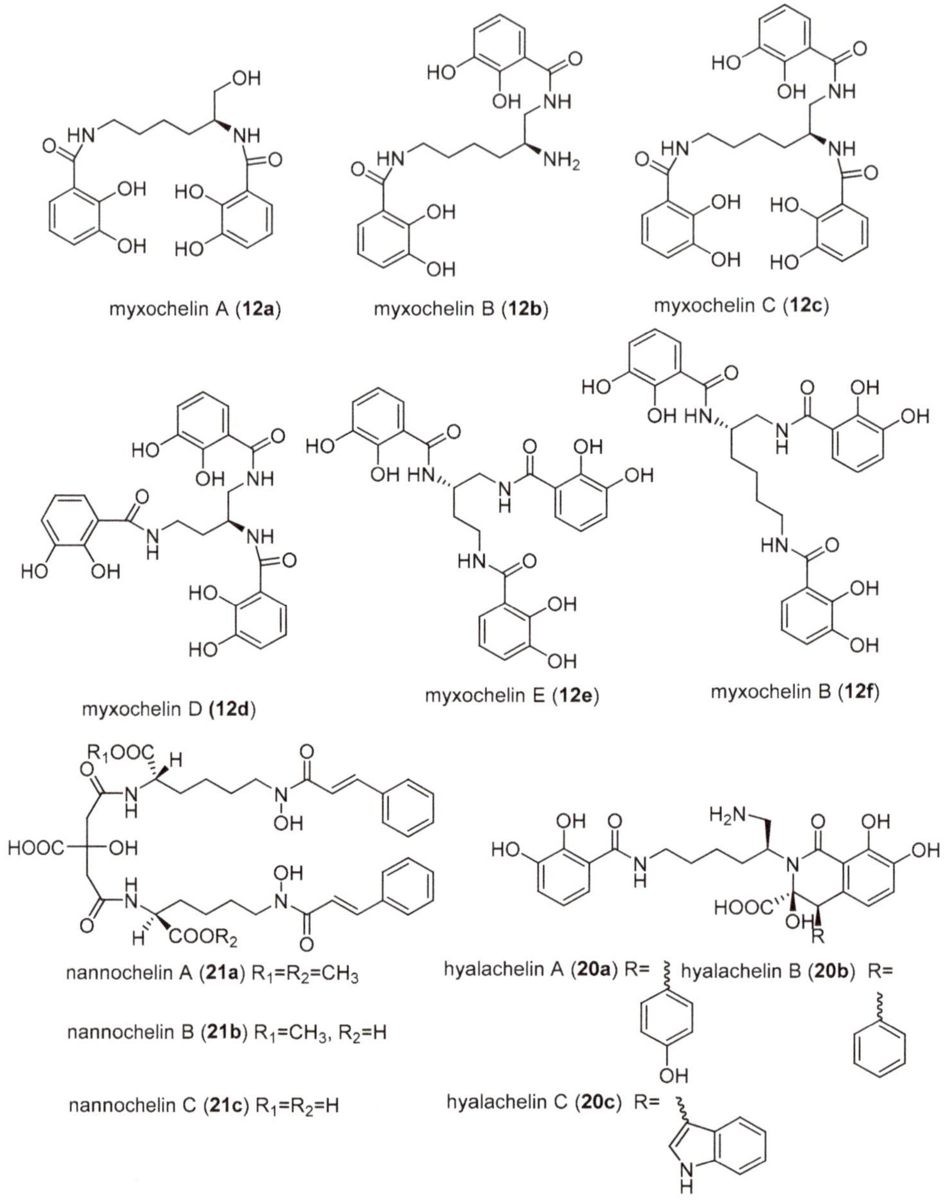

Figure 4. Myxobacterial-derived siderophores.

The re-emergence of Ebola occurred in West African countries causing 11,308 deaths leading to the WHO on 8 August 2014, declaring the epidemic to be an international public health emergency [66]. An experimental Ebola vaccine called rVSV-ZEBOV has been reported to show high protection against EVD [67]. No drug or licensed vaccine currently exists; hence there is an urgent need for drugs that inhibit entry or multiplication of EDV [62,67]. Developing an assay to test compounds for anti-EBOV poses a significant challenge because of the cost of equipment for the high risk, Biosafety S4 organism, involved [66]. However, various metabolites from myxobacteria were screened for EBOV inhibition by an assay with a surrogate system using Ebola envelope glycoprotein GP-pseudotyped lentiviral vectors (Figure 5) [67]. GP-pseudotype lentiviral vectors were used as tools to investigate the entry process of the viruses, enabling studies without the need of using the native Ebola virus reducing the safety level from the highest level 4 to level 2 [67]. The same analysis was conducted with vesicular stomatisis virus (VSV)-G-pseudotyped vectors to determine the EBOV-specificity of the inhibitory function of the compounds. Chondramides (**13a–13d**) were reported to inhibit EBOV-GP-mediated transduction with impressive IC_{50} values of 24–42 nM. The VSV-G-mediated transduction was less efficient, with an IC_{50} value of 55–111 nM [66]. Chondramides (**13a–13d**), a class of compounds known to interfere with actin, were isolated from a myxobacterium belonging to the genus *Chondromyces* [68]. Members of the genus *Chondromyces* belong to those myxobacteria known to synthesise two or more chemically unrelated secondary metabolites with different mechanisms of action [69]. Other promising hits were the noricumazoles, a family of polyketides from *Sorangium cellulosum*. Noricumazole A (**14a**) was found to inhibit EBOV-GP with an IC_{50} value of 0.33 µM. In fact, **14a** was found to be EBOV-GP specific and showed no significant inhibition against (VSV)-G-pseudotyped vectors [67]. Noricumazoles are known to be potassium channel blockers with **14a** known to be highly toxic while the derivatives, **14b** and **14c**, are equally active with lower toxicity [67,70]. The screening of myxobacterial natural compounds library resulted in the identification of inhibitors of EBOV-GP pseudotyped vectors, chondramides and noricumazole, whose mechanism of action is actin-stabilising and the channel blockers respectively [67]. These metabolites will give insights into the EBOV infection mechanism, rather than being used as drugs, because the modes of actions are expected to have side effects. However, the lower toxicity of **14b** and **14c**, which are derivatives of **14a**, is exciting and qualifies **14a** to be considered as a lead structure for the development of EBOV inhibitors.

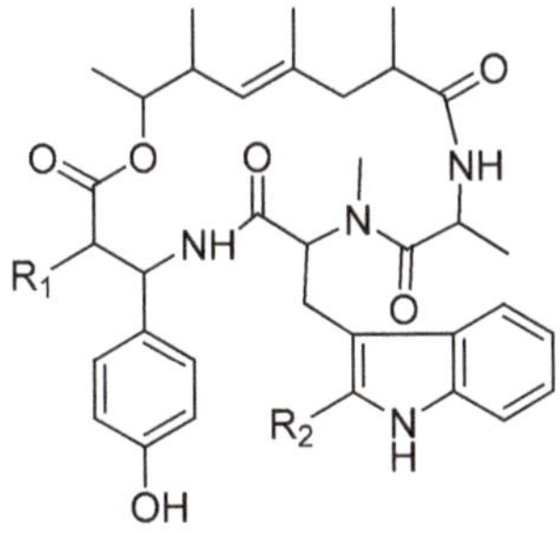

13

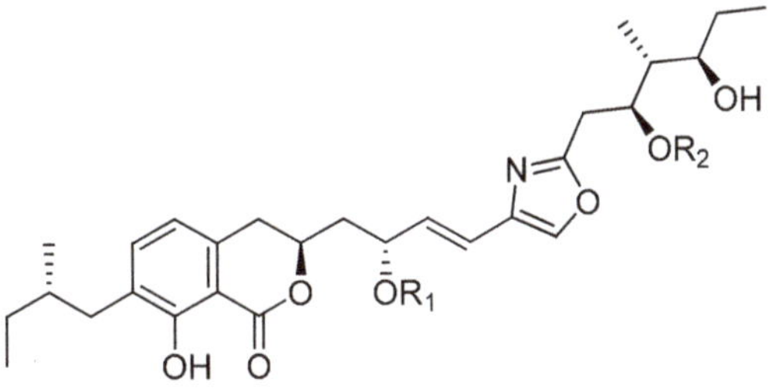

14

chondramide A (**13a**) R_1=OCH$_3$, R_2= H
chondramide B (**13b**) R_1=OCH$_3$, R_2= Cl
chondramide C (**13c**) R_1= H, R_2= H
chondramide D (**13d**) R_1= H, R_2= Cl

noricumazol A (**14a**) R_1= H, R_2 =H
noricumazol B (**14b**) R_1= H, R_2 = Alpha-D-arabino-Furanosyl
noricumazol C (**14c**) R_1= Beta-D-arabino-gluco-Pyranosyl, R_2 =H

Figure 5. Structures of myxobacterial-derived compounds with activity against Ebola virus (EBOV). Chondramides (**13a–d**) are known to be actin inhibitors, while noricumazols (**14a–c**) are known potassium channel inhibitors [70,71].

3.4. Hepatitis C Virus

HCV is an enveloped, single-stranded RNA virus with positive polarity (ss (+) RNA). HCV is transmitted by blood-to-blood contacts, such as through intravenous injections, blood transfusion, and various exposures to blood contaminants. It can also be transmitted by contact with bodily fluids including saliva or semen of an infected person [72]. By 2015, there were 71 million people infected with HCV globally [73]. HCV and hepatitis B virus (HBV) infection are the major causes of hepatocellular carcinoma (HCC), associated with cirrhosis [74]. Currently, no products are available to prevent HCV infection. There are some drugs available that can cure HCV infection [75]. However, treatment is complicated by HIV-HCV/HBC co-infections with drug-drug interactions between anti-HIV and anti-HCV drugs, resulting in serious side effects and can lead to the death of patient [8]. The discovery of broad-spectrum antivirals may play an essential role in overcoming this challenge.

The recently isolated compounds from *Labilithrix luteola*, labindoles A (**15a**), and B (**15b**) have been reported to have moderately inhibited HCV (Figure 6) [75]. Interestingly the labindoles were said to have no cytotoxicity, anti-bacterial or antifungal activities. 3-chloro-9H-carbazole (**17**) and 4-hydroxymethyl-quinoline (**18**) also isolated from *Labilithrix luteola* were reported to have a strong inhibition of HCV [75]. Soraphens are a family of polyketide-derived macrolactones comprising over 50 metabolites known for strong antifungal activity [76]. However, recent studies have suggested that soraphen A (**16**) inhibits HCV replication in HCV cell culture models expressing subgenomic and full-length replicons as well as a cell culture-adapted virus with an IC$_{50}$ value of 5 nM [11,77]. The HCV assay involved the development of subgenomic replicons that replicate autonomously in the human hepatoma cell line Huh-7 to be able to screen for anti-HCV activity. The subgenomic replicons are genetic materials from HCV, which represent the actual invasion and replication of HCV on the liver cells [75]. Furthermore, **16** is known to depolymerise the acetyl-CoA carboxylase (ACC) complexes into less active dimers [77]. The mechanism of action of **16** is a valuable probe to study the roles of ACC polymerisation and enzymatic activity in viral pathogenesis [77]. Various minor structure alterations of **16** did not affect the antiviral activity [11]. Owing to the fact that soraphens inhibit both HIV and HCV, it has been proposed that the broad-spectrum activity of **16** could be due to targeting commonly used host factors or pathways necessary for viral replication [78]. Another recently isolated myxobacteria-derived secondary metabolite, lanyamycin (**22**) from *Sorangium cellulosum* (strain Soce 481) moderately inhibited HCV with IC$_{50}$ value of 11.8 µM [78]. The macrolide, **22**, is closely related to the bafilomycins, a class of secondary metabolites from actinobacteria [79]. Interestingly bafilomycin A was previously shown to possess good activity (IC$_{50}$ value of 0.1 nM) against influenza A virus, which is below its cytotoxic levels [80]. Screening of **22** for activity against influenza A virus and other pathogenic viruses, and studying the mode of action would be interesting.

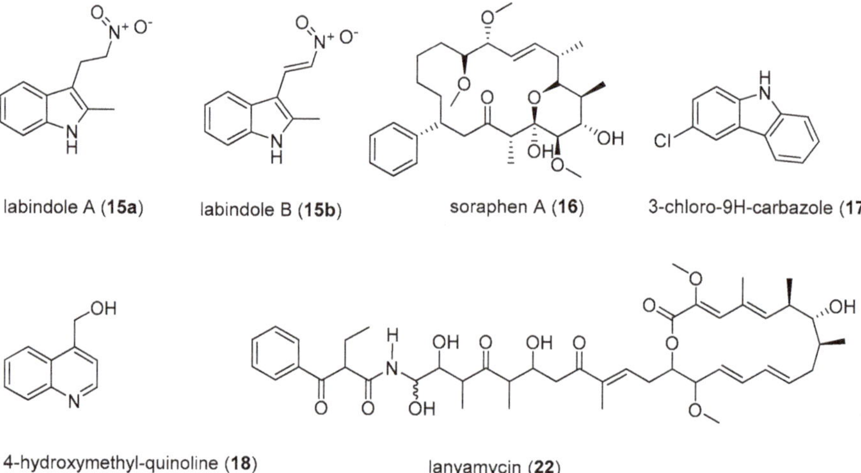

Figure 6. Compounds with activity against Hepatitis C Virus.

4. Conclusions

The reported antivirals discoveries call for more screening of myxobacterial-derived compounds, especially against other medically important viruses. Myxobacteria have demonstrated to be creative producers of molecules that can have valuable applications as possible lead structures for the development of antiviral drugs. Some broad-spectrum antivirals such as soraphen A could be of interest for the possibility of treating co-infection cases. Further, the possibility of using these myxobacteria derived secondary metabolites for treating both opportunistic infections and the HIV could be explored. The challenge of some compounds being toxic can be approached by structure modification for possibility of reducing toxicity while maintaining efficacy, as observed with the analogues of soraphen and noricumazoles. Moreover, some metabolites found to have potent antiviral activity may not be used as drugs themselves, due to toxicity, but they can serve as excellent tools to study and understand the viral invasion. This valuable information can be used to select other metabolites with similar mechanisms of action or structural modification of the compounds, to reduce their toxicity without substantially altering the activity. This ability of myxobacteria to produce such a vast number of secondary metabolites is most likely brought about by many years of evolution to adapt to survival in an ecological condition of competitive existence in the presence of competitors and invaders such as fungi, bacteriophages, and other bacteria. A recent comprehensive study involving molecular phylogeny with HPLC-MS profiling has revealed an unparalleled diversity of metabolites, along with strong correlations of the metabolite production to the phylogenetic position of the corresponding producer organisms [81]. Therefore, it appears promising to isolate more myxobacteria that represent novel genera and species from unexplored environments and screen them systematically for the production of further unique compounds with antiviral activities.

Author Contributions: Draft Preparation, L.S.M.; Editing of manuscript, M.S.

Funding: L.S.M. was supported by a Ph.D. grant from the DAAD-NACOSTI Scholarship funding program number (2014) 57139945 (91530969).

Acknowledgments: L.S.M. is highly indebted to DAAD-NACOSTI for the PhD Scholarship and TSC-Kenya for study leave. We are grateful to Joachim Wink and the technical staff of the HZI, Wera Collisi, Aileen Gollasch, Stephanie Schulz, Birte Trunkwalter and Klaus-Peter Conrad for the images of myxobacteria.

Conflicts of Interest: The authors declare no conflict of interest.

References

1. Nübel, U. Emergence and spread of antimicrobial resistance: Recent insights from bacterial population genomics. In *How to Overcome the Antibiotic Crisis—Facts, Challenges, Technologies & Future Perspective*; Stadler, M., Dersch, P., Eds.; Springer: Berlin, Germany, 2017; Volume 398, pp. 35–53.
2. Blair, W.; Cox, C. Current Landscape of Antiviral Drug Discovery. *F1000 Rev.* **2016**, *5*, 1–7. [CrossRef] [PubMed]
3. World Health Organization. Antimicrobial resistance. Available online: http://www.who.int/en/news-room/fact-sheets/detail/antimicrobial-resistance (accessed on 18 May 2018).
4. Dziuban, E.J.; DeVos, J.; Ngeno, B.; Ngugi, E.; Zhang, G.; Sabatier, J.; Wagar, N.; Diallo, K.; Nganga, L.; Katana, A.; et al. High prevalence of Abacavir-associated L74V/I mutations in Kenyan children failing antiretroviral therapy. *Pediatr. Infect. Dis. J.* **2017**, *36*, 758–760. [CrossRef] [PubMed]
5. U.S. Food & Drug Administration. Antiretroviral Drugs Used in the Treatment of HIV Infection. Available online: https://www.fda.gov/ForPatients/Illness/HIVAIDS/Treatment/ucm118915.htm (accessed on 22 May 2018).
6. National Institute of Allergy and Infectious Disease. Drugs That Fight HIV-1. Available online: https://aidsinfo.nih.gov/contentfiles/upload/HIV_Pill_Brochure.pdf (accessed on 22 May 2018).
7. DailyMed. Drug Listing Certification. Available online: https://dailymed.nlm.nih.gov/dailymed/index.cfm (accessed on 22 May 2018).
8. Reust, C.E. Common adverse effects of antiretroviral therapy for HIV disease. *Am. Fam. Phys.* **2011**, *83*, 1443–1451.

9.   Tantillo, C.; Ding, J.; Jacobo-Molina, A.; Nanni, R.G.; Boyer, P.; Hughes, S.H.; Pauwels, R.; Andries, K.; Jansen, P.A.; Arnold, E. Locations of anti-AIDS drug binding sites and resistance mutations in the three-dimensional structure of HIV-1 reverse transcriptase: Implications for mechanisms of drug inhibition and resistance. *J. Mol. Biol.* **1994**, *243*, 369–387. [CrossRef] [PubMed]

10.   Pawlotsky, J. Treatment failure and resistance with direct-acting antiviral drugs against Hepatitis C virus. *Hepatology* **2011**, *53*, 1742–1751. [CrossRef] [PubMed]

11.   Koutsoudakis, G.; Romero-Brey, I.; Berger, C.; Pérez-Vilaró, G.; Monteiro, P.P.; Vondran, F.W.R.; Kalesse, M.; Harmrolfs, K.; Müller, R.; Martinez, J.P.; et al. Soraphen A: A broad-spectrum antiviral natural product with potent anti-hepatitis C virus activity. *J. Hepatol.* **2015**, *63*, 813–821. [CrossRef] [PubMed]

12.   Herrmann, J.; Fayad, A.A.; Müller, R. Natural products from myxobacteria: Novel metabolites and bioactivities. *Nat. Prod. Rep.* **2017**, *34*, 135–160. [CrossRef] [PubMed]

13.   Hüttel, S.; Testolin, G.; Herrmann, J.; Planke, T.; Gille, F.; Moreno, M.; Stadler, M.; Brönstrup, M.; Kirschning, A.; Müller, R. Discovery and total synthesis of natural cystobactamid derivatives with superior activity against Gram-negative pathogens. *Angew. Chem. Int. Ed.* **2017**, *56*, 12760–12764. [CrossRef] [PubMed]

14.   Gerth, K.; Bedorf, N.; Irschik, H.; Höfle, G.; Reichenbach, H. The soraphens: A family of novel antifungal compounds from *Sorangium cellulosum* (Myxobacteria). *J. Antibiot.* **1994**, *47*, 23–31. [CrossRef] [PubMed]

15.   Reichenbach, H. Myxobacteria, producers of novel bioactive substances. *J. Ind. Microbiol. Biotechnol.* **2001**, *27*, 149–156. [CrossRef] [PubMed]

16.   Gerth, K.; Pradella, S.; Perlova, O.; Beyer, S.; Müller, R. Myxobacteria: Proficient producers of novel natural products with various biological activities—Past and future biotechnological aspects with the focus on the genus *Sorangium*. *J. Biotechnol.* **2003**, *106*, 233–253. [CrossRef] [PubMed]

17.   Dawid, W. Biology and global distribution of myxobacteria in soils. *FEMS Microbiol.* **2000**, *24*, 403–427. [CrossRef]

18.   Mohr, K.I.; Moradi, A.; Glaeser, S.P.; Kämpfer, P.; Gemperlein, K.; Nübel, U.; Schumann, P.; Müller, R.; Wink, J. *Nannocystis konarekensis* sp. nov., a novel myxobacterium from an Iranian desert. *Int. J. Syst. Evol. Microbiol.* **2018**, *68*, 721–729. [CrossRef] [PubMed]

19.   Wenzel, S.C.; Müller, R. Myxobacteria—'Microbial factories' for the production of bioactive secondary metabolites. *Mol. BioSyst.* **2009**, *5*, 567–574. [CrossRef] [PubMed]

20.   Weissman, K.J.; Müller, R. Myxobacterial secondary metabolites: Bioactivities and modes-of-action. *Nat. Prod. Rep.* **2010**, *27*, 1276–1295. [CrossRef] [PubMed]

21.   Korp, J.; Gurovic, M.S.V.; Nett, M. Antibiotics from predatory bacteria. *Beilstein J. Org. Chem.* **2016**, *12*, 594–607. [CrossRef] [PubMed]

22.   Goldman, B.S.; Nierman, W.C.; Kaiser, D.; Slater, S.C.; Durkin, A.S.; Eisen, J.A.; Ronning, C.M.; Barbazuk, W.B.; Blanchard, M.; Field, C.; et al. Evolution of sensory complexity recorded in a myxobacterial genome. *Prod. Natl. Acad. Sci. USA* **2006**, *103*, 15200–15205. [CrossRef] [PubMed]

23.   Wenzel, S.C.; Müller, R. The impact of genomics on the exploitation of the myxobacterial secondary metabolome. *Nat. Prod. Rep.* **2009**, *26*, 1385–1407. [CrossRef] [PubMed]

24.   Céspedes, A.D.; Hufendiek, P.; Crüsemann, M.; Schäberle, T.F.; König, G.M. Marine-derived myxobacteria of the suborder Nannocystineae: An underexplored source of structurally intriguing and biologically active metabolites. *Beilstein J. Org. Chem.* **2016**, *12*, 969–984. [CrossRef] [PubMed]

25.   Schneiker, S.; Perlova, O.; Kaiser, O.; Gerth, K.; Alici, A.; Altmeyer, M.O.; Bartels, D.; Bekel, T.; Beyer, S.; Bode, E.; et al. Complete genome sequence of the myxobacterium Sorangium cellulosum. *Nat. Biotechnol.* **2007**, *25*, 1281–1289. [CrossRef] [PubMed]

26.   Schäberle, T.F.; Lohr, F.; Schmitz, A.; König, G.M. Antibiotics from myxobacteria. *Nat. Prod. Rep.* **2014**, *31*, 953–972. [CrossRef] [PubMed]

27.   Bode, H.B.; Bethe, B.; Höfs, R.; Zeeck, A. Big effects from small changes: Possible ways to explore nature's chemical diversity. *ChemBioChem* **2002**, *3*, 619–627. [CrossRef]

28.   Seitz, R. Human Immunodeficiency Virus (HIV). *Transfus. Med. Hemother.* **2016**, *43*, 203–222.

29.   Curreli, F.; Kwon, Y.D.; Zhang, H.; Scacalossi, D.; Belov, D.S.; Tikhonov, A.A.; Andreev, A.I.; Altieri, A.; Kurkin, A.V.; Kwong, P.D.; et al. Structure-based design of a small molecule CD4-antagonist with broad spectrum anti-HIV-1 activity. *J. Med. Chem.* **2015**, *58*, 6909–6927. [CrossRef] [PubMed]

30. Asim, K.D.; Lin, R.; Shibo, J. Structure-based identification of small molecule antiviral compounds targeted to the gp41 core structure of the Human Immunodeficiency Virus type 1. *J. Med. Chem.* **1999**, *42*, 3203–3209.

31. Martinez, J.P.; Hinkelmann, B.; Fleta-Soriano, E.; Steinmetz, H.; Jansen, R.; Diez, J.; Frank, R.; Sasse, F.; Meyerhans, A. Identification of myxobacteria-derived HIV inhibitors by a high-throughput two-step infectivity assay. *Microb. Cell Fact.* **2013**, *12*. [CrossRef] [PubMed]

32. Zander, W.; Irschik, H.; Augustiniak, H.; Herrmann, M.; Jansen, R.; Steinmetz, H.; Gerth, K.; Kessler, W.; Kalesse, M.; Höfle, G.; et al. Sulfangolids, macrolide sulfate esters from *Sorangium cellulosum*. *Chem. Eur. J.* **2012**, *18*, 6264–6271. [CrossRef] [PubMed]

33. Available online: https://scifinder.cas.org/scifinder/view/scifinder/scifinderExplore.jsf (accessed on 10 July 2018).

34. Shen, Y.; Volrath, S.L.; Weatherly, S.C.; Elich, T.D.; Tong, L. A mechanism for the potent inhibition of eukaryotic acetyl-coenzyme A carboxylase by soraphen A, a macrocyclic polyketide natural product. *Mol. Cell* **2004**, *16*, 881–891. [CrossRef] [PubMed]

35. Mühlradt, P.F.; Sasse, F. Epothilone B stabilizes microtubuli of macrophages like taxol without showing taxol-like endotoxin activity. *Cancer Res.* **1997**, *57*, 3344–3346. [PubMed]

36. Goodin, S.; Kane, M.P.; Rubin, E.H. Epothilones: Mechanism of action and biologic activity. *J. Clin. Oncol.* **2004**, *22*, 2015–2025. [CrossRef] [PubMed]

37. Peterson, J.K.; Tucker, C.; Favours, E.; Cheshire, P.J.; Creech, J.; Billups, C.A.; Smykla, R.; Lee, F.Y.F.; Houghton, P.J. In vivo evaluation of Ixabepilone (BMS247550), A novel epothilone B derivative, against pediatric cancer models. *Clin. Cancer Res.* **2005**, *11*, 6950–6958. [CrossRef] [PubMed]

38. Reboll, M.R.; Ritter, B.; Sasse, F.; Niggemann, J.; Frank, R.; Nourbakhsh, M. The myxobacterial compounds spirangien A and spirangien M522 are potent inhibitors of IL-8 expression. *ChemBioChem* **2012**, *13*, 409–415. [CrossRef] [PubMed]

39. Jurkliewicz, E.; Jansen, R.; Kunze, B.; Trowitzsch-Klenast, W.; Porche, E.; Reichenbach, H.; Höfle, G.; Hunsmann, G. Three new potent HIV-1 inhibitors from myxobacteria. *Antivir. Chem. Chemother.* **1992**, *3*, 189–193. [CrossRef]

40. Trowitzsch-Kienast, W.; Forche, E.; Wray, V.; Reichenbach, H.; Jurkiewicz, E.; Hunsmann, G.; Höfle, G. Phenalamide, neue HIV-1-Inhibitoren aus *Myxococcus stipitatus* Mx s40. *Liebigs Ann. Chem.* **1992**, *1992*, 659–664. [CrossRef]

41. Garcia, R.; Stadler, M.; Gemperlein, K.; Müller, R. *Aetherobacter fasciculatus* gen. nov., sp. nov. and *Aetherobacter rufus* sp. nov., novel myxobacteria with promising biotechnological applications. *Int. J. Syst. Evol. Microbiol.* **2016**, *66*, 928–938. [CrossRef] [PubMed]

42. Plaza, A.; Garcia, R.; Bifulco, G.; Martinez, J.P.; Hüttel, S.; Sasse, F.; Meyerhans, A.; Stadler, M.; Müller, R. Aetheramides A and B, potent HIV-inhibitory depsipeptides from a Myxobacterium of the new genus "Aetherobacter". *Org. Lett.* **2012**, *14*, 2854–2857. [CrossRef] [PubMed]

43. Gerth, K.; Schummer, D.; Höfle, G.; Irschik, H.; Reichenbach, H. Ratjadon: A new antifungal compound from *Sorangium cellulosum* (Myxobacteria) Production, physico-chemical and biological properties. *J. Antibiot.* **1995**, *48*, 787–792. [CrossRef]

44. Fleta-Soriano, E.; Martinez, J.P.; Hinkelmann, B.; Gerth, K.; Washausen, P.; Diez, J.; Frank, R.; Sasse, F.; Meyerhans, A. The myxobacterial metabolite ratjadone A inhibits HIV infection by blocking the Rev/CRM1-mediated nuclear export pathway. *Microb. Cell Fact.* **2014**, *13*, 17. [CrossRef] [PubMed]

45. ICTV. Taxonomy. Available online: https://talk.ictvonline.org/taxonomy/ (accessed on 30 April 2018).

46. Britt, W.J.; Vugler, L.; Butfiloski, E.D.; Stevens, E.B. Cell surface expression of Human Cytomegalovirus (HCMV): Use of HCMV-recombinant vaccinia virus-infected cells in analysis of the human neutralizing antibody response. *J. Virol.* **1990**, *64*, 1079–1085. [PubMed]

47. Arvin, A.M.; Alford, C.A. Chronic Intrauterine and Perinatal Infections. In *Antiviral Agents and Viral Diseases of Man*; Galassov, G.J., Whitley, R.J., Merigan, T.C., Eds.; Raven Press: New York, NY, USA, 1990; Volume 3, pp. 497–580.

48. Field, A.K. Human cytomegalovirus: Challenges opportunities and new drug development. *Antivir. Chem. Chemother.* **1999**, *10*, 219–232. [CrossRef] [PubMed]

49. Castro, A.; Martinez, A. Novel agents for the treatment of human cytomegalovirus infection. *Exp. Opin. Ther. Pat.* **2000**, *10*, 165–177. [CrossRef]

50. Xiong, X.; Smith, J.L.; Chen, M.S. Effect of incorporation of Cidofovir into DNA by Human Cytomegalovirus DNA polymerase on DNA elongation. *Antimicrob. Agents Chemother.* **1997**, *24*, 594–599.
51. Boivin, G.; Chou, S.; Quirk, M.R.; Alejo, E.; Jordan, M.C. Detection of ganciclovir resistance mutations and quantitation of cytomegalovirus (CMV) DNA in leucocytes of patients with fatal disseminated CMV disease. *J. Infect. Dis.* **1996**, *173*, 523–528. [CrossRef] [PubMed]
52. Nassir, M.R.; Emerson, S.G.; Devivar, R.V.; Townsend, L.B.; Drach, J.C.; Taichman, R. Comparison of benzimidazole nucleosides ganciclovir on the *in vitro* proliferation and colony formation of human marrow progenitor cells. *Br. J. Haematol.* **1996**, *93*, 273–279. [CrossRef]
53. Kaul, D.R.; Stoelben, S.; Cober, E.; Ojo, T.; Sandusky, E.; Lischka, P.; Zimmermann, H.; Rübsamen-Schaeff, H. First report of successful treatment of multidrug-resistant cytomegalovirus disease with the novel anti-CMV compound AIC246. *Am. J. Transpl.* **2011**, *11*, 1079–1084. [CrossRef] [PubMed]
54. NEJM Journal Watch. Letermovir Approved to Prevent CMV Infection and Disease in Transplant Patients. Available online: https://www.jwatch.org/na45554/2017/12/07/letermovir-approved-prevent-cmv-infection-and-disease (accessed on 17 June 2018).
55. Schieferdecker, S.; König, S.; Koeberle, A.; Dahse, H.M.; Werz, O.; Nett, M. Myxochelins Target Human 5-Lipoxygenase. *J. Nat. Prod.* **2015**, *78*, 335–338. [CrossRef] [PubMed]
56. Nagoba, B.; Vedpathak, D. Medical applications of siderophores. *Eur. J. Gen. Med.* **2011**, *8*, 229–235. [CrossRef]
57. Miyanaga, S.; Obata, T.; Onaka, H.; Fujita, T.; Saito, N.; Sakurai, H.; Saiki, I.; Furumai, T.; Igarashi, Y. Absolute configuration and antitumor activity of myxochelin a produced by *Nonomuraea pusilla* TP-A0861. *J. Antibiot.* **2006**, *59*, 698–703. [CrossRef] [PubMed]
58. Korp, J.; König, S.; Schieferdecker, S.; Dahse, H.M.; König, G.M.; Werz, O.; Nett, M. Harnessing enzymatic promiscuity in myxochelins biosynthesis for the production of 5-lipoxygenase inhibitors. *ChemBioChem* **2015**, *16*, 2445–2450. [CrossRef] [PubMed]
59. Schieferdecker, S.; König, S.; Simona Pace, S.; Werz, O.; Nett, M. Myxochelin-inspired 5-lipoxygenase inhibitors: Synthesis and biological evaluation. *ChemMedChem* **2017**, *12*, 23–27. [CrossRef] [PubMed]
60. Miyanga, S.; Sakurai, H.; Saiki, I.; Onaka, H.; Igarashi, Y. Synthesis and evaluation of myxochelins analogues as antimetastatic agents. *Biol. Med. Chem.* **2009**, *17*, 2724–2732. [CrossRef] [PubMed]
61. Ambrosi, H.D.; Hartmann, V.; Pistorius, D.; Reissbrodt, R.; Trowitzsch-Kienast, W. Myxochelins B, C, D, E and F: A new structural principle for powerful siderophores imitating Nature. *Eur. J. Org. Chem.* **1998**, 541–551. [CrossRef]
62. Gaitatzis, N.; Kunze, B.; Müller, R. Novel insights into siderophore formation in myxobacteria. *Chembiochem.* **2005**, *6*, 365–374. [CrossRef] [PubMed]
63. Nadmid, S.; Plaza, A.; Lauro, G.; Garcia, R.; Bifulco, G.; Müller, R. Hyalachelins A–C, Unusual Siderophores Isolated from the terrestrial Myxobacterium *Hyalangium minutum*. *Org. Lett.* **2014**, *16*, 4130–4133. [CrossRef] [PubMed]
64. Kunze, B.; Trowitzsch-Kienast, W.; Hofle, G.; Reichenbach, H. Nannochelins A, B and C, new iron-chelating compounds from Nannocystis exedens (myxobacteria) production, isolation, physico-chemical and biological properties. *J. Antibiot.* **1991**, *45*, 147–150. [CrossRef]
65. Saha, M.; Sarkar, S.; Sarkar, B.; Sharma, B.K.; Bhattacharjee, S.; Prosun, T. Microbial siderophores and their potential applications: A review. *Environ. Sci. Pollut. Res.* **2016**, *23*, 3984–3999. [CrossRef] [PubMed]
66. World Health Organization. Ebola virus Disease. Available online: http://www.who.int/en/news-room/fact-sheets/detail/ebola-virus-disease (accessed on 5 May 2018).
67. Beck, S.; Henß, L.; Weidner, T.; Herrmann, J.; Müller, R.; Chao, Y.; Weber, C.; Sliva, K.; Schnierle, S. Identification of inhibitors of Ebola virus pseudotyped vectors from a myxobacterial compound library. *Antivir. Res.* **2016**, *132*, 85–91. [CrossRef] [PubMed]
68. Kunze, B.; Jansen, R.; Sasse, F.; Höflen, G.; Reichenbach, H.; Chondramides, A.D. New antifungal and cytostatic depsipeptides from *Chondromyces crocatus* (Myxobacteria). Production, physico-chemical and biological Properties. *J. Antibiot.* **1995**, *48*, 1262–1266. [CrossRef] [PubMed]
69. Reichenbach, H. Myxobacteria: A source of new antibiotics. *Trends Biotechnol.* **1988**, *6*, 115–121. [CrossRef]

70. Barbier, J.; Jansen, R.; Irschik, H.; Benson, S.; Gerth, K.; Böhlendorf, B.; Höfle, G.; Reichenbach, H.; Wegner, J.; Zeilinger, C.; et al. Isolation and total synthesis of icumazoles and noricumazoles—Antifungal antibiotics and cation-channel blockers from *Sorangium cellulosum*. *Angew. Chem. Int. Ed.* **2012**, *51*, 1256–1260. [CrossRef] [PubMed]

71. Sasse, F.; Steinmetz, H.; Heil, J.; Höfle, G.; Reichenbach, H. Tubulysins, new cytostatic peptides from Myxobacteria acting on microtubuli. Production, isolation, physico-chemical and biological properties. *J. Antibiot.* **2000**, *53*, 879–885. [CrossRef] [PubMed]

72. El-Serag, H.B. Epidemiology of viral hepatitis and hepatocellular carcinoma. *Gastroenterology* **2012**, *142*, 1264–1273. [CrossRef] [PubMed]

73. World Health Organization. Hepatitis C. Available online: http://www.who.int/en/news-room/factsheets/detail/hepatitis-c (accessed on 10 May 2018).

74. Shepard, C.W.; Finelli, L.; Alter, M.J. Global epidemiology of hepatitis C virus infection. *Lancet Infect. Dis.* **2005**, *5*, 558–567. [CrossRef]

75. Mulwa, L.; Jansen, R.; Praditya, D.; Mohr, K.; Wink, J.; Steinmann, E.; Stadler, M. Six heterocyclic metabolites from the Myxobacterium *Labilithrix luteola*. *Molecules* **2018**, *23*, 542. [CrossRef] [PubMed]

76. Bedorf, N.; Schomburg, D.; Gerth, K.; Reichenbach, H.; Höfle, G.; Bedorf, N.; Schomburg, D.; Gerth, K.; Reichenbach, H.; Höfle, G. Isolation and structure elucidation of soraphen $A_{1\alpha}$, a novel antifungal macrolide from *Sorangium cellulosum*. *Liebigs Ann. Chem.* **1993**, *1993*, 1017–1021. [CrossRef]

77. Singaravelu, R.; Desrochers, G.F.; Srinivasan, P.; O'Hara, S.; Lyn, R.; Müller, R.; Jones, D.M.; Russell, R.; Pezacki, J.P. Soraphen A: A probe for investigating the role of de novo lipogenesis during viral infection. *ACS Infect. Dis.* **2015**, *1*, 130–134. [CrossRef] [PubMed]

78. Gentzsch, J.; Hinkelmann, B.; Kaderali, L.; Irschik, H.; Jansen, R.; Sasse, F.; Frank, R.; Pietschmann, T. Hepatitis C virus complete life cycle screen for identification of small molecules with pro- or antiviral activity. *Antiviral Res.* **2011**, *89*, 136–148. [CrossRef] [PubMed]

79. Mulwa, L.S.; Jansen, R.; Praditya, D.F.; Mohr, K.I.; Okanya, P.W.; Wink, J.; Steinmann, E.; Stadler, M. Lanyamycin, a macrolide antibiotic from *Sorangium cellulosum*, strain Soce 481 (Myxobacteria). *Beilstein J. Org. Chem.* **2018**, *14*, 1554–1562. [CrossRef] [PubMed]

80. Yeganeh, B.; Ghavami, S.; Kroeker, A.L.; Mahood, T.H.; Stelmack, G.L.; Klonisch, T.; Coombs, K.M.; Halayko, A.J. Suppression of influenza A virus replication in human lung epithelial cells by noncytotoxic concentrations bafilomycin A1. *Am. J. Physiol. Lung Cell. Mol. Physiol.* **2015**, *308*, 270–286. [CrossRef] [PubMed]

81. Hoffmann, T.; Krug, D.; Bozkurt, N.; Duddela, S.; Jansen, R.; Garcia, R.; Gerth, K.; Steinmetz, H.; Müller, R. Correlating chemical diversity with taxonomic distance for discovery of natural products in myxobacteria. *Nat. Commum.* **2018**, *9*, 803. [CrossRef] [PubMed]

 *microorganisms*

Review

Chemical Elicitors of Antibiotic Biosynthesis in Actinomycetes

Anton P. Tyurin [1,*]**, Vera A. Alferova** [1] **and Vladimir A. Korshun** [1,2]

[1] Gause Institute of New Antibiotics, Bolshaya Pirogovskaya 11, 119021 Moscow, Russia;
    alferovava@gmail.com (V.A.A.); v-korshun@yandex.ru (V.A.K.)
[2] Shemyakin-Ovchinnikov Institute of Bioorganic Chemistry, Miklukho-Maklaya 16/10,
    117997 Moscow, Russia
*   Correspondence: ap2rin@gmail.com; Tel.: +7-499-246-6983

Received: 2 May 2018; Accepted: 6 June 2018; Published: 8 June 2018

Abstract: Whole genome sequencing of actinomycetes has uncovered a new immense realm of microbial chemistry and biology. Most biosynthetic gene clusters present in genomes were found to remain "silent" under standard cultivation conditions. Some small molecules—chemical elicitors—can be used to induce the biosynthesis of antibiotics in actinobacteria and to expand the chemical diversity of secondary metabolites. Here, we outline a brief account of the basic principles of the search for regulators of this type and their application.

Keywords: actinomycetes; antibiotic biosynthesis; silent biosynthetic pathways; γ-butyrolactones; HiTES; translation inhibitors

1. Introduction

Since the discovery of streptomycin by Selman A. Waksman, actinomycetes have become one of the most fruitful sources of new antibiotics. Most antibiotic classes in current clinical use were discovered during the "golden era", 1940–1960s, by phenotypic screening of soil microorganisms. Moreover, since the 1960s, a significant number of approved drugs has been designed using chemical modifications of natural scaffolds. Due to an "innovation gap" in this area, society is now facing an emerging threat of microbial drug resistance. An urgent need for new effective antimicrobials has become an important social and political issue [1,2].

On the other hand, achievements in genome sequencing of actinomycetes has revealed a difference between their potential and observed biosynthetic gene expression. Biosynthetic gene clusters (BGCs) are several generally contiguous genes encoding enzymes responsible for a stepwise assembly of complex bioactive molecules. According to data from the majority of published genomes, most BGCs remain "silent" under standard cultivation conditions. These silent, or cryptic, BGCs represent a potential source of new scaffolds for the discovery of novel antimicrobials [3–5]. Several techniques for activation of silent BGCs have been developed in recent decades, e.g., direct identification of BGCs and their expression in heterologous hosts [6–10], and systematic alteration of cultivation parameters (the "one strain—many compounds" (OSMAC) approach) [11–13]. These strategies are extremely powerful although still remaining laborious and resource-intensive, especially for large (>40 kb) BGCs. Further techniques comprise co-cultivation [14,15], ribosome engineering [16–18], and the use of chemical elicitors—compounds that induce the synthesis of antibiotics in actinomycetes [14,19–21]. The last approach is accounted and discussed succinctly in this review, covering literature up to December 2017. Here, we focus on small organic molecules capable in nanomolar to micromolar minimum effective concentrations to induce biosynthesis of secondary metabolite in actinomycetes.

2. γ-Butyrolactones (GBL) and Related (auto)Regulators

Historically, A-factor (**1**, Chart 1) (A for "autoregulation") was the first compound that revolutionized our views of the secondary metabolism and development cycle of actinomycetes. This γ-butyrolactone (GBL) derivative was discovered by Prof. Khokhlov and co-workers in 1967 [22].

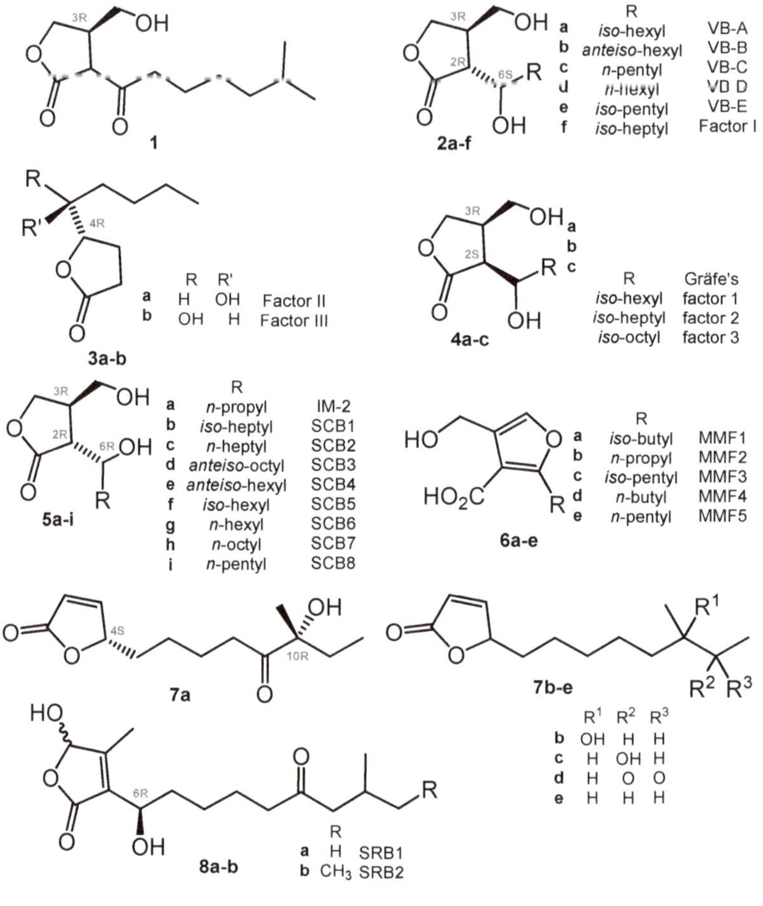

Chart 1. GBL and closely related regulators: butenolides (avenolide, *S. rochei* butenolides or SRB) and furans (methylenomycin furans or MMF).

A-factor acts as a pleiotropic regulator: it binds to the A-factor receptor protein (ArpA) and causes dissociation of this suppressor from DNA. This triggers the transcription of the *adpA* gene encoding the transcription activator AdpA, which in turn induces morphological differentiation, spore formation, and biosynthesis of secondary metabolites [23,24].

In further decades, many other closely related autoregulators were discovered, e.g., *Streptomyces virginiae* butanolides (VBs A-E, **2a–e**) [25–27], Factor I (**2f**) [28], Factors II and III (**3a,b**) [29], Gräfe's factors (**4a–c**) from *S. bikiniensis* and *S. cyaneofuscatus* [30], IM-2 (**5a**) [31,32], SCB1 (**5b**) [33,34], SCB2,3 (**5c,d**) [35] and SCB4–8 (**5e–i**) [36], methylenomycin furans (MMFs, **6**) [37], avenolide (**7a**) from *S. avermitilis* [38], related compounds (**7b–e**) from *S. albus* [39], and two *S. rochei* butenolides (SRBs, **8a,b**) [40]. The stereoconfiguration of avenolide analogues **7b–e** reported very recently has not been assigned [39]. All these compounds have a high degree of similarity in their chemical nature (butanolides, butenolides, or furans), biosynthesis pathways, and in mechanisms of action on actinomycetes. The transduction of the chemical signal starts from binding with a specific receptor protein (like ArpA in the A-factor signaling cascade) belonging to the TetR family of transcriptional regulators; the binding of compounds probably causes conformational changes, thus inducing the relocation of the DNA-binding domain and DNA release [23].

TetR transcriptional repressors are well represented and widespread in bacteria [41,42]. The recent study on the distribution of the GBL (IM-2, VB)-type and butenolide (avenolide)-type signaling showed these hormones to be common in actinomycetes. Production of regulators was screened using *S. lavendulae* FRI-5 (Δ*farX*), *S. virginiae* (Δ*barX*), and *S. avermitilis* (Δ*aco*) strains with disrupted essential biosynthetic genes for the corresponding hormone-like molecules. Twenty percent of the investigated strains produced **5a** (IM-2), another 20% produced VBs, and 24% of actinomycetes showed avenolide activity [43].

Thus, the potential for various applications of A-factor-like regulators is quite high: they induce biosynthesis of antibiotics of different classes, e.g., aminoglycosides (streptomycin), streptogramins (virginamycin), peptide ionophores (valinomycin) and others, including clinically significant antibiotics. For more detail see ref. [24].

3. Other Types of (auto)Regulators

However, GBLs and related compounds are not the only type of regulators found in actinomycetes. Hitherto, a few different types of autoregulators were reported (Chart 2), e.g., the pi-factor from *S. natalensis* (**9**) [44] and L-*N*-methylphenylalanyl-dehydrobutyrine diketopiperazine (**10**) from *S. globisporus* [45]. These molecules have no structural similarity to GBL-type compounds and their mechanisms of action remain unclear [46,47].

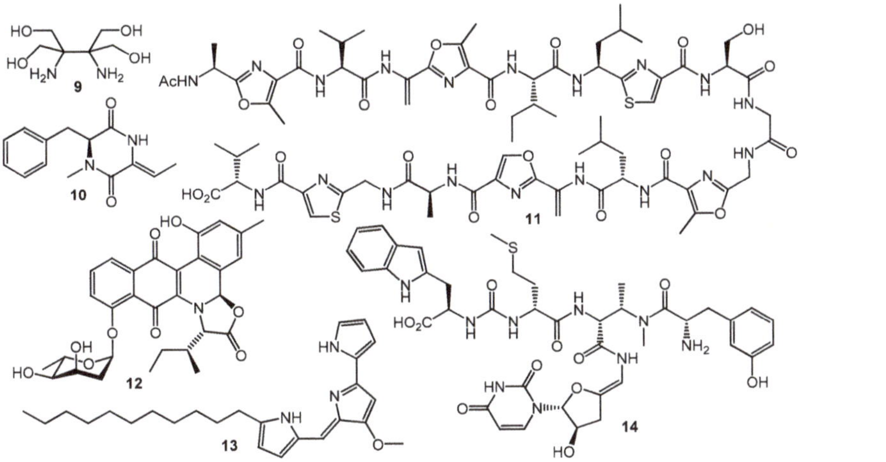

Chart 2. Other (non-GBL) types of regulators.

Goadsporin (**11**) [48,49], an active 19-aa peptide found in the culture broth of *Streptomyces* sp. TP-A0584, shows extremely broad elicitation activity on actinobacteria. Induction of sporulation and growth inhibition were observed in about 80% of *Streptomyces* sp. and *Micromonospora* sp. isolates after goadsporin treatment [50]. Possible mechanism of the goadsporin action on susceptible cells was postulated by Prof. H. Onaka and co-workers. Bioinformatics analysis of *godI*—self-resistance gene from the goadsporin BGC—showed a high degree of homology with *ffh* encoding the Ffh protein. The Ffh and 4.5S RNA form the signal recognition particles (SRPs) in prokaryotic cells. The SPRs recognize the signal peptide of membrane or secretory proteins and facilitate their translocation to the cytoplasmic membrane. Binding of Ffh protein with goadsporin causes the upregulation of protein translocation and maturation, growth inhibition, and alternation of secondary metabolism [51].

It was suggested that goadsporin and its analogs [51,52] could be useful in crop protection: *S. scabies*, which is known to cause potato scab, is effectively inhibited by the low concentrations of goadsporin.

Two "atypical" transcriptional regulators were discovered in *S. venezuelae*: JadR1 activates the transcription of jadomycin B (**12**) biosynthetic genes while repressing its own gene. Jadomycin B binds to the N-terminal receiver domain of JadR1, causing dissociation of JadR1 from target promoters. High expression of the biosynthetic enzyme genes leads to accumulation of the product in a concentration sufficient to inhibit the initial activation. Therefore, jadomycin acts as an autoregulator of its own synthesis. Similarly, the RedZ protein regulating production of undecylprodigiosine (**13**) is modulated by the end product. These findings demonstrate that end-product-mediated control of antibiotic biosynthesis may be widespread in nature, offering seminal basis for interspecies signaling and interaction [53,54]. Production of another class of antibiotics, uridyl peptides sansanmycins (sansanmycin A, **14**), active against multidrug-resistant *Mycobacterium tuberculosis*, is also controlled at the transcriptional level by the end-product-mediated mechanism [55].

Polyether antibiotics at sub-inhibitory concentrations can elicit the synthesis of secondary metabolites. Promomycin (**15**, Chart 3) and closely related compounds (salinomycin, monensin, and nigericin) promoted the synthesis of cryptic isonitrile antibiotic SF2768 (**16**) by *S. griseorubiginosus* [56,57]. This was the first example of isolation of a novel original compound using the chemical elicitation approach. However, the action of polyether antibiotics has not yet been studied on the molecular level.

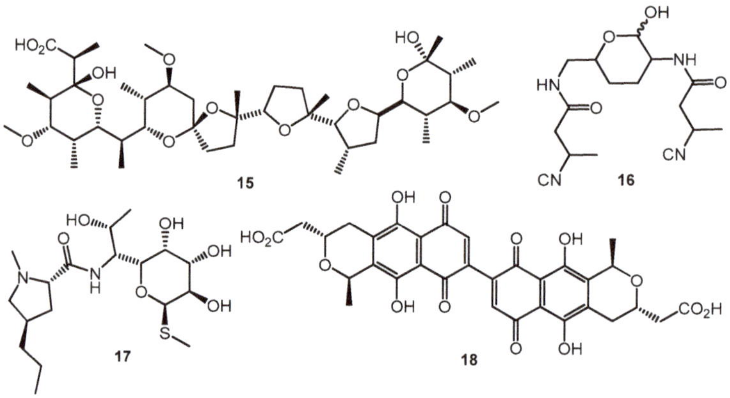

Chart 3. Antibiotic-antibiotic induction; elicitors and products.

Lincomycin (**17**) at one tenth of its MIC increased the expression of the pathway-specific regulatory gene actII-ORF4 in the enzyme gene cluster producing the blue-pigmented antibiotic actinorhodin (**18**) in *S. lividans*, thus resulting in actinorhodin overproduction [58]. Chloramphenicol activated biosynthetic genes at the transcriptional level and increased the amino acid pool 1.5- to 6-fold, enhancing the production of non-ribosomal peptide antibiotics [59]. Other ribosome-targeted antibiotics, especially thiosteptone and spectinomycin, at sub-inhibitory concentrations were also capable of altering colony morphology and antibiotics (**13** and **18**) production in *S. coelicolor* M145 [60].

The HiTES (High-Throughput Elicitor Screening) approach [61–63] was developed to discover small molecule elicitors of silent biosynthetic gene clusters and novel secondary metabolites using a simple fluorescent assay format. Three copies of *eGFP* (enhanced green fluorescent protein encoding gene) inserted in the *S. albus* J1074 genome both at a neutral site (*attB*) and inside the biosynthetic cluster of interest (*sur*, surugamide BGC). The difference in fluorescence between *attB::Psur-eGFPx3* (the *Psur* promoter region (~260 bp) upstream of *surE*) and *surE::eGFPx3* strains after testing compound treatment was used as an indicator of BGC's activation. Screening of a commercially-available natural products library (ca. 500 compounds) identified 6 potential elicitors, among these ivermectin b1a (**19**, Chart 4) and etoposide (**20**) had the maximum surugamide BGC activation effect and markedly changed the secondary metabolome profile of *S. albus* J1074—14 novel small molecules were isolated and characterized using these two inducers, e.g., surugamide J (**21**), albucyclone F (**22**), and surugamide F2 (**23**) [63].

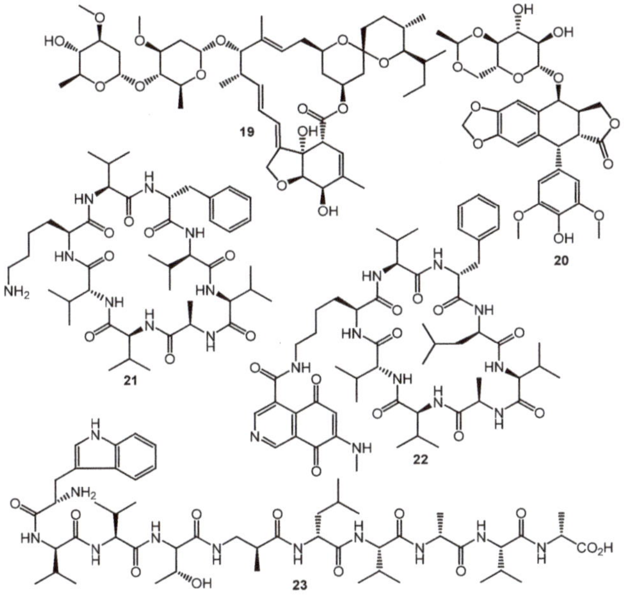

Chart 4. Natural products eliciting surugamide biosynthesis in *S. albus* J1074 (detected by HiTES).

4. Metabolism Remodeling and Epigenetic Control

Fully synthetic chemicals also can be utilized as elicitors. ARC2 (**24**, ARC from "Antibiotic-Remodeling-Compound") and similar diphenyl ethers (ARC3–5, **25–27**) (Chart 5) have become the first inducer examples discovered by systematic screening of 30,569 small molecules [64,65]. Taking into account the structural similarity of ARC compounds with triclosan (**28**), a well-known biocide and inhibitor of fatty acid biosynthesis, the mechanism of ARCs' action (Chart 5A) has been proposed. Generally, ARCs act as metabolism remodeling compounds, changing the balance between primary and secondary metabolic pathways. The mode of action was confirmed by the experimental data: ARCs changed fatty acid pool by inhibition of key biosynthetic enzymes like FabI. Remarkably, triclosan also can stimulate polyketide antibiotic synthesis at sub-inhibitory concentrations [59]. Further investigation of the different ARCs' effects on the streptomycetes biochemistry revealed the targets for different molecules (even structurally close) as not being the same; however, all of them influenced fatty acid metabolism [66]. Chlorinated analog Cl-ARC (**29**) was used to enhance the expression of cryptic biosynthetic genes, and the observed difference between growth on control and elicitor-containing media was studied using a metabolomics approach. More than 100 induced secondary metabolites were detected, including rare antibiotics [67]. Several identified structures are represented in Chart 6: oxohygrolidin (**30**), germicidins A–E (**31a–e**) 9-methylstreptimidone (**32**), nactins (**33**, **34**, **35**), desferrioxamines B,E (**36**, **37**), and arylomycin A$_4$ (**38**).

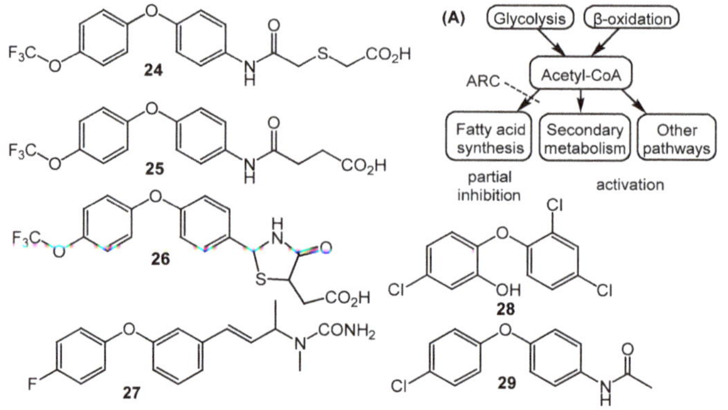

Chart 5. Synthetic diphenyl ethers (ARCs) as elicitors and their mode of action (**A**).

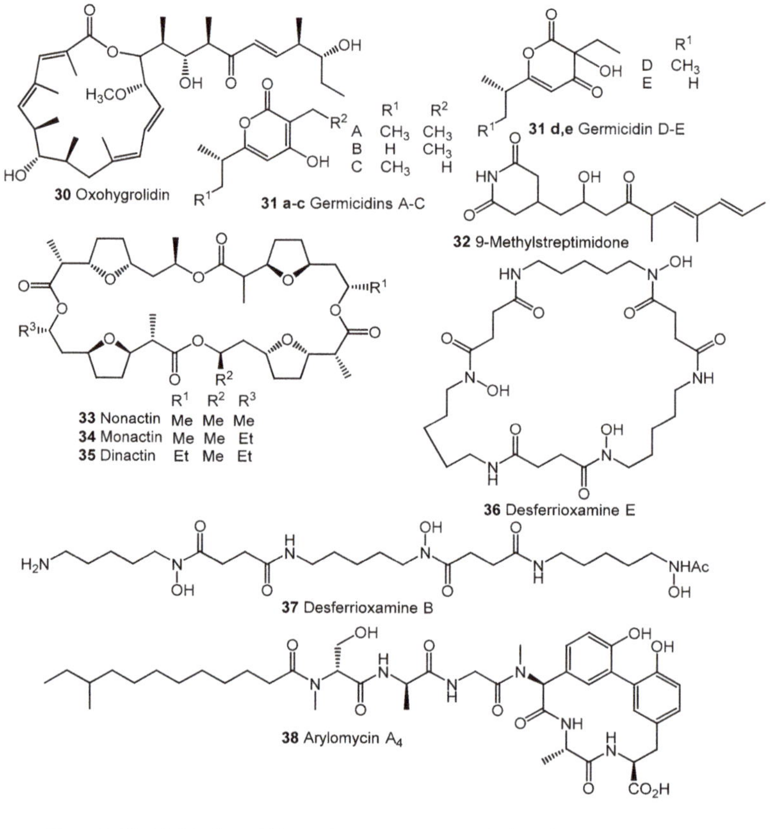

Chart 6. Cl-ARC-induced biosynthesis: examples of identified antibiotics [67].

Despite the significant difference in structural DNA organization between eukaryotes and actinomycetes, inhibitors of histone deacetylase (HDAC) like valproic acid (**39**) and suberanilohydroxamic acid (SAHA, **40**) (Chart 7) showed effect on antibiotic production by several *Streptomyces* strains, especially on nutrient-poor media. The finding was explained by the homology between several *S. coelicolor* enzymes (SCO0452, SCO6464, SCO3330) and human HDAC enzymes [68]. Epigenetic modification could be used for the control of antibiotic biosynthesis; the approach, however, is not well studied for actinobacteria.

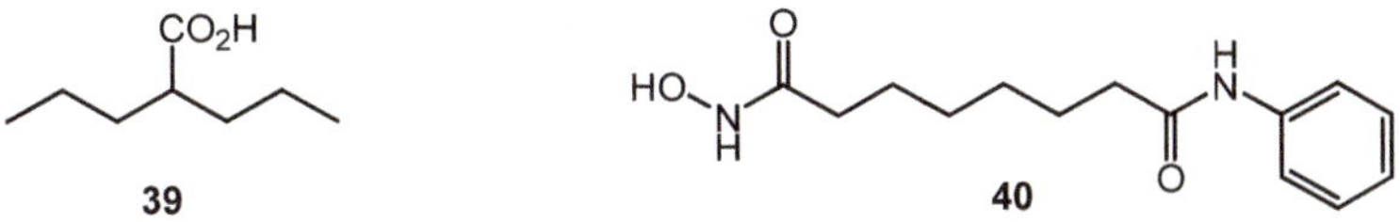

Chart 7. Histone deacetylase (HDAC) inhibitors.

5. Conclusions

Activation of antibiotics' biosynthesis in actinomycetes by using small-molecule elicitors seems to be a useful technique for drug discovery. The approach allows identification of new metabolite scaffolds and enhances synthesis of low yield secondary metabolites. Moreover, antibiotic-antibiotic induction relationships are believed to be the basis of chemical signaling between different actinobacterial strains and are, therefore, of interest for chemical ecology. Elicitation is an obvious cost-effective approach in biotechnology for increasing antibiotic yields and reducing fermentation time.

Acknowledgments: This work was supported in part by RFBR (project No. 18-33-00044) and by the budget funding of the topic 006 "Isolation and physico-phemical characterization of microbially produced antibiotic compounds" at the Gause Institute of New Antibiotics.

Conflicts of Interest: The authors declare no conflict of interest.

References

1. WHO Publishes List of Bacteria for Which New Antibiotics Are Urgently Needed. Available online: http://www.who.int/mediacentre/news/releases/2017/bacteria-antibiotics-needed/en/ (accessed on 27 February 2017).

2. The National Strategy to Prevent the Antimicrobial Resistance Spreading (2017–2030). Available online: http://government.ru/docs/29477/ (accessed on 3 October 2017).

3. Nett, M.; Ikeda, H.; Moore, B.S. Genomic basis for natural product biosynthetic diversity in the actinomycetes. *Nat. Prod. Rep.* **2009**, *26*, 1362–1384. [CrossRef] [PubMed]

4. Rutledge, P.J.; Challis, G.L. Discovery of microbial natural products by activation of silent biosynthetic gene clusters. *Nat. Rev. Microbiol.* **2015**, *13*, 509–523. [CrossRef] [PubMed]

5. Genilloud, O. Actinomycetes: Still a source of novel antibiotics. *Nat. Prod. Rep.* **2017**, *34*, 1203–1232. [CrossRef] [PubMed]

6. Yamanaka, K.; Reynolds, K.A.; Kersten, R.D.; Ryan, K.S.; Gonzalez, D.J.; Nizet, V.; Dorrestein, P.C.; Moore, B.S. Direct cloning and refactoring of a silent lipopeptide biosynthetic gene cluster yields the antibiotic taromycin A. *Proc. Natl. Acad. Sci. USA* **2014**, *111*, 1957–1962. [CrossRef] [PubMed]

7. Nah, H.-J.; Pyeon, H.-R.; Kang, S.-H.; Choi, S.-S.; Kim, E.-S. Cloning and heterologous expression of a large-sized natural product biosynthetic gene cluster in *Streptomyces* species. *Front. Microbiol.* **2017**, *8*, 394. [CrossRef] [PubMed]

8. Alberti, F.; Khairudin, K.; Venegas, E.R.; Davies, J.A.; Hayes, P.M.; Willis, C.L.; Bailey, A.M.; Foster, G.D. Heterologous expression reveals the biosynthesis of the antibiotic pleuromutilin and generates bioactive semi-synthetic derivatives. *Nat. Commun.* **2017**, *8*, 1831. [CrossRef] [PubMed]

9. Wei, Y.; Zhang, L.; Zhou, Z.; Yan, X. Diversity of gene clusters for polyketide and nonribosomal peptide biosynthesis revealed by metagenomics analysis of the Yellow Sea sediment. *Front. Microbiol.* **2018**, *9*, 295. [CrossRef] [PubMed]

10. Suroto, D.A.; Kitani, S.; Arai, M.; Ikeda, H.; Nihira, T. Characterization of the biosynthetic gene cluster for cryptic phthoxazolin A in *Streptomyces avermitilis*. *PLoS ONE* **2018**, *13*, e0190973. [CrossRef] [PubMed]

11. Bode, H.B.; Bethe, B.; Höfs, R.; Zeeck, A. Big effects from small changes: Possible ways to explore nature's chemical diversity. *ChemBioChem* **2002**, *3*, 619–627. [CrossRef]

12. Reen, F.J.; Romano, S.; Dobson, A.D.W.; O'Gara, F. The sound of silence: Activating silent biosynthetic gene clusters in marine microorganisms. *Mar. Drugs* **2015**, *13*, 4754–4783. [CrossRef] [PubMed]
13. Liu, M.; Grkovic, T.; Liu, X.; Han, J.; Zhang, L.; Quinn, R.J. A systems approach using OSMAC, Log P and NMR fingerprinting: An approach to novelty. *Synth. Syst. Biotechnol.* **2017**, *2*, 276–286. [CrossRef] [PubMed]
14. Abdelmohsen, U.R.; Grkovic, T.; Balasubramanian, S.; Kamel, M.S.; Quinn, R.J.; Hentschel, U. Elicitation of secondary metabolism in actinomycetes. *Biotechnol. Adv.* **2015**, *33*, 798–811. [CrossRef] [PubMed]
15. Van der Meij, A.; Worsley, S.F.; Hutchings, M.I.; van Wezel, G.P. Chemical ecology of antibiotic production by actinomycetes. *FEMS Microbiol. Rev.* **2017**, *41*, 392–416. [CrossRef] [PubMed]
16. Wang, G.; Hosaka, T.; Ochi, K. Dramatic activation of antibiotic production in *Streptomyces coelicolor* by cumulative drug resistance mutations. *Appl. Environ. Microbiol.* **2008**, *74*, 2834–2840. [CrossRef] [PubMed]
17. Tojo, S.; Tanaka, Y.; Ochi, K. Activation of antibiotic production in *Bacillus* spp. by cumulative drug resistance mutations. *Antimicrob. Agents Chemother.* **2015**, *59*, 7799–7804. [CrossRef] [PubMed]
18. Ochi, K. Insights into microbial cryptic gene activation and strain improvement: Principle, application and technical aspects. *J. Antibiot.* **2017**, *70*, 25–40. [CrossRef] [PubMed]
19. Zhu, H.; Sandiford, S.K.; van Wezel, G.P. Triggers and cues that activate antibiotic production by actinomycetes. *J. Ind. Microbiol. Biotechnol.* **2014**, *41*, 371–386. [CrossRef] [PubMed]
20. Okada, B.K.; Seyedsayamdost, M.R. Antibiotic dialogues: Induction of silent biosynthetic gene clusters by exogenous small molecules. *FEMS Microbiol. Rev.* **2017**, *41*, 19–33. [CrossRef] [PubMed]
21. Zarins-Tutt, J.S.; Barberi, T.T.; Gao, H.; Mearns-Spragg, A.; Zhang, L.; Newman, D.J.; Goss, R.J.M. Prospecting for new bacterial metabolites: A glossary of approaches for inducing, activating and upregulating the biosynthesis of bacterial cryptic or silent natural products. *Nat. Prod. Rep.* **2016**, *33*, 54–72. [CrossRef] [PubMed]
22. Khokhlov, A.S.; Tovarova, I.I.; Borisova, L.N.; Pliner, S.A.; Shevchenko, L.N.; Kornitskaia, E.I.; Ivkina, N.S.; Rapoport, I.A. The A-factor, responsible for streptomycin biosynthesis by mutant strains of *Actinomyces streptomycini*. *Dokl. Akad. Nauk SSSR* **1967**, *177*, 232–235. [PubMed]
23. Niu, G.; Chater, K.F.; Tian, Y.; Zhang, J.; Tan, H. Specialised metabolites regulating antibiotic biosynthesis in *Streptomyces* spp. *FEMS Microbiol. Rev.* **2016**, *40*, 554–573. [CrossRef] [PubMed]
24. Efremenkova, O.V. A-factor-like autoregulators. *Russ. J. Bioorg. Chem.* **2016**, *42*, 457–472. [CrossRef]
25. Yanagimoto, M.; Terui, G. Physiological studies on staphylomycin production II, formation of a substance effective in inducing staphylomycin production. *J. Ferment. Technol.* **1971**, *49*, 611–618.
26. Yanagimoto, M.; Yamada, Y.; Terui, G. Extraction and purification of inducing material produced in staphylomycin fermentation. *Hakko Kogaku Kaishi* **1979**, *57*, 6–14.
27. Yamada, Y.; Sugamura, K.; Kondo, K.; Yanagimoto, M.; Okada, H. The structure of inducing factors for virginiamycin production in *Streptomyces virginiae*. *J. Antibiot.* **1987**, *40*, 496–504. [CrossRef] [PubMed]
28. Gräfe, U.; Schade, W.; Eritt, I.; Fleck, W.F.; Radics, L. A new inducer of anthracycline biosynthesis from *Streptomyces viridochromogenes*. *J. Antibiot.* **1982**, *35*, 1722–1723. [CrossRef] [PubMed]
29. Gräfe, U.; Reinhardt, G.; Schade, W.; Krebs, D.; Eritt, I.; Fleck, W.F.; Heinrich, E. Isolation and structure of novel autoregulators from *Streptomyces griseus*. *J. Antibiot.* **1982**, *35*, 609–614. [CrossRef] [PubMed]
30. Gräfe, U.; Reinhardt, G.; Schade, W.; Eritt, I.; Fleck, W.F.; Radics, L. Interspecific inducers of cytodifferentiation and anthracycline biosynthesis from *Streptomyces bikinensis* and *S. cyaneofuscatus*. *Biotechnol. Lett.* **1983**, *5*, 591–596. [CrossRef]
31. Sato, K.; Nihira, T.; Sakuda, S.; Yanagimoto, M.; Yamada, Y. Isolation and structure of a new butyrolactone autoregulator from *Streptomyces* sp. FRI-5. *J. Ferment. Bioeng.* **1989**, *68*, 170–173. [CrossRef]
32. Hashimoto, K.; Nihira, T.; Sakuda, S.; Yamada, Y. IM-2, a butyrolactone autoregulator, induces production of several nucleoside antibiotics in *Streptomyces* sp. FRI-5. *J. Ferment. Bioeng.* **1992**, *73*, 449–455. [CrossRef]
33. Takano, E.; Nihira, T.; Hara, Y.; Jones, J.J.; Gershater, C.J.L.; Yamada, Y.; Bibb, M. Purification and structural determination of SCB1, a γ-butyrolactone that elicits antibiotic production in *Streptomyces coelicolor* A3(2). *J. Biol. Chem.* **2000**, *275*, 11010–11016. [CrossRef] [PubMed]
34. Takano, E.; Chakraburtty, R.; Nihira, T.; Yamada, Y.; Bibb, M. A complex role for the γ-butyrolactone SCB1 in regulating antibiotic production in *Streptomyces coelicolor* A3(2). *Mol. Microbiol.* **2001**, *41*, 1015–1028. [CrossRef] [PubMed]

35. Hsiao, N.-H.; Nakayama, S.; Merlo, M.E.; de Vries, M.; Bunet, R.; Kitani, S.; Nihira, T.; Takano, E. Analysis of two additional signaling molecules in *Streptomyces coelicolor* and the development of a butyrolactone-specific reporter system. *Chem. Biol.* **2009**, *16*, 951–960. [CrossRef] [PubMed]

36. Sidda, J.D.; Poon, V.; Song, L.; Wang, W.; Yang, K.; Corre, C. Overproduction and identification of butyrolactones SCB1–8 in the antibiotic production superhost *Streptomyces* M1152. *Org. Biomol. Chem.* **2016**, *14*, 6390–6393. [CrossRef] [PubMed]

37. Corre, C.; Song, L.; O'Rourke, S.; Chater, K.F.; Challis, G.L. 2-Alkyl-4-hydroxymethylfuran-3-carboxylic acids, antibiotic production inducers discovered by *Streptomyces coelicolor* genome mining. *Proc. Natl. Acad. Sci. USA* **2008**, *105*, 17510–17515. [CrossRef] [PubMed]

38. Kitani, S.; Miyamoto, K.T.; Takamatsu, S.; Herawati, E.; Iguchi, H.; Nishitomi, K.; Uchida, M.; Nagamitsu, T.; Omura, S.; Ikeda, H.; et al. Avenolide, a *Streptomyces* hormone controlling antibiotic production in *Streptomyces avermitilis*. *Proc. Natl. Acad. Sci. USA* **2011**, *108*, 16410–16415. [CrossRef] [PubMed]

39. Nguyen, T.B.; Kitani, S.; Shimma, S.; Nihira, T. Butenolides from *Streptomyces albus* J1074 act as external signals to stimulate avermectin production in *Streptomyces avermitilis*. *Appl. Environ. Microbiol.* **2018**, *84*, e02791-17. [CrossRef] [PubMed]

40. Arakawa, K.; Tsuda, N.; Taniguchi, A.; Kinashi, H. The butenolide signaling molecules SRB1 and SRB2 induce lankacidin and lankamycin production in *Streptomyces rochei*. *ChemBioChem* **2012**, *13*, 1447–1457. [CrossRef] [PubMed]

41. Ramos, J.L.; Martínez-Bueno, M.; Molina-Henares, A.J.; Terán, W.; Watanabe, K.; Zhang, X.; Gallegos, M.T.; Brennan, R.; Tobes, R. The TetR family of transcriptional repressors. *Microbiol. Mol. Biol. Rev.* **2005**, *69*, 326–356. [CrossRef] [PubMed]

42. Cuthbertson, L.; Nodwell, J.R. The TetR family of regulators. *Microbiol. Mol. Biol. Rev.* **2013**, *77*, 440–475. [CrossRef] [PubMed]

43. Thao, N.B.; Kitani, S.; Nitta, H.; Tomioka, T.; Nihira, T. Discovering potential *Streptomyces* hormone producers by using disruptants of essential biosynthetic genes as indicator strains. *J. Antibiot.* **2017**, *70*, 1004–1008. [CrossRef] [PubMed]

44. Recio, E.; Colinas, A.; Rumbero, A.; Aparicio, J.F.; Martin, J.F. PI factor, a novel type quorum-sensing inducer elicits pimaricin production in *Streptomyces natalensis*. *J. Biol. Chem.* **2004**, *279*, 41586–41593. [CrossRef] [PubMed]

45. Matselyukh, B.; Mohammadipanah, F.; Laatsch, H.; Rohr, J.; Efremenkova, O.; Khilya, V. *N*-methylphenylalanyl-dehydrobutyrine diketopiperazine, an A-factor mimic that restores antibiotic biosynthesis and morphogenesis in *Streptomyces globisporus* 1912-B2 and *Streptomyces griseus* 1439. *J. Antibiot.* **2015**, *68*, 9–14. [CrossRef] [PubMed]

46. Recio, E. Glycerol, ethylene glycol and propanediol elicit pimaricin biosynthesis in the PI-factor-defective strain *Streptomyces natalensis npi287* and increase polyene production in several wild-type actinomycetes. *Microbiology* **2006**, *152*, 3147–3156. [CrossRef] [PubMed]

47. Vicente, C.M.; Santos-Aberturas, J.; Guerra, S.M.; Payero, T.D.; Martín, J.F.; Aparicio, J.F. PimT, an amino acid exporter controls polyene production via secretion of the quorum sensing pimaricin-inducer PI-factor in *Streptomyces natalensis*. *Microb. Cell Fact.* **2009**, *8*, 8–33. [CrossRef] [PubMed]

48. Onaka, H.; Tabata, H.; Igarashi, Y.; Sato, Y.; Furumai, T. Goadsporin, a chemical substance which promotes secondary metabolism and morphogenesis in *streptomycetes*. I. Purification and characterization. *J. Antibiot.* **2001**, *54*, 1036–1044. [CrossRef] [PubMed]

49. Igarashi, Y.; Kan, Y.; Fujii, K.; Fujita, T.; Harada, K.-I.; Naoki, H.; Tabata, H.; Onaka, H.; Furumai, T. Goadsporin, a chemical substance which promotes secondary metabolism and morphogenesis in *streptomycetes*. II. Structure determination. *J. Antibiot.* **2001**, *54*, 1045–1053. [CrossRef] [PubMed]

50. Onaka, H. Novel antibiotic screening methods to awaken silent or cryptic secondary metabolic pathways in actinomycetes. *J. Antibiot.* **2017**, *70*, 865–870. [CrossRef] [PubMed]

51. Onaka, H.; Nakaho, M.; Hayashi, K.; Igarashi, Y.; Furumai, T. Cloning and characterization of the goadsporin biosynthetic gene cluster from *Streptomyces* sp. TP-A0584. *Microbiology* **2005**, *151*, 3923–3933. [CrossRef] [PubMed]

52. Ozaki, T.; Yamashita, K.; Goto, Y.; Shimomura, M.; Hayashi, S.; Asamizu, S.; Sugai, Y.; Ikeda, H.; Suga, H.; Onaka, H. Dissection of goadsporin biosynthesis by in vitro reconstitution leading to designer analogues expressed in vivo. *Nat. Commun.* **2017**, *8*, 14207. [CrossRef] [PubMed]

53. Wang, L.; Tian, X.; Wang, J.; Yang, H.; Fan, K.; Xu, G.; Yang, K.; Tan, H. Autoregulation of antibiotic biosynthesis by binding of the end product to an atypical response regulator. *Proc. Natl. Acad. Sci. USA* **2009**, *106*, 8617–8622. [CrossRef] [PubMed]

54. Wang, W.; Ji, J.; Li, X.; Wang, J.; Li, S.; Pan, G.; Fan, K.; Yang, K. Angucyclines as signals modulate the behaviors of *Streptomyces coelicolor*. *Proc. Natl. Acad. Sci. USA* **2014**, *111*, 5688–5693. [CrossRef] [PubMed]

55. Li, Q.; Wang, L.; Xie, Y.; Wang, S.; Chen, R.; Hong, B. SsaA, a member of a novel class of transcriptional regulators, controls sansanmycin production in *Streptomyces* sp. strain SS through a feedback mechanism. *J. Bacteriol.* **2013**, *195*, 2232–2243. [CrossRef] [PubMed]

56. Amano, S.; Morota, T.; Kano, Y.; Narita, H.; Hashidzume, T.; Yamamoto, S.; Mizutani, K.; Sakuda, S.; Furihata, K.; Takano-Shiratori, H.; et al. Promomycin, a polyether promoting antibiotic production in *Streptomyces* spp. *J. Antibiot.* **2010**, *63*, 486–491. [CrossRef] [PubMed]

57. Amano, S.; Sakurai, T.; Endo, K.; Takano, H.; Beppu, T.; Furihata, K.; Sakuda, S.; Ueda, K. A cryptic antibiotic triggered by monensin. *J. Antibiot.* **2011**, *64*, 703. [CrossRef] [PubMed]

58. Imai, Y.; Sato, S.; Tanaka, Y.; Ochi, K.; Hosaka, T. Lincomycin at subinhibitory concentrations potentiates secondary metabolite production by *Streptomyces* spp. *Appl. Environ. Microbiol.* **2015**, *81*, 3869–3879. [CrossRef] [PubMed]

59. Tanaka, Y.; Izawa, M.; Hiraga, Y.; Misaki, Y.; Watanabe, T.; Ochi, K. Metabolic perturbation to enhance polyketide and nonribosomal peptide antibiotic production using triclosan and ribosome-targeting drugs. *Appl. Microbiol. Biotechnol.* **2017**, *101*, 4417–4431. [CrossRef] [PubMed]

60. Wang, H.; Zhao, G.; Ding, X. Morphology engineering of *Streptomyces coelicolor* M145 by sub-inhibitory concentrations of antibiotics. *Sci. Rep.* **2017**, *7*, 13226. [CrossRef] [PubMed]

61. Seyedsayamdost, M.R. High-throughput platform for the discovery of elicitors of silent bacterial gene clusters. *Proc. Natl. Acad. Sci. USA* **2014**, *111*, 7266–7271. [CrossRef] [PubMed]

62. Okada, B.K.; Wu, Y.; Mao, D.; Bushin, L.B.; Seyedsayamdost, M.R. Mapping the trimethoprim-induced secondary metabolome of *Burkholderia thailandensis*. *ACS Chem. Biol.* **2016**, *11*, 2124–2130. [CrossRef] [PubMed]

63. Xu, F.; Nazari, B.; Moon, K.; Bushin, L.B.; Seyedsayamdost, M.R. Discovery of a cryptic antifungal compound from *Streptomyces albus* J1074 using high-throughput elicitor screens. *J. Am. Chem. Soc.* **2017**, *139*, 9203–9212. [CrossRef] [PubMed]

64. Craney, A.; Ozimok, C.; Pimentel-Elardo, S.M.; Capretta, A.; Nodwell, J.R. Chemical perturbation of secondary metabolism demonstrates important links to primary metabolism. *Chem. Biol.* **2012**, *19*, 1020–1027. [CrossRef] [PubMed]

65. Ochi, K.; Okamoto, S. A magic bullet for antibiotic discovery. *Chem. Biol.* **2012**, *19*, 932–934. [CrossRef] [PubMed]

66. Ahmed, S.; Craney, A.; Pimentel-Elardo, S.M.; Nodwell, J.R. A synthetic, species-specific activator of secondary metabolism and sporulation in *Streptomyces coelicolor*. *ChemBioChem* **2013**, *14*, 83–91. [CrossRef] [PubMed]

67. Pimentel-Elardo, S.M.; Sørensen, D.; Ho, L.; Ziko, M.; Bueler, S.A.; Lu, S.; Tao, J.; Moser, A.; Lee, R.; Agard, D.; et al. Activity-independent discovery of secondary metabolites using chemical elicitation and cheminformatic inference. *ACS Chem. Biol.* **2015**, *10*, 2616–2623. [CrossRef] [PubMed]

68. Moore, J.M.; Bradshaw, E.; Seipke, R.F.; Hutchings, M.I.; McArthur, M. Use and discovery of chemical elicitors that stimulate biosynthetic gene clusters in *Streptomyces* bacteria. *Meth. Enzymol.* **2012**, *517*, 367–385. [CrossRef] [PubMed]

MDPI

St. Alban-Anlage 66

4052 Basel

Switzerland

Tel. +41 61 683 77 34

Fax +41 61 302 89 18

www.mdpi.com

Microorganisms Editorial Office

E-mail: microorganisms@mdpi.com

www.mdpi.com/journal/microorganisms